Great Salt Lake Biology

Bonnie K. Baxter • Jaimi K. Butler
Editors

Great Salt Lake Biology

A Terminal Lake in a Time of Change

Springer

Editors
Bonnie K. Baxter
Great Salt Lake Institute
Westminster College
Salt Lake City, UT, USA

Jaimi K. Butler
Great Salt Lake Institute
Westminster College
Salt Lake City, UT, USA

ISBN 978-3-030-40351-5 ISBN 978-3-030-40352-2 (eBook)
https://doi.org/10.1007/978-3-030-40352-2

This Springer imprint is published by the registered company Springer Nature Switzerland AG.
The registered company address is: Gewerbestrasse 11, 6330 Cham, Switzerland

To our children and our students: the future generations who will continue the work to keep Great Salt Lake great.

Preface

Dear Readers,

Are you expecting a typical story about lake ecology? You may be disappointed as there is nothing typical about Great Salt Lake nor this time in Earth's history. This book aims to reframe and update our knowledge base, include unheard voices, and dispense with incomplete understandings. It is a time of change and reckoning.

Jaimi met Great Salt Lake for the first time in the mid-1980s as a young girl. Locals thought the rising lake would drown the entire Salt Lake Valley. The lake was at its highest level ever recorded, and she plunked rocks directly into the water from the shoulder of the highway. All of the islands of Great Salt Lake were true islands, and their animals were isolated. Winters at this time were dramatic, with deep snowfall that shut the city down for days. This is in stark contrast to now. It is rare for snow to accumulate in the valley, plunking rocks into the water of the lake is impossible from the shoulder of the highway, and mammalian predators are walking across land bridges to invade pelican breeding grounds. Change dominates Jaimi's relationship with Great Salt Lake over her 20-year career.

Bonnie grew up in North Carolina, playing on Atlantic beaches, boating and otherwise in the water. In the intermountain west, she felt landlocked and was drawn to the shores of Great Salt Lake. It was not enough to play there; she began to work at the lake, studying the tiny microorganisms and their place in this otherworldly landscape. In the two decades of her Utah career, Great Salt Lake shrunk dramatically. Bonnie watched salt playa appear where she used to canoe with students. Instead of studying life in the water, she began studying the ability of life to survive, desiccated in salt.

Great Salt Lake, Utah, USA, is one of the largest bodies of water in North America and is unique in that this terminal lake is adjacent to a major metropolitan area. The lake impacts the millions of people living there, and they impact the lake. Great Salt Lake is at a crossroads. Piecemeal management, gaps in oversight, inapplicable air and water quality standards, and misconceptions create hurdles. Do we want this lake to remain a vibrant ecosystem, or let it dry into a dusty ecological and economic catastrophe? We have an opportunity to decide the future of the lake, and science should light the way. A more coordinated effort on all aspects of Great Salt Lake management could make the difference.

Driven by our experiences, a dozen years ago we built Great Salt Lake Institute together, to connect people to the lake through research and education. Through this work, we discovered scientists and managers who are passionate about their work on the lake. This groundswell of people and new data precipitated the idea of this book. Our goal is to build an understanding of the biology of this immense system, its various salinity gradients, and its significance in the hemisphere. But we must situate our knowledge in the context of this time of change where the lake will be impacted by human population growth, water diversions, climate change, and pollution.

Books on Great Salt Lake have been few, and we are indebted to those authors and editors, from Stansbury to Gwynn. We also express our gratitude to the chapter authors of this volume; thank you for getting salty with us! *Great Salt Lake Biology: A Terminal Lake in a Time of Change* is not only an updated view; it is the very first book to focus on the biology of the lake.

Salt Lake City, UT, USA

Bonnie K. Baxter
Jaimi K. Butler

Contents

1 **Water Development, Consumptive Water Uses, and Great Salt Lake** . . . 1
Sarah E. Null and Wayne A. Wurtsbaugh

2 **Climate Change and Great Salt Lake** . . . 23
Bonnie K. Baxter and Jaimi K. Butler

3 **Relationships Between Humans and Great Salt Lake: Dynamics of Change** . . . 53
Carla Koons Trentelman

4 **Microbialites of Great Salt Lake** . . . 87
Melody R. Lindsay, Eric C. Dunham, and Eric S. Boyd

5 **Unexpected Complexity at Salinity Saturation: Microbial Diversity of the North Arm of Great Salt Lake** . . . 119
Swati Almeida-Dalmet and Bonnie K. Baxter

6 **Invertebrates and Phytoplankton of Great Salt Lake: Is Salinity the Driving Factor?** . . . 145
Katherine L. Barrett and Gary E. Belovsky

7 **Great Salt Lake *Artemia*: Ecosystem Functions and Services with a Global Reach** . . . 175
Brad Marden, Phil Brown, and Thomas Bosteels

8 **Importance of Great Salt Lake to Pelagic Birds: Eared Grebes, Phalaropes, Gulls, Ducks, and White Pelicans** . . . 239
Michael R. Conover and Mark E. Bell

9 **Great Salt Lake Shorebirds, Their Habitats, and Food Base** . . . 263
Ella Dibble Sorensen, Heidi Morrill Hoven, and John Neill

10 American White Pelicans of Gunnison Island, Great Salt Lake, Utah 311
Ashley M. Kijowski, John Neill, Adam Wickline, Jessica Swift, Jaimi K. Butler, David A. Kimberly, Jim Van Leeuwen, John Luft, and Kyle Stone

11 Amphibians and Reptiles of Antelope Island, Great Salt Lake, Utah 345
David A. Kimberly and Chloe L. Fender

12 Shoreline Plants of Great Salt Lake 369
Ember S. Bradbury and David L. Parrott Jr.

13 Invasive Plants of Great Salt Lake Wetlands: What, Where, When, How, and Why? 397
Karin M. Kettenring, Chad R. Cranney, Rebekah Downard, Keith R. Hambrecht, Emily E. Tarsa, Diane R. Menuz, and Christine B. Rohal

14 Mercury Bioaccumulation and Biomagnification in Great Salt Lake Ecosystems 435
Abigail F. Scott and Frank J. Black

15 The Rozel Point Tar Seeps and Their Impact on the Local Biology at Great Salt Lake, Utah 463
Kara L. Kornhauser, H. Gregory McDonald, Rebecca S. Dennis, and Jaimi K. Butler

16 Great Salt Lake as an Astrobiology Analogue for Ancient Martian Hypersaline Aqueous Systems 487
Scott M. Perl and Bonnie K. Baxter

Index 515

Chapter 1
Water Development, Consumptive Water Uses, and Great Salt Lake

Sarah E. Null and Wayne A. Wurtsbaugh

Abstract Great Salt Lake (GSL) covers 5500 km^2 (2100 mi^2) at its unimpacted elevation and is the eighth largest saline lake in the world. Its highly productive food web supports millions of migratory birds and the economic value of the lake is estimated at $1.5 billion US dollars in 2019. Droughts and wet cycles have caused huge fluctuations in lake level, area, and salinities, and this variation has masked anthropogenic impacts. Recent work, however, has determined that consumptive water uses in the watershed have depleted inflows by approximately 39%, with 63% used by agriculture, 11% by cities, 13% by solar ponds, and 13% by other uses. This has lowered the lake by 3.4 m, decreased its area by 51%, and reduced its volume by 64% as of 2019. Projected water development of the lake's primary tributary could lower the lake approximately 1.5 m more. Climate change, to date, has not noticeably influenced lake level. Per capita water use in Utah is the second highest in the nation and is 2.6-fold higher than other semiarid nations. Potential solutions exist to reduce consumptive water uses and stabilize or increase the GSL water level. Water conservation is likely the most economical solution, with permanently mandated water cutbacks costing $14–96 million ($5–32 per person). Water conservation paired with water markets reduce costs further, costing between $2 and $16 per person. Descriptions of potential solutions to reduce consumptive water uses and stabilize GSL level are a starting point to encourage discussion. Strategies have yet to be prioritized or thoroughly evaluated. Quantifying water diversions from rivers that feed GSL and consumptive water uses will allow Utahns to make defensible decisions to manage water resources and the lake's biology for long-term ecological, recreational, and economic benefit.

Keywords Great Salt Lake · Water level · Lake elevation · Depletion · Terminal lake · Salinity · Utah

---

S. E. Null (✉) · W. A. Wurtsbaugh
Department of Watershed Sciences, Utah State University, Logan, UT, USA
e-mail: sarah.null@usu.edu

B. K. Baxter, J. K. Butler (eds.), *Great Salt Lake Biology*,
https://doi.org/10.1007/978-3-030-40352-2_1

1.1 Introduction

Utah's Great Salt Lake (GSL) is the eighth largest terminal lake in the world by surface area and is one of Utah's most recognizable features. It supports a highly productive food web with microbial mats, phytoplankton, and macroinvertebrates (Belovsky et al. 2011; Pace et al. 2016), provides wetland habitat for millions of migratory birds (Aldrich and Paul 2002; Downard et al. 2014), and substantially contributes to the state economy through mineral extraction, commercial brine shrimp harvest, and recreation (Bioeconomics Inc. 2012). However, water diversions and consumptive water uses from rivers that feed GSL have reduced the lake level by about 3.4 m (11 ft) and lake area by over 50%. Planned development of water supplies in the watershed threatens to lower the lake further and reduce its ecological, cultural, and economic value (Wurtsbaugh et al. 2017).

Diking of the shallow system (mean depth 4.5 m) and the disproportionate inflow of freshwater into some sections has resulted in four bays with salinities ranging from freshwater to 34%. These salinity regimes, in turn, support very different biological communities, ranging from freshwater fishes and invertebrates of the estuarine-line bays, to brine shrimp, and to only halo-tolerant Archaea, bacteria, and some algae in the most saline area.

Lake elevation affects the biology of GSL by influencing salinity, nutrients, water temperature, depth, lake habitat area, and exposed lake bed (Barrett and Belovsky 2020). However, to understand GSL's water level, it is necessary to understand the hydrology of the lake and upstream water uses that reduce streamflow to the lake and thus alter lake levels. Here we summarize the effects on GSL elevation from water development, diversions, and consumptive uses. We include long-term records of streamflows to Great Salt Lake to demonstrate that lake level decline is primarily from water development and consumptive uses. We discuss how GSL elevation decline affects biology of the lake and its surrounding wetlands. The chapter ends with a discussion of the economic benefits from saline lakes, opportunities to increase GSL's elevation, and potential future water development changes along Utah's Wasatch Front and their implications for lake level.

1.2 Great Salt Lake Hydrology

1.2.1 Water and Salt Balance

Streams flow into terminal lakes, but water leaves only through evaporation. In other words, terminal lakes have no stream outlet. Major streams feeding GSL include the Bear, Weber, and Jordan Rivers, which drain the west-slope of the Wasatch Range and collectively make up approximately 65% of the lake's inflows (Fig. 1.1). Direct

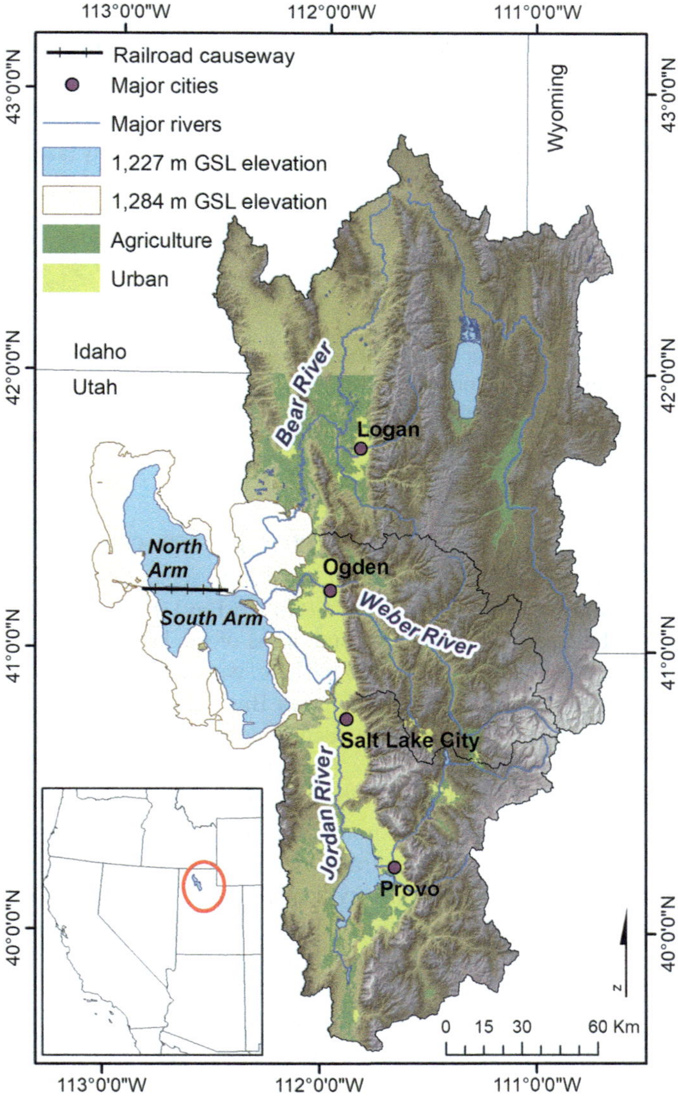

Fig. 1.1 Great Salt Lake, watersheds, and major rivers, with recent high and low lake elevations

precipitation to the lake surface accounts for about 33% of inflows, with the remainder from groundwater and ephemeral West Desert streamflows (Bedford 2005). Of the 65% of streamflow contributions to the lake, the Bear River provides 58%, the Weber River provides 15%, the Jordan River provides 22%, and the remainder is from small streams.

Prior to construction of a railroad causeway that divided the lake in half, the GSL was typical of a terminal lake, where dissolved salt concentration varied

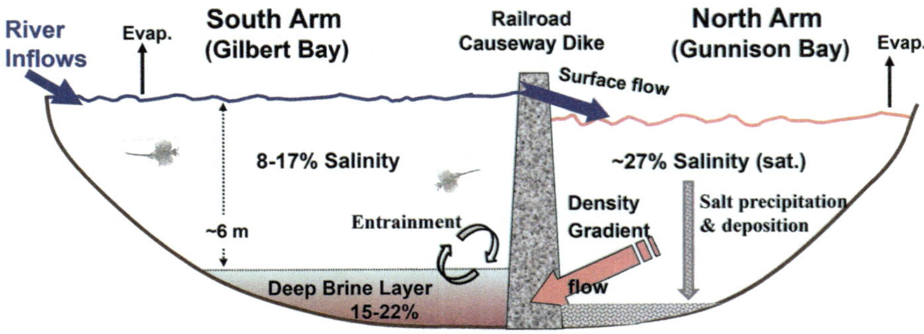

Fig. 1.2 Bidirectional water and salt transport between the south and north arms of Great Salt Lake. At most lake levels, evaporation brings the north arm to saturation and NaCl precipitates to the bottom. The deep density flows from the north to the south arm creates a semi-stable deep brine layer at a depth of about 6 m. Wind mixing entrains a portion of this layer into the surface layer of the south arm

inversely with lake volume (Loving et al. 2000). In 1959, GSL was bisected by Union Pacific Railroad Company's railroad causeway, separating the lake into the north arm (Gunnison Bay) and the south arm (Gilbert Bay) (Fig. 1.1). The exchange of water and salt between the north and south arms of the lake was forever altered. Additional causeways have further divided the lake, with an automobile causeway to Antelope Island partially separating Farmington Bay from Gilbert Bay, and a salt flume partially restricting interchange between Bear River Bay and Gilbert Bay.

The Bear, Weber, and Jordan Rivers flow into the south arm (Gilbert Bay), providing freshwater to the south arm of the lake and keeping the surface elevation approximately 0.27 m higher than the north arm (Mohammed and Tarboton 2012) (Fig. 1.1). Surface inflow to the north arm is nearly all saline water from the south arm. The south arm has lower salinity (typically 8–17%) and higher biodiversity since it receives nearly all of the streamflow. With little freshwater, but high evaporation, the north arm is often completely saturated with salt (~27%) (Johnson et al. 2020).

Some water and salt flows through the railroad causeway at a breach and through the causeway fill. Two rectangular culverts were originally built for boater access between the north and south arms of the lake. They provided bidirectional flow between the north and south arms of the lake, but were closed in 2012 and 2013 because they were subsiding into the soft lake bed sediment (White et al. 2015). After 3 years of negligible water and salt exchange, Union Pacific breached the causeway in 2016, which allowed water exchange between the north and south arms of GSL. This brought the elevations of the north and south arms closer to each other. Nevertheless, elevation and salinity differences between the north and south arms remain, with surface flows moving less salty water from the south arm into the north arm. However, there is a countercurrent flow near the lake bed, as very dense, saline water moves from the north to the south arm (Fig. 1.2). This dense water does not mix readily with the more buoyant water in the south arm, creating a deep brine layer, or monimolimnion. Decomposition of organic matter in this layer makes it anoxic with high concentrations of toxic hydrogen sulfide. Brine shrimp and brine flies cannot survive in this layer. Wind events create turbulence that erodes the deep brine layer, bringing its volume into equilibrium with the inflowing brine from the north. Jones and Wurtsbaugh (2014) roughly estimated that 40% of the deep brine layer is entrained into the surface layer each year.

1.2.2 Fluctuating Lake Levels Through Time

Lakes are integrators of droughts, floods, land use, and water use (Schindler 2009), so the elevation of GSL varies through time (Baxter and Butler 2020) (Fig. 1.3). Recorded lake elevations have ranged by over 6 m (20′) in the past 170 years, with lower lake elevations approximately halving lake volume and area (Wurtsbaugh et al. 2017) (Fig. 1.1). The shallow bays on the east shore of the lake are impacted even more by water diversions and drought. In 2016, when the lake reached its lowest recorded elevation, about 75% of Bear River Bay and Farmington Bay were dry. These bays have fresher water than the north and south arms, so low water levels greatly reduce important bird habitat and the biodiversity of the lake (Wurtsbaugh et al. 2017).

Fig. 1.3 Weighted mean elevation of Great Salt Lake measured at USGS gages 10010000 (Saltair Boat Harbor) and 10010100 (Near Saline, Utah). The lowest elevation of 1277.5 m (4191.2 feet) was reached in September, 2016

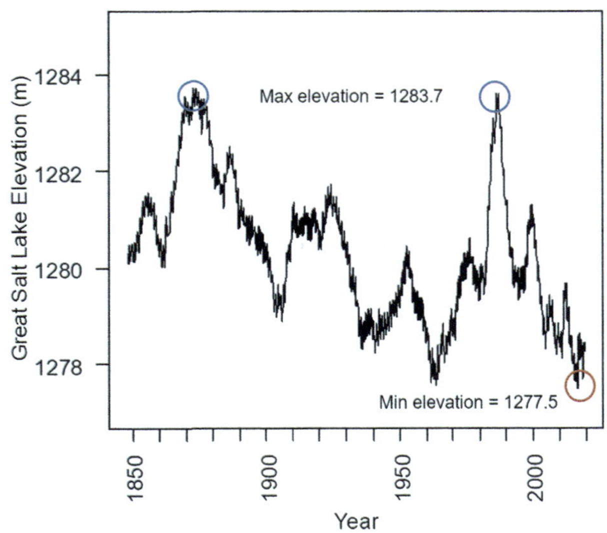

1.2.3 Constant River Flows Through Time

Despite droughts and pluvials that cause marked changes in lake levels, there has not been a significant long-term streamflow trend in the measured record from watersheds that drain into GSL (Fig. 1.4). In fact, a tree-ring reconstruction of streamflow identified medieval droughts that persisted for decades, but identified no long-term, climate-driven changes to precipitation or streamflows in the past 150 years (DeRose et al. 2015). Gillies et al. (2012) documented an overall slight increase in precipitation in Utah over the 1950–2003 period. Consequently, papers that attribute GSL's decline to climate change have been debunked (Wine et al. 2019).

1.3 Water Development and Consumptive Water Uses

While GSL elevation responds to precipitation changes from droughts and floods, those natural events do not have a persistent trend on lake elevation. On the contrary, GSL decline has coincided with water development since pioneers inhabited Salt Lake Valley and the Wasatch Front. Utah had a pronounced dam-building era from the 1930s to the 1990s (Fig. 1.5), which largely coincided with water development throughout the American West (Reisner 1993). Today, Utah has an elaborate system of water infrastructure (Fig. 1.6), including major federal projects like the Central Utah Project that transfers water from the Colorado Basin to the Salt Lake Watershed, the Weber River Project, and the Provo River Project. Pumps and diversion structures take water directly from rivers to irrigate fields.

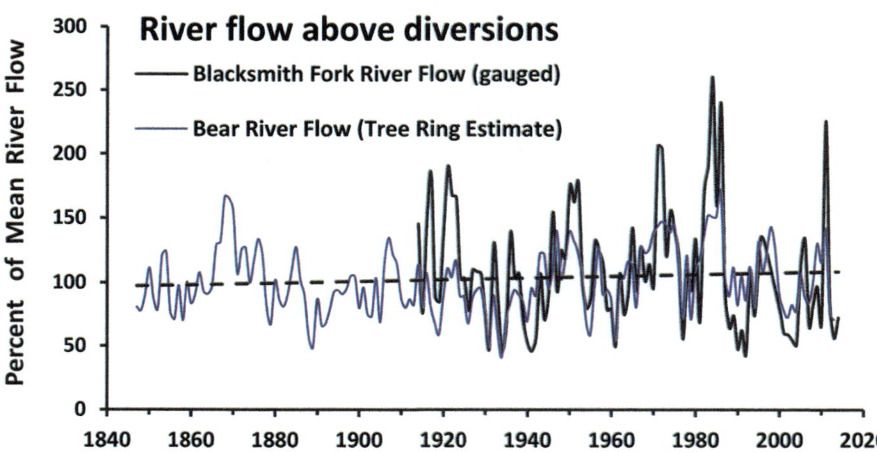

Fig. 1.4 Estimated streamflows in Great Salt Lake headwater streams upstream of diversions. Flows in the Bear River are based on tree-ring reconstructions (figure reprinted from Wurtsbaugh et al. 2017)

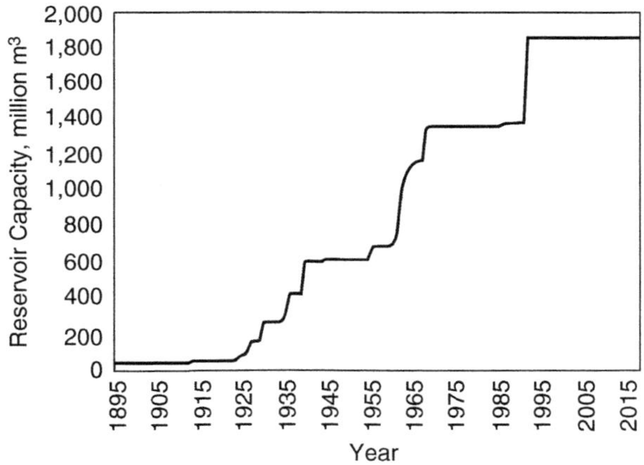

Fig. 1.5 Cumulative reservoir capacity in the Bear, Weber, and Jordan watersheds, 1895–2006 (data from National Atlas 2006)

Diversions redirect water away from streams to other uses, usually agricultural or urban uses. However, some diverted water eventually finds its way back to the river downstream and flows to the GSL. A small but growing literature has explored water pathways from inefficient irrigated distribution systems (Jensen 2007; Boelens and Vos 2010). This water is not truly "lost" from the system if it eventually returns to downstream ground and surface water bodies. True consumptive water uses, sometimes called depletions, include water that is consumed, evaporated, or transpirated—water that will not return to GSL or the streams that feed it. It is consumptive water uses that are important for considering relationships between water development, water use, and GSL elevation decline.

Consumptive water uses were calculated by the Utah Division of Water Resources since 1847, when record keeping began in this area (Wurtsbaugh et al. 2017). Consumptive water use is split among multiple groups. Irrigated agriculture uses 63% of water, mineral extraction from GSL uses 13%, cities and industry use 11%, impounded wetlands use 10%, and evaporation from reservoirs use 3% of water (Fig. 1.7a) (Wurtsbaugh et al. 2017). Hydrologic modeling has shown that if no diversions or consumptive water extractions had occurred, then GSL level would be 3.4 m (11 ft) higher than it is today (Fig. 1.7b). Overall, water development and consumptive uses of water have reduced streamflows by 39% (Wurtsbaugh et al. 2017). Consumptive water uses have decreased the area of the lake by 51% and lake volume by 64%.

Variable lake elevations, caused by water depletions, droughts, pluvials, and salt extraction, have had a large effect on the salinity of GSL. The lake contained approximately 5 billion metric tonnes of salt. During an unusually wet cycle in the mid-1980s, the rising lake level threatened infrastructure encroaching around the shoreline, and a pumping project was undertaken to move water to the desert west of the lake. Along with the water, approximately 0.5 billion tonnes of salt were deposited in the west desert. Despite an investment of $72 million, the pumps

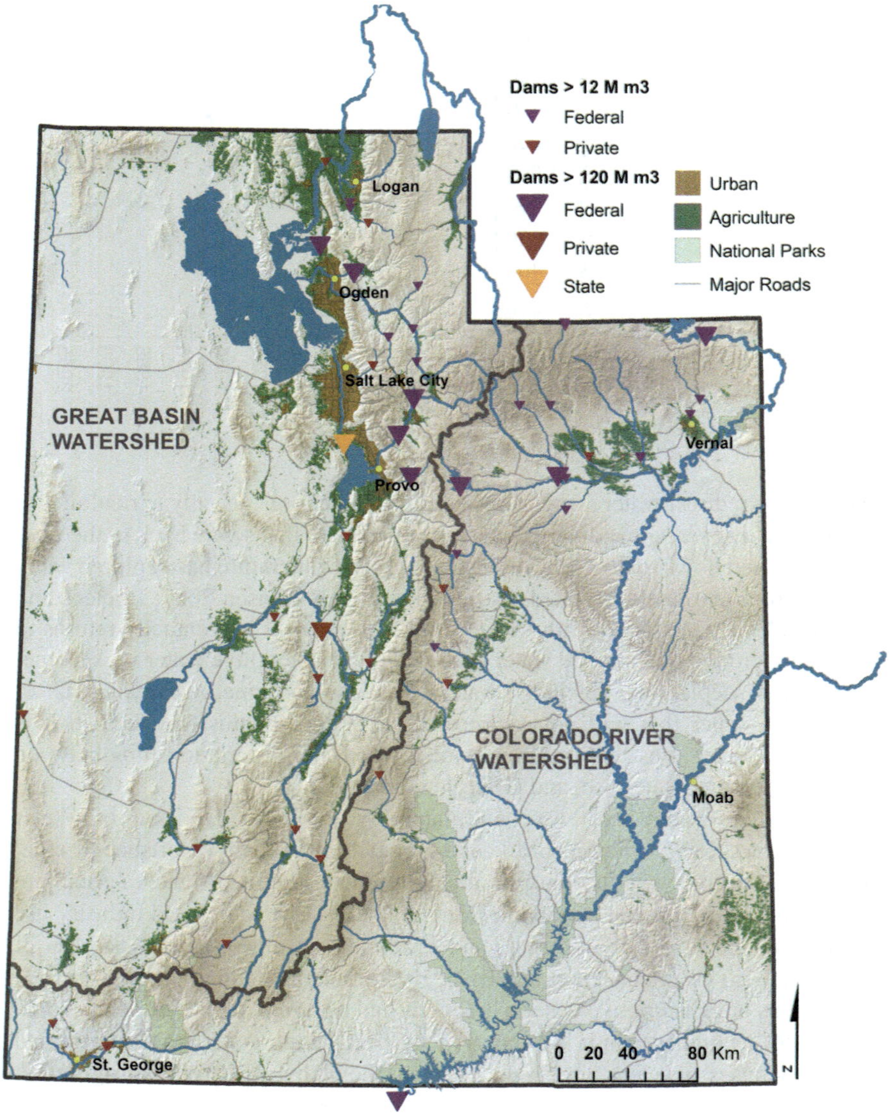

Fig. 1.6 Utah watersheds and water infrastructure

were only used for about 24 months until June 1989 because the pluvial ended, naturally lowering the lake (White et al. 2015).

The natural wet and dry cycles have a large influence on the salinity. When the lake reached a high level in 1985, salinities in the south arm decreased to 5.8%. In contrast, when the lake reached its near lowest level in 1961, salts were concentrated to 28%.

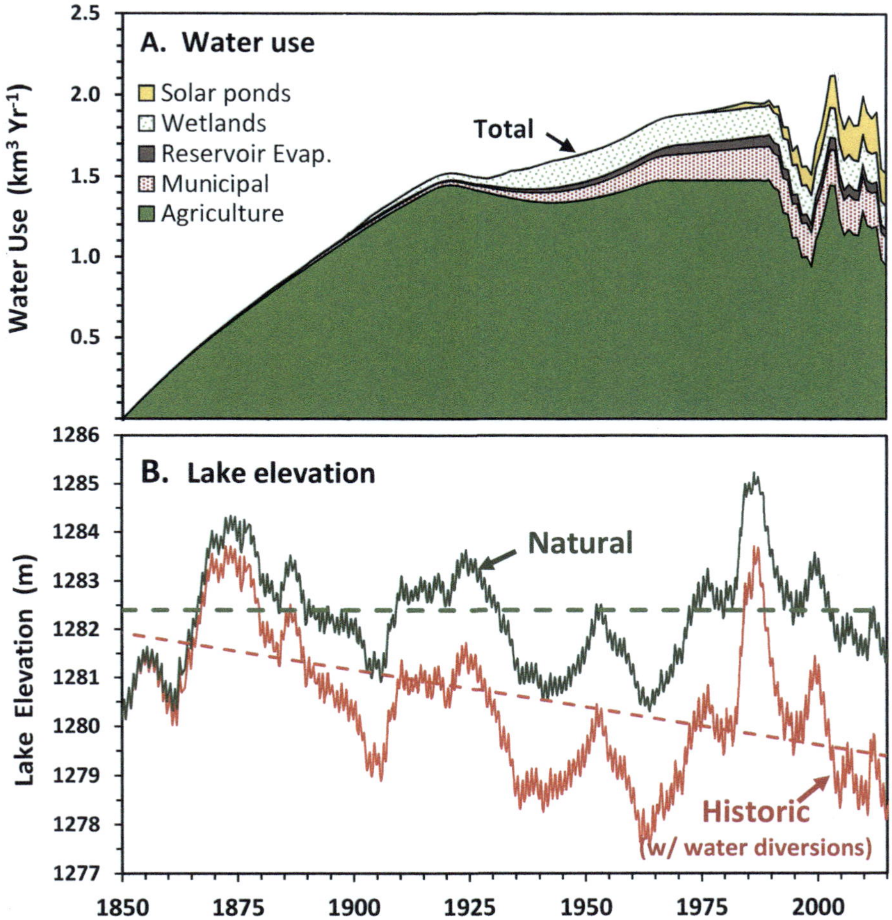

Fig. 1.7 (**a**) Estimated consumptive water use by user group from 1850 to 2013. (**b**) Measured USGS lake level (red) and modeled lake level had consumptive water uses not occurred (green) (figure modified from Wine et al. 2019)

1.4 Great Salt Lake Elevation and Biology

1.4.1 Habitat Connectivity with Bays and Wetlands

GSL elevation affects lake and wetland biology (Fig. 1.8). Farmington and Bear River Bays, the large bays on the east side of the lake, function as estuaries, with salinity gradients from freshwater near inflows, to hypersaline conditions near their connections with Gilbert Bay. Salinities also vary greatly with droughts and pluvials (Wurtsbaugh et al. 2012). Increasing salinities reduce biodiversity (Hammer 1986), but overall, the bays are highly productive and diverse. Under hypersaline

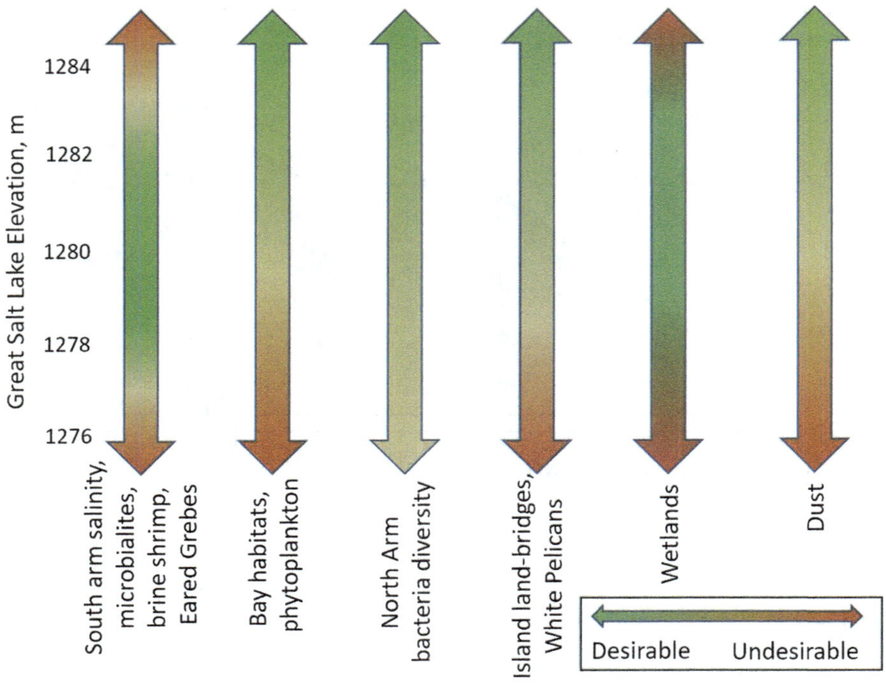

Fig. 1.8. General relationships between Great Salt Lake elevation and biological parameters (UDNR FFSL 2013)

conditions, the invertebrate community is dominated by brine shrimp (*Artemia franciscana*) and brine flies (*Ephydra* spp.). As salinities drop below 6–7%, macroinvertebrates like *corixids* (water boatmen) can flourish and prey on brine shrimp, and the community becomes dominated by cladocerans, copepods, and chironomids in the benthic zone. Near river inflows, the bays contain a variety of fish species (Armstrong and Wurtsbaugh 2019). Both bays are important habitat for shorebirds, migratory waterfowl, and other species (Paul and Manning 2002; Wurtsbaugh 2018), and waterfowl hunting is an important component of the $136 million spent on recreation at the lake (Bioeconomics Inc. 2012; Aldrich and Paul 2002).

Due to water use and drought, over 75% of Farmington and Bear River Bays lake bottoms has been exposed in recent years, causing vast playas that are a source of dust for the millions of residents along the Wasatch Front (Hahnenberger and Nicoll 2014; Perry et al. 2019). This is especially important for Bear River Bay, where low water levels in 2018 resulted in a 15 km dry section between it and Gilbert Bay (Fig. 1.9).

Fig. 1.9 Abandoned sampling device on the desiccated portion of Bear River Bay caused by water use and drought. Photo: Sept. 2019 when over 150 km^2 of the bay was dry

1.4.2 Salinity and Great Salt Lake Biota

Salinity has an inverse relationship with lake level, so as streamflows decrease, salinity increases (Fig. 1.10). Microbial mats surround the perimeter of GSL and are sensitive to increased salinity. They contribute to the formation of microbialites, which develop when cyanobacteria and periphytic algae reduce the pH and cause limestone structures to form (Lindsay et al. 2020). They are nearly the only solid substrate in the lake, and as such, are important habitat for larval brine flies (*Ephydra* spp.) that feed on the microbial community (Pace et al. 2016; Wurtsbaugh et al. 2011; Collins 1980). Ongoing research is investigating the salinity thresholds of GSL microbial mats and the health of microbiolites.

When the north arm is saturated, conditions are intolerable for most phytoplankton and macroinvertebrates; however, a diversity of bacteria and Archaea thrive in the hypersaline water (Baxter and Zalar 2019; Almeida-Dalmet and Baxter 2020).

The moderate salinity of the south arm, which averages 13%, supports large populations of macroinvertebrates like brine shrimp (*Artemia franciscana*) and brine fly (*Ephydra cinera*). When GSL salinity exceeds approximately 12%, brine shrimp become physiologically stressed and their production begins to decline, but they are abundant at least up to salinities of 20% (Barnes and Wurtsbaugh 2015). Brine flies have higher salinity tolerances, but their growth also declines as salinities climb above 12%. Brine shrimp and brine flies are the most important food resource for birds inhabiting the south arm. Consequently, increasing salinity due to water

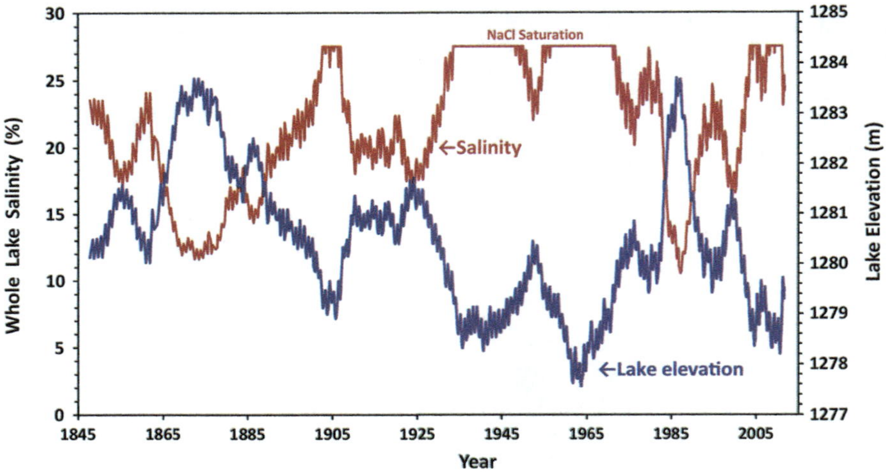

Fig. 1.10 Estimated whole lake elevations and salinities if Great Salt Lake had not been divided by a railroad causeway. Derived from Null et al. (2013)

development is a potential threat to the Great Salt Lake food web. For example, 0.5–5.5 million Eared Grebes (*Podiceps nigricollis*) migrate to GSL to feed on brine shrimp and brine flies (Conover and Bell 2020; Wurtsbaugh et al. 2011). While doubling their weight at GSL, Eared Grebes loses the ability to fly. Thus, if the lake became too saline and brine shrimp production was unable to support the Eared Grebe population, they would be stranded and unable to fly to other lakes with more abundant food sources.

The solid-fill railroad causeway complicates biological effects of salinity (Fig. 1.2). As mentioned above, streams flow into the south arm, while the north arm typically remains saturated. While unnatural, this maintains a range of salinities in GSL. Under normal conditions, salinities in the south arm support high densities of brine shrimp and brine flies, a diversity of phytoplankton, and large populations of Eared Grebes, Phalaropes, and other birds that feed on the macroinvertebrates. During exceptionally wet years, such as those in the mid-1980s, salinity in the south arm becomes too low (6%) to support brine shrimp. However, during these events the salinity in the north arm declines to around 21% and high densities of brine shrimp may be present (Wurtsbaugh and Berry 1990; Wurtsbaugh 1992). A range of salinities also supports phytoplankton biodiversity, as phytoplankton species vary with salinity though time and by lake depth (Belovsky et al. 2011).

1.4.3 Land Bridges and Bird Rookeries

Lake islands begin to connect to land when lake levels drop below 1280.8 m (4202 ft). When lake level falls to 1278.6 m (4195 ft), all islands are accessible by land or are separated only by shallow water (Fig. 1.7) (UDNR FFSL 2013). When land bridges form, predators and people can reach bird rookeries. For example, a large population of White Pelicans (*Pelecanus erythrorhynchus*) nest on Gunnison Island in the north arm of GSL. Shallow, wadeable water separates the island from the mainland at about 1279.3 m (4197 ft) and a land bridge form when lake levels drops below about 1278.0 m (4193 ft) (UDNR FFSL 2013). The GSL level has consistently been lower than 1279.3 m (4197 ft) since 2012 (Fig. 1.2), threatening one of the largest White Pelican rookeries in North America (Kijowski et al. 2020).

1.4.4 Dust

Water development in the basin has exposed 2100 km^2 of lake bed. Desiccated saline lake beds generate fine dust that harms human health (Griffin and Kellogg 2004) and agriculture (Micklin 2007). Impacts have been well-studied at the Aral Sea in Central Asia where 12,700 km^2 of lake bed was exposed due to water development for agriculture (Crighton et al. 2011; Micklin 2007; Indoitu et al. 2015). In California's small (285 km^2) Owens Lake, dust from the dried lake bed has exceeded US air quality standards for large particulate particles (PM10) (Ramboll Environ US Corporation 2016) and allegedly increased the incidence of lung infections, asthma, and other respiratory diseases in the area (Kittle 2000). To mitigate this dust problem the City of Los Angeles will spend US$3.6 billion over 25 years (Ramboll Environ US Corporation 2016). The area of exposed Great Salt Lake sediments is over 7 times that of Owens Lake, and the population near the GSL is 85 times higher than the sparse population near Owens Lake. Consequently, the potential impact of dust for the Wasatch Front is of concern.

Studies on dust emissions from the exposed bed of GSL are just beginning. There are no epidemiological studies of the impact of playa dust on human health, but the potential risks are high given results from the dried Aral Sea and Owens Lake. Hahnenberger and Nicoll (2014) found that dust that originated from GSL and reached Salt Lake City was important, but other dust sources west of the city were problematic for the city more frequently. A recent study by Perry et al. (2019) found that only about 9% of the currently exposed lake bed is likely to produce dust during wind events. Surface crusts and vegetation protect other areas from wind scour. However, Perry found that if all the protective crusts were destroyed by rains, natural erosion, and human activities, 22% or 460 km^2, of the exposed lake bed would produce dust. Exposed lake bed area would increase or decrease by 23–46% for each meter change in lake level, depending on the bay. Consequently, any additional water development in the basin will increase the potential for dust production.

Perry also analyzed lake bed sediments for heavy metals and found that nine elements exceeded the Residential Regional Screening Levels established by the US Environmental Protection Agency (EPA 2019). Four of these elements (arsenic, lanthanum, lithium, and zirconium) had some values that exceeded the EPA's Industrial Screening Levels. However, mercury concentrations were well below either screening level. These metals will not necessarily cause a health risk depending on exposure to the dust. Regardless, site-specific exposure assessments should be done for the most problematic metals (Perry et al. 2019).

1.5 Potential Solutions to Reduce Consumptive Water Uses and Stabilize Great Salt Lake Level

Saline lakes are valuable. In 2012, GSL generated approximately \$1.3 billion of Utah's gross domestic product (Bioeconomics Inc. 2012), or about \$1.48 billion in 2019 dollars, assuming an average inflation rate of 1.62%. Of that amount, about 85.5% is generated from mineral extraction, 10.3% from recreation, and 4.3% from the brine shrimp industry (Bioeconomics Inc. 2012). In addition to the quantifiable economic value, GSL is immensely valuable ecologically for the millions of birds that utilize the ecosystem. Conversely, the vast lake bed, when dried, becomes a source of dust that can cause costly health problems for millions of residents in greater metropolitan Salt Lake City (Perry et al. 2019).

1.5.1 Mechanisms to Preserve Saline Lakes Globally

Elevations to maintain desired ecosystem services of GSL have been estimated (UDNR FFSL 2013); however, minimum streamflow requirements or mechanisms to maintain lake elevation have not been developed. A wide array of strategies have been implemented to preserve other saline lakes (Table 1.1). For example, litigation-driven water conservation at Mono Lake (Hart 1996), water purchases from willing sellers at Walker Lake (Elmore et al. 2016), and an interbasin water transfer at the Dead Sea (Gavrieli et al. 2011) are varied mechanisms to maintain lake elevations. For other saline lakes, substantial changes have been tolerated to preserve a remnant of the lake or to maintain select ecosystem services. Diking has reduced Aral Sea area to about 5% the size of the original lake, allowing salinities to remain low enough in the remnant lake to support a fish community (Micklin 2016). Shallow flooding, managed wetlands, and gravel mitigate for airborne dust at Owens dry lake bed (Gutrich et al. 2016), while shallow flooding, ponds, and berms are used at Salton Sea to maintain some minimal habitat and reduce airborne dust (California Natural Resources Agency 2015).

Table 1.1 Mechanisms to restore saline lakes

Restoration goal	Mechanism	Lake (Location)
Increase lake elevation	Litigation and water conservation	Mono Lake (California, USA)
	Environmental water purchases	Walker Lake (Nevada, USA)
	Interbasin water transfer	Dead Sea (Jordan, Israel, and Palestine)
Reduce lake area	Diking	Aral Sea (Kazakhstan and Uzbekistan)
Mitigate dust and preserve habitat	Shallow flooding, managed wetlands, and gravel cover	Owens Lake (California, USA)
	Habitat ponds, berms, shallow flooding	Salton Sea (California, USA)

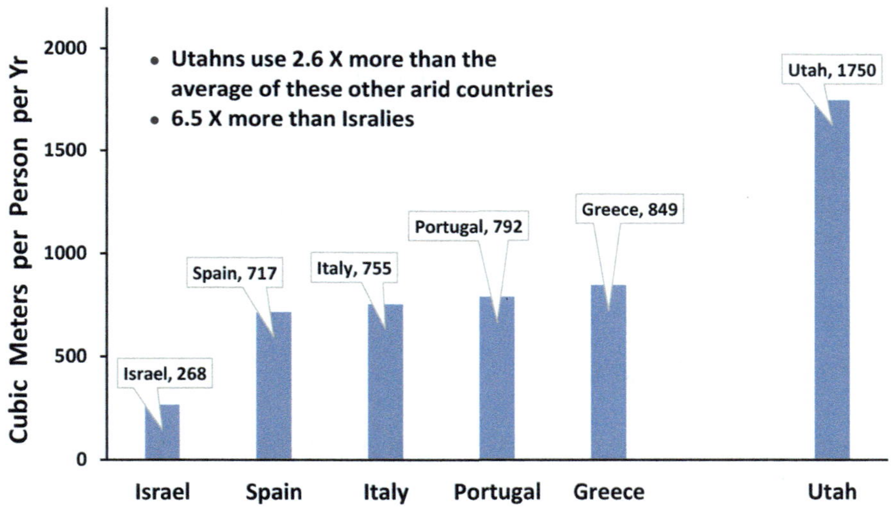

Fig. 1.11 Per capita, water use in Utah compared to that other arid regions. Data derived from Pacific Institute (2013)

1.5.2 Opportunities to Preserve Great Salt Lake Level

Utah has over 3 million people with about 80% of the population living in the metropolitan Wasatch Front (Fig. 1.1). Utahns have the second highest municipal and industrial per capita water use of the United States at 1109 L per person per day (293 gallons per person per day) (Office of the Legislative Auditor General 2015). Agriculture is also a major use of water, and when it is added, total water use far exceeds that of other arid regions in the world (Fig. 1.11). Permanently implementing water cutbacks to urban and agricultural water users could cost between $14 and 96 million ($5–32 per person), depending on upper and lower cost estimates (Edwards and Null 2019). However, with a water conservation market between water users and watersheds, costs drop substantially to $6–48 million ($2–16 per person). Water conservation measures are varied and could include low water use toilets, showers, and washing machines, urban and agricultural water scheduling, turf conversion, rain barrels, and more (Edwards et al. 2017). These costs are inexpensive, although opportunity costs given by lost benefits of consumptive water use could also generate supply curves for water to GSL (Génova et al. 2018). Proposed state legislation to enable water banking in Utah may facilitate water trading in the future. Water banking allows farmers or other water users to forego their water use without forfeiting water rights. It enables water trading between willing water sellers and buyers and may have the potential to facilitate dedicated streamflows to GSL.

Some urban water districts are considering implementing green infrastructure such as rain barrels, retention ponds, permeable pavement, or bioswales to recharge groundwater and baseflows (Prudencio and Null 2018). However, this approach is unlikely to offset anticipated drying from climate change (York et al. 2015). A menu of potential strategies to provide water to GSL has been developed, including water conservation, groundwater management, reducing vegetation around the lake, applying the Public Trust Doctrine used to preserve Mono Lake, removing dams, enlarging dams, and interbasin water transfers (SWCA Environmental Consultants 2017; Clyde 2016). The list is a starting point to encourage discussion and strategies that have yet to be prioritized or thoroughly evaluated.

1.5.3 Future Changes

Consumptive water uses have caused GSL elevation to decline by 3.4 m (11 ft) since pioneers colonized Salt Lake Valley (Wurtsbaugh et al. 2017). Utah's population is anticipated to double by 2060 and 80% of the population lives along the Wasatch Front in the GSL watershed. This suggests that water development and diversions to urban and agricultural users will change and evolve in the coming decades.

In fact, considerable water development of the Bear River for urban and agricultural use is being considered. The Bear River Compact between the States of Utah, Idaho, and Wyoming envisions the development of 1600 million m^3 (1.3 million af) of water. If all of this water is developed, it would lower the lake more than 1.6 m (5.4 ft) beyond what has already occurred. This would raise salinity to over 22% and brine shrimp populations would be greatly diminished. Idaho and Wyoming have not yet funded projects to develop their water. But in 1991, the Utah legislature passed the Bear River Development Act, which directs the state to develop 270 million m^3 (220,000 af) of surface water in the Bear River and its tributaries through construction of reservoirs and associated facilities. Hydrologic modeling suggests that Utah's Bear River Development Project, which is estimated to cost $1.5 billion, would lower the GSL by an additional 20 cm (8 in.) (Wurtsbaugh et al. 2017).

Utah has not fully developed its share of water from the Colorado River according to the Colorado River Compact and transfers of water have been proposed (SWCA Environmental Consultants 2017). Cost estimates are unavailable, although the proposed Lake Powell Pipeline to southwest Utah is estimated to cost $1.1–1.8 billion (LPP 2019). While increasing water supply appears to be a costly, but possible, solution to maintain GSL elevation, previous research has shown that interbasin transfers temporarily mask water supply problems, but do not address underlying problems of unsustainable water use and development (AghaKouchak et al. 2015). In other words, interbasin water transfers have been shown to be a temporary fix that backfires in the long run (Gohari et al. 2013).

1.6 Concluding Remarks

Population growth and development in a semiarid climate elicits questions like: How will water development and operations affect flows to the GSL and surrounding wetland habitats? What is the role of water conservation, water markets, stormwater management, water infrastructure, and coordinated management of existing facilities to simultaneously maintain human benefits and preserve ecosystems?

Restoration of other terminal lakes has shown that it is more costly to restore lakes and the ecosystem services they provide than to preserve them from the outset. For instance, Libecap (2009) estimated that the costs of litigating the out-of-basin water transfers for California's Mono Lake over 20 years likely exceeded the actual value of the water. Mitigating for airborne dust in Owens dry lake bed has already cost over $1 billion, and that does not include restoration of the lake (Gutrich et al. 2016). It is estimated to cost $3.6 billion over 25 years (Ramboll Environ US Corporation 2016). Even securing water for terminal lakes is costly. As of 2016, $57 million had been spent purchasing about 24 million m^3 (20,000 acre-ft) of water from willing sellers to increase the elevation of Walker Lake (Null et al. 2017).

Large declines in lake level and salinity threaten the unique biology of GSL. Quantifying water diversions from rivers that feed GSL, consumptive water uses, and total streamflow that reaches GSL will allow Utahns to make defensible decisions to manage water resources and GSL biology for long-term ecological and economic benefit. Utah has potential opportunities and multiple alternatives to improve water management and maintain water supply to GSL.

Acknowledgments This work was partially supported from a National Science Foundation CAREER Award (#1653452). Any opinions, findings, and conclusions or recommendations expressed in this chapter are those of the authors and do not necessarily reflect the views of the National Science Foundation. Craig Miller of the Utah Department of Water Resources shared data and guided the development of the hydrological depletion model for GSL. David Tarboton shared hypsographic data allowing us to predict exposed lake bed area. Peter Wilcock, Justin DeRose, Maura Hahnenberger, and Jonnie Moore participated in the development of an earlier paper documenting the decline of GSL.

References

AghaKouchak A, Norouzi H, Madani K, Mirchi A, Azarderakhsh M, Nazemi A, Nasrollahi N, Farahmand A, Mehran A, Hasanzadeh E (2015) Aral Sea syndrome desiccates Lake Urmia: call for action. J Great Lakes Res 41(1):307–311

Aldrich TW, Paul DS (2002) Avian ecology of Great Salt Lake. In: Gwynn JW (ed) Great Salt Lake: an overview of change. Utah Department of Natural Resources, Salt Lake City, UT, pp 343–374

Almeida-Dalmet S, Baxter BK (2020) Unexpected complexity at salinity saturation: microbial diversity of the north arm of Great Salt Lake, Utah. In: Baxter BK, Butler JK (eds) Great Salt Lake biology: a terminal lake in a time of change. Springer, Cham

Armstrong T, Wurtsbaugh WA (2019) Impacts of eutrophication on benthic invertebrates & fish prey of birds in Farmington and Bear River Bays of Great Salt Lake. Final Report to the Utah Division of Forestry, Fire & State Lands, Salt Lake City, UT, 41 p. https://digitalcommons.usu.edu/wats_facpub/1099/

Barnes BD, Wurtsbaugh WA (2015) The effects of salinity on plankton and benthic communities in the Great Salt Lake, Utah, USA: a microcosm experiment. Can J Fish Aquat Sci 72(6):807–817. https://doi.org/10.1139/cjfas-2014-0396

Barrett KL, Belovsky GE (2020) Invertebrates and phytoplankton of Great Salt Lake: is salinity the driving factor? In: Baxter BK, Butler JK (eds) Great Salt Lake biology: a terminal lake in a time of change. Springer, Cham

Baxter BK, Butler JK (2020) Climate change and Great Salt Lake. In: Baxter BK, Butler JK (eds) Great Salt Lake biology: a terminal lake in a time of change. Springer, Cham

Baxter BK, Zalar P (2019) The extremophiles of great salt lake: complex microbiology in a dynamic hypersaline ecosystem. In: Model ecosystems in extreme environments. Elsevier, Amsterdam, pp 57–99

Bedford D (2005) Utah's Great Salt Lake: a complex environmental-societal system. Geogr Rev Wiley Online Library 95(1):73–96

Belovsky GE, Stephens D, Perschon C, Birdsey P, Paul D, Naftz D, Baskin R, Larson C, Mellison C, Luft J, Mosley R, Mahon H, Van Leeuwen J, Allen DV (2011) The Great Salt Lake ecosystem (Utah, USA): long term data and a structural equation approach. Ecosphere 2 (3):1–40

Bioeconomics Inc. (2012) Economic significance of the Great Salt Lake to the State of Utah. Great Salt Lake Advisory Council, Salt Lake City, UT

Boelens R, Vos J (2010) The danger of naturalizing water policy concepts: water productivity and efficiency discourses from field irrigation to virtual water trade. Agric Water Manag 108:16–26. https://doi.org/10.1016/j.agwat.2011.06.013

California Natural Resources Agency (2015) Salton sea species conservation habitat monitoring and adaptive management plan. Prepared by Cardno Inc. and Environmental Science Associates, 64 p. https://water.ca.gov/-/media/DWR-Website/Web-Pages/Programs/Engineering-And-Construction/Files/Design-Build/Salton-Sea-Reports/07_Monitoring_Adaptive_Mgmt_Plan.pdf

Clyde SE (2016) Water rights for Great Salt Lake: is it the impossible dream? Utah Water Law, CLE International, Salt Lake City, UT, pp 1–29. http://www.clydesnow.com/images/Articles%2D%2DGreat-Salt-Lake-Paper-2016-01090010xB165B.pdf

Collins N (1980) Population ecology of Ephydra cinerea Jones (Diptera, Ephydridae), the only benthic metazoan of the Great Salt Lake, USA. Hydrobiologia 68:99–112

Conover MR, Bell ME (2020) Importance of Great Salt Lake to Pelagic Birds: Eared Grebes, Phalaropes, Gulls, Ducks, and White Pelicans. In: Baxter BK, Butler JK (eds) Great Salt Lake biology: a terminal lake in a time of change. Springer, Cham

Corporation REUS (2016) Great Basin Unified Air Pollution Control District: 2016 Owens Valley Planning Area PM10 State implementation plan. Great Basin Unified Air Pollution Control District. Bishop, California, p 1494. http://www.gbuapcd.org/Air%20Quality%20Plans/OVPA_SIP_2016/2016_SIP_FINAL_20160413.pdf

Crighton EJ, Barwin L, Small I, Upshur R (2011) What have we learned? A review of the literature on children's health and the environment in the Aral Sea area. Int J Public Health 56:125–138. https://doi.org/10.1007/s00038-010-0201-0

DeRose RJ, Bekker MF, Wang S-Y, Buckley BM, Kjelgren RK, Bardsley T, Rittenour TM, Allen EB (2015) A millennium-length reconstruction of Bear River stream flow, Utah. J Hydrol 529:524–534

Downard R, Endter-Wada J, Kettenring K (2014) Adaptive wetland management in an uncertain and changing arid environment. Ecol Soc 19(2)

Edwards EC, Null SE (2019) The cost of addressing saline lake level decline and the potential for water conservation markets. Sci Total Environ 651(1):435–442

Edwards EC, Bosworth RC, Adams P, Baji V, Burrows A, Gerdes C, Jones M (2017) Economic insight from Utah's water efficiency supply curve. Water 9(3):214

Elmore LR, Null SE, Mouzon NR (2016) Effects of environmental water transfers on stream temperatures. River Res Appl 32(7):1415–1427

EPA (2019) Regional Screening Levels (RSLs)--Generic Tables. https://www.epa.gov/risk/regional-screening-levels-rsls-generic-tables.

Gavrieli I, Lensky NG, Abelson M, Ganor J, Oren A, Brenner S, Shahar H (2011) Red Sea to Dead Sea water conveyance (RSDSC) study: Dead Sea research team, Rep. GSI/10/2011

Génova PA, Null SE, Olivares MA (2018) An economic-engineering method to allocate water between instream and off-stream uses. J Water Resour Plann Manag 145(3):04018101. https://doi.org/10.1061/(ASCE)WR.1943-5452.0001026

Gillies RR, Wang S-Y, Booth MR (2012) Observational and synoptic analyses of the winter precipitation regime change over Utah. J Clim 25:4679–4698

Gohari A, Eslamian S, Mirchi A, Abedi-Koupaei J, Bavani AM, Madani K (2013) Water transfer as a solution to water shortage: a fix that can backfire. J Hydrol 491:23–39

Griffin DW, Kellogg CA (2004) Dust storms and their impact on ocean and human health. EcoHealth 1:284–295

Gutrich JJ, Gigliello K, Gardner KV, Elmore AJ (2016) Economic returns of groundwater management sustaining an ecosystem service of dust suppression by alkali meadow in Owens Valley, California. Ecol Econ 121:1–11

Hahnenberger M, Nicoll K (2014) Geomorphic and land cover identification of dust sources in the eastern Great Basin of Utah, U.S.A. Geomorphology 204:657–672. https://doi.org/10.1016/j.geomorph.2013.09.013

Hammer UT (1986) Saline lake ecosystems of the World. Dr W. Junk, Dordrecht

Hart J (1996) Storm over Mono: the Mono Lake battle and the California water future. University of California Press, Berkeley, CA. 0520203682

Indoitu R, Kozhoridze G, Batyrbaeva M, Vitkovskaya I, Orlovsky N, Blumberg D, Orlovsky L (2015) Dust emission and environmental changes in the dried bottom of the Aral Sea. Aeol Res 17:101–115. https://doi.org/10.1016/j.aeolia.2015.02.004

Jensen ME (2007) Beyond irrigation efficiency. Irrig Sci 25:233–245. https://doi.org/10.1007/s00271-007-0060-5

Johnson WP, Wurtsbaugh WA, Belovsky GE, Baxter BK, Black F, Angeroth C, Jewell P, Yang S (2020) Geochemistry of Great Salt Lake. Encyclopedia of water: Science, Technology, and Society (in Press)

Jones EF, Wurtsbaugh WA (2014) The Great Salt Lake's monimolimnion and its importance for mercury bioaccumulation in brine shrimp (*Artemia franciscana*). Limnol Oceanogr 59:141–155

Kijowski AK, Neill J, Wickline A, Swift J, Butler JK, Kimberly DA, VanLeeuwen J, Luft J, Stone K (2020) American White Pelicans of Gunnison Island. In: Baxter BK, Butler JK (eds) Great Salt Lake biology: a terminal lake in a time of change. Springer, Cham

Kittle S (2000) Survey of reported health effects of Owens Lake particulate matter. Great Basin Unified Air Pollution Control District. Bishop, California. 10 p. http://www.gbuapcd.org/Information/SurveyofReportedHealthEffectsofOwensLakeParticulateMatter.pdf

Libecap GD (2009) The tragedy of the commons: property rights and markets as solutions to resource and environmental problems. Aust J Agric Resour Econ 53(1):129–144

Lindsay MR, Dunham E, Boyd ES (2020) Microbialites of Great Salt Lake. In: Baxter BK, Butler JK (eds) Great Salt Lake biology: a terminal lake in a time of change. Springer, Cham

Loving BL, Miller CW, Waddell KM (2000) Water and salt balance of Great Salt Lake, Utah, and simulation of water and salt movement through the causeway, 1987-98. Water-Resources Investigations Report. U.S. Geological Survey, Washington, DC

LPP (Lake Powell Pipeline) (2019) Washington County's repayment options for the Lake Powell Pipeline fact sheet. https://lpputah.org/wp-content/uploads/2019/01/LLPFactSheet-Repayment-Options-012819.pdf

Micklin P (2007) The Aral Sea disaster. AREPS 35:47–72. https://doi.org/10.1146/annurev.earth.35.031306.140120

Micklin P (2016) The future Aral Sea: hope and despair. Environ Earth Sci 75(9):844

Mohammed IN, Tarboton DG (2012) An examination of the sensitivity of the Great Salt Lake to changes in inputs. Water Resour Res 48(11). https://doi.org/10.1029/2012WR011908

Null SE, Wurtsbaugh WA, Miller C (2013) Can the causeway in the Great Salt Lake be used to manage salinity Friends of the Great Salt Lake Newsletter 19(1), 2 p. http://works.bepress.com/wayne_wurtsbaugh/143/

Null SE, Mouzon NR, Elmore LR (2017) Dissolved oxygen, stream temperature, and fish habitat response to environmental water purchases. J Environ Manag 197(15):559–570

Office of the Legislative Auditor General (2015) A performance audit of projections of Utah's water needs. Report to the Utah Legislature Number 2015-01, 82 p. https://water.utah.gov/1AuditProjectionsUtahWaterNeeds.pdf

Pace A, Bourillot R, Bouton A, Vennin E, Galaup S, Bundeleva I, Patrier P, Dupraz C, Thomazo C, Sansjofre P, Yokoyama Y (2016) Microbial and diagenetic steps leading to the mineralisation of Great Salt Lake microbialites. Sci Rep 6:31495

Pacific Institute (2013) The World's water, Vol. 7. http://worldwater.org/wp-content/uploads/2013/07/ww8-table2.pdf

Paul DS, Manning AE (2002) Great Salt Lake Waterbird Survey Five-Year Report (1997–2001). Publication Number 08-38. Utah Division of Wildlife Resources, Salt Lake City, UT

Perry KD, Crosman ET, Hoch SW (2019) Results of the Great Salt Lake dust plume study (2016-2018). Final Report to Utah Forestry, Fire and State Lands, Salt Lake City, UT, 305 p. https://d1bbnjcim4wtri.cloudfront.net/wp-content/uploads/2019/12/10101816/GSL_Dust_Plumes_Final_Report_Complete_Document.pdf

Prudencio L, Null SE (2018) Stormwater management and ecosystem services: a review. Environ Res Lett 13(3):033002. https://doi.org/10.1088/1748-9326/aaa81a

Reisner M (1993) Cadillac desert: the American West and its disappearing water. Penguin Books, New York, NY

Schindler DW (2009) Lakes as sentinels and integrators for the effects of climate change on watersheds, airsheds, and landscapes. Limnol Oceanogr 54(6, Part 2):2349–2358

SWCA Environmental Consultants (2017) Water for Great Salt Lake. Prepared for Great Salt Lake Advisory Council. Salt Lake City, UT, 54 p

UDNR FFSL (Utah Department of Natural Resources Division of Forestry, Fire and State Lands) (2013) Final Great Salt Lake Comprehensive Management Plan and Record of Decision. Prepared by SWCA Environmental Consultants. Salt Lake City, UT, 391 p. https://ffsl.utah.gov/wp-content/uploads/OnlineGSL-CMPandROD-March2013.pdf

White JS, Null SE, Tarboton DG (2015) How do changes to the railroad causeway in Utah's Great Salt Lake affect water and salt flow? PloS One 10(12):e0144111

Wine ML, Null SE, DeRose JR, Wurtsbaugh WA (2019) Climatization—negligent attribution of Great Salt Lake desiccation: a comment on Meng (2019). Climate 7(5):67

Wurtsbaugh WA (1992) Food web modifications by an invertebrate predator in the Great Salt Lake (USA). Oecologia 89:168–175

Wurtsbaugh WA (2018) Effects of eutrophication on birds in three bays of Great Salt Lake: a comparative analysis with Utah DWR Waterbird Survey Data. Final Report. Utah Division of Forestry, Fire and State Lands. https://digitalcommons.usu.edu/wats_facpub/1012/

Wurtsbaugh WA, Berry TS (1990) Cascading effects of decreased salinity on the plankton chemistry, and physics of the Great Salt Lake (Utah). Can J Fish Aquat Sci 47(1):100–109. https://doi.org/10.1139/f90-010

Wurtsbaugh WA, Gardberg J, Izdepski C (2011) Biostrome communities and mercury and selenium bioaccumulation in the Great Salt Lake (Utah, USA). Sci Total Environ 409:4425–4434. https://doi.org/10.1016/j.scitotenv.2011.07.027

Wurtsbaugh WA, Marcarelli AM, Boyer GL (2012) Eutrophication and metal concentrations in three bays of the Great Salt Lake (USA). 2009 Final report to the Utah Division of Water Quality, Salt Lake City, UT, 70 p. http://digitalcommons.usu.edu/cgi/viewcontent.cgi?article=1548&context=wats_facpub

Wurtsbaugh WA, Miller C, Null SE, DeRose RJ, Wilcock P, Hahnenberger M, Howe F, Moore J (2017) Decline of the world's saline lakes. Nat Geosci 10:816–821

York C, Goharian E, Burian SJ (2015) Impacts of large-scale stormwater green infrastructure implementation and climate variability on receiving water response in the Salt Lake City area. Am J Environ Sci 11(4):278

Chapter 2
Climate Change and Great Salt Lake

Bonnie K. Baxter and Jaimi K. Butler

Abstract Terminal lakes are highly susceptible to climate change impacts since water that enters through precipitation, runoff, and groundwater must be balanced with water that leaves through evaporation. A change in this equation can lead to a decline in elevation, which can be tragic for the ecosystem, particularly if the closed basin is shallow. Great Salt Lake faces many threats that will impact the volume of water in the depression of the Bonneville Basin where it resides. If the lake's level declines, salinity increases, and wetlands are altered. Salinity is a driver of microbial diversity and, as this foundation of the ecosystem is altered, so will be the rest of the food web, affecting large numbers of avian migrators along the Pacific and Central fly-ways. Human population growth and water diversions for agriculture have put a strain on Great Salt Lake, resulting in a terminal lake whose trajectory is downward in surface area. How might anthropogenic climate change impact this scenario? Alterations in temperature can influence the timing of snowmelt and change evapotranspiration. As temperatures increase and droughts persist, climate change will amplify the decline in lake elevation, creating more dust from the exposed lakebed. Dust blowing into inhabited valleys will worsen air quality with particulates and may be laden with the pollutants collected by the lake. Early melting of the snowpack in the Wasatch Mountains due to higher temperatures would be further impacted as airborne dust from the dry shorelines is deposited during storms and can reduce the albedo of snow, altering groundwater recharge of the watershed. The current status of Great Salt Lake, with no water rights of its own and increasing pressures for water use upstream, does not bode well for the survival of this critical ecosystem given climate change predictions for the southwestern United States.

Keywords Great Salt Lake · Climate change · Terminal Lake · Thalassohaline Lake

B. K. Baxter (✉) · J. K. Butler
Great Salt Lake Institute, Westminster College, Salt Lake City, UT, USA
e-mail: bbaxter@westminstercollege.edu

B. K. Baxter, J. K. Butler (eds.), *Great Salt Lake Biology*,
https://doi.org/10.1007/978-3-030-40352-2_2

2.1 The Nature of Terminal Lakes

Nearly 27% of the Earth's land-surface area is characterized by systems lacking an outlet for water drainage (de Martonne 1927). In some cases, the watershed will deliver water to a large single internal basin, forming an endorheic terminal lake (Williams 1993, 1996). These closed basins are by nature dynamic, responding to changes in input and output of water; water enters through direct precipitation, runoff, or groundwater seeps, and it leaves only through evaporation. They are also typically hypersaline with respect to marine water, which is 3.5% total salts, since the water flowing into the lake brings minerals that cannot leave, and evaporation results in a higher concentration of dissolved salts (Litchfield 2011; McGenity and Oren 2012).

Terminal lakes are sensitive to climate variation, both natural fluctuations and anthropomorphic climate change, as well as modifications to the land-surface hydrology such as damming and diverting flow (Williams 1996, 2002). In addition, these lakes, particularly larger ones in arid regions such as Great Salt Lake (GSL), present a great opportunity to study hydroclimatic change as they are superb recorders of regional conditions over time (Bowen et al. 2019). Data on the long-term record of precipitation and evaporation cycles can be inferred from such a closed basin (Mann et al. 1995).

2.1.1 *Thalassohaline Lakes*

Some terminal lakes such as GSL are thalassohaline, which refers to their origin being related to the sea, either in their chemical nature, e.g., the proportions of the salts present, or evidence for a prehistoric connection to the ocean (Ventosa and Arahal 2009). These lakes are characterized by a predominance of Na^+ and Cl^- ions but may feature other ions in lower concentrations (especially Mg^{2+}, SO_4^{2-}, K^+, Ca^{2+}, Br^-, HCO_3^-, and F^-) (Ventosa and Arahal 2009; Litchfield 2011). The geochemistry of GSL is consistent with the ion content of thalassohaline lakes (Johnson et al. 2019) and reflects solute inputs from weathering of a diversity of rocks in its drainage basin (Jones et al. 2009). Other terminal lakes are athalassohaline, meaning they are of continental origin, such as the Dead Sea where divalent cations such as Mg^{2+} and Ca^{2+} predominate over monovalent ions like Na^+ and Cl^- (Oren 1993).

As the water evaporates the salts concentrate. Each salt reaches its individual saturation point and precipitates out of the water, dependent on temperature, in a predictable order (Ventosa and Arahal 2009): first some carbonates, such as calcium carbonate, start to precipitate at around 10%. GSL carbonates have been described by Lindsay et al. (2019, 2020). NaCl (halite) precipitates above 30% salinity, of which we see evidence along the GSL north arm shoreline in the warm summer months (Perl and Baxter 2020). This provides the geochemistry backdrop for the ecosystem. Climate change, diversions, and other threats to water can result in an increase of mineral-rich shorelines that impact the terminal basin.

2.1.2 *Bonneville Basin over Time*

Modern GSL sits in the bottom of the Bonneville Basin, one of the lowest depressions in the largest contiguous inland watershed (Fig. 2.1) on the North American continent, the Great Basin (Cohenour and Thompson 1966). Over the last several million years, this area primarily held shallow lakes such as GSL, or mudflats and playa (Atwood et al. 2016). But the Bonneville Basin had several deep lake episodes during the last 780 thousand years, including late Pleistocene Lake Bonneville, 30–12 thousand years ago, covering an enormous swath of Utah and into other western states (Oviatt et al. 1999; Shroder et al. 2016) (Fig. 2.1).

During its existence, Lake Bonneville experienced dramatic fluctuations in climate. The examination of calcium carbonate deposits and stratigraphic columns of the shorelines demonstrated millennial-scale lake level fluctuations with amplitudes of 50 m between 30 and 10 thousand years ago (Oviatt 1997). Independent studies on vegetation and glacial records support this climate model during the transition between the Pleistocene and the Holocene (Godsey et al. 2005). Climatic events that impacted Lake Bonneville are consistent with climate records that describe other locations on Earth during the late Pleistocene, suggesting that millennial-scale climate change was global (Oviatt 1997; Godsey et al. 2005). The formation of GSL from the expansive Lake Bonneville occurred as a quick transition, geologically speaking, over just a few thousand years (Atwood et al. 2016; Shroder et al. 2016). As the Pleistocene thawed, the planet warmed, resulting in evaporation and leaking of Bonneville waters, sometimes in dramatic bursts. The margins of modern GSL lake were formed following these events, about 13,000 years ago (Fig. 2.1).

Today the Bonneville Basin is home to GSL, which is a vibrant ecosystem. It is the largest lake in the western United Sttates and one of the largest terminal saline lakes in the world (Keck and Hassibe 1979; Arnow and Stephens 1990; Johnson et al. 2019). The first white explorer to describe GSL, while mapping the river systems of the west, was John C. Frémont (1845). Captain Howard Stansbury led the earliest robust mapping expedition, for the US government, which captured the geology, chemistry, and biology of this immense lake in the nineteenth century (Stansbury 1855). His cartographic efforts were followed by the acclaimed work of G.K. Gilbert, who was the first to suspect the shelves visible on the mountains surrounding GSL were shorelines of deep lakes (Fig. 2.2) (Gilbert 1890). Gilbert proposed the existence of Lake Bonneville, and he recorded in remarkable detail the geology that supported this prehistoric lake event.

Current work on the Bonneville Basin has utilized a combination of geologic mapping and sophisticated techniques to provide a model of the sequence of lake events (Oviatt and Shroder 2016). The recorded history of the Bonneville Basin gives us a foundational baseline on which to understand current impacts on GSL and to build future climate projections of this closed interior basin system.

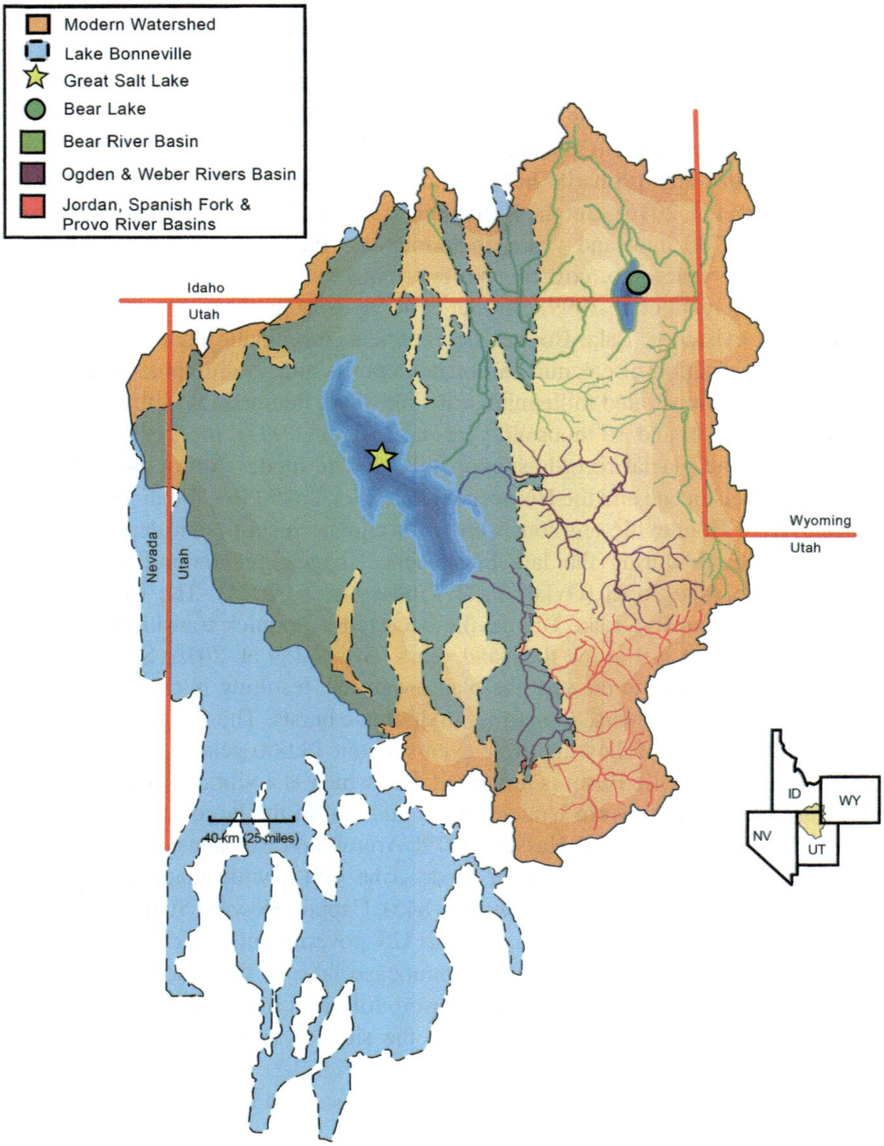

Fig. 2.1 Pleistocene Lake Bonneville and modern Great Salt Lake (GSL), including the current watershed. The area of Lake Bonneville is indicated, an immense freshwater lake during the last ice age that covered the Bonneville Basin in the western US states: Utah, Nevada, Wyoming, and Idaho. In the modern watershed, rivers flow into GSL, the bottom of the basin for approximately the last 13 thousand years. Image credit: Johanna Bossart, adapted from Atwood et al. (2016)

Fig. 2.2 Drawing by G.K. Gilbert's party of hillsides in the Wasatch Mountain Range, highlighting the shorelines of Lake Bonneville, which show climate change in the subsequent drops in elevation to the Provo, Stansbury, and Gilbert shorelines. Image is public domain

2.1.3 Great Salt Lake Historic Elevation Fluctuation

As a terminal lake, the elevation of GSL is dependent on its land-surface hydrology factors, which result in natural fluctuations (Fig. 2.3). The earliest recorded surface level of GSL was measured by Frémont in 1843 (Frémont 1845). Using several barometers, he recorded 4200 ft (1280.16 m) above sea level, which is likely accurate within 0.3 m (Arnow and Stephens 1990). This is considered the "average" elevation for GSL, as it is a typical mean cited in elevation studies over more than two centuries of measurements (Atwood et al. 2016). Stansbury took barometric readings as well during his expedition years (Stansbury 1855), and Gilbert estimated lake elevation during the time period of 1847–1874 using oral reports, observations of Antelope Island and Stansbury Island bars, and gauge readings near Black Rock and Farmington (Gilbert 1890). From 1875 to 1938, there is a decent record of gauge recordings from various sites, readings taken weekly or monthly (Arnow 1984; Arnow and Stephens 1990). Since that time until today, the US Geological Survey, in cooperation with the Utah Division of Water Rights, has kept accurate records and reliable numbers showing lake fluctuations (USGS 2019). In addition, the riverine input is measured by gauges placed in the Bear River (since 1902), the Weber River (since 1907), and the Jordan (since 1949) (Loving et al. 2000; Mohammed and Tarboton 2012).

Thus, for the better part of two centuries, GSL's elevation has been monitored, and there are a few events that are worth discussing. The historic low was recorded in 1963 after a century-long decline in elevation (Arnow and Stephens 1990; Gwynn 2007). The high point since monitoring began was 1986, following some interesting weather conditions causing early snowmelt and subsequent flooding (Karl and Young 1986). The flooding caused damage and financial losses, including 250 million US dollars in flood mitigation and 52 million for a brine (saline water) pumping station,

Fig. 2.3 Great Salt Lake elevation fluctuations. Landsat satellite imagery of Great Salt Lake at two time points. The left panel is 1985 and the right is 2010, image credit: NASA and USGS Landsat and Water Program, public domain

which could be used to prevent future flooding (Jellison et al. 2008). These elevation vacillations probably mimic that of GSL since its formation 13 thousand years ago (Atwood et al. 2016). Karl and Young (1986) hypothesized that the 1980s "wet period" of GSL flooding would occur with a periodicity of 100 years, but recent measurements indicate that the overall current trajectory is downward, without anticipated dramatic rises, most of which is likely due to freshwater diversions upstream of GSL (Wurtsbaugh et al. 2017). Climate change may also be contributing to this decline (Wang et al. 2018), but there is some debate about the proportion (Meng 2019; Wine et al. 2019).

Lake elevation data tracking is critical to understanding climate impacts, as temperature warming can increase evaporation, and this affects the salt concentrations of the brine directly. The salinity of GSL at any given time differs across the various regions of the lake from freshwater where rivers empty into the lake to 34% in the salt-saturated north arm (Baxter et al. 2005; Naftz et al. 2011, Johnson et al. 2019; USGS 2019). In the pelagic bays, the north arm and south arm open water, the salinity is measurably different from season to season, and from year to year (Fig. 2.3) (Gwynn 2007). As GSL elevation changes, the salts are either concentrated or diluted; therefore, salinity in these bays changes inversely to the lake elevation, though there is no longer a direct relationship due to anthropomorphic factors (Jones et al. 2009). The GSL relative water budget (Arnow and Stephens 1990) is written as:

$$\text{Inflow} = \text{Outflow} \pm \text{Storage Change}$$

This budget, along with salt concentrations of the water, impacts surface area/ volume and ultimately evaporation. The major inflow to GSL is from three rivers that drain the Wasatch Range snowpack and empty into the south arm of the lake (Fig. 2.1), but also from direct precipitation and minor groundwater inflow (Jones et al. 2009). Outflow is by evaporation. In sum, shifting climate parameters may cause vacillation of the GSL elevation and alter the salinity, which can have consequences for the ecosystem (discussed below in Sect. 2.3.4).

Because of the elevation shifts and patterns, GSL is a useful model with which we may learn about climate change. Climate does appear to influence the historical variation of the GSL volume; quasi-decadal variations create high amplitude oscillations in lake level (Fig. 2.3) (Mann et al. 1995), and the shallow bathymetry (average depth of ~5 m) makes GSL sensitive to climate variability (Jones et al. 2009; Tweed et al. 2011). Studies that investigate this sensitivity of GSL's response to changes in precipitation, streamflow, evaporation, surface area, volume, and salinity will be invaluable in building our understanding (Mohammed and Tarboton 2012).

2.2 Anthropomorphic Impacts on Great Salt Lake

Humans have likely been in Utah since the Pleistocene, between 20,000 and 15,000 years ago, the time of the high point of Lake Bonneville (Raghavan et al. 2015; Madsen 2015, 2016). Lake Bonneville and its watershed would have given humans a food source of freshwater fish. As the water level oscillated over time, responding to a changing climate, humans would have moved to follow the changing shorelines (Madsen 1999). As the water evaporated over a couple thousand years, and the lake got saltier, human interactions with the lake would have changed, perhaps pickleweed or salt collecting instead of fishing. As modern GSL was formed, ample remaining groundwater sources for freshwater springs existed until the sources were significantly depleted about 9500 years ago (Madsen 2015; Oviatt et al. 2015). The significance of the GSL wetlands to the native people is evident in that the Fremont buried dozens of their dead in this region, around the wetlands of GSL circa AD 400–1000 (Parr et al. 1996; Coltrain and Leavitt 2002).

On more recent timescales, approximately one thousand years ago until the late nineteenth century, Shoshonean-Paiute peoples with shared dialects occupied the Great Basin, and many lived around the margins of GSL (Cuch 2000; Baxter 2018). The Goshute people in particular were known for their connection to climate and cycles that influence the fauna and flora of their desert homeland, including the precise capacity of the land for grazing animals. Friction arose when white people moving into the area from the 1847 Mormon emigration began farming and raising livestock on land belonging to the Goshute, essentially starving the indigenous people who had to relocate. Thus began the settlement of the GSL area from descendants of Europeans (Deseret News 1907). From 1847 forward, humans had a different relationship with GSL that the species had previously, and they began to intervene with the ecosystem. This appears in stark contrast to indigenous peoples who lived alongside the shifting shorelines for thousands of years (Baxter 2018). However, in the GSL wetlands, fluctuating water availability and instability of climatic conditions did impact the behavior of native people, some of whom abandoned agriculture in this inhospitable area (Coltrain and Leavitt 2002). And humans have likely always influenced the lake (Trentelman 2020).

Residents of the Salt Lake valley in the late nineteenth century until the present have had an enormous impact on the GSL hydrology and, thus, the biology. They built resorts, only to have the shorelines recede or swell (Nichols 2014). They responded to flooding by building giant pumps (Karl and Young 1986), only to find the lake waters had receded before the expensive machines were completed. Smells of sulfate-reducing bacteria (Boyd et al. 2017) and sewage dumping into GSL (Smith 1936) have added to the fractious relationship Utahns have with the lake. In fact, the majority of people surveyed in the state consider it a waste for freshwater to enter the lake instead of being utilized by people directly (Bedford 2009).

The anthropomorphic interventions described below were intended to maximize the lake for industry, recreation, and transport. But they have inadvertently created a new GSL. In thinking about the future of the lake and the realities of climate change, it is this altered setting that we must consider, where water flow is restricted in various ways.

2.2.1 Damming and Causeways

GSL has been compartmentalized into various bays by the creation of structures that limit the flow of water (Arnow 1984; Arnow and Stephens 1990; Gwynn 2007). In the late 1950s, construction of a rock-filled railroad causeway effectively bisected GSL, segmenting the north arm away from the majority of freshwater input (Fig. 2.4a) (Cannon and Cannon 2002). This created an artificial salinity gradient between the largest bays of the lake (Adams 1964; Madison 1970). The north arm is at salinity saturation, which in the summer temperatures is around 34% salt. This limits the algae, invertebrates, and birds, which make up the south arm GSL food web, but the microbial diversity of pigmented halophilic microorganisms is immense, causing the pink hues seen in this part of the lake (Fig. 2.4a) (Baxter et al. 2005; Almeida-Dalmet et al. 2015; Baxter 2018). In contrast, the south arm receives the freshwater input from the watershed and is less saline, currently around 15%, resulting in a productive aquatic ecosystem teeming with phytoplankton, brine fly larvae and brine shrimp (Wurtsbaugh and Gliwicz 2001; Belovsky et al. 2011). Recent adjustments to the causeway have changed this disparate salinity gradient very little, given the haloclines and density differences that prevent mixing, though models suggest that they may balance the salt over time (White et al. 2015). Another prominent causeway, to allow vehicular access to Antelope Island, was built to the north end of the Island in the 1960s (Gwynn 2007). It effectively separates Farmington Bay to the south from the water north of this road.

Other damming events created critical bird habitats or salt evaporation ponds (Fig. 2.4b), but they also diverted water from the bottom of the basin, GSL. In the early twentieth century, the Bear River was diked and dammed for avian habitat

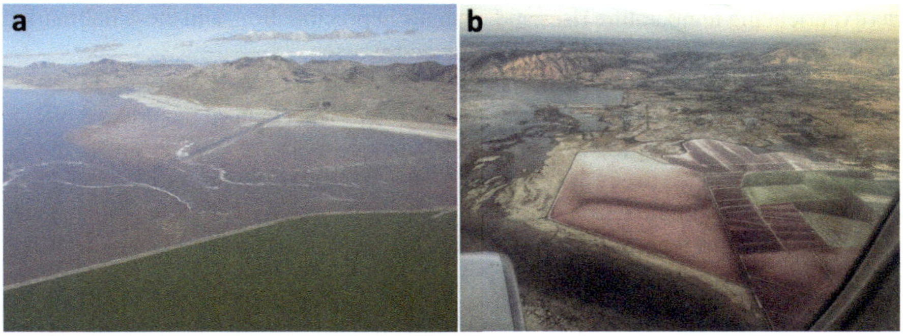

Fig. 2.4 Damming and causeways. (**a**) The railroad causeway separating the north arm from the south arm of Great Salt Lake. The rose-colored halophilic archaea and bacteria in the north arm are contrasted with a diverse group of algae species that color the south arm green. A trench, used to deliver high-density brine from the western salt ponds to a processing facility, can be seen in the middle of the photograph, image credit: Jaimi Butler. (**b**) Compass Minerals salt ponds at Great Salt Lake, Ogden, Utah. This view from an aircraft is eastward toward the Wasatch Mountains. In addition to the constructed evaporative salterns, you can also see containment of freshwater in Willard Bay in the top left of the photograph. This water is prevented from entering the basin but provides a reservoir for fishing and recreation, image credit: Tracy Christopherson

creating the Bear River Migratory Bird Refuge (USBR 1962; USDFW 2019). The structures are on the margin of the lake and have been maintained over time, preventing inflow of this water to GSL. For similar reasons, the Farmington Bay Wildlife Management Area was constructed beginning in 1935, which created a freshwater bay for bird habitat (State of Utah 2019a). Mineral extraction ponds are numerous around the lakeshore, further segmenting the lake brines (Fig. 2.4b) (Behrens 1980; Bingham 1980; State of Utah 2019b).

Lake level fluctuates in these various compartments depending on their connections to riverine and groundwater input. Also, the surface area, local salinity, and depths can impact evaporation. Therefore, one would predict some bays of GSL will be impacted more directly from activities that alter water input or from climate change.

2.2.2 *Diversions*

Land use changes upstream in a watershed, such as development for housing, industry, or agriculture, will change the freshwater input for a lake and can compound the impacts of climate change (Taylor et al. 2013). The level of GSL has been dramatically altered by the removal of freshwater before it has a chance to reach the lake (Null and Wurtsbaugh 2020). Diversions are estimated to have reduced the elevation of lake water by 3.4 m, decreased the volume of GSL by 48%, and exposed around 50% of the lakebed (Wurtsbaugh et al. 2017). This trend began before Utah was experiencing temperature rises and changes in precipitation, and these hallmarks of climatic change are expected to continue and worsen.

Population growth projections for the State of Utah indicate a doubling by 2065 (Ewert 2014; Perlich et al. 2015). Most of this growth will occur in the urban areas in the watershed of GSL. Thus, we expect to encounter more pressure to divert freshwater for anthropogenic uses (Yidana et al. 2010). This puts GSL in a perilous position where the effects of climate change can be devastating. A thorough discussion of the impact of diversions on GSL appears in this volume (Null and Wurtsbaugh 2020).

2.2.3 Industry

The most prevalent industries that work on GSL are the brine shrimp, *Artemia franciscana,* cyst harvesting companies, and the mineral extraction companies that produce various salts. The *Artemia* are an integral part of the GSL pelagic food web (Conover and Bell 2020). For this reason, the impacts of harvesting must be closely monitored, and this is done by a technical advisory group including industry partners, scientists, and the State of Utah Division of Wildlife Resources Great Salt Lake Ecosystem Program (Marden et al. 2020; Belovsky et al. 2011; Belovsky and Perschon 2019; State of Utah 2019c). This combination of stakeholders, and an open meeting format, has resulted in a well-managed industry that values brine shrimp as a species and not just a commodity, leaving a small footprint on the ecosystem.

Although culinary salts are no longer produced at this site, the GSL mineral extraction industry produces sodium chloride for road and water softener salt, magnesium chloride primarily for steel production, and potassium sulfate for fertilizer (Behrens 1980; Bingham 1980; Jones et al. 2009; State of Utah 2019b). The first step of production involves bringing lake water into constructed ponds, evaporating the water, and collecting the salts as each type reaches its individual saturation points and precipitates (Ventosa and Arahal 2009). The process removes desirable salts for GSL brine, but also returns unwanted salts to the lake. Each company's procedure is approved by, and land is leased through the State of Utah Department of Natural Resources (Bioeconomics, Inc. 2012; State of Utah 2019b). The footprint of mineral extraction as a whole may be more impactful than the brine shrimp industry due to water sequestration, but attempts are made to balance the removal of minerals with the input from the watershed, and the relative ion composition is not impacted (Gwynn 2007). In addition, bird habitat loss is a concern when open lake water is traded for dammed regions with higher salt concentrations (Conover and Bell 2020; Sorenson et al. 2020).

If the watershed delivers less water and evaporation increases, as climate change ensues, and a higher population in the region uses more upstream water, more lakebed will become exposed (Wurtsbaugh et al. 2017) and the brine in the south arm will become more concentrated. This could endanger the *Artemia* industry if the south arm waters become too saline (Marden et al. 2020). Also, more exposed lakebed will likely create difficulties for the mineral extraction industry around the lake margins. Although low lake elevation and the resulting higher salinity are great for mineral extraction process of evaporative collection, it becomes more challenging to deliver brine long distances from the lake to the salt ponds (Wurtsbaugh et al. 2016).

2.2.4 Pollution

Salt lakes are not protected by regulations on freshwater that govern human health, and therefore, they have a history of being dumping grounds for solid waste or effluents (Jellison et al. 2008). The US Clean Water Act is defined largely by the impact on drinking water or on fish, as pollutants that bioaccumulate in animals may impact humans who eat them (Hawkins 2015). Rivers that feed into GSL are protected by this federal Act, but this does not take into account that they are flowing into a closed basin and may accumulate toxins. Additionally, since the open waters of GSL do not contain fish, it has been difficult to assign standards for various pollutants that might prove toxic to, or be unhealthy for, the ecosystem (Adams et al. 2015).

Most important in discussing contaminants, GSL is a terminal lake and, therefore, a closed hydrological system. Any pollutant that enters the basin may accumulate, and some have the potential for biomagnification or for modification of the toxicity levels, such as mercury (Naftz et al. 2008a; Scott and Black 2020). Coal burning and mining in the west has led to atmospheric deposition or direct discharge of heavy metals into GSL, including mercury but also arsenic, copper, and selenium (e.g., Naftz et al. 2008b; Beisner et al. 2009; Adams et al. 2015). Reactive halides (e.g., Cl^-) are associated with salt lakes, including GSL, and can work to make metals more soluble (e.g., HgCl) in the aquatic system (Stutz et al. 2002). Agriculture and urban runoff from wastewater effluent bring excess nutrients into GSL and can cause eutrophic eruptions (Marcarelli et al. 2006). What is more, in saline lakes, the halophilic microorganisms do not degrade pollutants as readily as the microbial communities of freshwater systems (Oren et al. 1992).

Pollutants, then, remain in the GSL basin even as water evaporates, but they do have the means to escape through the biota of the lake. Contaminants may leave the GSL system through transfer from the aquatic to the terrestrial sphere through the food chain (e.g., Conover and Vest 2009; Saxton et al. 2013). And once bioaccumulated in birds, the pollutant may travel, as many of the GSL avian population migrate, taking the toxins with them (Conover and Vest 2009; Dicataldo et al. 2010). Significant in the context of climate change and a trend in declining lake elevations, newly exposed shorelines may contain pollutants that can become airborne, blowing into the populated centers along the Wasatch Front (Silcox et al. 2012; Whiteman et al. 2014).

2.3 Anthropogenic Climate Change

The Pleistocene to Holocene transition exemplified by Lake Bonneville (Fig. 2.1; Sect. 2.1.2) demonstrates natural climatic change as the Earth warmed up from the last ice age. However, the climatic variation recorded in recent decades reflects both anthropogenic and natural factors. Anticipated global trends expected from climate change have been well summarized by the Intergovernmental Panel on Climate Change (IPCC), a body representing multiple (~40) countries through the United Nations, whose role involves assessing the related science (IPCC 2018). This recent report represents a review of more than 6000 scientific studies, and the major conclusion was that the Earth will warm 1.5 °C, above pre-industrial levels, by 2040 if carbon emissions continue at the current rate. This will impact our planet with erratic weather, rising coastlines, droughts, and deluges.

One climate change impact discussed in the literature is alteration of precipitation, either in quantity or type. For example, some regions of the planet, such as the northeastern United States, will likely experience extreme rain (Witze 2018). This is due to warming, as warm air can hold more moisture. Also, more hurricanes and more extreme monsoons may be a common hallmark of some locales, instead of a rare characteristic. Various climate models disagree when this idea is applied regionally. Local systems, such as GSL, would affect the way moisture is deposited. Many models suggest dry areas could become more drought ridden (Cook et al. 2015), but some studies point to increased precipitation (Donat et al. 2016; Witze 2018). When looking at climate change over land masses, most areas get wetter, and previously dry areas may be at risk of flooding events (Donat et al. 2016). Donat and coworkers processed data over a 59-year period, looking at rain and snowfall in global regions that were wetter and drier than average land areas. Their analysis indicated that annual precipitation and extreme precipitation increased by 1–2% per decade in dry regions, including the western United States. Indeed, global warming has already increased the frequency of record-breaking rainfall events over most land areas (Lehmann et al. 2015).

Climate variability has also altered watershed groundwater systems, which then cannot be replenished by recharge (Anderson 2012; Treide 2012; Taylor et al. 2005, 2013; Meixner et al. 2016). Related to this, warmer temperatures change snowpack in regions of high elevation, affecting how snow is distributed or its accumulation. Global warming can cause the earlier melting of snow or rain in place of snow. This may impact recharge in groundwater significantly. Watersheds that depend on retention of water and slow delivery from melting snow will lose water more rapidly; this bad timing ultimately diminishes runoff and soil moisture (Cayan et al. 2010; Taylor et al. 2005, 2013).

All ecosystems have developed under the selection pressures of the past envelope of climate conditions, including temperature and precipitation variations. When change is slow over geologic time, the biodiversity adapts and evolves, but such sudden changes as the Earth is currently experiencing can have profound outcomes on the ecology of a system (USGCRP 2017).

2.3.1 Climate Projections for the Southwestern United States

Global warming does not heat all landmasses and oceans evenly; there are regions that will be more impacted than others. Early climate studies identified the southwestern United States (Utah, Nevada, New Mexico, Arizona, and parts of California) as a major "hotspot" for global warming sites (e.g., Thomson et al. 2005; Diffenbaugh et al. 2008; Dominguez et al. 2010; MacDonald 2010). A recent survey of NOAA data concerning projected temperature rises of 2.0 °C continues to rank this region as a hotspot (Mufson et al. 2019). Predictions for Utah are in the second tier of highest hotspot rankings and are not as dire as for the states southward in this region. However, topography matters, and intensity is increased in the high elevations of the Rocky Mountains in Utah and Colorado.

The US Global Change Research Program recently issued the Fourth National Climate Assessment Report, which collected data across federal agencies and provided predictions for climate change repercussions in the United States (USGCRP 2017). Although modeling predicts higher amounts of rainfall in some areas (Donat et al. 2016; Witze 2018), the southwestern United States is likely to maintain and increase drought conditions (Kunkel et al. 2013; USGCRP 2017). Snowpack and streamflow amounts are projected to decline, decreasing surface water supply reliability for cities, agriculture, and ecosystems.

In fact, the southwestern United States has experienced prolonged drought since 2001, evident in the annual precipitation being 20–40% lower than the mean across the area (Cayan et al. 2010; Woodhouse et al. 2010). Higher average temperatures in this region have contributed to longer warm seasons, resulting in less surface water and an increase in wildfires. Simulations (IPCC 2018) and paleoclimate soil hydrology study predictions (Fritz 1996; Cook et al. 2015) suggest that the southwestern states, including Utah, are likely to experience more severe droughts in the coming years.

The scientific rationale for drought conditions in the southwestern United States involves the linkage between this area and both the Pacific and Atlantic Ocean sea surface temperatures (SST) (Wagner et al. 2010). Pacific Ocean SST affect winter precipitation of the southwest in part by the positioning of westerly storm tracks. During El Niño conditions, when the Pacific SST are warm, the storm track is shifted toward the south and more moisture is delivered to these southwestern states. The hydrological cycle of the Great Basin is characterized by a quasi-decadal oscillation (QDO), a cyclic period of about 12 years, which is related to the Pacific QDO based on measurements of these SST changes (Wang et al. 2010). Alternatively, warm North Atlantic Ocean SST result in storm tracks in the northern parts of both the Atlantic and Pacific Oceans, which move toward the North Pole and weaken, and these conditions bring less winter precipitation to the Southwest United States. Winter is the primary season of hydrologic recharge in the southwest (Kunkel et al. 2013). Without winter storms, despite precipitation during hot months when runoff and rapid evaporation occur, groundwater does not recharge, and drought continues (Taylor et al. 2005, 2013).

Lack of moisture in drought conditions, coupled with high temperatures, can have cascading effects on ecosystems. Wildfires in the southwestern United States have increased dramatically in recent years (Hurteau et al. 2014). This climate-related flammability drives structural changes in plant biodiversity and communities and can allow for the accumulation of cheat grass or other invasives which make great tinder for future fires. This in turn affects all ecosystems downwind. In Utah, increasing air quality problems are due to fires in the west (Mallia et al. 2015).

2.3.2 Climate Projections for the Wasatch Front

Climate projections are not completely characterized in Utah and the larger Great Basin region; there is more work to do in understanding this area (Kunkel et al. 2013), especially along the front of the Wasatch Mountain Range where GSL resides. An analysis of data over the last century indicated a significant increasing trend in air temperature in Utah (dos Santos et al. 2011). One study of winter precipitation noted that snowfall has decreased by 9% in the last half century, reflecting not only less precipitation but a shift in the type or precipitation; a higher percentage of the moisture fell as rain instead of snow (Gillies et al. 2012). Snow depth and storm systems that bring snow have decreased all across Utah.

Regional features are important to consider, for example, mountains can greatly enhance precipitation from the creation of low-pressure systems, and they collect and store water in the form of snow and ice. As stated, this heavy snowpack is critical for water resources in the intermountain west and is an important factor of the hydrological cycle (Gillies et al. 2012). The Wasatch Range of the Rocky Mountains are high in elevation and deliver water to the valleys below. For this reason, northern Utah is likely to experience fewer dry spells than other areas of the southwest, but serious droughts nonetheless (Cayan et al. 2010). Mountain snowmelt preserves soil moisture as the snow melts slowly over the warm seasons, so the timing of the melting action is critical (Painter et al. 2018). Any alteration in snowmelt can influence water availability downstream in the watershed and can affect potential flooding, water scarcity, and water quality. In the framework of climate change, projected decreases in snowpack and increases in evapotranspiration due to early melting (and more precipitation falling as rain versus snow) will certainly impact the downstream ecosystems of the Wasatch Front (Garfin et al. 2014; Scalzitti et al. 2016).

As the human populations have grown along the valleys of the Wasatch Range, air pollution and dust have increased. These particles are interacting with snow in the mountains, increasing absorbed solar radiation by reducing the reflective capacity (albedo), and increasing snow melt (Gillies et al. 2012; Crosman and Horel 2016; Painter et al. 2018; Skiles et al. 2018; Goodman et al. 2019). Even when temperatures are not rising locally, this reduced albedo effectively causes early snowmelt and increase in stream flow, moving water through the watershed too quickly to recharge groundwater (Skiles et al. 2018). Population growth in Utah is expected to continue, particularly in the northern part of the state (reviewed in Trentelman 2020), which will only increase this effect.

2.3.3 *Great Salt Lake and a Warming Earth*

Considering anthropogenic climate change, the largest concern for GSL is quantity of water. As discussed above, GSL is a terminal lake, and its elevation responds dramatically to seasonal or annual fluctuations in water input. Also, the lake itself contributes to the hydroclimate of GSL and the Wasatch Front with lake-effect precipitation (Yeager et al. 2013). Rising air temperatures affect lake water temperatures (Crosman and Horel 2009), which increases evaporation (Strong et al. 2014). Global warming events or shifts that impact precipitation will have large-scale effects on this lake. The characteristic volume changes make GSL a useful case study for climate change (Mann et al. 1995), similar to other terminal lakes such as Lake Urmia in Iran which is rapidly disappearing (Abbaspour et al. 2012).

GSL is already at risk of water depletion from diversions and damming (Wurtsbaugh et al. 2017; Null and Wurtsbaugh 2020). The shorelines have been exposed 48% since the historic high of 1986. Considering the discussion in Sect. 2.3.2, the future GSL watershed will likely experience less water input due to the regional shift in snowfall to rainfall, higher rates of evapotranspiration, and the timing of snowmelt (due to increased air temperatures or decreased albedo) (Skiles et al. 2018). This is similar to projections made regarding Mono Lake in California in an extensive modeling study (Ficklin et al. 2013). Applying this model to GSL, the system will very likely experience a feedback loop. Less water results in more shoreline exposure, which can create more dust from the dry playa, which can be deposited on the Wasatch Mountain snow, which can decrease the water in the watershed. In this scenario, this cycle would cause a rapid decline in the elevation of this shallow lake as the faster rate of snowmelt would be compounded by dust forcing (Painter et al. 2018).

An important structural attribute to consider in climate change scenarios: GSL is very shallow and has a greater surface-area-to-volume ratio relative to similar bodies of water (Fig. 2.5). Utah's inland sea sits at the bottom of a wide, flat basin. Other terminal lakes are smaller but are much deeper. For example, Pyramid Lake (Nevada, USA) is a tenth the size of GSL considering surface measurements, but it holds 25% more water (Fig. 2.5). The shallow nature of GSL, plus the high salinity content, makes GSL more susceptible to evaporation and, thus, dramatic elevation fluctuations (Atwood et al. 2016; USGS 2019).

A terminal lake is an excellent recorder of climatic change, and the lakes that have existed in the Bonneville Basin over time have left a sedimentary record of the lakes that preceded GSL (Oviatt 1997; Atwood et al. 2016). This gives researchers a foundation from which to build and data to reinterpret as methods get more

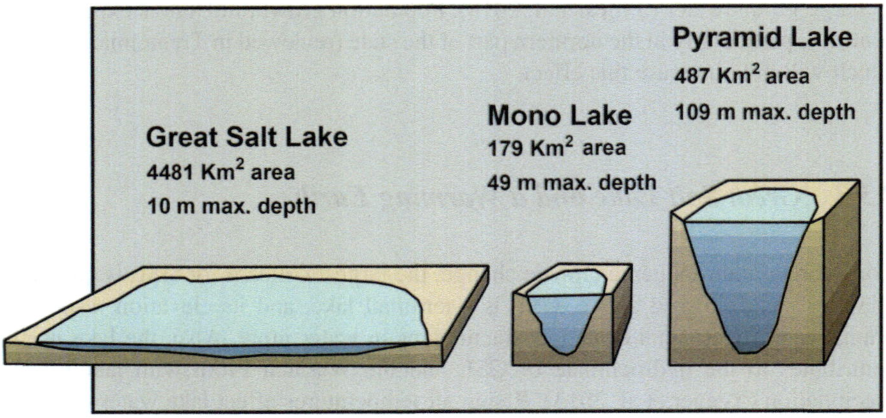

Fig. 2.5 Great Salt Lake surface area and volume relative to other terminal lakes in the Great Basin of the southwestern United States. Image credit: Johanna Bossart, adapted from the Genetics Science Learning Center, University of Utah

refined allowing them to predict the future of GSL levels. Reconstruction of paleoenvironmental conditions from the GSL sediments can help us understand future changes. A recently published GSL model that employs such paleoenvironmental data (Bowen et al. 2019) will be a useful tool for more precise prediction of future climate fluctuations. Also, a mass balance modeling study evaluating the sensitivity of GSL to input changes and predicting future lake elevations (Mohammed and Tarboton 2012) demonstrated how we can use historic GSL elevation data to make climate predictions. The authors of this work confirmed that fluctuation in streamflow is the predominating impact factor in GSL elevation, which is consistent with the modeling study on Mono Lake (Ficklin et al. 2013). Other current ecological models (Barrett and Belovsky 2020) may have different assumptions regarding the pressure of human populations, but all will be key for projecting the ecosystem dynamics and response to changes by the biological system.

2.3.4 *Impact on Biology of Great Salt Lake*

Sudden changes do not give life time to adapt and evolve especially in the context of an entire ecosystem. Anthropomorphic climate change may produce enormous outcomes for the ecology of a system (USGCRP 2017). Decreases in water input in the context of rising temperature will certainly affect the biota of GSL. Much of this is driven by changes in salinity as organisms are adapted to particular salinity ranges, and the higher salinity predicted in drought conditions may have major effects on the plankton as well as the higher trophic levels of the lake (Williams 1996, 2002).

2.3.4.1 Microbial Consortia and Primary Producers

Salinity rises are correlated with a reduction in microbial diversity in other saline lakes, especially phytoplankton (Ficklin et al. 2013; Moore 2016; Smith et al. 2010; Tweed et al. 2011). At the foundational level, the primary producers in mats associated with the microbialite structures in the benthic region thrive at current south arm concentrations of salts but are absent in the north arm hypersaline water (Lindsay et al. 2017, 2019). The other factor that affects microbialites is exposure. These carbonate structures layered with microbial mats are relatively close to the shoreline, which makes them vulnerable to being out of water if the lake level and surface area decline (Lindsay et al. 2020). Other phototroph communities in the water column, which also power the lake with the energy of photosynthesis, have documented shifts in community structure across salinity gradients (Meuser et al. 2013). Dilution of the south arm during the floods of the mid-1980s resulted in only 6% salinity in the south arm (compared to around 15% currently), and the diversity of phototrophs increased (Stephens 1990).

2.3.4.2 Invertebrates

If GSL elevation continues to decline as models predict, the photosynthetic primary producers will decrease in both number and species composition. Since these microorganisms serve as food for the pelagic invertebrates, brine shrimp (*Artemia franciscana*) and brine fly larvae (*Ephydra* spp.), changes in the phototroph assemblages will directly impact the success of these secondary consumers. In fact, high water years at GSL resulted in higher numbers of shrimp correlating with increased phototrophs in the water column (Stephens 1990). This positive correlation holds true for low water years and increasing salinity as well; both phototrophs and brine shrimp populations are reduced (Belovsky et al. 2011). However, species-specific blooms of algae may counter this model: large diatoms, which were difficult for *Artemia* to ingest, increased in numbers in 1999 during a relatively high water year and impacted the shrimp population negatively. Considering the wetlands, the mudflats and playa around the lake would also be altered dramatically with shifting shorelines and changes in salt concentrations, impacting other invertebrates' availability to the higher trophic levels (Sorenson et al. 2020).

There are also salinity impacts on invertebrates not related to food sources. The historic high for GSL was correlated with a change in the buoyancy of *Artemia* cysts (Stephens 1990), making them harder to harvest, and the predation of juveniles by corixids (Fig. 2.6a) (Wurtsbaugh 1992). These insects only live at marine salinities or just above and are usually restricted to the zones where freshwater groundwater seeps meet GSL brine. Though brine shrimp tolerate a broad range of salinities, up to salt saturation, depending on their phase of life cycle, at the higher salinities they are stressed and may not reproduce well (Stephens and Birdsey Jr 2002). Brine fly larvae have a narrower range, and a reduction in biomass is observed when salinity is increased above 20% (Barnes and Wurtsbaugh 2015).

Fig. 2.6 Climate change impacts on the biology of Great Salt Lake. (**a**) Corixids are predators for brine shrimp when salinity of brine allows their habitats to overlap. Image credit: Jaimi Butler. (**b**) The PELIcam project (Kijowski et al. 2020), studying a breeding colony of American white pelicans on Gunnison Island, has documented coyotes on the island following development of a land bridge. On the left of this image, you see a heron nest that was attacked by a coyote (seen on the right). The camera also caught the simultaneous disturbance of other birds in the area, which is another impact beyond predation. Future shrinking shorelines may be devastating to bird populations that depend on isolation from predators. Image credit: The PELIcam project (Utah Division of Wildlife Resources and Great Salt Lake Institute at Westminster College)

Temperature rises will certainly affect invertebrates in the GSL ecosystem. Currently, the adult *Artemia* die off as the winter approaches (Marden et al. 2020). Prior to this time, in the autumn, decreasing temperatures signal a shift from live birth to dormant encysted embryos. Air temperature changes, predicted by climate change models, will warm the lake water (Crosman and Horel 2016). A warmer GSL could cause the overwintering of adults and less cyst production, changing the life cycle. To our knowledge, work on temperature-sensitive gene regulation has not been done in *Artemia*, but this type of genetic switch likely exists. A common cellular response to warmer environs is the production of Heat Shock Proteins which help stabilize other proteins (Kregel 2002). Also, some genes are differentially expressed at different temperatures, and this has been documented in crustaceans (Whiteley et al. 1997). Although we do not yet understand the specific impacts of temperature rise on GSL invertebrates, it is likely to negatively affect the population, especially considering the added impact of salinity stress.

2.3.4.3 Avian Species

Ten million birds represented by 338 species depend on GSL and its diverse and extensive wetland assemblages as an enriching stopover during migration (Bellrose 1980; Paul and Manning 2002; Aldrich and Paul 2002; Neill et al. 2016; Oring et al. 2019; Sorenson et al. 2020). The invertebrates of the open water, brine shrimp and brine fly larvae/pupae, combined with the invertebrates in the wetlands, provide diverse food sources for a variety of avian feeders. Salinity impacts on the invertebrates as described above will certainly affect the next trophic level, primarily birds.

A shrinking GSL with lower water levels has physical impacts as well as creating zones of higher salinity; lower water levels can change feeding or nesting habitat, forming land bridges to islands which permit predator species access (Fig. 2.6b) (Kijowski et al. 2020) or altering depth. Wetland water depth is correlated with the leg length of the shorebirds that feed and nest there (Sorenson et al. 2020). Decreases in elevation impact the pelagic birds as well, resulting in less open water habitat (Conover and Bell 2020). The impact of drought on GSL birds can already be observed with some important species, such as North America's Wilson's phalaropes (*Phalaropus tricolor*) and red-necked phalaropes (*Phalaropus lobatus*) (Frank and Conover 2019). Water depth was a critical factor in the density of phalarope flocks at GSL sites studied.

The National Audubon Society recently sounded the alarm on how declining GSL levels may impact birds worldwide since the lake system contains five designated "Globally Important Bird Areas" and is an important site within the Western Hemisphere Shorebird Reserve Network (Shoop 2019). Millennium Ecosystem Assessment Predictions pointed at the combined effect of land use changes and climate change which will imperil avian species (Jetz et al. 2007). Even ignoring shifting migration patterns due to climate change, this study projected the loss of 950–1800 species of birds by the year 2100. Avian-rich sites such as GSL have an important role to play in preventing species loss. Already, bird surveys reveal a loss in numbers of critical species in the GSL system (Conover and Bell 2020; Sorenson et al. 2020).

2.4 The Future of Great Salt Lake

The tools and models for managing lakes are built on a system of freshwater lake management, which will not work for a saline lake. As we look forward for GSL, we must work in the context of a closed terminal basin with water that accumulates solutes that enter. Also, threats to water inputs may be ignored since the lake water is not utilized for human consumption nor fish.

2.4.1 Anthropogenic Climate Change Combined with Other Watershed Pressures

A recent GSL climate change publication created a debate because of the invalid assumptions that water diversions occurred after water had entered the lake as opposed to upstream of the lake (Meng 2019; Wine et al. 2019). This discussion, however, enlightens all of us. Clearly both land use changes, such as an increase in diversions for agriculture (Sect. 2.2.2), and climate change (Sect. 2.3), will have the same disastrous effect of less water making its way to GSL, at the bottom of the Bonneville Basin. Streamflow into GSL is the dominant factor impacting lake levels (Mohammed and Tarboton 2012), and it is impacted by both climate change (Ficklin et al. 2013) and water diversions (Null and Wurtsbaugh 2020) in ways that are more than additive given the positive feedback loop resulting from dust creation and its impact on the timing of snowmelt.

Water development may currently be the largest threat, and citizens and governments may intervene to change this impact to the watershed (Grimm et al. 1997). However, watershed managers may move policy in the wrong direction if they blame a shrinking GSL entirely on global warming as this framing may make citizens feel as if the water loss is out of their control (Wine et al. 2019). How do we move forward with realistic projections of both water needs and climatic effects?

Pushing back against the more pressing anthropomorphic threats, such as upstream diversions, will serve GSL well, but we should also incorporate expectations of more extreme climatic conditions. Fragile terminal lake ecosystems, so dependent on water input, should be managed in the context of global climate change. As the temperatures rise in the coming decades (Cook et al. 2015), we will likely see a synergistic effect, compounding the water loss in the system (Jetz et al. 2007; Taylor et al. 2013; Huybers et al. 2016). Therefore, it is critical to consider both forces, climate change impacts and water diversions, when making predictions about the future of GSL. Ignoring climate in water projections puts GSL and its wetlands in peril.

2.4.2 Water and Air Pollution in a Dry Basin

Lower GSL elevations will change the lake's interplay with accumulation and processing of pollutants, described in Sect. 2.2.4. Considering heavy metals, less lake surface area could result in reduced atmospheric deposition, for example (Scott and Black 2020), which is considered to be the most significant source of Hg contamination (Peterson and Gustin 2008; Naftz et al. 2009). Given the significant levels of Hg and its methylated forms currently in the GSL waters, lessened deposition may be too little too late.

Exposure of shoreline is a concern as water levels recede since pollutants of the lake will then become contaminants of the salt playa, which can enter the air in dust clouds (Goodman et al. 2019). California's Owens Lake is a harbinger of the dust that can be generated when a lake system becomes dry (Cahill et al. 1996; Gill et al. 2002). Dust from GSL playa would contain such heavy metals and the remnants of anything that has been dumped or drained or discharged into GSL since the human population upswell began in the late nineteenth century.

The air quality problem along the Wasatch Range could be tremendous as climate change amplifies water loss, even considering the dust in absence of any toxins. The significance of dust is overlooked with respect to its influence on ecological systems and human health (Field et al. 2010). The inhabited valleys of the Wasatch Front currently experience long periods of temperature “inversions,” characterized by a stable boundary layer that holds in pollution, suppressing vertical mixing (Silcox et al. 2012). Especially in the winter, dangerous pollution has been measured under the inversion envelope, including long periods of $PM_{2.5}$ (particulate matter with a diameter of less than 2.5 μm) concentrations that exceed the National Ambient Air Quality Standard and threaten human health (Silcox et al. 2012; Whiteman et al. 2014). More shoreline will certainly lead to more dust, likely year-round instead of seasonally, and an increase in air quality problems for the citizenry of Utah. As discussed above, this problem becomes cyclical as an increase in dust deposition on snow changes the timing of melt, thereby reducing water inputs into GSL (Skiles et al. 2018). Therefore, less water to the lake leads to conditions that favor even less water to the lake, resulting in poorer air quality.

2.4.3 Protecting Great Salt Lake

GSL is a unique terminal lake in the southwestern United States, a critical bird habitat, a protector of regional air quality, and an economic driver for the region. It is in danger of disappearing as global climate change puts pressure on an already fragile system. What can be done? Understanding such elements as the management of the lake, water rights, and water pricing structure in the watershed are seminal points to discuss concerning action.

2.4.3.1 Government Oversight

The US Federal Government established a public trust doctrine, after the Revolutionary War, which mandated that individual states would hold navigable water bodies and related resources “in trust” for the benefit of the people of that state (Koehler 1995; Ryan 2015). This doctrine has been employed to preserve groundwater, most notably in the case of Mono Lake, California (Spiegel 2010). The US Supreme Court ruled in favor of the National Audubon Society, an outcome that resulted in a guarantee of a minimum amount of water delivered to Mono Lake. GSL is similarly a terminal lake in the west that has been deemed navigable, and it is held in public trust. This doctrine has not yet been used in legal argument to preserve inflow to GSL.

While GSL is considered to be held in public trust, it has not been assigned a water right. During the westward US expansion in the nineteenth century, the practice of resource allocation emerged, first governing land and minerals but eventually this ethic applied to the scarce sources of water (Shupe et al. 1989). Senior water rights are still respected in the arid western states, and they have become commodities that can be bought and sold. Since GSL was not filled with freshwater that could be used for drinking, irrigating, or fishing, it never was assigned a water right of its own.

The management of GSL by the State of Utah is accomplished through several agencies, with the primary one being the Utah State Division of Forestry, Fire, and State Lands, which oversees a Comprehensive Management Plan mandated by the Utah Legislature (State of Utah 2000). Brine shrimp and phytoplankton are monitored and managed by the Great Salt Lake Ecosystem Program, a special project within the Division of Wildlife Resources (State of Utah 2019c). The Division of Water Quality monitors criteria that measure pollutant inputs (State of Utah 2019d). The Division of Water Resources is disconnected from the other management agencies and does not manage GSL per se. It is "tasked with planning, conserving, developing and protecting Utah's water resources" (State of Utah 2019e), but the job of both conserving and developing can be juxtaposed. Currently, this agency is engaged in the Bear River Development project which may threaten the lake's water input (Utah Rivers Council 2017).

2.4.3.2 Water Conservation

Water conservation is an environmental ethic, especially concerning ecosystems and their biological components (Wilson 2017; Campbell et al. 2004). But it is also smart policy for dry places. Utah is one of the most arid landscapes in North America. However, the water use is inefficient (Ewert 2014). Perhaps this is related to water pricing as Utah has the cheapest water in the west, which does not encourage sustainable practices (Utah Rivers Council 2017). Cities in the western United States that have increased the price of water have reduced water usage by homes and agriculture. Indeed, this has been shown to be an effective strategy in longitudinal studies (Campbell et al. 2004). The cost of not conserving should also be weighed. Owens Lake was famously sucked dry by water piped to Los Angeles, CA (Cahill et al. 1996), and now residents pay monthly, as part of their water bill, to mitigate the dust of the dry Owens valley. Municipalities in Utah should be looking at increasing the price of water to encourage conservation of water, and this would bring more water to GSL as we approach the pressures of climate change.

2.4.3.3 Action

The Utah State Legislature, with the support of many GSL stakeholders, recently adopted a resolution that was signed by Utah Governor Gary Herbert, "HCR010: Concurrent Resolution to Address Declining Water Levels of the Great Salt Lake" (State of Utah 2019f). Though this resolution does not mandate action, it does serve to recognize the critical importance of adequate water flows to the GSL ecosystem. A prior legislative action, House Bill 343, which created the Great Salt Lake Advisory Council, was passed in 2010 (State of Utah 2019g). This group of stakeholders and officials meets monthly and works to advise management agencies on sustainable use, protection, and development of the lake. In both cases, these actions were not associated with funding nor oversight, but could serve to elevate the conversation of the health of GSL in the future.

State managers, advocacy groups, academics, industry representatives, recreation and tourism officials, and the public have been meeting formally and informally for the last two decades (e.g., FRIENDS of Great Salt Lake 2018) and have formed a passionate GSL community that is informed by science. Vibrant discussions about water rights for the lake, public trust doctrine, and water pricing structures dominate many meetings of stakeholders. Collaboration is key for managing the future of GSL as anthropogenic climate change descends upon Utah's important saline lake.

Acknowledgements The authors are grateful for input and edits from Erik Crosman and helpful discussions about water diversions with Sarah Null and Wayne Wurtsbaugh.

References

Abbaspour M, Javid AH, Mirbagheri SA, Givi FA, Moghimi P (2012) Investigation of lake drying attributed to climate change. Int J Environ Sci Technol 9(2):257–266

Adams TC (1964) Salt migration to the northwest body of Great Salt Lake, Utah. Science 143 (3610):1027–1029

Adams WJ, DeForest DK, Tear LM, Payne K, Brix KV (2015) Long-term monitoring of arsenic, copper, selenium, and other elements in Great Salt Lake (Utah, USA) surface water, brine shrimp, and brine flies. Environ Monit Assess 187(3):118

Aldrich TW, Paul DS (2002) Avian ecology of Great Salt Lake. In: Gwynn JW (ed) Great Salt Lake: an overview of change. Utah Department of Natural Resources, Salt Lake City, UT, p 584

Almeida-Dalmet S, Sikaroodi M, Gillevet PM, Litchfield CD, Baxter BK (2015) Temporal study of the microbial diversity of the north arm of Great Salt Lake. Microorganisms 3:310–326

Anderson RB (2012) Quantity and quality of groundwater discharge in a hyper-saline lake environment, Great Salt Lake, Utah, USA. Thesis. The University of Utah, Salt Lake City, UT

Arnow T (1984) Water-level and water-quality changes in Great Salt Lake, Utah, 1847-1983, vol 913. US Department of the Interior Geological Survey, Reston, VA

Arnow T, Stephens DW (1990) Hydrologic characteristics of the Great Salt Lake, Utah, 1847-1986 (No. 2332). US Government Printing Office, Washington, DC

Atwood G, Wambeam TJ, Anderson NJ (2016) The present as a key to the past: Paleoshoreline correlation insights from Great Salt Lake. In: Oviatt CG, Shroder JF (eds) Lake Bonneville a scientific update, 1st edn. Elsevier, Amsterdam, pp 1–25

Barnes BD, Wurtsbaugh WA (2015) The effects of salinity on plankton and benthic communities in the Great Salt Lake, Utah, USA: a microcosm experiment. Can J Fish Aquat Sci 72(6):807–817

Baxter BK (2018) Great Salt Lake microbiology: a historical perspective. Int Microbiol:1–17

Baxter BK, Litchfield CD, Sowers K, Griffith JD, DasSarma PA, DasSarma S (2005) Microbial diversity of Great Salt Lake. In: Gunde-Cimerman N, Oren A, Plemenitaš A (eds) Adaptation to life at high salt concentrations in archaea, bacteria, and eukarya. Cellular origin, life in extreme habitats and astrobiology, 9th edn. Springer, Dordrecht, pp 9–25

Bedford D (2009) The Great Salt Lake America's Aral Sea? Environ Sci Policy Sustain Devel 51:8–21

Behrens P (1980) Industrial processing of Great Salt Lake brines by Great Salt Lake Minerals and Chemicals Corporation. In: Gwynn JW (ed) Great Salt Lake: a scientific, historical and economic overview. Utah Geological Survey, Salt Lake City, UT, pp 223–228

Beisner K, Naftz DL, Johnson WP, Diaz X (2009) Selenium and trace element mobility affected by periodic displacement of stratification in the Great Salt Lake, Utah. Sci Total Environ 407 (19):5263–5273

Bellrose FC (1980) Ducks, geese & swans of North America. Stackpole Books, Harrisburg, PA

Belovsky GE, Perschon WC (2019) A management case study for a new commercial fishery: brine shrimp harvesting in Great Salt Lake, Utah, USA. Ecol Appl 29(3):e01864. https://doi.org/10.1002/eap.1864

Belovsky GE, Stephens D, Perschon C, Birdsey P, Paul D, Naftz D, Baskin R, Larson C, Mellison C, Luft J, Mosley R (2011) The Great Salt Lake ecosystem: long term data and a structural equation approach. Ecosphere 2:1–40. https://doi.org/10.1890/ES10-00091.1

Bingham CP (1980) Solar production of potash from the brines of the Bonneville Salt Flats. In: Gwynn JW (ed) Great Salt Lake: a scientific, historical and economic overview. Utah Geological Survey, Salt Lake City, UT, pp 229–242

Bioeconomics, Inc. (2012) Economic significance of the Great Salt Lake to the State of Utah. Missoula, MT. https://documents.deq.utah.gov/water-quality/standards-technical-services/great-salt-lake-advisory-council/Activities/DWQ-2012-006864.pdf. Accessed 19 Aug 2019

Bowen GJ, Nielson KE, Eglinton TI (2019) Multi-substrate radiocarbon data constrain detrital and reservoir effects in Holocene sediments of the Great Salt Lake, Utah. Radiocarbon:1–22

Boyd ES, Yu R-Q, Barkay T, Hamilton TL, Baxter BK, Naftz DL, Marvin-DiPasquale M (2017) Effect of salinity on mercury methylating benthic microbes and their activities in Great Salt Lake, Utah. Sci Total Environ. https://doi.org/10.1016/j.scitotenv.2016.12.157

Cahill TA, Gill TE, Reid JS, Gearhart EA, Gillette DA (1996) Saltating particles, playa crusts and dust aerosols at Owens (dry) Lake, California. Earth Surf Process Landf 21(7):621–639

Campbell HE, Johnson RM, Larson EH (2004) Prices, devices, people, or rules: the relative effectiveness of policy instruments in water conservation 1. Rev Policy Res 21(5):637–662

Cannon JS, Cannon MA (2002) The Southern Pacific Railroad trestle - past and present. In: Gwynn JW (ed) Great Salt Lake: an overview of change. Special Publication of the Department of Natural Resources, Salt Lake City, UT, pp 283–294

Cayan DR, Das T, Pierce DW, Barnett TP, Tyree M, Gershunov A (2010) Future dryness in the southwest US and the hydrology of the early 21st century drought. Proc Natl Acad Sci USA 107 (50):21271–21276

Cohenour RE, Thompson KC (1966) Geologic setting of Great Salt Lake. Utah Geological and Mineralogical Survey, Salt Lake City, UT

Coltrain JB, Leavitt SW (2002) Climate and diet in Fremont prehistory: economic variability and abandonment of maize agriculture in the Great Salt Lake Basin. Am Antiq 673:453–485

Conover MR, Bell ME (2020) Importance of Great Salt Lake to pelagic birds: eared grebes, phalaropes, gulls, ducks, and white pelicans. In: Baxter BK, Butler JK (eds) Great Salt Lake biology: a terminal lake in a time of change. Springer, Cham

Conover MR, Vest JL (2009) Selenium and mercury concentrations in California gulls breeding on the Great Salt Lake, Utah, USA. Environ Toxicol Chem 28(2):324–329

Cook BI, Ault TR, Smerdon JE (2015) Unprecedented 21st century drought risk in the American Southwest and Central Plains. Sci Adv 1(1):e1400082

Crosman ET, Horel JD (2009) MODIS-derived surface temperature of the Great Salt Lake. Remote Sens Environ 113(1):73–81

Crosman ET, Horel JD (2016) Winter lake breezes near the Great Salt Lake. Bound Layer Meteorol 159(2):439–464

Cuch FS (2000) History of Utah's American Indians. University Press of Colorado, Louisville, CO

De Martonne E (1927) Regions of interior-basin drainage. Geogr Rev 17(3):397–414

Dicataldo G, Hayes DF, Miller TG, Scanlan L (2010) Selenium speciation and distribution in a wetland system of the Great Salt Lake, Utah. Environ Eng Sci 27(9):777–788

Diffenbaugh NS, Giorgi F, Pal JS (2008) Climate change hotspots in the United States. Geophys Res Lett 35(16)

Dominguez F, Cañon J, Valdes J (2010) IPCC-AR4 climate simulations for the Southwestern US: the importance of future ENSO projections. Clim Change 99(3–4):499–514
Donat MG, Lowry AL, Alexander LV, O'Gorman PA, Maher N (2016) More extreme precipitation in the world's dry and wet regions. Nat Clim Change 6(5):508
dos Santos CA, Neale CM, Rao TV, da Silva BB (2011) Trends in indices for extremes in daily temperature and precipitation over Utah, USA. Int J Climatol 31(12):1813–1822
Ewert EC (2014) The coming challenge: population growth and water decline. In: Crimmel H (ed) Desert water: the future of Utah's water resources. University of Utah Press, Salt Lake City, UT, pp 13–28
Ficklin DL, Stewart IT, Maurer EP (2013) Effects of projected climate change on the hydrology in the Mono Lake Basin, California. Clim Change 116:111–131
Field JP, Belnap J, Breshears DD, Neff JC, Okin GS, Whicker JJ, Painter TH, Ravi S, Reheis MC, Reynolds RL (2010) The ecology of dust. Front Ecol Environ 8(8):423–430
Frank MG, Conover MR (2019) Threatened habitat at Great Salt Lake: importance of shallow-water and brackish habitats to Wilson's and Red-necked phalaropes. The Condor Ornithol Appl 21(2): duz005. https://doi.org/10.1093/condor/duz005
Frémont JC (1845) Report of the exploring expedition to the Rocky Mountains in the year 1842 and to Oregon and North California in the years 1843-44: Printed by order of the Senate of the United States, vol 174. Gales & Seaton, Washington, DC
Friends of Great Salt Lake (2018). https://www.fogsl.org/programs/great-salt-lake-issues-forum. Accessed 11 Sept 2019
Fritz SC (1996) Paleolimnological records of climatic change in North America. Limnol Oceanogr 41(5):882–889
Garfin G, Jardine A, Merideth R, Black M, LeRoy S (2014) Assessment of climate change in the Southwest United States, National Climate Assessment Regional Technical Input Report Series. Island Press, Washington, DC
Gilbert GK (1890) Lake Bonneville. U. S. Geological Survey Monograph 1, 438 p
Gill TE, Gillette DA, Niemeyer T, Winn RT (2002) Elemental geochemistry of wind-erodible playa sediments, Owens Lake, California. Nucl Instrum Methods Phys Res B Beam Interact Mater Atoms 189(1-4):209–213
Gillies RR, Wang SY, Booth MR (2012) Observational and synoptic analyses of the winter precipitation regime change over Utah. J Clim 25(13):4679–4698
Godsey HS, Currey DR, Chan MA (2005) New evidence for an extended occupation of the Provo shoreline and implications for regional climate change, Pleistocene Lake Bonneville, Utah, USA. Quat Res 63(2):212–223
Goodman MM, Carling GT, Fernandez DP, Rey KA, Hale CA, Bickmore BR, Nelson ST, Munroe JS (2019) Trace element chemistry of atmospheric deposition along the Wasatch Front (Utah, USA) reflects regional playa dust and local urban aerosols. Chem Geol 530:119317
Grimm NB, Chacón A, Dahm CN, Hostetler SW, Lind OT, Starkweather PL, Wurtsbaugh WW (1997) Sensitivity of aquatic ecosystems to climatic and anthropogenic changes: the Basin and Range, American Southwest and Mexico. Hydrol Process 11(8):1023–1041
Gwynn JW (2007). Great Salt Lake brine chemistry databases and reports, 1966–2006. Utah Geological Survey
Hawkins CP (2015) The clean water rule: defining the scope of the clean water act. Freshw Sci 34 (4):1585–1587
Hurteau MD, Bradford JB, Fulé PZ, Taylor AH, Martin KL (2014) Climate change, fire management, and ecological services in the southwestern US. For Ecol Manag 327:280–289
Huybers K, Rupper S, Roe GH (2016) Response of closed basin lakes to interannual climate variability. Clim Dyn 46:3709–3723
Jellison R, Williams WD, Timms B, Alcocer J, Aladin NV (2008) Salt lakes: values, threats and future. In: Polunin NVC (ed) Aquatic ecosystems. Cambridge University Press
Jetz W, Wilcove DS, Dobson AP (2007) Projected impacts of climate and land-use change on the global diversity of birds. PLoS Biol 5(6):e157

Johnson WP, Wurtsbaugh W, Belovsky GE, Baxter BK, Black F, Angeroth C, Jewell P, Yang S (2019) Geochemistry of Great Salt Lake. In: Maurice PA (ed) Encyclopedia of water, science, technology and society. Wiley, Hoboken, NJ

Jones BF, Naftz DL, Spencer RJ, Oviatt CG (2009) Geochemical evolution of Great Salt Lake, Utah, USA. Aquat Geochem 15(1–2):95–121

Karl TR, Young PJ (1986) Recent heavy precipitation in the vicinity of the Great Salt Lake: just how unusual? J Clim Appl Meteorol 25(3):353–363

Keck W, Hassibe W (1979) The Great Salt Lake. U.S. Geological Survey 25, Salt Lake City

Kijowski AK, Neill J, Wickline A, Swift J, Butler JK, Kimberly DA, VanLeeuwen J, Luft J, Stone K (2020) American White Pelicans of Gunnison Island. In: Baxter BK, Butler JK (eds) Great Salt Lake biology: a terminal lake in a time of change. Springer, Cham

Koehler CL (1995) Water rights and the public trust doctrine: resolution of the Mono Lake controversy. Ecol LQ 22:541

Kregel KC (2002) Heat shock proteins: modifying factors in physiological stress responses and acquired thermotolerance. J Appl Physiol 92(5):2177–2186

Kunkel KE, Stevens LE, Stevens SE, Sun L, Janssen E, Wuebbles D, Redmond KT, Dobson JG (2013) Regional climate trends and scenarios for the U.S. National Climate Assessment NOAA Technical Report NESDIS 142-5 Part 5 Climate of the Southwest US

Lehmann J, Coumou D, Frieler K (2015) Increased record-breaking precipitation events under global warming. Clim Change 132(4):501–515

Lindsay MR, Anderson C, Fox N, Scofield G, Allen J, Anderson E, Bueter L, Poudel S, Sutherland K, Munson-McGee JH, Van Nostrand JD, Zhou J, Spear JR, Baxter BK, Lageson DR, Boyd ES (2017) Microbialite response to an anthropogenic salinity gradient in Great Salt Lake, Utah. Geobiology 15(1):131–145

Lindsay MR, Johnson RE, Baxter BK, Boyd ES (2019) Effects of salinity on microbialite-associated production in Great Salt Lake, Utah. Ecology 100(3):e02611

Lindsay MR, Dunham EC, Boyd ES (2020) Microbialites of Great Salt Lake. In: Baxter BK, Butler JK (eds) Great Salt Lake biology: a terminal lake in a time of change. Springer, Cham

Litchfield CD (2011) Saline lakes. In: Encyclopedia of geobiology. Springer, Heidelberg, pp 765–769

Loving BL, Waddell KM, Miller CW (2000) Water and salt balance of Great Salt Lake, Utah, and simulation of water and salt movement through the causeway, 1987–98. US Geol Surv Water Resour Invest Rep 2000-4221

MacDonald GM (2010) Water, climate change, and sustainability in the southwest. Proc Natl Acad Sci USA 107(50):21256–21262

Madison RJ (1970) Effects of a causeway on the chemistry of the brine in Great Salt Lake, Utah. In: Utah Geological and Mineralogical Survey Water-Resources Bulletin 14

Madsen DB (1999) The nature of Great Basin environmental change during the Pleistocene/Holocene transition and its possible impact on human populations. In: Beck C (ed) Models for the millennium: the current status of Great Basin anthropological research. University of Utah Press, Salt Lake City, UT, pp 75–82

Madsen DB (2015) A framework for the initial occupation of the Americas. PaleoAmerica 1:217–250

Madsen DB (2016) The early human occupation of the Bonneville Basin. In: Oviatt CG, Shroder JF (eds) Lake Bonneville: a scientific update, Developments in earth surface processes, vol 20. Elsevier, Amsterdam, pp 504–520

Mallia DV, Lin JC, Urbanski S, Ehleringer J, Nehrkorn T (2015) Impacts of upwind wildfire emissions on CO, CO2, and PM2. 5 concentrations in Salt Lake City, Utah. J Geophys Res Atmos 120(1):147–166

Mann ME, Lall U, Saltzman B (1995) Decadal-to-centennial-scale climate variability: insights into the rise and fall of the Great Salt Lake. Geophys Res Lett 22(8):937–940

Marcarelli AM, Wurtsbaugh WA, Griset O (2006) Salinity controls phytoplankton response to nutrient enrichment in the Great Salt Lake, Utah, USA. Can J Fish Aquat Sci 63:2236–2248

Marden B, Brown P, Bosteels T (2020) Great Salt Lake *Artemia*: ecosystem functions and services with a global reach. In: Baxter BK, Butler JK (eds) Great Salt Lake biology: a terminal lake in a time of change. Springer, Cham

McGenity TJ, Oren A (2012) Hypersaline environments. In: Bell EM (ed) Life at extremes: environments, organisms and strategies for survival. CAB International, Wallingford, UK, pp 402–437

Meixner T, Manning AH, Stonestrom DA, Allen DM, Ajami H, Blasch KW, Brookfield AE, Castro CL, Clark JF, Gochis DJ, Flint AL (2016) Implications of projected climate change for groundwater recharge in the western United States. J Hydrol 534:124–138

Meng Q (2019) Climate change and extreme weather drive the declines of saline lakes: a showcase of the Great Salt Lake. Climate 7(2):19

Meuser JE, Baxter BK, Spear JR, Peters JW, Posewitz MC, Boyd ES (2013) Contrasting patterns of community assembly in the stratified water column of Great Salt Lake, Utah. Microb Ecol 66 (2):268–280

Mohammed IN, Tarboton DG (2012) An examination of the sensitivity of the Great Salt Lake to changes in inputs. Water Resour Res 48(11)

Moore JN (2016) Recent desiccation of western great basin saline lakes: lessons from Lake Abert, Oregon, U.S.A. Sci Total Environ 554–555:142–154. https://doi.org/10.1016/j.scitotenv.2016.02.161

Mufson S, Mooney C, Eilperin J, Muyskens J (2019) 2°C Beyond the limit: extreme climate change has arrived in America. Washington Post, Aug. 13, 2019

Naftz D, Angeroth C, Kenney T, Waddell B, Darnall N, Silva S, Perchon C, Whitehead J (2008a) Anthropogenic influences on the input and biogeochemical cycling of nutrients and mercury in Great Salt Lake, Utah, USA. Appl Geochem 23(6):1731–1744

Naftz DL, Johnson WP, Freeman ML, Beisner K, Diaz X, Cross VA (2008b) Estimation of selenium loads entering the south arm of Great Salt Lake, Utah. US Geological Survey scientific investigations report, 5069

Naftz DL, Fuller C, Cederberg J, Krabbenhoft D, Whitehead J, Garberg J, Beisner K (2009) Mercury inputs to Great Salt Lake, Utah: reconnaissance-phase results. Nat Resour Environ Iss 15(1):5

Naftz DL, Millero FJ, Jones BF, Green WR (2011) An equation of state for hypersaline water in Great Salt Lake, Utah, USA. Aquat Geochem 17:809–820

Neill J, Leite B, Gonzales J, Sanchez K, Luft J (2016) 2015 Great Salt lake eared grebe aerial photo survey: Annual report. Utah Division of Wildlife Resources, Salt Lake City, UT

Nichols J (2014) No lasting resorts: the shifting shoreline of Great Salt Lake. In: Proceedings of world conference on environmental history, Guimarães, Portugal

Null SE, Wurtsbaugh WA (2020) Water development, consumptive water uses, and Great Salt Lake. In: Baxter BK, Butler JK (eds) Great Salt Lake biology: a terminal lake in a time of change. Springer, Cham

Oren A (1993) The Dead Sea—alive again. Experientia 49(6-7):518–522

Oren A, Gurevich P, Azachi M, Henis Y (1992) Microbial degradation of pollutants at high salt concentrations. Biodegradation 3(2–3):387–398

Oring LW, Neel L, Oring KE (2019) Intermountain West Regional Shorebird Plan. https://www.shorebirdplan.org/wp-content/uploads/2013/01/IMWEST4.pdf. Accessed 11 Sept 2019

Oviatt CG (1997) Lake Bonneville fluctuations and global climate change. Geology 25(2):155–158

Oviatt CG, Shroder JF (eds) (2016) Lake Bonneville a scientific update, 1st edn. Elsevier, Amsterdam

Oviatt CG, Thompson RS, Kaufman DS, Bright J, Forester RM (1999) Reinterpretation of the Burmester Core, Bonneville Basin, Utah. Quat Res 52:180–184

Oviatt CG, Madsen DM, Miller DM, Thompson RS, McGeehin JP (2015) Early Holocene Great Salt Lake, USA. Quat Res 84:57–68

Painter TH, Skiles SM, Deems JS, Brandt WT, Dozier J (2018) Variation in rising limb of Colorado River snowmelt runoff hydrograph controlled by dust radiative forcing in snow. Geophys Res Lett 45(2):797–808

Parr RL, Carlyle SW, O'Rourke DH (1996) Ancient DNA analysis of Fremont Amerindians of the Great Salt Lake wetlands. Am J Phys Anthropol 99:507–518
Paul DS, Manning AE (2002) Great Salt Lake Waterbird Survey Five-Year Report (1997–2001); Publication Number 08-38. Utah Division of Wildlife Resources, Salt Lake City, UT
Perl S, Baxter BK (2020) Great Salt Lake as an astrobiology analogue for ancient Martian hypersaline aqueous systems. In: Baxter BK, Butler JK (eds) Great Salt Lake biology: a terminal lake in a time of change. Springer, Cham
Perlich PS, Hollingshaus M, Harris ER, Tennert J, Hogue MT (2015) Utah's long-term demographic and economic projections summary. Research brief. Kem C. Gardner Policy Institute, University of Utah, Salt Lake City, UT
Peterson C, Gustin M (2008) Mercury in the air, water and biota at the Great Salt Lake (Utah, USA). Sci Total Environ 405(1–3):255–268
Raghavan M, Steinrücken M, Harris K, Schiffels S, Rasmussen S, DeGiorgio M, Albrechtsen A, Valdiosera C, Ávila-Arcos MC, Malaspinas AS, Eriksson A (2015) Genomic evidence for the Pleistocene and recent population history of Native Americans. Science 349:aab3884. https://doi.org/10.1126/science.aab3884
Ryan E (2015) The public trust doctrine, private water allocation, and Mono Lake: The historic saga of National Audubon Society v Superior Court. Environ Law 45:561
Saxton HJ, Goodman JR, Collins JN, Black FJ (2013) Maternal transfer of inorganic mercury and methylmercury in aquatic and terrestrial arthropods. Environ Toxicol Chem 32(11):2630–2636
Scalzitti J, Strong C, Kochanski A (2016) Climate change im-pact on the roles of temperature and precipitation in western U.S. snowpack variability. Geophys Res Lett 43:5361–5369
Scott AF, Black FJ (2020) Mercury bioaccumulation and biomagnification in Great Salt Lake ecosystems. In: Baxter BK, Butler JK (eds) Great Salt Lake biology: a terminal lake in a time of change. Springer, Cham
Shoop M (2019) Watershed moment for addressing Great Salt Lake's declining water levels. https://www.audubon.org/news/watershed-moment-addressing-great-salt-lakes-declining-water-levels. Accessed 11 Sept 2019
Shroder JF, Cornwell K, Oviatt CG, Lowndes TC (2016) Landslides, alluvial fans, and dam failure at Red Rock Pass: the outlet of Lake Bonneville. In: Oviatt CG, Shroder JF (eds) Lake Bonneville a scientific update, 1st edn. Elsevier, Amsterdam, pp 75–85
Shupe SJ, Weatherford GD, Checchio E (1989) Western water rights: the era of reallocation. Nat Resour J 29:413–417
Silcox GD, Kelly KE, Crosman ET, Whiteman CD, Allen BL (2012) Wintertime PM2. 5 concentrations during persistent, multi-day cold-air pools in a mountain valley. Atmos Environ 46:17–24
Skiles SM, Mallia DV, Hallar AG, Lin JC, Lambert A, Petersen R, Clark S (2018) Implications of a shrinking Great Salt Lake for dust on snow deposition in the Wasatch Mountains, UT, as informed by a source to sink case study from the 13–14 April 2017 dust event. Environ Res Lett 13(12):124031
Smith WW (1936) Evidence of a bacterial flora indigenous to the Great Salt Lake. M.S. Thesis. University of Utah, Salt Lake City, UT
Smith MD, Goater SE, Reichwaldt ES, Knott B, Ghadouani A (2010) Effects of recent increases in salinity and nutrient concentrations on the microbialite community of Lake Clifton (Western Australia): are the thrombolites at risk? Hydrobiologia 649:207–216
Sorenson ED, Hoven HM, Neil J (2020) Great Salt Lake shorebirds, their habitats and food base. In: Baxter BK, Butler JK (eds) Great Salt Lake biology: a terminal lake in a time of change. Springer, Cham
Spiegel D (2010) Can the public trust doctrine save western groundwater. NYU Envtl L J 18:412
Stansbury H (1855) Exploration of the Valley of the Great Salt Lake: including a reconnaissance of a new route through the Rocky Mountains. Lippincott, Gramabo, Philadelphia
State of Utah (2000) Great Salt Lake comprehensive management plan and decision document. Utah Department of Natural Resources, Salt Lake City, UT

State of Utah (2019a) https://wildlife.utah.gov/habitat/farmington_bay.php. Accessed 15 Aug 2019
State of Utah (2019b) https://wildlife.utah.gov/gsl/industries/index.php. Accessed 15 Aug 2019
State of Utah (2019c) https://wildlife.utah.gov/gsl/. Accessed 15 Aug 2019
State of Utah (2019d). https://deq.utah.gov/water-quality/great-salt-lake-water-quality-strategy. Accessed 15 Aug 2019
State of Utah (2019e) https://water.utah.gov/. Accessed 15 Aug 2019
State of Utah (2019f) https://le.utah.gov/~2019/bills/static/HCR010.html. Accessed 15 Aug 2019
State of Utah (2019g) https://deq.utah.gov/water-quality/hb-343-great-salt-lake-advisory-council. Accessed 15 Aug 2019
Stephens DW (1990) Changes in lake levels, salinity and the biological community of Great Salt Lake (Utah, USA), 1847–1987. Hydrobiologia 197(1):139–146
Stephens DW, Birdsey PW Jr (2002) Population dynamics of the brine shrimp, *Artemia franciscana*. Great Salt Lake, and regulation of commercial shrimp harvest:327–336
Strong C, Kochanski AK, Crosman ET (2014) A slab model of the Great Salt Lake for regional climate simulation. J Adv Model Earth Syst 6:602–615
Stutz J, Ackermann R, Fast JD, Barrie L (2002) Atmospheric reactive chlorine and bromine at the Great Salt Lake, Utah. Geophys Res Lett 29(10):18–11
Taylor RG, Scanlon B, Döll P, Rodell M, Van Beek R, Wada Y, Longuevergne L, Leblanc M, Famiglietti JS, Edmunds M, Konikow L, Green TR, Chen J, Taniguchi M, Bierkens MFP, MacDonald A, Fan Y, Maxwell RM, Yechieli Y, Gurdak JJ, Allen DM, Shamsudduha M, Hiscock K, Yeh PJ, Holman I, Thomson AM, Brown RA, Rosenberg NJ, Srinivasan R, Izaurralde RC (2005) Climate change impacts for the conterminous USA: an integrated assessment. Clim Change 69(1):67–88
Taylor RG, Scanlon B, Döll P, Rodell M, Van Beek R, Wada Y, Longuevergne L, Leblanc M, Famiglietti JS, Edmunds M, Konikow L (2013) Ground water and climate change. Nat Clim Change 3(4):322
The Deseret News (1907, July 27) Weary path trodden by intrepid band, to the shores of America's Dead Sea. pp 4–5
The Intergovernmental Panel on Climate Change (IPCC) (2018) Summary for policymakers. In: Masson-Delmotte V, Zhai P, Pörtner H-O, Roberts D, Skea J, Shukla PR, Pirani A, Moufouma-Okia W, Péan C, Pidcock R, Connors S, Matthews JBR, Chen Y, Zhou X, Gomis MI, Lonnoy E, Maycock T, Tignor M, Waterfield T (eds) Global warming of 1.5°C. An IPCC special report on the impacts of global warming of 1.5°C above pre-industrial levels and related global greenhouse gas emission pathways, in the context of strengthening the global response to the threat of climate change, sustainable development, and efforts to eradicate poverty. IPCC, Geneva
Thomson AM, Brown RA, Rosenberg NJ, Srinivasan R, Izaurralde RC (2005) Climate change impacts for the conterminous USA: an integrated assessment. Clim Change 69(1):67–88
Treide H (2012) Ground water and climate change. Nat Clim Change 3(4):1–8
Trentelman CK (2020) The relationship between humans and Great Salt Lake: dynamics of Change. In: Baxter BK, Butler JK (eds) Great Salt Lake biology: a terminal lake in a time of change. Springer, Cham
Tweed S, Grace M, Leblanc M, Cartwright I, Smithyman D (2011) The individual response of saline lakes to a severe drought. Sci Total Environ 409:3919–3933
United States Bureau of Reclamation (1962) Bear River Project, Part I, Feasibility Report, Oneida Division, Idaho and Utah. Part II, Reconnaissance Report, Blacksmith Fork Division, Utah. United States Bureau of Reclamation, Salt Lake City, p 86
United States Division of Fish and Wildlife (2019) https://www.fws.gov/Refuge/Bear_River_Migratory_Bird_Refuge/about.html. Accessed 15 Aug 2019
United States Geologic Survey (2019) http://ut.water.usgs.gov/greatsaltlake/elevations Accessed 18 Aug 2019
United States Global Climate Research Program (2017) In: Wuebbles DJ, Fahey DW, Hibbard KA, Dokken DJ, Stewart BC, Maycock TK (eds) Climate science special report: Fourth national climate assessment, Vol I. U.S. Global Change Research Program, Washington, DC, 470 p

Utah Rivers Council (2017) Alternatives to Bear River Development. http://static1.squarespace.com/static/5a46b200bff2007bcca6fcf4/5a4d26a4448c8a3e3c8810dd/5a4d26b6448c8a3e3c881341/1515005622382/Bear-River-Alternatives.pdf?format=original. Accessed 11 Sept 2019.

Ventosa A, Arahal DR (2009) Physico-chemical characteristics of hypersaline environments and their biodiversity. Extremophiles 2:247–262

Wagner JD, Cole JE, Beck JW, Patchett PJ, Henderson GM, Barnett HR (2010) Moisture variability in the southwestern United States linked to abrupt glacial climate change. Nat Geosci 3(2):110

Wang SY, Gillies RR, Jin J, Hipps LE (2010) Coherence between the Great Salt Lake level and the Pacific quasi-decadal oscillation. J Clim 23(8):2161–2177

Wang J, Song C, Teager JT, Yao F, Famiglietti JS, Sheng Y, MacDonald GM, Brun F, Schmied HM, Marston RA, Wada Y (2018) Recent global decline in endorheic basin water storges. Nat Geosci 11(12):926–932

White JS, Null SE, Tarboton DG (2015) How do changes to the railroad causeway in Utah's Great Salt Lake affect water and salt flow? PloS one 10(12):e0144111

Whiteley NM, Taylor EW, El Haj AJ (1997) Seasonal and latitudinal adaptation to temperature in crustaceans. J Therm Biol 22(6):419–427

Whiteman CD, Hoch SW, Horel JD, Charland A (2014) Relationship between particulate air pollution and meteorological variables in Utah's Salt Lake Valley. Atmos Environ 94:742–753

Williams WD (1993) Conservation of salt lakes. Hydrobiologia 267:291–306

Williams WD (1996) What future for saline lakes? Environment 38(13–20):38–39

Williams WD (2002) Environmental threats to salt lakes and the likely status of inland saline ecosystems in 2025. Environ Conserv 29:154–167

Wilson EO (2017) Biophilia and the conservation ethic. In: Evolutionary perspectives on environmental problems. Routledge, London, pp 263–272

Wine ML, Null SE, DeRose RJ, Wurtsbaugh WA (2019) Climatization—negligent attribution of Great Salt Lake desiccation: a comment on Meng (2019). Climate 7(5):67

Witze A (2018) Why extreme rains are gaining strength as the climate warms. Nature 563 (7732):458–460. https://doi.org/10.1038/d41586-018-07447-1

Woodhouse CA, Meko DM, MacDonald GM, Stahle DW, Cook ER (2010) A 1,200-year perspective of 21st century drought in southwestern North America. Proc Natl Acad Sci 107 (50):21283–21288

Wurtsbaugh WA (1992) Food-web modification by an invertebrate predator in the Great Salt Lake (USA). Oecologia 89(2):168–175

Wurtsbaugh WA, Gliwicz ZM (2001) Limnological control of brine shrimp population dynamics and cyst production in the Great Salt Lake, Utah. Hydrobiologia 466:119–132. https://doi.org/10.1023/A:101450251

Wurtsbaugh WA, Miller C, Null S, Wilcock P, Hahnenberger M, Howe F (2016) Impacts of water development on Great Salt Lake and the Wasatch Front. Utah State University, White Paper. https://works.bepress.com/wayne_wurtsbaugh/171/

Wurtsbaugh WA, Miller C, Null SE, DeRose RJ, Wilcock P, Hahnenberger M, Howe F, Moore J (2017) Decline of the world's saline lakes. Nat Geosci 10(11):816. https://doi.org/10.1038/ngeo3052

Yeager KN, Steenburgh WJ, Alcott TI (2013) Contributions of lake-effect periods to the cool-season hydroclimate of the Great Salt Lake basin. J Appl Meteorol Climatol 52(2):341–362

Yidana SM, Lowe M, Emerson RL (2010) Wetlands in northern Salt Lake Valley, Salk Lake County, evaluation of the threats posed by groundwater development and drought. Utah Geological Survey Report of Investigation 268, Salt Lake City, UT

Chapter 3
Relationships Between Humans and Great Salt Lake: Dynamics of Change

Carla Koons Trentelman

Abstract The relationships between humans and Great Salt Lake (GSL) have always involved change and adaptation, for both humans and the lake. Humans have changed the lake in numerous ways, and the lake has changed humans, frequently in recursive processes. Further, forces external to both humans and GSL have affected both. Some key themes of change and the need for adaptation have included the elevation and related size of GSL, and humans' attempts to manage that; technology, both lake-related and related to broader social trends; and finally, development, of population and of communities. Using sociological research findings, this chapter addresses gaps in our understanding of some of the many and drastic changes occurring with the lake, listening to people's perceptions of and experiences with GSL and the transitions they have observed. Challenges for the future are outlined, including the complexities of needing to meet humans' needs without further compromising the health of the lake.

Keywords Great Salt Lake · Natural resource sociology · Humans and terminal lakes · Humans/lake relationships · Humans/lake interactions

3.1 Great Salt Lake and Humans

It may seem curious to see a sociological chapter in a book focused on the biology of Great Salt Lake (GSL), but given the substantial effect humans have had on the lake's ecosystem, it is an important inclusion for a volume considering changes to that ecosystem. Humans have lived near GSL and used lake resources from prehistoric times through the present. Our relationships with the lake have always involved change and adaptation, for both humans and the lake. Particularly since the European settlement of northern Utah, humans have changed the lake, and the lake has changed humans. Forces external to both have brought changes to both, whether

C. K. Trentelman (✉)
Department of Sociology, Weber State University, Ogden, UT, USA
e-mail: Carlatrentelman@weber.edu

B. K. Baxter, J. K. Butler (eds.), *Great Salt Lake Biology*,
https://doi.org/10.1007/978-3-030-40352-2_3

due to weather and climate, technology, economic forces, or other phenomena that affected both humans and the lake they lived near. Some key recurring themes of change and the need for adaptation have included the elevation and related size of GSL and humans' attempts to manage that; human development, both in population and in the growth of communities; and technology, some related directly to the lake and attempts to manage it, but also related to broader trends, for example transportation.

Humans have had relationships with GSL for a long time. Its eastern wetlands were used by Native Americans for over 10,000 years (Simms and Stuart 2002; Baxter 2018). The earliest mapping by European Americans was conducted by Howard Stansbury and his team, mapping the full lake in 1849 (Stansbury 1855). Yet the social and sociological aspects of GSL can be characterized as understudied, unappreciated, and unusual, similarly to a theme throughout this book. This researcher found GSL nearly virgin territory in terms of a social science informed understanding of the human part of the lake ecosystem. Yet human interaction has brought about tremendous change in that ecosystem. While my research found considerable evidence that the lake is unappreciated by many, I also found that, at least among those who live near the lake, there is a good deal of appreciation of the lake, with a proportion of lake neighbors feeling quite attached to it (Trentelman 2009). As for the lake's unusualness, many people living close to GSL enjoy its uniqueness.

This chapter examines various dynamics of the relationships between humans and GSL. First, the impacts of technology, economics, development, and lake elevation on these relationships are explored. The perceptions of people who live nearby or have other connections to GSL are considered, drawing particularly from qualitative research to allow their voices to be heard. I focus on the lake's human neighbors, in some ways, more vulnerable to changes in GSL than others in northern Utah. They get to know GSL on a more intimate level than many, experiencing both positive and negative aspects of that relationship. The chapter then addresses some of the more daunting challenges for the future. While there is cause for hope, there are no simple solutions for the complexities in our relationships with GSL, particularly for meeting humans' needs without further compromising the health of the lake.

3.1.1 Great Salt Lake

GSL is a shallow, terminal lake at the bottom of the Great Basin. It settled into its current footprint 13,000 years ago after a process of formation that took hundreds of thousands of years (Atwood et al. 2016). One of the largest terminal lakes in the world, at its historically common lake elevation of about 1280 m (4200 ft) above sea level, GSL is roughly 121 km (75 miles) long and 45 km (28 miles) wide, covering 2736 km^2 (1700 square miles) (USGS 2019b). The lake's elevation determines salinity levels, typically three to five times saltier than the ocean in the open water

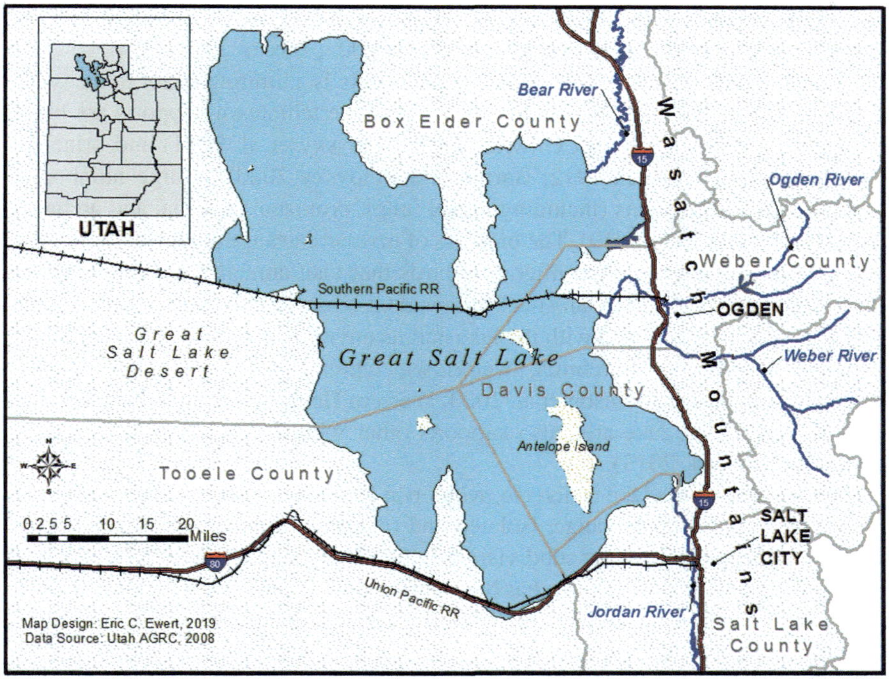

Fig. 3.1 A map of Great Salt Lake (GSL), Utah, in its geographic surroundings. Noted are the five counties that contain the lake, the city of Ogden and Salt Lake City, as well as the three rivers that feed the lake, Bear River, Weber River (with its tributary, Ogden River), and Jordan River. The largest islands are visible, with Antelope Island labeled (the sizable island north of Antelope Island is Fremont Island). The railroad causeway is shown crossing the lake west of Ogden. The inset figure shows GSL within the state of Utah. Map design by Eric C. Ewert, Weber State University, Ogden, UT

of the south arm, the isolated north arm reaching up to ten times ocean salinity. With an average depth of only 4.27 m (14 ft),[1] relatively small changes in elevation affect the size of this shallow lake dramatically. At the historic low point of 1277.4 m (4191 ft) above sea level set in 1963, GSL covered 1529 km^2 (950 square miles). At the historic high of 1283.77 (4211.85), reached in both 1986 and 1987, the lake covered 3701.5 km^2 (2300 square miles), creating substantial problems with flooding around the lake's shoreline. Annually, GSL swells with spring runoff, delivered primarily by Bear River, Weber River (joined by Ogden River not far from GSL), and Jordan River (Fig. 3.1). The lake shrinks back down under the summer sun. If there is heavy precipitation, the year ends with more water than at its beginning; if water is scarce, the elevation dips lower than the start point (Bedford 2014; USGS 2019b). While the variation between high and low elevation refreshes the wetlands, it has greatly affected local populations, creating settlement problems

[1]GSL has a maximum depth of 10.7 m (35 ft) deep (USGS 2019b).

since the Europeans came 150 years ago (Bedford 2006), and long before that for Native American settlements (Simms and Stuart 2002; Baxter 2018).

Compared to freshwater lakes, GSL's high salinity is a limiting factor to the types of aquatic life in the lake. While the only macroinvertebrates it supports are brine shrimp (e.g., Wurtsbaugh and Gliwicz 2001; Belovsky et al. 2011) and brine fly larvae and pupae (Aldrich 1912; Barrett and Belovsky 2020), a large number of halophilic microorganisms (including algae, fungi, protozoa, bacteria, and archaea) thrive (Baxter and Zalar 2019). The biomass of invertebrates is nonetheless immense and supports some ten million migratory birds that visit annually, making GSL an internationally significant avian site. Add resident birds, and the number represents 338 avian species associated with the lake and its environs that have been identified, including waterbirds, shorebirds, waterfowl, raptors, and marsh and upland-associated songbirds (Sorenson et al. 2020; Western Hemisphere Shorebird Reserve Network 2019). The lake area also supports other wildlife, including a variety of mammals (UDFFSL 2013).

The shoreline of the lake is diverse, with open lake, bays, and wetlands bordering a desert military bombing range, isolated and remote rural areas, small rural communities, and rapidly growing subdivisions in five Utah counties (Fig. 3.1). Much of the settled land nearest to the lake has been agricultural (Trentelman 2009), although this has been changing (UDFFSL 2013). Both near GSL and higher in the watershed, agriculture competes with municipalities for water that feeds the lake (Jenkins 2002). With extremes of desert on GSL's west side, and estuary areas, large wetlands, and high-water tables around much of the rest, only in Weber and Davis counties do many people live near the lake (Trentelman 2009). These communities have been growing rapidly, and growing closer to the lake than they have before (UDFFSL 2013).

3.1.2 Great Salt Lake Research and a Sociological Study

As noted earlier, GSL has a history of being unappreciated as well as understudied. A newspaper piece observed, "Most of the 1.6 million people who live along the Wasatch Front rarely think about their vast neighbor… When the topic of the lake comes up, the first image that flashes through many minds is of a stinking, brine-fly-infested cesspool that occasionally floods highways and low-lying subdivisions" (Woolf 1999, reprinted in Topping 2002, p 274). In *Utah Historical Quarterly*'s recent special issue on GSL, the introduction noted, "There's a sturdy cliché about the Great Salt Lake: most people think the lake is too much. Too flat, too shallow, too salty, too stinky to love—or even to refrain from fouling" (Nichols 2019, p 3). Some local scientists have speculated that these more negative views of GSL may be partially responsible for the paucity of scientific research until recently, citing a "historic disgust" with a lake that has been "viewed as a cesspool" (Hummel 2006). However, the lake is also difficult and expensive to study, due to its large size, extreme salinity, and extreme weather (e.g., high winds). The methodology and equipment are not designed to withstand the conditions present in this lake system (personal communication with Jaimi Butler, 9/15/19). Whatever the reason, much about GSL had not been studied systematically until fears of a possible collapse of the lucrative brine shrimp industry led to the formation of the Utah Division of Wildlife Resource's Great Salt Lake Ecosystem Program (GSLEP) in 1996. Of note, this program was created when the industry itself requested increased regulation of the harvest and higher fees from brine shrimp companies, in order to protect the viability of the resource. The fees fund GSLEP; the program's objective is methodically researching and monitoring and managing the wildlife of the GSL ecosystem (GSLEP 2019; Belovsky et al. 2011). Now considerable research has been conducted on GSL and its environs, greatly improving the understanding of ecosystem dynamics. However, much work remains (UDFFSL 2013). As recently as 2012, a report on the health of the lake began, "The Great Salt Lake is one of the most important and least understood ecosystems in Utah, and possibly North America" (Gaddis et al. 2012, p 1).

This historic scarcity of research included the social sciences. This researcher worked to fill some of that need by conducting a study of perceptions of GSL, focusing particularly on residents living near the lake. The study utilizes qualitative data collected from 18 interviews of residents, resource management professionals, and county commissioners in Box Elder, Weber, and Davis Counties, and two focus groups of Weber and Davis county residents. A survey was also conducted with Weber and Davis county residents living within 2.45 km (1.5 miles) of the lake. The response rate of 83.7% brought in 381 completed questionnaires[2] (Trentelman 2009).

How local people feel about GSL is important, since they are part of the lake's ecosystem, and among those most affected by problems within that ecosystem. Knowing their perceptions also allows social scientists to gauge their degree of protectiveness and willingness to behave in a conservation-oriented manner. Research on sense of place, place attachment, and caring for a place has found these dynamics can be associated with valuing the environmental traits of the place, environmentally responsible behavior and concern, sensitivity to environmental impacts, and increased commitment to the place (Williams et al. 1992; Kaltenborn 1998; Vaske and Kobrin 2001; Vorkinn and Riese 2001; Kaltenborn and Williams 2002). I draw from the findings of this study throughout the remainder of this paper.[3]

[2]The survey has a sampling error of plus or minus 4.23%.

[3]The reader will notice a lack of percentages and numbers related to the qualitative portions of this research, as I do not attempt to quantify the findings from the qualitative data.

3.2 Relationships with Great Salt Lake and Change

There are many aspects to humans' relationships with GSL, with a number of intervening factors. These have included lake elevation and related size of GSL, lake management strategies, and human development, both in population size and density, and in the development of communities. Additionally, both people and the lake have been affected by technology, economics, and other social phenomena. These all comprise the dynamics of change in the relationships between people and the lake, where lake and people adapt to and change each other, and both are changed by other external forces.

Given the diversity of the long shoreline of GSL, with its open lake, bays, and wetlands, it makes sense that people who live in lake-adjacent areas are not uniform in their relationships with the lake. Differences in the lay of the land mean that, while some of GSL's closest neighbors watch the sun set over its waters from lawn chairs in their yards, others cannot see the lake from their property. Additionally, public access to the open lake is quite limited. As a result, some of GSL's nearest neighbors do not think of themselves as living near the lake. These differences were correlated with people's sense of the lake, and strong divergence in what GSL means to them (Trentelman 2009).

It could be argued that, unlike owners of more typical lakeside property, those living closest to GSL are situated to experience fewer of the more positive aspects of the lake while suffering more of the negative. Some of the best views of the lake are seen kilometers away from the shores, on the mountain benches to the east, where some homes have large picture windows that capture the beauty of GSL. Meanwhile, the closest neighbors experience the brunt of ever-shifting lake elevations and resultant building regulations, with briny waters at times threatening to flood and at others, receding from view, leaving dust from the lake bed to blow in the wind (Trentelman 2009). A strong hydrogen sulfide (or "rotten egg gas") odor is associated with GSL (Bedford 2006; Hummel 2006). The clouds of brine flies, mosquitoes, gnats, and other insects are another nuisance. Both odor and insects can be more bothersome the closer one is to the lake. This constellation of challenges has historically affected land uses and property values near GSL.

There are a number of lake-related recreation activities, and those living nearest have close access to wetlands, public waterfowl management areas, and other wildlife refuges that ring much of the east and north sides of GSL. Birdwatchers and waterfowl hunters find this a positive aspect of living nearby. However, there are few points of public access to the open lake, so living close to GSL does not necessarily give one close access to it; in fact, it can be at least 30 min away for many of these nearest neighbors. Additionally, many are not aware of GSL's recreation amenities (Brunson and Nicholson 1999; Trentelman 2009). These challenges with access and lack of awareness, along with GSL's negative reputation, may account for an unusual dynamic for people living within less than 2.5 km (1.5 miles) of such a large lake: nine percent of survey respondents had never visited the lake, its islands, bays, shorelands, or affiliated refuges and preserves. These were not just newcomers. Of respondents who had never visited GSL, 39% had lived near GSL for at least 10 years, and 27% had lived there 20 years or more (Trentelman 2009).

To develop more understanding of residents' perceptions of GSL, qualitative data were used to consider the meanings the lake held for these nearest neighbors. A set of themes emerged from the diverse meanings expressed. Some participants gave rich descriptions of strongly held, very positive senses of GSL, some providing evidence of feeling particularly connected or attached to the lake. On the other hand, some research participants were among those unappreciative of GSL. These meanings are presented in more depth later, but it is clear that people's relationships with GSL are diverse (Trentelman 2009).

The relationships between people and GSL are not static, however. Over the years, they have changed many times, affected by technology, economic forces, the ever-changing lake elevation and people's attempts to manage it, and increasing development.

3.2.1 Technology

Throughout history, technology has brought many changes that have affected both the lake and humans. Improvements in transportation technology are one example of forces that have affected and changed both people and lake, and the relationships between them.

3.2.1.1 The Coming of the Railroad

Shortly after the establishment of the Transcontinental Railroad, in 1902 the Central Pacific Railroad built the Lucin Cutoff to improve its route west (Peterson 2001). The Cutoff included a 12-mile wooden trestle which cut straight across the northern portion of GSL, from Ogden to Lucin, at the time, the world's longest bridge over open water. This created a significant short-cut for the railroad since traveling around the lake was 69.2 km (43 miles) longer, with significant grades and curves (Cannon and Cannon 2002). In 1959, a sand, gravel, and rock-fill causeway replaced the then aging wooden trestle, leading to a recursive relationship between the lake and the causeway. Of note, the arm of the lake north of the causeway has no substantial source of freshwater, as all three rivers feed the lake area south of the railroad causeway. When GSL rose, the causeway had to be built up to keep the tracks above water. It eventually became a dam as it settled and filled in with brine, a barrier to water and salt circulation between the saltier north arm and the fresher south arm, which substantially changed the salinity levels. When the lake elevation rose in 1984, the damming effect led to a three- to four-foot elevation difference between the two arms, and to reduce flooding by the south arm, a breach was built into the causeway. The breach also redistributed the salt loads between the arms (Gwynn

2002; UDFFSL 2013). Additional water circulation problems have occurred since, requiring further intervention, e.g., another breach was constructed in the causeway in 2016 after a just over 1-m (3.3-ft) elevation difference (USGS 2016; Larsen 2018). These breaches are also necessary for maintaining the navigability of the lake.

An example of human activity creating changes in GSL, the causeway has affected lake elevation, circulation, and salinity to the point of substantially altering the environmental conditions. It has left the north arm hypersaline, and of note, a pinkish color due primarily to the halophilic archaea, creating a sharp visual contrast between the two arms (Fig. 3.2). This, in turn, has created "great challenges" for GSL managers (UDFFSL 2013, pp 2–10).

Another effect on people and lake from the new railroads was an increase in GSL tourism and people's familiarity with the lake. A number of successful lake resorts opened between 1870 and the early 1900s, several of which were developed by the railroad corporations (Peterson 2001; Nichols 2014). The resorts offered a variety of activities, but the main draw was the opportunity to bathe, swim, and float in the

Fig. 3.2 Great Salt Lake as captured by the International Space Station June 19, 2019. The color differences are indicative of the salinity differences between the lake's north arm and south arm, caused by the railroad causeway reducing flow between the arms. The differences are due to halophilic microbiota in the hypersaline north arm, which give the water a pinkish to purple hue from the pigmentation of the cells (NASA 2019)

briny, mineral-rich waters touted as restorative and as having health benefits for a number of illnesses. The resorts had lunch stands and restaurants, and some had saloons; they offered hotel accommodations and one had summer residences for rent. They had dance floors and employed dance orchestras and big bands, two offered steamboat excursions, and at least one had amusement rides, midway games, and concessions. The resorts on the east side of GSL had easy access by way of a railroad line between Ogden and Salt Lake City, and eventually two resorts on the south shore also had railroad access. This access gave them a considerable advantage over other GSL resorts. For example, Lake Park resort, located on the rail line east of the lake, was less attractive than resorts on the south shore, yet this railroad access afforded it a large share of the tourism business (Travous 1980).

Visiting the lake via these resorts was quite popular, until a combination of lake and social conditions eventually closed nearly all of them. A majority of the resorts succumbed to problems from either receding lake levels leaving them high and dry, or rising levels resulting in flooding. Additionally, large fires did substantial damage to the wooden structures of some of them. The Depression made things more difficult by drying up people's expendable income. Further damage was done when World War II rerouted local train lines (Travous 1980; Peterson 2001; Nichols 2014).

3.2.1.2 The Rise in Automobile Travel

Along with these changes in railroad accessibility, the increase in automobile travel by the late 1940s also affected the appeal of GSL resorts. Recreation interests shifted to mountain canyons that were now easy to reach by car, taking people away from the lake and further drying up traffic to its resorts (Travous 1980; Nichols 2014).

Eventually this shifted again, with cars allowing easy visitation to GSL via the Antelope Island causeway. The state purchased a portion of the island for a state park in the 1960s, and the causeway was built from the mainland west of the city of Syracuse to the island in 1969 (Travous 1980; Trentelman 2009).

This causeway gave people much more access to GSL, particularly since Antelope Island is one of the only public points of access to the open lake. The island currently sees more than 500,000 visitors annually (personal communication with Jeremy Shaw, Park Manager, Antelope Island State Park, 9/17/19). Park staff, organizations such as FRIENDS of Great Salt Lake, and nearby schools and universities use the island for educational programming. Recreation opportunities include birding and other wildlife viewing, swimming or floating, boating, and camping, among others. In interviews and focus groups, participants often talked about the island and the lake as though they were synonymous. Antelope Island likely makes the biggest difference for local people as well as tourists being able to experience GSL.

> I think my kids, most of their experience with the Great Salt Lake would have been from Antelope Island where they was actually able to go out and bob like a cork in the water... Antelope Island's probably one of the greatest things to happen to the State of Utah. 'Course, we know how many visitors visit that a year, but it really gives our children—the schools go

> out there and they have a chance to really see it's one of the wonders of the world. . . .That's a fascinating thing, for them to be able to see that. People travel for thousands of miles to see what we have in our backyard.—Davis County resident

> We spent a lot of time riding our horses out on Antelope Island... The trails are pretty accessible and you can see a mixture of animals most people don't get to see—bighorn sheep, the buffalo, the deer, the coyotes... So that's the memories we have is spending a lot of time out there.—Davis County resident

Research participants talked about having participated in a number of recreation activities on the island, from bicycling and hiking to birding, other wildlife watching, and stargazing. Many talked about Antelope Island being their introduction to GSL, or the place their children first experienced the lake. On the other hand, some participants complained that there really was not anything to do on the island, e.g., "there's nothing really to go out there for," and "it's like you go and see it, and then you're done."

While Antelope Island State Park has provided benefits for people, and for GSL in terms of the lake advocacy these positive interactions foster, there have also been disruptions to the lake. Like the northern railroad causeway, the Antelope Island causeway has created a barrier to water circulation, cutting off Farmington Bay from GSL's open waters, with the reverse effect on salinity as the railroad causeway. Farmington Bay receives all the water from Jordan River, so the causeway keeps Farmington Bay far fresher than the rest of the lake (salinity typically around five percent), and the rest of the lake benefits little from this freshwater input (UDFFSL 2013). Additionally, human-caused nutrient conditions in Farmington Bay, exacerbated by the bay being closed off, are responsible for much of the odor associated with GSL (Wurtsbaugh et al. 2002).

3.2.1.3 Summary

These examples illustrate how the relationships between humans and GSL impact both entities. Also, they illuminate how external forces, for example transportation technology, have affected not only both humans and lake but also the relationships between the two. These technologies have had profound impacts on the lake system, and also on humans who are part of that ecosystem, impacts that continue to affect both people and GSL. This demonstrates how long-lasting the effects of historical actions have been, which should serve as a cautionary tale for future interactions. While these examples draw on historical relationships between people and GSL, they demonstrate how long-lasting the effects of the interplay can be, as both are still being affected by the changes from the interaction outlined here.

3.2.2 Economics

GSL has provided people with ways to make a living since before the European settlement of the area (Simms and Stuart 2002; Baxter 2018). Harvesting salt is likely the oldest economic use of the lake still in practice. While it continues to be a lucrative industry, albeit a changed one, other economic pursuits from earlier times have faded. GSL helped many local people get through financially lean times, including a few research participants and their predecessors. One participant described his family harvesting salt to get through the Depression. They would go out to the dry lake bed with a wagon:

> During the Depression we fixed up a wagon with large tires and we'd go out there... with horses and load up a ton [of salt] every day. . . .That was pretty good money for Depression years. Truckers from Montana and Wyoming would come down and buy it, to sell to the ranchers I guess.—Davis County resident

Participants talked about family members who settled near GSL in the late 1800s to early 1900s, and did lake-related work. One grandfather ran steam pumps for salt works; another hauled cedar posts from nearby Promontory Mountains by boat, selling them at the lakeshore. A participant described his grandfather hunting in the GSL wetlands, where "the sky would just be black with ducks." He would bring a boat full of ducks home, family members would clean and pluck them, and then, "He'd take them to Ogden the next day, in a wagon—a whole wagon load of ducks—and he'd sell them, 25 cents apiece for 'em. That was good money in those days."

Some research participants had made money from hunting and trapping, too.

> I courted my wife trapping muskrats. I was outta high school, I couldn't get a job. So, I trapped muskrats. I would travel three or four miles [five or six km] a day, set trap lines out, and all my buddies, they were penniless but I always had a few dollars.
>
> Well, we all taught our kids to do that. My boys trapped muskrats for a few extra bucks for Christmas, and we taught 'em all how to trap muskrats 'cause they could get some pretty good money out of them. . . .And it all comes back to the lake and the drainages going out to it. . .—Two Weber County residents

More recently, though, many lake neighbors knew little about GSL's economic benefits, despite a large mineral extraction industry (salts and other metals and minerals), substantial aquaculture revenue from the harvesting of brine shrimp cysts, and a varied recreational sector bringing money into the state. Only 39% of survey respondents indicated they thought GSL added to the area's economic health (Trentelman 2009). This is not surprising though, since GSL's economic value had been both understudied and unappreciated. Until recently, estimates of the lake's economic value had "largely been piecemeal" (Bioeconomics 2012, p 9), and were rarely accessible or publicized. This was remedied in 2012 when a comprehensive study of the economic significance of GSL to the state of Utah found that industry, aquaculture, recreation, and other economic values provided by the lake and its ecosystem total $1.32 billion in economic output, including $375 million in total labor income, providing 7700 full- and part-time jobs annually (Bioeconomics 2012).

The argument could be made that, in addition to GSL being an economic resource for humans, some aspects benefit the lake as well. Fees from the brine shrimp industry fund important lake research through GSLEP, noted earlier (GSLEP 2019). Not only has this research program managed the brine shrimp harvest, setting parameters at sustainable levels that allow harvest only of the excess needed for feeding of the avian population and reproduction, it has also greatly increased understanding of the ecosystem needs for the sustainability of the lake and its environs as a whole (Trentelman 2009). GSL-related recreation and tourism not only bring an economic boost to the area, they are responsible for much of the positive exposure people have to the lake, and play a substantial role in lake neighbors having a sense of place for the lake, and even feeling attached to GSL. Additionally, the findings from the economic study have given resource managers and lake advocates empirical data that can inform decision-making and legislation regarding the lake and its ecosystem.

The economic aspects also create some tensions in the relationships between humans and the lake. The mineral industry makes changes to the lake, as that is what extraction does, and some of these changes are more consequential than others (Trentelman 2009; UDFFSL 2013). At times environmental concerns have been raised, for example, about expansions of evaporation ponds occurring too close to protected nesting areas, with the potential to impact nesting and facilitate predation; about air pollution due to chlorine emissions; and about resource extraction-related hazards that have required EPA intervention in clean-up efforts (Trentelman 2009).

Aspects of lake-related recreation and tourism can also create problems for the lake's ecosystem, for example, unauthorized off-highway vehicle use on exposed lakebed or the islands can damage habitat and increase risk of nesting predation (UDFFSL 2013). One research participant talked about her son's Boy Scout troop canoeing and kayaking very close to Egg Island, important habitat for colony nesters, during nesting season (Trentelman 2009), an activity that could easily disrupt nesting patterns. Nesting populations being affected by the presence of humans in the GSL system have raised concerns at least since Alfred Lambourne's 1909 book, *Our Inland Sea*. Behle (1958) noted a number of ways human presence has unintentional negative effects on nesting success, including increased risk of predation due to animals following human trails. Lambourne noted that human activities have resulted in some species abandoning their historic nesting places, at times even abandoning entire islands, for example, Hat Island (Lambourne 1909).

The complexities in the relationships between humans and the lake can be further illustrated by considering one of the less obvious economic benefits to the local area, that of waste assimilation. Partially because the lake system is seen as being able to process these nutrients (at least partly due to hungry brine shrimp), public sewage treatment facilities are able to discharge higher nutrient levels into GSL than they could if the lake water was used for drinking water or other consumptive uses. The lower level of sewage treatment results in substantial avoided costs for local communities, totaling somewhere between 10 million and nearly 60 million dollars annually (Bioeconomics 2012). Yet the lake being used in this way has affected the lake system, particularly in Farmington Bay, where approximately 50% of the water inflow is from municipal treatment plants (Wurtsbaugh et al. 2012). The resultant eutrophic conditions, algal blooms, and high cyanotoxin levels may threaten birds and other wildlife, and at times pose health concerns for people recreating in the bay. Additionally, these conditions have been directly linked to the odor associated with the lake, giving GSL the reputation of being a stinky place. Unfortunately, local residents are unaware of their own contribution to the smell, and the economic benefits they reap from this. Yet, these odor issues account for a substantial portion of the negative perceptions of the lake (Trentelman 2009).

3.2.3 *Lake Elevation and Size*

3.2.3.1 High Lake Elevations

As noted earlier, lake elevation has a dramatic effect on the size of GSL. While this great lake covers 2736 km^2 (1700 square miles) at its common elevation of 1280 m (4200 ft) above sea level, during the 1980s when the lake swelled to its historic high, it covered 3701.5 km^2 (2300 square miles), resulting in flood damage that totaled over $240 million, including costs incurred by transportation, industry, public lands and facilities, private lands, and habitat. Flooding occurred in rural and residential areas to the south, east, and northeast of GSL (UDFFSL 2013) (Fig. 3.3).

Humans have attempted to manage lake elevation during its highest levels. In fact, in 1982 the Utah State Legislature passed a bill mandating that GSL be

Fig. 3.3 Left: Great Salt Lake (GSL), September 1972. This satellite image provides a baseline shot of GSL at an average elevation and size. Right: GSL, September 1987. This satellite image captures the lake near its historic high elevation of 1283.77 m (4211.85 ft) above sea level. The size difference compared to the 1972 image gives some indication of the areas that flooded. This image also shows the evaporation basin in the West Desert where water from GSL was being pumped. At this point the pumping had been going on for 6 months. Source: USGS Earthshots (image is a false-colored-composite satellite image)

managed at an elevation below 4202 (UDFFSL 2013). The folly of this legislation was realized rather quickly as the lake rose 3.66 m (12 ft) from 1982 through 1986 to its historic high level. At that point, attempts to manage the elevation led to the development of the West Desert Pumping Project at a cost of $60 million. The pumps were built, and for 26 months pumped lake water into the West Desert, lowering the lake level 0.13 m (15 in.) in addition to the natural decline which had already begun. Due to concerns over cost and construction speed, the pumps pulled water from GSL's north arm, affecting the overall salinity of the lake. The Pumping Plant has not been used since, but is maintained for future use and inspected monthly at an annual cost of about $9000. This is another example of the recursive relationship between humans and GSL: material, natural events lead to weather- and climate-driven lake rise, which affects local humans. The humans respond in ways that affect the lake, which then further affects the humans.

People's relationships with GSL are so affected by lake elevation that, when looking at their interactions with and perceptions of the lake at various historical times, it is important to consider the elevation at that time. For the research project referred to in this chapter, data were collected 2005–2007, when GSL had been below average in elevation for some time. Many participants had lived near GSL long enough to experience varying elevations, e.g., 44% of survey respondents had lived near GSL for at least 20 years (Trentelman 2009).

Some participants who had lived near the lake during the flooding years of the 1980s remember well what it was like, and report varying experiences with the flooding.

> . . .the lake was in our yard. That's how high it was. . . .It washed out all the dikes, it was right at the edge of the yard. The neighbors came and helped us build a sand barrier, but it took out all the plants that we had. And it was here for a long time, I mean that was the thing, I think it was here like 2 years and it was very stressful.—Davis County resident

> When that lake came up, that was quite a shock to our community—we never realized we lived that close and were that vulnerable. It forced carp right up into our lawns and our yards, they came up the drainages to try and get away from the salt water. We could sit or stand out on our decks and see the lake, right there, and it inundated some farm land and we found out that the lake can be a treacherous thing as well as a beautiful thing to see. And it was quite a shock to us when that happened.—Weber County resident

A Weber County resident reported the flooding killed 21,000 acres of marsh in Ogden Bay, and the saline lake water flooded into Weber River:

> And the salt came clear back [into] the river, clear to the river bridge... it killed all the vegetation along the edge of the river. So we got that rotten vegetation smell really bad in those years when it was killing all that vegetation.–Weber County resident

The possibility of more high-water years made some people nervous about moving into the area even after GSL had been back to average and below-average elevations for well over a decade.

Briny flood waters affected farms all along GSL's east side. Participants who farmed in Davis and Box Elder counties described "rinsing off" or flushing their land with freshwater, and then being able to return to farming; one farmer said it took a couple of years for the ground to recover. A Box Elder farmer had to replace kilometers of fence that had been destroyed. Due to high clay content in their soil, some Weber County farmers and residents were still plagued with salt in their soil 20 years later. In addition to the financial and emotional costs, the flooding of the 1980s also resulted in building and development limitations based on the high-water mark, due to county moratoriums, changes in FEMA flood plain maps, or both (Trentelman 2009).

3.2.3.2 Low Lake Elevations

Low elevations also affect the area, with people living nearest to GSL experiencing more negative consequences than most. The most significant problem is lake dust from the exposed lake bed, an issue of increasing concern due to other local air quality issues (Larsen 2015).

> . . .With the lake being down the last several years, there's been a lot of exposed lake bed. . . .When the wind blows, especially when you get a big, heavy south wind, the area on the mainland has been picking up a lot of lake dust, which has been a bit of a problem. We've heard a lot of complaints—actually had people call us and tell us we needed to water it down so it wouldn't do that anymore.—GSL resource manager

> The soil grows the garden really good, the thing that's bad is if it's a drought year and the lake's down and there's a big wind and it blows the salt onto your plants. . .you better wash it off! 'Cause they'll croak—they don't like it.—Davis County resident

> . . .the salt etching from the stormy breezes that come—and now, like my grill on my patio that's stainless steel [is marred]—Davis County resident

Low elevations challenge GSL's positive qualities, for example, changing aesthetics as the lake shrinks into the horizon, and reducing recreational opportunities in water ever more distant from beaches and too shallow for boating. Of course, low elevations affect the lake ecosystem, raising salinity, reducing habitat available for many species such as shorebirds, and threatening protection smaller islands provide as predators can more easily access nesting colonies. This also affects humans, as it limits opportunities for ecotourism, educational research, and hunting, for example (UDFFSL 2013).

For some lake neighbors, the cause of anxiety is the variability in elevation and lake size, as it makes long-term planning difficult. A rancher who used the privately owned Fremont Island for grazing talked about this difficulty.

> That's the problem we have with. . .Fremont Island. There's been several ranchers that tried to make it work, and I don't know if we're going to be able make it work, because the Great Salt Lake fluctuates so much. . . .the one year we had to leave our calves out there, we couldn't get them in to sell 'em because the lake was so low we couldn't get a barge in to load them and so the next year we had to bring in calves and yearlings and, I mean, it was a job. . . .[Ranchers] just have such a hard time making it work because of the elevation, the fluctuation.—Weber County resident

Lake elevation issues were important to survey respondents. An open-ended survey question asked what respondents saw as the most important issue or concern related to GSL. Of the 213 who answered, 26% named water level issues (14% of all survey respondents), the highest concern raised. Of these, 17 respondents specified low elevation, and smaller numbers were concerned about future flooding possibilities or about elevation variability (Trentelman 2009).

GSL elevation has continued its decline, coming very close to the 1963 historic low level (1277.4 m/4191 ft) several times in the last 5 years (e.g., Larsen 2015; Deamer 2016; O'Donoghue 2018) (see Fig. 3.4). While weather and climate conditions (e.g., drought) have played a part, human diversions are responsible for much of the decline. As increasing amounts of water are diverted and consumed for agricultural, industrial, and urban uses, water supply to GSL has decreased, resulting in a lower lake level. In fact, diversions of the water supply to GSL have decreased the lake's elevation by a full 3.4 m (11 ft) (Wurtsbaugh et al. 2017). Importantly, 63% of the water in the GSL Basin is used by agriculture.

Fig. 3.4 Great Salt Lake, September 2016, elevation 1277.84 m (4192.40 ft) above sea level, very near the historic low of 1277.42 (4191) set in 1963. In 2016, 48% of the lakebed was exposed (Wurtsbaugh et al. 2017). Source: USGS Earthshots (image is a false-colored-composite satellite image)

In northern Utah, there is some tension in the competition for water between agriculture and municipal and industrial uses, with ongoing increases in population and related growth of communities. However, for GSL, the bigger issue is competition for water between agriculture and the lake. Box Elder County is primarily an agricultural area, and a large proportion of the farms and ranches get their irrigation water from Bear River, a primary source of freshwater for the lake. There, water going to GSL is viewed negatively. A resource manager observed,

> [S]ome of what I hear is that water that goes into the Great Salt Lake is wasted. And I've heard that a lot, there's still a lot of folks out there that don't understand that there's a value to the lake. So 'let's dam up the Bear completely and pump every single gallon...'—GSL resource manager

A Box Elder County public official validated this observation.

> Where [GSL is] not a fresh body of water, it's not really something that's useful in the agriculture world. . . .nope, it's not useful to agriculture.—Box Elder official

A rancher from the area, describing what he saw as the benefits of damming Bear River, commented, "Once the Bear River crosses there in Corrinne, the bridge right there, it's gone." This is the sentiment the resource manager was describing, that water going into GSL is seen as wasted—once it crosses the bridge, "it's gone."

Among research participants, this sentiment was fairly specific to Box Elder County, where farming has a long history of using Bear River for irrigation. Participants who farmed and ranched in western Weber County, closer to the lake than Box Elder County farmers, and where Bear River is not the source of irrigation water, saw things substantially differently, raising concerns about potential harm to GSL's ecosystem from damming Bear River.

3.2.3.3 Summary

The extremes of lake elevation have greatly affected the lake's nearest neighbors, northern Utah, and the state more broadly. During high-water times, attempts were made to manage the elevation, from legislative mandates to building pumps. These attempts have been costly, but not as costly as the flooding that occurred in the 1980s. For the last 15 years, the bigger concern has been low elevations and related damage. Interestingly, thus far there have been no real attempts to manage low elevations. While some of the lower elevation is due to weather and climate, much of it is from diverting waters that would otherwise feed GSL, for irrigation and other uses. Wurtsbaugh and coauthors argue that to "significantly reduce water use, a balanced conservation ethic needs to consider all uses, including agriculture. . ." (Wurtsbaugh et al. 2016, p. 1).

3.2.4 Development

As long as people have lived near GSL, they have had to adjust to the lake's elevation. This has affected settlement and land use patterns, both historically and more recently. The areas near GSL have been rural and agricultural. Where these areas fall within the boundaries of an incorporated city, they have been on the rural outskirts, with large lots and some farm animals. Other areas are in unincorporated rural communities, and there are some isolated farms or ranches as well.

While the flooding during the 1980s was hard on these agricultural lands, the resulting changes to the FEMA flood plain maps and establishment of building moratoriums served to maintain the rural feel. Additionally, a high-water table and being seen as too remote have slowed down residential development. Despite these drawbacks, recently in Davis and Weber counties the communities closest to the lake have been growing increasingly suburban, with new housing developments creating more densely populated neighborhoods near the shoreline (McKitrick 2007; UDFFSL 2013; Vandenack 2019).

3.2.4.1 Rural Place

For a majority of the research participants, the ruralness of the lake area was an attraction. Many lived there primarily because it was a rural area, and did not want it altered by development. While some farmed the land, others liked how near they were to nature, and especially the seclusion, quiet, and open space. When talking about their land on the east side of the lake, participants made comments such as "There's a wide, open space to the west and no housing developments will go there," "No development will happen west of our home," and referred to "a western barrier." While it may seem this has little to do with the lake, many were aware that GSL and its protected wetlands limit the potential for development and growth.

The closest neighbors of GSL often talked about the ruralness and the lake almost interchangeably. When asked about their experience with the lake, they often talked about the rural, when asked if the ruralness was the attraction to the area rather than the lake, many said they liked both (Trentelman 2009). A few made it clear the ruralness was what mattered, like one participant who said, "We moved out here not because of the lake, we moved out here because of the seclusion. And when that goes, lake or no lake, we'll be gone." A Box Elder County farmer described the lake as providing a shelter from development:

> [A]nd so you watch all that [city life and development] in the distance, knowing that the Great Salt Lake and its niche provides, I guess a shelter, if you want to call it that—some people see it as a barrier, I actually see it as a shelter, from too many neighbors.

> It's a lifestyle. . .it's a rural thing, and I think the marsh and the Great Salt Lake, they provide that because they're not going to let in a lot. . .—Box Elder County resident

The lake has played a role in the degree of ruralness of the area. Yet as development increasingly encroaches on these rural aspects so important to the lake's neighbors, it is also encroaching on GSL and its environs. Many look to GSL for shelter from this development; however, there is concern that not even the lake can slow down these changes (Trentelman 2009).

3.2.4.2 Population Development

Utah has experienced a surge in population projected to continue well into the future. Between 2000 and 2010, the state added more than half a million people, a nearly 24% increase. Most of this growth was in northern Utah, within the GSL Basin. Further, the state's population is projected to nearly double between 2015 and 2065 (Ewert 2014; Perlich et al. 2015).

This increase in population is of concern for GSL and its neighbors for two reasons: an increased need for places for people to live, and for water for their use. In terms of places to live, in Weber, Davis, and Salt Lake counties, agricultural land is being converted to residential and commercial uses, with single family subdivisions constructed on what was formerly farm or ranch land (UDFFSL 2013). This development is occurring ever closer to GSL.

Weber County's population is projected to rise from under 250,000 in 2015 to nearly 390,000 by 2065 (Perlich et al. 2015), and county planners anticipate most of that growth will be in the western portion of the county ever closer to GSL, continuing a decade-long trend (Vandenack 2018a). This makes sense, since there is more undeveloped open space there, partially due to the difficulties of building closer to GSL (Trentelman 2009). Now to address flood plain restrictions, developers are bringing in fill to raise the elevation of areas for subdivisions. The increase in development has some farmers and longtime residents of the area worried (Vandenack 2018b). Residents of townships and other unincorporated areas are working toward possible incorporation of a large swath of western Weber County into one municipality. They have accepted that continued growth appears to be inevitable, but want some control over how development proceeds (Vandenack 2019).

With double-digit growth between 2010 and 2016, Davis County has been growing faster than Weber County, and accelerated growth near GSL has been occurring for over 20 years (McKitrick 2007; Vandenack 2017). For example, Syracuse, the last city before the Antelope Island causeway, grew from about 9000 in 2001 to 24,000 in 2007 (McKitrick 2007), and was estimated at 30,400 in 2018 (US Census 2019). Davis County's population is projected to increase from about 336,000 in 2015 to nearly 545,000 in 2065 (Perlich et al. 2015).

These population changes will affect water use and availability, including waters that feed GSL. It seems intuitive that this increase in population would necessitate an increase in water for human use. With the state's projected gain being nearly double the current population, many argue that even with redoubled attempts to conserve, it is unlikely the state will be able to provide water for the increased population without developing additional water sources (Utah Foundation 2014; GWSAT 2017).

Toward this end, the State of Utah is doing preparatory work to eventually build the Bear River Development, a dam project on Bear River that would divert water for municipal and industrial use in both the northern and southern parts of the Great Salt Lake Basin. The concern for GSL is its dependence on Bear River for water. According to estimates, the Bear River Project would decrease lake elevation about 0.2 m (8.5 in.). Wurtsbaugh et al. (2017) estimate this would expose another 80 square kilometers (30 square miles) of lakebed.

One counterargument is that Utah's water use is wasteful and inefficient. The nation's second most arid state, Utah's per capita water use is considerably higher than other arid places, including others in the western United States. There is much more that can be done for reducing water use in all sectors, including agriculture (e.g., Ewert 2014). Another counterargument is that, as development increases, lands that have been used for irrigated agriculture are being converted to municipal and industry use, freeing up considerable water. For example, the general manager of one of the water districts slated to use water from the Bear River Development has reported that in his district, urbanization creates a 20% surplus of water compared to when that land was used for agriculture (URC 2019). Thus, water use projections based on current and recent past use of water in these areas greatly overestimate the amount of water needed as farmland shrinks and municipalities grow. The downside to this equation for GSL is the reduction in water that recharges the lake's wetlands that occurs with agricultural uses of these lands; however, the building of the Bear River Development has the potential to affect the whole ecosystem more negatively than this.

These issues are being considered at the state level by a number of entities (Utah Foundation 2014; GWSAT 2017). Conservation efforts have been made, with some success, although not enough: by 2016, water conservation had reduced urban per capita use by 18%, yet overall municipal water use increased by five percent because of Utah's growing population and urbanization (Wurtsbaugh et al. 2016). Importantly, this outcome could be used as evidence that conservation alone cannot supply the water needs of a growing state. It could also be used as evidence that the state is not doing enough and greater conservation measures must be taken.

Other complexities for GSL accompany the growth of cities fed by increasing population. As water use shifts from agriculture to municipal and industrial uses, less water feeds the subirrigation groundwater and lands adjacent to GSL receive reduced return flows. Additionally, development encroaches on uplands, reducing the buffer for GSL wetlands. Together these things affect the wetland and GSL ecosystem (UDFFSL 2013). Substantial acreage of lake-adjacent lands is in holdings for habitat protection, held by state and federal Wildlife Management Areas, by groups like the Nature Conservancy and the National Audubon Society, and by private landowners, with additional acreage owned and managed by private hunting clubs; however, these preservation efforts are not enough to protect lake or rurality from advancing development.

In addition to the concerns about development disrupting the rural nature of the areas near GSL where the lake's neighbors live, one participant also raised this concern:

> You know what I wonder? What is the effect on the lake of all the people moving in, 'cause the water gets treated and dumped into the lake. What happens to the lake when all these people move here?—Davis County resident

3.2.4.3 Summary: Development

Humans accommodated the lake and its ecosystem in their settlement patterns. They adapted to high lake levels, and the potential for future high levels. However, increases in their population and development are now affecting the lake ecosystem, with the potential to affect it far more substantially if the Bear River Development is realized. These human-related changes to the lake are also affecting the lifestyle preferences of the lake's nearest neighbors.

3.2.5 Appreciated! Meanings of Great Salt Lake

Thus far readers have been introduced to ways GSL has affected residents economically, as a protection from development, and as it adapted to human technology, as well as the ways lake elevations have affected residents. But how do they feel about the lake more broadly? What does GSL mean to its nearest neighbors? Research participants were asked if GSL held any particular meaning for them, and if so, what that meaning was. Several themes emerged, including the lake as a part of nature, or as a place to recreate. For some, GSL was meaningful because of social connections, for others, because of its unique traits. Some participants felt quite negative about GSL, and others really did not think about the lake at all (Trentelman 2009).

3.2.5.1 Attachment

When asked how they felt about living close to GSL, some participants expressed a substantial positive connection with the lake. Sixteen percent of survey respondents, and a good number of qualitative participants, felt attached to GSL (Trentelman 2009).

> It's a privilege, it's a tremendous privilege.—Davis County resident

> Why a lot of us are as close as we are is because of the memories and the nostalgia that comes with the land and being close to the lake, enjoying the benefits that the lake actually offered us.—Weber County resident

> I can't imagine being without it... I can't visualize living out here and not having this beauty.—Weber County resident

> I told my husband I want to be buried in West Point, right there on the top where you can look down on the lake...there's a big enough connection that I want to end up there... Yeah, that's pretty connected, isn't it? If you're gonna put your bones there.—Davis County resident

3.2.5.2 The natural aspects of Great Salt Lake

For many, GSL was perceived as a natural place with its own beauty. Participants described sunsets over the lake, and taking time to enjoy them. Some enjoyed the night skies over the lake.

> How many times do you come home and someone will say, 'Did you see the sunset tonight?' And what do you do? You haul your lawn chair to some certain spot and you look at the sunset.—Davis County resident

> There's two times a year the sun sets on the water—when it comes off the tail of that Fremont Island and when the sun hits that water, the light just shoots across the water...it's pretty neat. ...It only lasts for a few days, maybe a week or so.—Davis County resident

> ...The moon is really pretty too. ...it's so dark, sometimes we get the pretty effect of the moon kind of glowing off the lake. I mean sometimes it seems like it's not even dark out here.—Weber County resident

Many participants noted the closeness to wildlife. This included numerous avian references, from Great Blue Herons sitting on top of a neighbor's shed to large numbers of wintering bald eagles. They also described bighorn sheep, bison, deer, elk, coyotes, foxes, turtles, and others, found on the shores, in the basins where the rivers meet the lake, and on the islands.

> We've seen pelicans close to the house and that's been awesome because my oldest daughter likes birds so we're always out, seeing what birds we can find.—Davis County resident

> ...we are so lucky...to be able to sit in your backyard and watch the geese and the ducks and hear them... and all the other birds. We live right out in a waterfowl area. ...I was watching the cranes today and it's just so much fun...but I've always loved birds, so I really appreciate it when I can go out on my balcony and watch them birds. ...Not only that, but they have deer now. Deer and elk...—Weber County resident

Participants talked about lake-related weather, including watching the weather on the lake.

> I get an enjoyment out of watching the news and seeing what the prediction is, then watching the fronts move in on the lake...—Weber County resident

> I remember one night...a storm was coming. ...the wind coming in and blowing my hair, and I remember looking over that direction and the redness of the sunset...with the grey...those clouds will come in over, and they're really low...and they're really, really black on top, and I thought, 'This is a very, very cool thing that I'm experiencing right now.'—Davis County resident

3.2.5.3 Recreation Aspects of Great Salt Lake

Many participants indicated that GSL meant recreation to them. The activities discussed included boating, swimming, and floating in the lake, photography, bird-watching, and hunting. People fished at two of the fresher bays. Other activities were not directly lake-related, but the lake area provided a place to do the activities, such as horseback riding, bicycling, jogging and walking, going for drives, and riding four-wheelers.

> I grew up there and, you know, I just loved it. I really don't understand what other kids did for a childhood, because we just played out in the fields and rode horses and every day we was on our horses bareback and went down in the swimming hole, swimming, and then hunting. It just was the childhood that we had, and it was fun.—Weber County resident

> It was always a mysterious place to go. It was like going to the ocean. It was a fun place to go. I like the lake and of course, being a hunter, the lake furnished us a lot of opportunities to hunt waterfowl. I can remember many stories that my father and my grandfather used to tell about hunting on the entrance to the lake where the water, the rivers, empty into it.—Weber County resident

3.2.5.4 Social Aspects of Great Salt Lake

Some people's perceptions of GSL were related to their social world rather than to the lake's features: their heritage, a sense of community, family traditions and stories, or childhood.

Nearly one-fifth of survey respondents indicated they lived on, or very near, property that had been in their family for multiple generations, some for three to five generations, and some since before 1900. All the interview and focus group participants with intergenerational family roots described feeling very connected to the area, to their "family ground." One older man talked about having lived there, where his ancestors had homesteaded, since he was 3 years old.

> ...my father was originally from here...and all the relatives, my ancestors, lived [there] at that time. ...we came down here as a family in the old homestead that my grandfather owned... [My father] eventually bought the old homestead and then I bought it from him. ...Before I got married, I made my wife sign a contract that that's where we was going to live. ...I'd said, 'If you marry me, we're going to live out there next to the lake.' That's where I was born and raised, rode my horses bareback for years, and so that's why I'm there. And I've already bought a burial plot in the local cemetery, that's where my remains will probably go as well.—Weber County resident

> There's some really great pioneer history for some of us, so, you know, that really adds to your connection, I think.—Davis County resident

Some participants described important family occasions and events being tied to the lake.

> I got engaged on Antelope Island.—Survey response

> As kids...we all swam, we all floated. I took my kids out there probably just last year... You know you walk through leg-deep sand and you walk through the bugs, and they thought that was pretty cool... They did basically what we did as kids, although they really couldn't float as well and I don't think they had showers in those days, but now you can clean yourself off. ...It was kind of letting kids know this was where we hung out as kids.—Weber County resident

For some, GSL was important because it is where they raised their families.

> How I'm connected is because I have lived here in the bulk of raising my five children as they're growing up... just enjoyin' that—and the times that they've had there [at lake sites], and all of that nature...—Davis County resident

> I think my kids all feel a connection. . . . this has been their growing-up years and they love it. I've heard my kids say they'd love to have a spot along the lakeshore.—Davis County resident

3.2.5.5 Great Salt Lake as a Unique Place

Some participants focused on the unusual aspects of the lake. For them, GSL was unique, and living near the lake meant living in a unique place. For several, this meant the uniqueness of the natural aspects, for example, the oolitic sand, or even the unusualness of a salt lake.

> . . .I remember when I was a little boy going out with my parents and relatives, and they would build little houses out of wire and they'd put them in the lake and then they'd bring in these little crystal houses. In fact, I dropped one of my cowboy hats in the lake about a month ago and I hurried and fished it out, and by the time we got back to Promontory, my hat was just white. So I'm just going to take an old hat out there and making it a crystal hat—the lake crystallizes everything, makes a really beautiful thing.—Weber County resident

Several participants felt the lake being well known made it special and unique.

> It's really easy to find the lake on a map, even on a globe. 'See that little point there? That's where we are.' Not many people can say that.—Davis County resident

> That's our claim to fame. If someone says, 'where do you live?' and I say, 'Have you heard of the Great Salt Lake?' I don't care if I'm in Nebraska, or where I'm at, 'Have you ever heard of the Great Salt Lake?' 'Yes.' 'I live about three miles [five km] from the Great Salt Lake.' They know where you're at.—Weber County resident

> It's a famous lake, it's the biggest lake west of the Mississippi. It's the only salt lake in the nation. If you say, 'I live by the Great Salt Lake' someone knows where you're at, it's the only lake in Utah that's really on the map. . . It's just something important.—Weber County resident

3.2.5.6 Those with a Negative Sense of Great Salt Lake, or No Sense At All

Compared to how positively people tend to feel about freshwater lakes, another somewhat unusual aspect of GSL is that for some people, it garners a negative response, or even no response at all. Even for some of the lake's nearest neighbors, GSL is unappreciated.

GSL has a number of aspects seen as negative, everything from odor and bugs to lake dust, the salt, and the perception that "there's nothing out there." For some participants (e.g., 4.7% of all survey respondents, as well as a few interviewees and focus group members), the lake, on the whole, was an irritating place with few, if any, redeeming qualities.

> [Commenting on the area near the Antelope Island causeway] They got that long sweeping stinking, slimy, brine shrimp-infested, rotten—oh, that's where it stinks, that's where the lake stinks! That lake has a distinct smell. . .—Weber County resident

> Stinky, smelly, mosquito infested. . .—Davis County resident

Negative characterizations by the broader public make it difficult for GSL's nearest neighbors at times. One participant commented,

> You get visitors out there, there's nothing—they'd don't like anything about it. It's a dead lake to them, it's a—you know, it stinks, there's bugs.—Weber County resident

There is also a perception from others that it is a difficult place to live. Some focus group members described being discouraged from moving into the area.

> Everybody in our [church] was making fun of us, 'oh no, not the sewer, down by that lake.'—Davis County resident

> . . .And some people said, 'Oh, well, what about the smell? What about the mosquitoes?'—Davis County resident

> I hear, 'you poor people that have to live in the most miserable place in Syracuse'.—Davis County resident

A Weber County official assumed people likely lived near GSL only because of affordable land:

> I'd imagine the property value out there when they bought was pretty low, and probably no one really wanted to be out by the lake.—Weber County official

This negativity about the area near GSL has likely existed a long time. A man in his late 70s described it when he was a boy in school, when the area west of Syracuse near the lake was called "the bottoms."

> We was kind of treated like second-class citizens because we were from 'the bottoms' and the elite were up here [in Syracuse]. . . .There was some pretty wealthy families moved in up here. . .their kids didn't have to work and everybody down there had to work. . . .They'd have problems at school, maybe, a lot of fistfights. . .but the ol' principal we had said, 'well it's always those kids from the bottoms causing troubles, you know.'—Davis County resident

A proportion of participants (e.g., 6.8% of all survey respondents) indicated the lake had no particular meaning; they did not really think about it. This, despite living very near a lake that, at its common elevation, covers 2414 km^2 (1500 square miles).

> The lake's irrelevant... The lake's just there. . . Never thought about the lake much, actually.—Weber County resident

> The lake has nothing to do about how I feel about living here.—Survey response

> Don't really pay any attention to it.—Survey response

> I don't think about it.—Survey response

GSL was not part of these residents' lives, but rather an undistinguished, unimportant part of the landscape. A resource manager speculated about this dynamic.

> The reason being, I think, is they just don't realize what's out here. And when I do find new people that come out here and see this, they're just kind of 'Wow, I never knew this was out here, I just had no idea.' And some of these are people who lived within a mile of the lake their entire lives, 30, 40 years. And they just don't have a clue, they just never think about it.—GSL resource manager

Some residents are ignorant of basic information about GSL and its environs. A resource manager in Davis County said with a sigh,

> I've talked to people out here who don't even know that that's an island out there. They're just like, 'What are those mountains?' Antelope Island. 'Oh, there's an island out there?'—GSL resource manager

3.2.5.7 Summary of Perceptions

This exploration of how GSL's nearest neighbors experience the lake has demonstrated a variety of perceptions. While there is evidence of the lake being unappreciated, with some participants holding negative perceptions and the lake not mattering to others, a substantial majority indicated GSL held meaning for them. Most participants felt positively about the lake, including feeling protective toward it. This constellation of GSL meanings held by those living nearby is indicative of its complexity. With the foci varying from traits of GSL itself to more social aspects of the lake area, it is evident that people's relationships with GSL are diverse (Trentelman 2009).

These data demonstrate varying ways GSL affects the lives of people who live nearby. For example, the qualitative research revealed that those who live closest to the lake can experience lake-related challenges others do not, including in some areas, more potent and pervasive odors, damage from lake dust and from salt, and even stigma from living near GSL. Yet lake neighbors also identified positive things others do not have the opportunity to enjoy (Trentelman 2009).

3.3 Challenges for the Future

This volume's title refers to GSL being in an "age of change." While change and adaptation have always part of the relationships between GSL and humans, the current challenges seem more daunting than they have been at least since the flooding in the 1980s. Some may argue things are even more daunting now. Looking forward, challenges to GSL, nearby humans, and the relationships between them include continued population growth and development and its effect on the lake, and related pressures on GSL elevation and its effect on the ecosystem, including the humans who are part of it. Climate change is the next external force (although human-caused) affecting both. GSL's nearest neighbors have raised concerns related to their relationships with the GSL area. Despite these challenges, there is some cause for hope.

3.3.1 Challenges Related to Development

As noted earlier, Utah's population is projected to continue growing substantially well into the future. After hitting 3 million in 2013, it is projected to reach 4 million just 17 years later, in 2032, and 5.8 million in 2065 (Perlich et al. 2015). Considerable growth is projected in counties within the GSL Basin. Both Weber and Davis counties are projected to continue growing rapidly, and it is expected that their growth will occur in the western parts of the counties, increasing the population density near the lake (Vandenack 2018a, 2017). Additionally, both Salt Lake and Tooele counties are continuing to grow, and are developing industrial areas increasingly near GSL (UDFFSL 2013), for example the development of a large distribution hub or "inland port," the boundaries of which include 16,147 acres of land in Salt Lake County's Northwest Quadrant immediately south of the lake (Utah Inland Port 2019). Much of this area is wetlands, 4000 acres according to one Salt Lake City Council member (Nixon 2019), raising concerns for the ecosystem and the birds it supports. An industrial example from Box Elder County is the development of a commercial landfill on the southern tip of Promontory Point, the landmass that extends into the lake from the north, creating GSL's north arm. The 2000-acre site is permitted to open as soon as a contract with a municipal or county client can be developed, although there are no contract negotiations going on as of August 2019 (Larsen 2019).

These increases in population and development bring additional pressure and tension over the availability of water for GSL. Any expansion in the diversion of the lake's water supply, e.g., diverting more water from Bear River, will increasingly affect the lake elevation.

3.3.2 Challenges Related to Great Salt Lake Elevation and Water for the Lake

A considerable challenge for both lake ecosystem and humans is that GSL elevation is already low, and has been for much of the last decade and a half (SWCA 2017; USGS 2019a). As noted earlier, research has found that water diversions are responsible for a substantial proportion of this decline. GSL is already experiencing these reductions now, before any additional increase in population, urbanization of the areas closest to the lake, or further development of lake-bound water. In 2016, 48% of the lakebed was dry (Wurtsbaugh et al. 2017). It has been argued that the lake's ecosystem cannot afford any further impact on GSL elevation.

Yet, in addition to increases in population and urbanization, climate change is highly likely to affect the amount of water available for GSL. Some of this change is still difficult to predict, but there is agreement among climate scientists and other experts that, while actual precipitation may not decline, more of it will fall as rain than snow, thus reducing Utah's important snowpack which feeds streams, rivers, and eventually GSL. Projected snowpack declines will lead to lower GSL elevations and increased salinity (UDFFSL 2013; Bedford 2014; Utah Foundation 2014; GWSAT 2017; Baxter and Butler 2020).

Of course, climate change will not just affect the lake. These effects on rivers and streams will affect all water users in the GSL Basin, from agriculture to cities and industry. This has the potential to substantially exacerbate tension over water.

These lake elevation challenges can create tremendous problems for the GSL ecosystem, due to increased salinity, exposed lakebed, dried out wetlands, and other issues. Additionally, they can create problems for Utah's economy by affecting the brine shrimp industry, the mineral extraction industry, and the amount and types of recreation the lake can support. Further, the potential for health consequences from lake dust increases (UDFFSL 2013). Finally, these issues can negatively affect people's relationships with GSL, relationships that, among other things, are important for continuing advocacy for the lake (Trentelman 2009).

3.3.3 Concerns for the Future from the Lake's Nearest Neighbors

While able to experience positive things others miss out on, the people who live closest to GSL are also most affected by some of the negative aspects of GSL. Hence, their concerns about the lake and lake management should be given consideration. When asked what they saw as the most important issue or concern related to GSL, as noted earlier, the highest concern expressed was water elevation issues. The second ranking concern was preservation concerns, including "saving the quality of lake waters/shoreline," "conserving and protecting its fragile environment and wildlife," and "saving areas for migratory birds and wild animals" (Trentelman 2009).

As noted earlier, participants raised a number of concerns about the increasing development they were watching. While they were certainly anxious about losing their rural way of life, they were also concerned about the lake. They worried about how the growing population might affect GSL habitat and water quality, about whether increasing industry in the area is leading to the pollution of GSL, and about increasing population resulting in further diversion of the freshwater that feeds the lake.

3.3.4 Cause for Hope

While there are numerous and sobering causes for concern for GSL, its ecosystem, and the people who live nearby, there is also cause for hope in a number of areas. There is evidence that at least the nearest neighbors of the lake feel protective of it; increasing research is being conducted on the various consequences of a shrinking lake, with efforts being made to develop solutions; and it appears that awareness of these issues may be increasing.

The sociological research found evidence that the nearest neighbors feel protective of GSL. In addition to the preservation concerns raised by a substantial proportion of participants noted above, there were a number of other indications of strong support for protecting GSL. For example, the majority of survey respondents agreed that GSL must be taken care of, so future generations can enjoy it (69% agreed, of which 32% strongly agreed), that GSL is important for providing habitat for wildlife (77% agreed, 34% strongly), and that more should be done to protect the lake and its wetlands (58% agreed, 25% strongly). Further, as noted previously, among those who live nearest to GSL, a substantial number of people feel attached to the lake, and a majority of participants had positive perceptions of it (Trentelman 2009). This is hopeful; again social science research has found that sense of place and place attachment have been associated with increased commitment to the place, preservation efforts, and support for resource management (Williams et al. 1992; Kaltenborn and Williams 2002).

Research is now underway to examine health issues from lake dust (Larsen 2015), and the economic and societal effects of a shrinking lake (AECOM 2019). Entities such as Utah Foundation are also conducting research on solutions to the water conundrums that exist in Utah. Rather than large development projects, Utah Foundation's report (2014) appears to privilege substantial increases in conservation, arguing to address water law and restructure water rates statewide to accomplish this end. Further, policy makers and politicians at both state and local levels appear to be paying attention. During the 2019 session of the Utah State Legislature, bills and resolutions were proposed that would support GSL in several ways. For example, a concurrent resolution to address declining water levels of GSL passed both houses unanimously; monies were appropriated for GSL-related research; and while the passage of an amended version of a bill to meter irrigation water reduced the intended effectiveness of the original bill, it still moves the needle forward (De Freitas 2019).

Efforts are being made at political and policy levels to increase awareness of the potential consequences of a shrinking lake, and also devise policy solutions to the current trend of ever smaller shares of water making it to GSL. For example, when Utah's governor convened a Water Strategy Advisory Team to make recommendations for a state water strategy, conservationists and lake advocates were among the stakeholders included. While the advisory team's final recommendations continue to include water development projects (such as the Bear River Development), they also recommend strategies such as water recycling and reuse. They argue any water development projects must be weighed, carefully looking at all benefits and costs, including to the environment:

> The challenge for the future is how to assure that reasonable human needs are met, while—through improved science, water efficiency, and thoughtful policy and legal changes—adequate water is also available for the environment (GWSAT 2017).

Environmental needs have not always been a priority in Utah. These examples and others appear to provide anecdotal evidence that awareness of these issues is increasing, but we need empirical confirmation that this is the case. This speaks to the need for further social science research on perceptions of the lake, and on the degree of willingness that exists to support policy changes.

3.3.5 Summary

The challenges for the future of the relationships between GSL and humans are clearly daunting, particularly with ongoing tensions between increasing development and the difficulties for both lake and humans related to lake elevation. GSL's neighbors are concerned for the future of the lake. Fortunately, there appears to be increasing concern about the fate of GSL, including by decision makers.

3.4 Conclusions

This chapter has examined the relationships between humans and GSL, and a variety of intervening factors including technology, economics, the variable lake elevation, and human development. With each, the ways in which humans have affected the lake and the lake affected humans have been outlined. Research has demonstrated these relationships are varied: while some of the lake's nearest neighbors appear to be unappreciative of GSL, the majority of participants find it meaningful. Some indicate feeling very connected, or attached, to the lake. This matters, since those with place attachment or sense of place are more likely to feel protective about that place, and supportive of conservation-oriented management practices.

In the process of making progress for human society, such as improving transportation technology and building agriculture in an arid state, actions taken have had substantial consequences for GSL's ecosystem, e.g., building two causeways, and developing infrastructure to divert water for farming. While acknowledging the positive outcomes from these efforts, we are now paying attention to the negative consequences as well, and investing more in scientific investigation, both of these past actions and of potential effects of those in the future. Efforts are being made to incorporate these concerns into policy and management decisions. It is critical for this work to continue, to make gains for the health of the lake, and relatedly, for the humans.

Importantly, while humans have attempted to manage the lake during the crises (for humans) of high elevation years, thus far we have not made similar attempts with problematically low elevations. The difference is that with high elevations, we could focus on "fixing" the lake, whereas with low elevations, we need to focus on the humans, on our own practices. We clearly have more control over the latter, even with the incredible messiness of managing human behavior, compared to the degree of control we have over the natural forces that drive the lake side of the equation (see, e.g., Ewert 2014; Wurtsbaugh et al. 2016, 2017). This is one of several areas where further social science work would be beneficial.

While there is some cause for hope, there are no simple solutions for the complexities here. In one way or another, humans will continue to affect the lake and the lake will continue to affect humans in the recursive ways that have been outlined here. Further, these relationships will also continue to affect the larger GSL ecosystem, including the biological processes addressed in the other chapters of this volume. This drives home the importance of the humans continuing to work toward real solutions for their need for water without further compromising the health of the lake—and their own health in the process.

References

AECOM (2019) Consequences of drying lake systems around the world. Research report prepared for State of Utah Great Salt Lake Advisory Council

Aldrich JM (1912) The biology of some western species of the dipterous genus *Ephydra*. J N Y Entomol Soc 20:77–102

Atwood G, Wambeam TJ, Anderson NJ (2016) The present as a key to the past: paleoshoreline correlation insights from Great Salt Lake. In: Oviatt CG, Shroder JF (eds) Lake Bonneville a scientific update. Elsevier, Amsterdam

Barrett KL, Belovsky GE (2020) Invertebrates and phytoplankton of Great Salt Lake: is salinity the driving factor? In: Baxter BK, Butler JK (eds) Great Salt Lake biology: a terminal lake in a time of change. Springer, Cham

Baxter BK (2018) Great Salt Lake microbiology: a historical perspective. Int Microbiol 21 (3):79–95

Baxter BK, Butler JK (2020) Climate change and Great Salt Lake. In: Baxter BK, Butler JK (eds) Great Salt Lake biology: a terminal lake in a time of change. Springer, Cham

Baxter B, Zalar P (2019) Model ecosystems in extreme environments. In: Seckbach J, Rampelotto P (eds) The extremophiles of Great Salt Lake: complex microbiology in a dynamic hypersaline ecosystem. Elsevier, Amsterdam, pp 57–99

Bedford DP (2006) Utah's Great Salt Lake: a complex environmental-societal system. Geogr Rev 95(1):73–96

Bedford DP (2014) Climate change and the future of Great Salt Lake. In: Crimmel H (ed) Desert water: the future of Utah's water resources. University of Utah Press, Salt Lake City, UT, pp 80–97

Behle WH (1958) The bird life of Great Salt Lake. University of Utah Press, Salt Lake City, UT

Belovsky GE, Stephens D, Perschon C, Birdsey P, Paul D, Naftz D, Baskin R, Larson C, Mellison C, Luft J, Mosley R (2011) The Great Salt Lake Ecosystem: long term data and a structural equation approach. Ecosphere 2:1–40. https://doi.org/10.1890/ES10-00091.1

Bioeconomics (2012) Economic significance of the Great Salt Lake to the State of Utah. Research report prepared for State of Utah Great Salt Lake Advisory Council. https://documents.deq.utah.gov/water-quality/standards-technical-services/great-salt-lake-advisory-council/Activities/DWQ-2012-006864.pdf. Accessed 7/5/19

Brunson MW, Nicholson B (1999) Use, attitudes, and beliefs about the Great Salt Lake among Davis County residents. Professional report 1999-02, Institute for Outdoor Recreation and Tourism, Utah State University, Logan, UT

Cannon JS, Cannon MA (2002) The Southern Pacific railroad trestle—past and present. In: Gywnn JW (ed) Great Salt Lake: an overview of change. Utah Department of Natural Resources, Salt Lake City, UT, pp 283–291

Deamer K (2016) Utah's Great Salt Lake is shrinking. Live Science, December 1. https://www.livescience.com/57055-utah-great-salt-lake-shrinking.html. Accessed 7/9/19

De Freitas L (2019) The Great Salt Lake geopolitical landscape, water, and the shapes of things to come. Executive Director's Message blog, FRIENDS of Great Salt Lake. https://www.fogsl.org/news-and-archives/executive-directors-message/item/765-the-great-salt-lake-geopolitical-landscape-water-and-the-shapes-of-things-to-come. Accessed 7/19/19

Ewert EC (2014) The coming challenge: population growth and water decline. In: Crimmel H (ed) Desert water: the future of Utah's water resources. University of Utah Press, Salt Lake City, UT, pp 13–28

Gaddis E, Hornbeck JH, McCulley E, Nicholson B, Crowley P, Stutz A, Burfitt L, Pecorelli J, Low G (2012) Definition and assessment of Great Salt Lake health, prepared for State of Utah Great Salt Lake Advisory Council. https://documents.deq.utah.gov/water-quality/standards-technical-services/great-salt-lake-advisory-council/Activities/DWQ-2012-006862.pdf. Accessed 7/5/19

Governor's Water Strategy Advisory Team (GWSAT) (2017) Recommended state water strategy. Report. https://www.envisionutah.org/images/FINAL_Recommended_State_Water_Strategy_7.14.17_5b15d.pdf. Accessed 7/10/19

Great Salt Lake Ecosystem Program (GSLEP) (2019) GSLEP website, Utah Department of Natural Resources. https://wildlife.utah.gov/gsl/. Retrieved 7/7/2019

Gwynn JW (2002) The railroads proximate to Great Salt Lake, Utah. In: Gywnn JW (ed) Great Salt Lake: an overview of change. Utah Department of Natural Resources, Salt Lake City, UT, pp 273–281

Hummel D (2006) Great Salt Lake organisms fuel scientific hopes: Utah colleges toiling to unlock their mysteries. Deseret Morning News, May 30. https://www.deseretnews.com/article/635211428/Great-Salt-Lake-organisms-fuel-scientific-hopes.html. Accessed 7/4/2019

Jenkins M (2002) Suburbanites compete for the lake's fresh water. High Country News, April 29, p 11

Kaltenborn BP (1998) Effects of sense of place on responses to environmental impacts: a study among residents in Svalbard in the Norwegian High Arctic. Appl Geogr 18(2):169–189

Kaltenborn BP, Williams DR (2002) The meaning of place: attachments to Femundsmarka National Park, Norway, among tourists and locals. Norweg J Geogr 56:189–198

Lambourne A (1909) Our inland sea: the story of a homestead. The Deseret News, Salt Lake City, UT

Larsen L (2015) As Great Salt Lake dries up, Utah air quality concerns blow in. Standard-Examiner, October 11. https://www.standard.net/news/environment/as-great-salt-lake-dries-up-utah-air-quality-concerns/article_104452df-6fd1-5f2a-8601-0432cc1f5ad5.html. Accessed 7/9/19

Larsen L (2018) 1 year after the breach, Great Salt Lake causeway creates mixed feelings. Standard-Examiner, January 22. https://www.standard.net/news/environment/year-after-the-breach-great-salt-lake-causeway-creates-mixed/article_c7e0d88f-3748-51d4-b177-79ec136ee88b.html. Accessed 7/6/19

Larsen L (2019) A landfill on the tip of the Promontory cape on the Great Salt Lake is on the verge of opening for business. Salt Lake Tribune, August 19. https://www.sltrib.com/news/politics/2019/08/19/landfill-tip-promontory/. Accessed 8/29/19

McKitrick C (2007) Davis County officials seek to steer future growth to a happy balance. The Salt Lake Tribune, June 18. https://archive.sltrib.com/article.php?id=6166751&itype=NGPSID. Accessed 7/10/19

NASA (2019) Utah's Great Salt Lake. Space Station image iss059e099093 (June 10, 2019). https://www.nasa.gov/image-feature/utahs-great-salt-lake. Accessed 7/25/19

Nichols J (2014) No lasting resorts: the shifting shoreline of Great Salt Lake. In: Conference paper, World conference on environmental history, Portugal

Nichols J (2019) Introduction. Utah Hist Q 87(1):3–7

Nixon N (2019) SLC council approves $28m tax break for inland port development. KUER NPR Utah, August 27. https://www.kuer.org/post/slc-council-approves-28m-tax-break-inland-port-development#stream/0. Accessed 8/29/19

O'Donoghue AJ (2018) Great Salt Lake: a lake no more? Deseret News, May 10. https://www.deseretnews.com/article/900018277/great-salt-lake-a-lake-no-more.html. Accessed 7/9/19

Perlich PS, Hollingshaus M, Harris ER, Tennert J, Hogue MT (2015) Utah's long-term demographic and economic projections summary. Research brief, Kem C. Gardner Policy Institute, University of Utah, Salt Lake City, UT

Peterson D (2001) Tale of the Lucin: a boat, a railroad and the Great Salt Lake. Old Waterfront, Trinidad, CA

Simms SR, Stuart ME (2002) Ancient American Indian life in the Great Salt Lake wetlands: archaeological and biological evidence. In: Gywnn JW (ed) Great Salt Lake: an overview of change. Utah Department of Natural Resources, Salt Lake City, UT, pp 71–83

Sorenson ED, Hoven HM, Neil J (2020) Great Salt Lake shorebirds, their habitats and food base. In: Baxter BK, Butler JK (eds) Great Salt Lake biology: a terminal lake in a time of change. Springer, Cham

Stansbury H (1855) Exploration of the valley of the Great Salt Lake: including a reconnaissance of a new route through the Rocky Mountains. Lippincott, Gramabo, Philadelphia

SWCA Environmental Consultants (SWCA) (2017) Water for Great Salt Lake. Research report prepared for Great Salt Lake Advisory Council, Salt Lake City, UT

Travous KE (1980) Recreation on the Great Salt Lake. In: Gwynn JW (ed) Great Salt Lake: a scientific, historical and economic overview. Utah Geological and Mineral Survey, Utah Department of Natural Resources, Salt Lake City, UT, pp 33–45

Trentelman CK (2009) "Big, smelly, salty lake that I call home": sense of place with a mixed amenity setting. Ph.D. Dissertation. Utah State University, Logan, UT

United States Census Bureau (US Census) (2019) QuickFacts: Syracuse City, Utah. https://www.census.gov/quickfacts/syracusecityutah. Accessed 7/10/19

United States Geological Survey (USGS) (2016) New breach allows flow on the Great Salt Lake. USGS website post. https://www.usgs.gov/media/videos/new-breach-allows-flow-great-salt-lake. Accessed 7/6/19

United States Geological Survey (USGS) (2019a) Great Salt Lake at Saltair Boat Harbor, UT. National Water Information System: Web Interface (live data). July 9. https://waterdata.usgs.gov/ut/nwis/uv/?site_no=10010000&PARAmeter_cd=62614. Accessed 7/9/19

United States Geological Survey (USGS) (2019b) Great Salt Lake overview, results. USGS website post. https://www.usgs.gov/centers/ut-water/science/great-salt-lake-0?qt-science_center_objects=0#qt-science_center_objects. Accessed 7/6/19

Utah Division of Forestry, Fire & State Lands (UDFFSL) (2013) Great Salt Lake comprehensive management plan and record of decision. Utah Department of Natural Resources, Division of Forestry, Fire & State Lands, Salt Lake City, UT

Utah Foundation (2014) Flowing toward 2050: Utah's water outlook. Research report, Utah Foundation, Salt Lake City, UT. http://www.utahfoundation.org/reports/flowing-toward-2050-utahs-water-outlook/. Accessed 7/10/19

Utah Inland Port (2019) Background. https://www.utahinlandport.org/inland-port-background-1. Accessed 8/29/19

Utah Rivers Council (URC) (2019) Alternatives to Bear River water development: implementing inexpensive alternatives to the west's largest new water project, 4th edn. Salt Lake City, UT

Vandenack T (2017) Davis County cities boom, easily outpace growth of Weber County cities. Standard-Examiner, June 10. https://www.standard.net/news/local/davis-county-cities-boom-easily-outpace-growth-of-weber-county/article_35094aa0-8559-5b09-8796-b5a6129d478c.html. Accessed 7/10/19

Vandenack T (2018a) Rural Weber County residents bristle at development, but growth is coming. Standard-Examiner, January 11. https://www.standard.net/news/local/rural-weber-county-residents-bristle-at-development-but-growth-is/article_8be57bc7-7002-551e-9b30-57d1ac38700a.html. Accessed 7/10/19

Vandenack T (2018b) West Haven, western Weber Co. cites register fastest growth, new data shows. Standard-Examiner, May 24. https://www.standard.net/news/government/west-haven-western-weber-co-cities-register-fastest-growth-new/article_8fc36e00-bb99-5390-a5c8-25e8313b21c4.html. Accessed 7/10/19

Vandenack T (2019) Western Weber County residents mull incorporation, turning zone into city. Standard-Examiner, March 17. https://www.standard.net/news/government/western-weber-county-residents-mull-incorporation-turning-zone-into-city/article_b702a047-d4fe-5cdc-8cfd-7c9f8c66cb1d.html. Accessed 7/10/19

Vaske JJ, Kobrin KC (2001) Place attachment and environmentally responsible behavior. J Environ Educ 32(4):16–21

Vorkinn M, Riese H (2001) Environmental concern in a local context: the significance of place attachment. Environ Behav 33(2):249–263

Western Hemisphere Shorebird Reserve Network (2019). https://www.whsrn.org/great-salt-lake. Accessed 6 Aug 2019

Williams DR, Patterson ME, Roggenbuck JW, Watson AE (1992) Beyond the commodity metaphor: examining emotional and symbolic attachment to place. Leisure Sci 14:29–46

Woolf J (1999) What do we want to do with the lake? Salt Lake Tribune, November 25–28. Reprinted in Topping G (ed) (2002) Great Salt Lake: an anthology, Utah State University Press, Logan, UT, pp 274–276

Wurtsbaugh WA, Gliwicz ZM (2001) Limnological control of brine shrimp population dynamics and cyst production in the Great Salt Lake, Utah. Hydrobiologia 466:119–132. https://doi.org/10.1023/A:101450251

Wurtsbaugh WA, Marcarelli M, Christison C, Moore J, Gross D, Bates S, Kircher S (2002) Comparative analysis of pollution in Farmington Bay and the Great Salt Lake, Utah. Aquatic Ecology Laboratory class project 2001, report to the Utah Division of Water Quality. College of Natural Resources, Utah State University, Logan, UT

Wurtsbaugh WA, Marcarelli AM, Boyer GL (2012) Eutrophication and metal concentrations in three bays of the Great Salt Lake (USA). Watershed Sciences Faculty Publications. Paper 550. http://digitalcommons.usu.edu/wats_facpub/550

Wurtsbaugh W, Miller C, Null S, Wilcock P, Hahnenberger M, Howe F (2016) Impacts of water development on Great Salt Lake and the Wasatch Front. Watershed Sciences Faculty Publications, Paper 875. https://digitalcommons.usu.edu/wats_facpub/875/. Accessed 7/9/19

Wurtsbaugh WA, Miller C, Null SE, DeRose RJ, Wilcock P, Hahnenberger M, Howe F, Moore J (2017) Decline of the world's saline lakes. Nat Geosci 10(11):816. https://doi.org/10.1038/ngeo3052

Chapter 4
Microbialites of Great Salt Lake

Melody R. Lindsay, Eric C. Dunham, and Eric S. Boyd

Abstract Fossilized organo-sedimentary structures (microbialites) have been identified in sedimentary rocks dated to 3.5 Ga, with some reports of putative microbialite structures in rocks that are even older. These findings have spurred significant interest in understanding the role of biology in the formation of microbialites and the role of microbialites in sustaining biodiverse contemporary and non-contemporary ecosystems. Microbialites in Great Salt Lake (GSL) form reef-like structures that cover an estimated 20% of the lake bottom and thus represent the most extensive assemblage of extant microbialites on Earth. GSL microbialites are colonized by complex photosynthetic microbial mats consisting of both Cyanobacteria and algae (diatoms) that contribute fixed carbon supporting a diversity of heterotrophic microorganisms also within these mats. These diverse microbial communities are also thought to be involved in the formation of carbonate minerals that can then lithify and preserve microbialite structure. Biomass produced by these complex microbial communities supports a variety of higher forms of life, including brine flies and brine shrimp that themselves serve as food sources for a diverse array of shore and migratory birds. Consequently, the microbialites and associated mat communities represent integral components of the aquatic ecosystem at GSL and represent useful analogs for understanding microbialite ecology in past Earth environments. This chapter overviews the key microbial taxa that comprise microbialite mat communities and the metabolic processes that support them, highlighting the importance of these "living fossils" and their linkages with the health of the greater GSL ecosystem and their significance as analogs for understanding ecosystem function on early Earth.

Keywords Microbialite · Cyanobacteria · Diatoms · Primary production · Secondary production · *Artemia* · Salinity · Trophic · Food web · Archean · Dolomite · Aragonite

---

M. R. Lindsay · E. C. Dunham · E. S. Boyd (✉)
Department of Microbiology and Immunology, Montana State University, Bozeman, MT, USA
e-mail: eboyd@montana.edu

B. K. Baxter, J. K. Butler (eds.), *Great Salt Lake Biology*,
https://doi.org/10.1007/978-3-030-40352-2_4

4.1 Introduction

Lithified organo-sedimentary structures have been discovered in rocks dating back to 3.481 billion years ago (bya), representing the oldest identifiable fossil assemblages and the earliest accepted evidence for life on early Earth (Allwood et al. 2006; Sugitani et al. 2015; van Kranendonk et al. 2008; Walter et al. 1980). Interactions between microbial life and the environment can result in the formation of contemporary organo-sedimentary structures, termed microbialites (Burne and Moore 1987; Riding 1991). Benthic microbial communities, oftentimes in the form of a mat or a biofilm that colonize lacustrine and marine environments, can bind and trap detrital sediment and serve as the locus of mineral precipitation. Lithification of these organo-sedimentary structures can lead to their preservation in the rock record (Dupraz and Visscher 2005; Riding 2011).

Modern microbialites can display structures similar to those present in ancient rocks, a finding that (along with other lines of evidence) has led to the interpretation that these early structures are biogenic (Laval et al. 2000). Therefore, studies at the intersection of biology and geology as they relate to modern microbialites can potentially inform on how microbial communities, their activities, and interactions with their environment influence the formation and preservation of microbialites (Bosak et al. 2013; Jahnert and Collins 2011; Pepe-Ranney et al. 2012). To the extent that such links can be made between the formation of these structures in modern-day environments and those preserved in the rock record, such studies can provide insight into the physiology and ecology of early microbial life and its potential role in sustaining biodiverse ecosystems (Edgcomb et al. 2014).

Microbialites are found in limited locations around the world today, such as in the saline marine environments of Hamelin Pool of Shark Bay, Western Australia (Collins and Jahnert 2014; Logan 1961; Pages et al. 2014), and Highborne Cay, Bahamas (Myshrall et al. 2010), and in lacustrine environments including Pavilion Lake, Canada (Theisen et al. 2015), Lake Tanganyika, Africa (Cohen et al. 1997), Lake Salda Golu, Turkey (Braithwaite and Zedef 1994), and Lake Alchichica, Mexico (Gerard et al. 2013), among many others. Of these, the microbialite "reefs" in Great Salt Lake (GSL), Utah, USA, represent the largest expanse of microbialites on modern Earth (Baskin 2014). As such, the microbialites of GSL not only represent extensive modern analogs to examine the role of biology in the formation of such structures, with application to our understanding of early forms of microbial life, they also represent an opportunity to develop an understanding of their role in supporting the biodiverse ecosystem of GSL.

The most common metabolic process associated with microbialites in modern environments is autotrophic, oxygenic photosynthesis, catalyzed by a variety of Cyanobacteria or algal species (Dupraz and Visscher 2005; Riding 2011). As such, microbialites and their associated photosynthetic microbial assemblages contribute a large amount of biomass in the form of primary production to the local ecosystem. This is certainly the case for the extensive microbialites in GSL (Lindsay et al. 2017), which have been estimated to contribute ~9 times more chlorophyll (primary productivity) per meter^2 than lake phytoplankton (Wurtsbaugh et al. 2011) that, in turn, directly or indirectly supports a variety of higher trophic level taxa of economic and ecological importance (Belovsky et al. 2011; Wurtsbaugh 2009). In this chapter, we examine microbialites of GSL to better understand the biology of the microbial mats that grow in association with these benthic structures and their contribution to lake productivity. We discuss the potential impacts of natural and anthropogenic changes on the functioning of GSL microbialites, in particular changes in salinity that are predicted to accompany climate change. Finally, we review the processes that lead to mineral formation in microbialite-associated microbial mats in GSL and processes involved in mat lithification.

4.2 Microbialites of Great Salt Lake

4.2.1 Geologic Setting

GSL occupies one of the lowest depressions in the Great Basin Province which is bounded by the north-south trending Wasatch fault zone to the east (Baskin 2014; Chidsey et al. 2015; Cohenour and Thompson 1966). The Sevier orogenic system produced thrust faults and folds during the Cretaceous that would become the GSL basin (Mohapatra and Johnson 1998). Then, from the middle Eocene to the early Miocene (49–20 Ma), crustal extension resulted in collapse, and subsequent extension from the middle Miocene to the present (17 Ma–present) produced the current GSL basin and range structural architecture (Mohapatra and Johnson 1998). Other than the most extensive Northern Hemisphere glaciations which formed the larger Lake Bonneville, the current GSL area (Fig. 4.1a, b) (Lindsay et al. 2017) and its characteristics are likely similar to what they have been for the last 780 ka (Oviatt et al. 1999, 2015) or even longer. The basin occupied by the modern GSL is a half graben produced by extensional faults, and the older bedrock beneath the lake has been subject to multiple phases of brittle deformation and is highly fractured (Baskin 2014; Cohenour and Thompson 1966; Gwynn 1996; Jones et al. 2009; Mohapatra and Johnson 1998).

GSL is the largest lake in the western United States and the fourth largest terminal lake in the world (Keck and Hassibe 1979). It is a shallow, meromictic lake that exhibits a maximum and a mean depth of ~9.0 and 4.3 m, respectively. The major source of freshwater inflow to GSL comes from three rivers: the Bear, Weber, and Jordan Rivers (Fig. 4.1a). These three rivers all flow into the southern end of the lake and account for 95% of total water input to GSL (Baskin 2014; Belovsky et al. 2011; Chidsey et al. 2015; Gwynn 1996; Jones et al. 2009; Naftz et al. 2011). Between 1900 and 1959, the average salinity in GSL ranged from 20 to 27% (Stephens 1990). In 1959, a rock and gravel railroad causeway was constructed, dividing the lake into two "arms" termed the north arm (NA) and the south arm (SA) (Cannon and Cannon 2002) (Fig. 4.1). Differences in where freshwater enters GSL (largely on the southern end) result in differing geochemical landscapes in the SA versus the NA; the salinity fluctuates in the SA while the salinity of the NA is maintained at near the

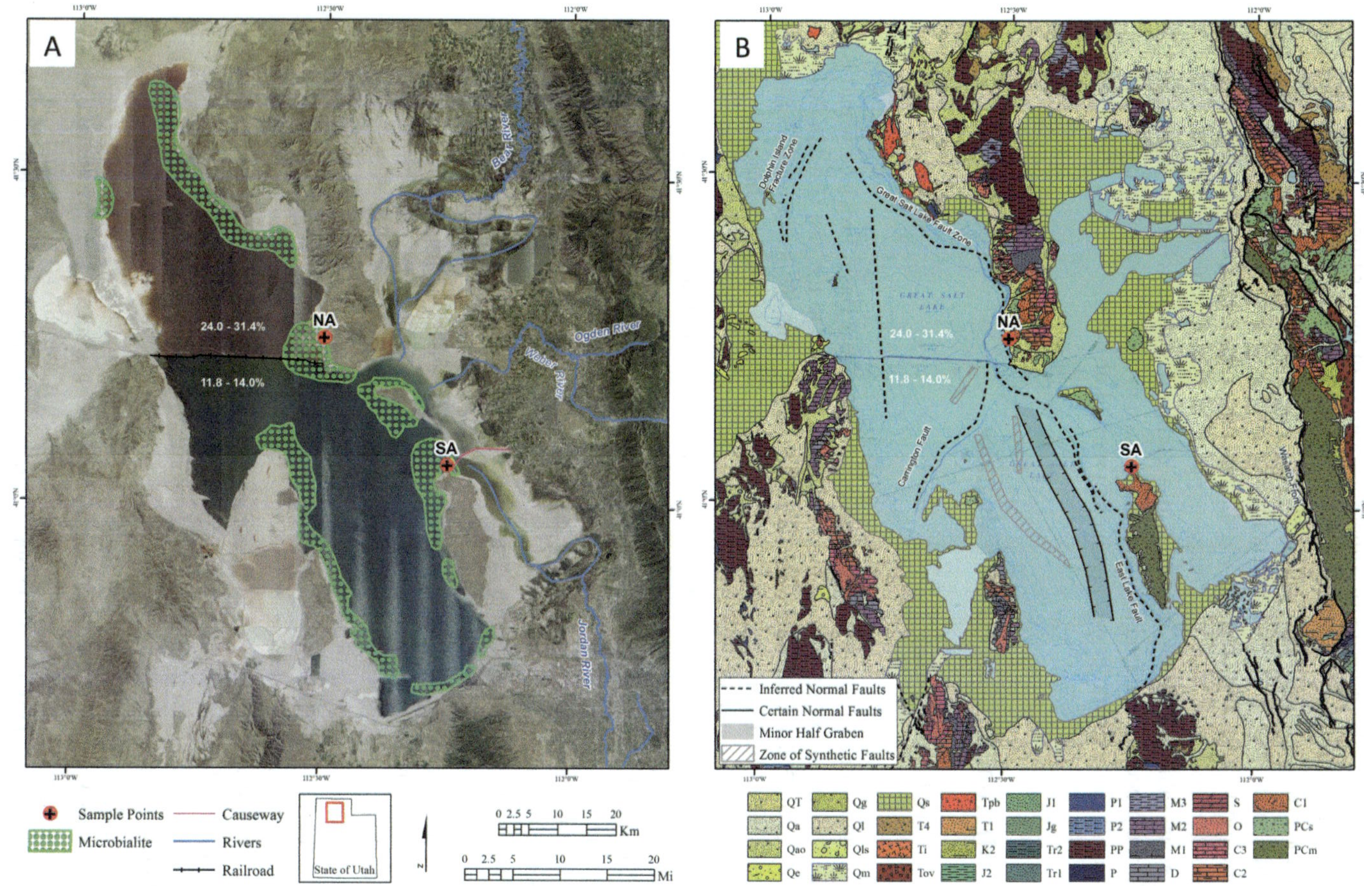

Fig. 4.1 (**a**) Map of Great Salt Lake (GSL) constructed with salinity (%), structural, stream, transportation, and microbialite reference layers. Imagery from *National Agriculture Imagery Program* (NAIP) 1 m resolution orthoimagery (2014) accessed from the Utah Geological Survey (UGS) server. Salinity values for

saturation of sodium chloride (Gwynn 1996; Naftz et al. 2011; Oren 2006a). While the abundance of ions is higher in the NA than in the SA due to dilution in the SA by freshwater, their composition is similar (Domagalski et al. 1989) and both "arms" are dominated by sodium and chloride (Oren 2006a). Mirabilite (sodium sulfate) deposits can also form seasonally on the lake bottom and have been found in sediment layers at depth in GSL (Bouton et al. 2016). These deposits also contribute to the salinity of GSL and the nutrient supply in the lake ecosystem (Anderson et al. 2014; Oviatt et al. 2015).

4.2.2 Geographic Location of Microbialites in Great Salt Lake

Microbialites of GSL are estimated to cover ~20% of the lake bottom (Fig. 4.1a) (Baskin 2014; Eardley 1938), with the extent of their aerial exposure varying depending on lake level (Wurtsbaugh et al. 2017). The extent of microbialite coverage in GSL was first mapped and described in detail during the droughts of 1933 and 1934 (Eardley 1938), and they were shown to be extensive in shallow areas of the lake (Fig. 4.1a, b). The limited distribution of microbialites to the shallow margins of GSL is likely due, at least in part, to these areas being more conducive to sunlight penetration throughout the entire shallow water column, since solar radiation is necessary to drive the metabolism of the presumed architects of microbialites, photosynthetic diatoms and Cyanobacteria (Chidsey et al. 2015; Lindsay et al. 2017).

In addition to the shallow margins of GSL that allow for light penetration and photosynthetic organisms to flourish, recent mapping suggests that the underlying structural geology of the GSL basin likely plays a key role in determining the distribution of microbialites (Fig. 4.1). Maps (Eardley 1938) and high-resolution seismic reflection data (Baskin et al. 2012, 2013) indicate that microbialites are associated with microtopographic highs that result from faults, folds, or monoclinal features on the lake floor of GSL. The key fault that is thought to have the greatest effect on the distribution of microbialites in GSL is the East Lake Fault, which defines the eastern and southeastern edge of GSL (Fig. 4.1b) (Bouton et al. 2016). This fault, in addition to other minor faults, may facilitate entry of freshwater into these locations that could promote the formation of microbialites through delivery of ions and nutrients. Moreover, oil seeps and associated brine and nutrients are also found along the north and north-east margins of GSL and are likely to contribute nutrients to the local environment, thereby promoting growth of microbialite-associated phototrophic microbial mats (Sei and Fathepure 2009).

Fig. 4.1 (continued) the south arm (SA) and the north arm (NA) were reported previously (Lindsay et al. 2017). The microbialite layer is modified from Eardley (1938). (**b**) Geologic and bathymetric map of GSL. Data layers and references for parts **a** and **b** adapted from Lindsay et al. (2017)

4.2.3 Structure and Mineralogical Composition of Microbialites

The morphology of the microbialites sampled from several sites around GSL has been described previously (Carozzi 1962). Microbialites collected offshore of Promontory Point have been categorized into four different morphotypes: (1) subparallel festooned ridges, (2) tongue-like festooned ridges, (3) composite rings and flat-topped mounds, and (4) isolated mounds. Cross sections indicate that these different reported microbialite morphologies likely represent the exaggerated expression of underlying topographical differences (Carozzi 1962). Domal stromatolitic-microbialite structures have also been described at Bridger Bay on the northwestern tip of Antelope Island (Chidsey et al. 2015). These low-profile domes exhibit a raised outer ring morphology (Fig. 4.2a), and it is likely that their shapes were created by wave erosion of material from around the base of the microbialite mound (Chidsey et al. 2015).

The inner microstructure of microbialites from GSL has also been characterized, with the earliest study finding them to be a mix between laminated and non-laminated forms (Halley 1976). Most microbialites in GSL form on a base of oolitic grains or sand. Ooids in GSL are small spherical or elongated grains composed of radial aragonite crystalline fabrics surrounding a core of detrital mineral grains or even brine shrimp fecal pellets (Chidsey et al. 2015; Eardley 1938; Gwynn 1996). SA microbialite hand samples also show that oolitic sand grains as well as brine shrimp cysts can be incorporated into the microbialite fabric, and photosynthetic biomass is often present in the outer portions of the structures as indicated by the visual presence of pigments (Fig. 4.2c). SA microbialites have also been shown to comprise patches of carbonate that are fused to form a clotted thrombolitic fabric; the deeper sections of microbialites have a more stromatolitic fabric (Lindsay et al. 2017; Pace et al. 2016). In contrast, microbialites from the NA (Fig. 4.2b) exhibit differences from those from the SA, including being much harder and more difficult to break apart (Lindsay et al. 2017). Like SA microbialites, those of the NA also have sand grains integrated into the fabric, but there is no clear visual presence of photosynthetic pigments (Fig. 4.2d). Unlike SA microbialites those from the NA also comprise abundant salt precipitates.

The mineralogy of microbialites from different locations around GSL has been determined by X-ray diffraction (XRD) as well as elemental spot analyses by field-emission microscopy (FEM) (Lindsay et al. 2017). XRD analyses revealed the most abundant mineral in both SA and NA microbialites to be calcium carbonate, specifically aragonite, with other non-crystalline minerals or phases below the detection limit (1–2 weight %) (Fig. 4.2e, f). Micritic aragonite crystals were also detected on the faces of larger aragonite crystals in SA microbialites (Fig. 4.2e) but were absent from NA microbialites (Lindsay et al. 2017). Micritic aragonite has been suggested to be produced through biological activities in the GSL (Pedone and Folk 1996). FEM analyses indicated that the other mineral components in the NA structures included quartz and clay (Fig. 4.2f), and consistent with this observation,

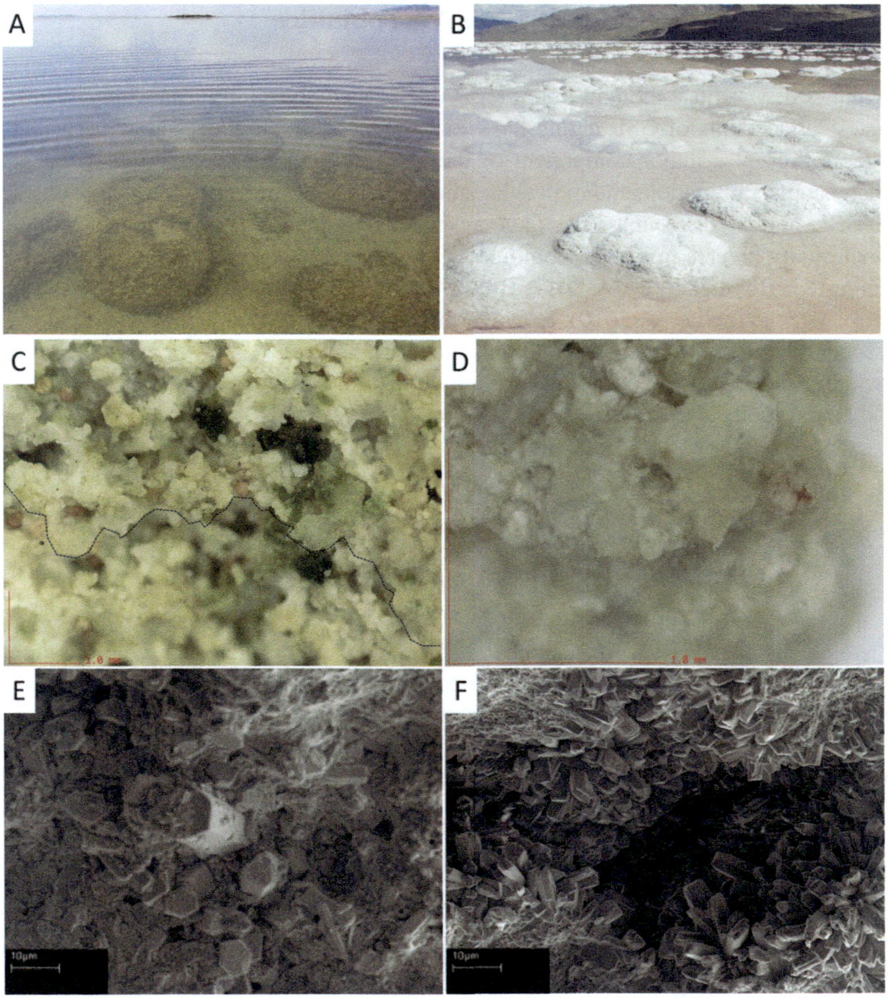

Fig. 4.2 (**a**) Submerged microbialites at the northern tip of Antelope Island in the south arm (SA) of Great Salt Lake (GSL), near Bridger Bay. Image taken August 12, 2017. (**b**) Partially exposed microbialites off the south-western tip of Promontory Peninsula in the north arm (NA) of GSL. Image taken May 9, 2016. (**c**) Dino-light image of a SA microbialite hand sample, with magnification to ×68 showing fenestral fabric and a pore roughly 5 mm in diameter (indicated by dotted black line. (**d**) Image of a NA microbialite hand sample, with magnification to ×200. (**e**) Field-emission microscopy (FEM) image of SA orthorhombic aragonite crystals overlain with micritic aragonite precipitates. (**f**) FEM image of a pore space in a NA microbialite filled with crystalline aragonite lacking micritic aragonite precipitate. Panels **c**–**f** adapted from Lindsay et al. (2017)

sand grains were observed in the microstructure of NA hand samples (Fig. 4.2d). In addition to aragonite, XRD analyses of SA microbialites also reveal the presence of dolomite and a magnesium-silica rich mineral phase (Pace et al. 2016).

4.3 Biology of Great Salt Lake Microbialites

While GSL was once thought to be sterile due to its high salinity (Stansbury 1855), historical accounts of GSL dating back to 1870 refute this notion and suggest that algae represent a major component of the lake microbiota (Tilden 1898). Recent microscopic, cultivation, and molecular studies have confirmed the presence of algae in GSL and at the same time have revealed a much more taxonomically and functionally diverse microbial community in waters and sediments than previously suggested (Boyd et al. 2014, 2017; Brandt et al. 2001; Lindsay et al. 2017; Meuser et al. 2013; Parnell et al. 2010). However, microbiological analyses of microbialite-associated microbial mat communities have only recently been performed and include detailed perspectives from both macro/microscopic (Wurtsbaugh et al. 2011) and molecular analyses (Lindsay et al. 2017, 2019).

4.3.1 Productivity of Great Salt Lake Microbialites in the South Arm and North Arm

Visibly, there are clear differences in the planktonic microbial communities inhabiting the waters of the SA and NA of GSL. The NA is at times visually pink purple in color, whereas the SA is green brown in color, a dichotomy so stark that it is even visible from space (Fig. 4.1a). The difference in color can be attributed to the types of microbial plankton in waters in each of these areas, with those in the NA comprising non-photosynthetic halophilic Archaea (Almeida-Dalmet et al. 2015; Kemp et al. 2018; Tazi et al. 2014) that often dominate planktonic communities (Lindsay et al. 2017). These halophilic Archaea produce carotenoid pigments which accumulate in the cell membranes and impart a distinctive pink-purple color to the NA (Jones and Baxter 2017; Post 1980). In contrast, planktonic communities in the SA are dominated by phototrophic algae and Cyanobacteria (Barnes and Wurtsbaugh 2015; Belovsky et al. 2011; Lindsay et al. 2017; Wurtsbaugh et al. 2011), which give the SA a green-brown color. These differences are similarly apparent from a visual inspection of microbialites sampled from the SA and NA of GSL, with those from the SA showing a green-brown color (Fig. 4.2a) consistent with the presence of chlorophylls and oxygenic phototrophs. Periphyton (microbial mats) associated with microbialites has been estimated to represent 70% of the phytoplankton in the SA of GSL (Wurtsbaugh 2009). Indeed, abundant chlorophyll has been extracted from SA microbialites (Wurtsbaugh 2009), and microscopic analyses of the communities reveal the presence of abundant Cyanobacteria (Belovsky et al. 2011). In contrast, microbialites from the NA show a pink-purple color (Fig. 4.2b) consistent with carotenoid pigments and the presence of halophilic Archaea (Post 1980).

Using qPCR of archaeal, bacterial, and eukaryal small subunit (SSU) rRNA genes as a proxy for the abundance of cells and their biomass, it was shown that

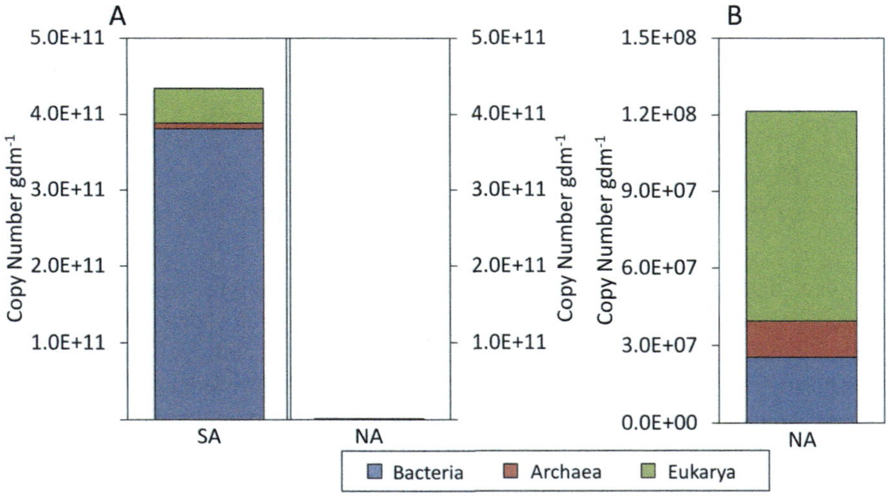

Fig. 4.3 (**a**) Abundance of archaeal, bacterial, and eukaryl small subunit (SSU) rRNA genes per gram dry mass (gdm) associated with microbialites sampled from the south arm (SA) and north arm (NA) of Great Salt Lake as determined by quantitative PCR. (**b**) Rescaled figure showing the abundance of SSU rRNA genes gdm^{-1} in NA microbialites

microbialites in the SA host orders of magnitude more SSU rRNA templates than NA microbialites (Fig. 4.3a) (Lindsay et al. 2017). When all SSU rRNA gene templates from each of the three taxonomic domains were combined, a SA microbialite was shown to host three orders of magnitude greater SSU rRNA gene templates than a NA microbialite. Moreover, the composition of the microbialite-associated communities varied markedly, with those sampled from the SA being dominated by Bacteria (80% of total SSU rRNA gene templates) (Fig. 4.3a) and those from the NA being dominated by Eukarya (55% of total SSU rRNA gene templates) (Fig. 4.3b). Archaeal 16S rRNA gene templates were surprisingly not abundant in either the SA or NA microbialites but were relatively more common in the NA than the SA microbialite-associated community (8.8% and 2.4% of total SSU rRNA gene templates in the NA and SA, respectively) (Lindsay et al. 2017). Collectively, these data suggest that the microbialite-associated community in the SA is more productive than that in the NA.

In other aquatic systems, a positive relationship exists between the productivity of a microbial community and its biodiversity (as reviewed in Smith 2007). Indeed, SSU rRNA gene sequences from each of the three domains (Archaea, Bacteria, and Eukarya) sampled from SA microbialite-associated communities exhibited higher species diversity and richness than those from the NA (Lindsay et al. 2017). As will be discussed below, the low productivity and/or diversity associated with the NA microbialite communities is likely to be a consequence of stress imposed by elevated salinity on microbial inhabitants (Oren 2006a). It follows that the lower biomass associated with microbialite communities in the NA (~1000 times less SSU rRNA gene templates) likely places constraints on the extent that nutrients can be transferred from this ecological compartment to higher trophic level compartments in the NA food web.

4.3.2 Biological Inhabitants of Microbialites in the South Arm and North Arm of Great Salt Lake

The first detailed characterization of microbial communities associated with microbialites in the SA of GSL was conducted using microscopy (Wurtsbaugh 2009), which revealed that a cyanobacterium designated as *Aphanothece* sp. represented over 99% of the cells. To the author's knowledge, there are no published microscopic analyses of the composition of NA microbialites. Here, we review a recent molecular-based study that characterized the composition of SSU rRNA gene sequences associated with microbialites collected from the SA and the NA of the GSL (Lindsay et al. 2017). At the time of sample collection (July 2011), waters overlying the SA microbialite had a temperature of 18 °C, a pH of 7.2, and a salinity of 11.8%. In contrast, waters overlying the NA microbialite had a temperature of 20 °C, a pH of 7.9, and salinity of 31.4%.

Archaea The most abundant archaeal 16S rRNA gene operational taxonomic unit (OTU) in the SA microbialite community was most closely related to *Halorubrum* sp. S26-1 of the order Haloferacales (Fig. 4.4a). This taxon represented 7.7% of the total archaeal 16S rRNA gene templates recovered (Lindsay et al. 2017). At the order level taxonomic rank, the most abundant OTUs in the SA belonged to the orders Halobacteriales and Haloferacales, which are well-characterized halophilic organisms that are often detected in hypersaline environments (as reviewed in Oren 2006b; Oren and Ventosa 2017). Likewise, numerous sequences affiliated with the order Methanosarcinales, one of the seven described taxonomic orders that comprise methanogens (Kendall and Boone 2006), were identified in the SA. Methanogens affiliated with the Methanosarcinales are typically acetoclastic (disproportionate acetate to methane and carbon dioxide; Kendall and Boone 2006), and these organisms have been identified in numerous high salinity environments (Oren 2006a).

The taxonomic composition of archaeal 16S rRNA genes associated with NA microbialites was similar to that of SA microbialites and comprised numerous sequences affiliated with the Haloferacales and Halobacteriales (Fig. 4.4a) (Lindsay et al. 2017). Like the SA microbialite, the dominant OTU in the NA microbialite community was closely affiliated with *Halorubrum* sp. S26-1. Additionally, the NA microbialite community comprised an OTU most closely related to *Haloquadratum walsbyi* (within the order Haloferacales). However, unlike the SA microbialites, sequences belonging to the order Methanosarcinales were not detected in the NA microbialites. This is consistent with previous evidence suggesting that acetoclastic methanogens are not tolerant of high salinity conditions (Oren 2006a), such as those associated with the NA (24.0–31.4%; Fig. 4.1a).

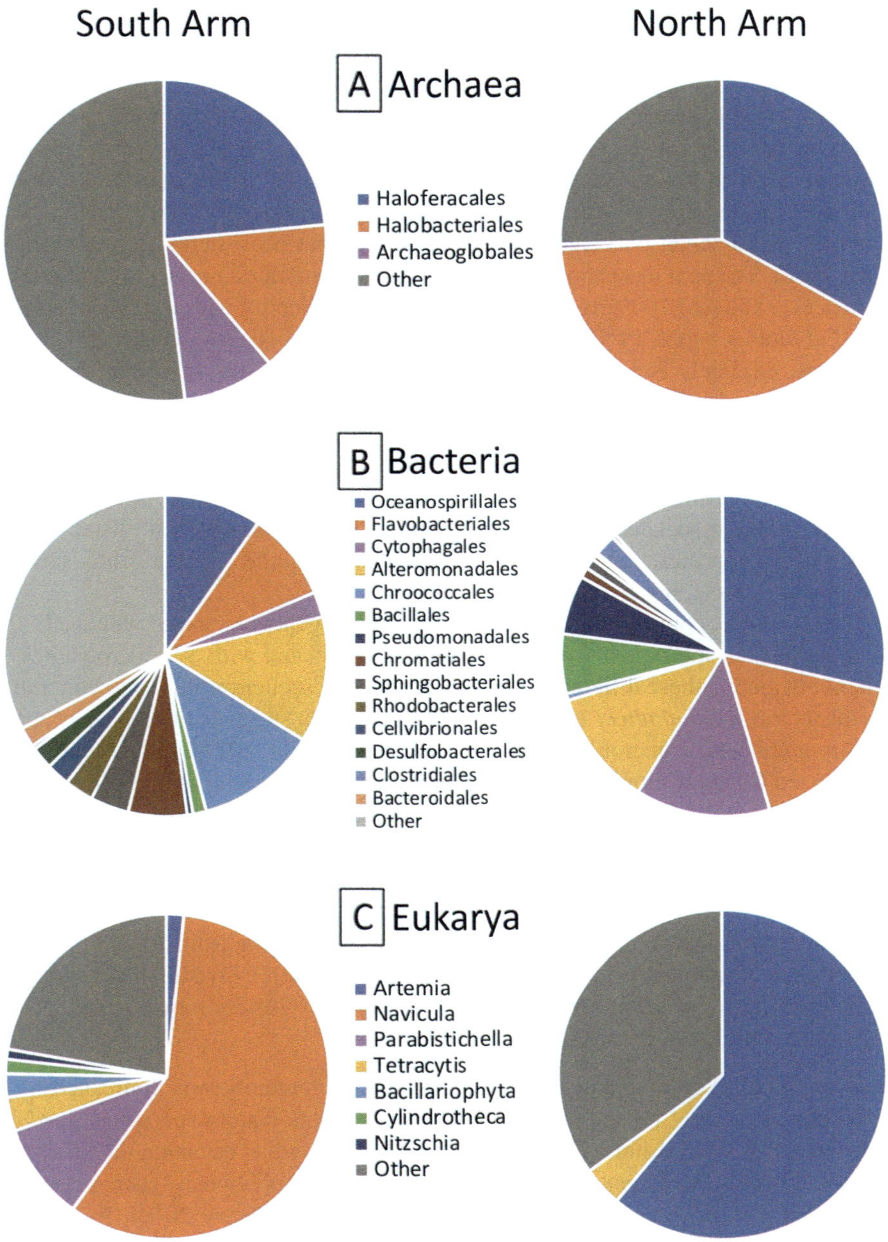

Fig. 4.4 Composition of small subunit (SSU) rRNA genes recovered from microbialites sampled from the south arm and north arm of Great Salt Lake. Representative operational taxonomic units (OTUs) for each library were binned at the order level for archaeal SSU rRNA genes (**a**), at the order level for bacterial SSU rRNA genes (**b**), and at the genus level for eukaryl SSU rRNA genes (**c**). Taxonomic bins (at the same three taxonomic levels) which represented $<1.0\%$, $<1.0\%$, and $<0.1\%$ of the total sequences from each assemblage, respectively, were pooled and depicted as "other." Figure adapted from Lindsay et al. (2017)

Bacteria The most abundant bacterial 16S rRNA gene OTU within SA microbialite community (12.7% of bacterial reads) was most closely related (100% 16S rRNA gene sequence identity) to the heterotroph *Marinimicrobium haloxylanilyticum* of the order Alteromonadales. This organism was originally isolated from GSL and was shown to degrade simple and complex polysaccharides including xylan, starch, carboxymethyl cellulose, and galactomannan (Møller et al. 2010), compounds that might be expected to be produced and/or excreted by phototrophs that co-inhabit the mat such as has been observed in other microbial mat ecosystems (Bateson and Ward 1988). The next most abundant OTU in the SA communities (at 12.0% of total sequences) exhibited 100% 16S rRNA gene sequence identity with the cyanobacterium *Euhalothece* sp. MPI 96N304 within the order Chroococcales (Fig. 4.4b). The strain was originally isolated from a coastal lagoon located near Guerrero Negro, Baja California Sur, Mexico (Garcia-Pichel et al. 1998). Strain MPI 96N304 is an obligate halophile and a moderate thermophile, characteristics that likely allow the strain to cope with elevated temperatures in the shallow depths of GSL marginal environments during periods of high solar radiation. The majority of the remaining SA microbialite-associated microbial mat community members were affiliated with heterotrophic species, including a moderate halophile within the order Oceanospirillales (Lindsay et al. 2017).

The taxonomic composition of bacterial SSU rRNA genes associated with the NA microbialite was substantially different than that associated with SA microbialites. Paramount among these differences was the absence of sequences affiliated with the cyanobacterium *Euhalothece* (Fig. 4.4b), which has been suggested to be one of the primary architects of microbialites in the SA (Lindsay et al. 2017). In fact, none of the detected OTUs at significant abundances were affiliated with putative photosynthetic organisms. Rather, the most abundant member of the NA microbialite assemblage was most closely related to the extremely halophilic, obligately aerobic, and chemoorganotrophic *Salinibacter ruber* within the order Sphingobacteriales (Antón et al. 2002) (Fig. 4.4b). The type strain was isolated from a saltern crystallizer pond in Spain and was shown to grow optimally at salinities ranging from 20 to 30% and did not grow at salinities of less than 15% (Antón et al. 2002). Overall, there was an absence of bacterial phototrophic species within the NA microbialite and NA planktonic communities (Lindsay et al. 2017).

Eukarya The SA microbialite-associated eukaryal community was dominated by an OTU most closely related to the benthic phototrophic diatom species *Navicula salinicola* (Fig. 4.4c) (Lindsay et al. 2017). Consistent with its detection in GSL SA microbialite-associated mats, *N. salinicola* strains have previously been isolated from or detected in brackish areas or inland salt springs (Guiry 2019). It is also worth noting that silica, a requirement for frustrule development in diatoms (Darley and Volcani 1969), was detected in the SA microbialite albeit at low amounts (<4 weight %) (Lindsay et al. 2017). This observation was further supported by a microscopic analysis of a SA microbialite indicating the presence of sand grains. Another abundant OTU within the SA microbialite and planktonic communities was closely affiliated with the phototrophic green alga *Tetracytis texensis* (Fig. 4.4c)

(Lindsay et al. 2017). Thus, although cyanobacterial phototrophs appear to numerically dominate SA microbialite mat communities when compared to algal phototrophs, both are likely to contribute photosynthate and biomass for heterotrophic secondary consumers especially since the eukaryal species are typically larger in size (Mogany et al. 2018; Phyto'pedia 2012).

Interestingly, 18S rRNA gene sequences affiliated with the brine fly *Ephydra* were not detected in microbial mat communities associated with the SA microbialites (Lindsay et al. 2017). *Ephydra* species are common inhabitants of SA microbialite mats (Wurtsbaugh et al. 2011) and can be visually seen in association with these structures, including in the microbialite hand sample depicted in Fig. 4.5a. Their lack of detection might have been due to the size of the sample that was used for extraction of DNA for PCR and sequencing, which might have been too small to include ephydrid larvae and their DNA. While sequences affiliated with *Ephydra* were not detected in DNA extracted from SA microbialites, those affiliated with the brine shrimp *Artermia* were detected (Fig. 4.5b), likely due to the presence of cysts in the fabric of the structure (Fig. 4.2c).

Based on Lindsay et al. (2017), as well as other literature published on GSL ecology (e.g., Belovsky et al. 2011; Wurtsbaugh 2009), the key biological inhabitants of the SA of GSL were identified and detailed in Fig. 4.5c as a food web schematic illustrating the flow of biomass and nutrients. Here, photosynthetic Cyanobacteria and algae represent the base of the food web and generate photosynthate and biomass to support secondary consumers, including both heterotrophic Bacteria and Archaea, as well as *Ephydra* and *Artemia*. Microbial consumers are likely supported by photosynthate (e.g., carbohydrates, organic acids) produced by the primary producers, whereas higher order secondary consumers such as *Artemia* and *Ephydra* likely feed on phototrophic cells. *Artemia* and *Ephydra* serve as food for a diversity of shorebirds and migratory birds. Further details of the GSL SA food web are presented below.

The NA microbialite eukaryal community was dominated by two OTUs most closely related to *Artemia* (Fig. 4.4c) (Lindsay et al. 2017). While *Artemia* can tolerate high salinities (Dana and Lenz 1986), the salinity in the NA (31.4%) at the time that the microbialite was collected for molecular analyses would not have been conducive to its survival. Thus, it is not clear if the *Artemia* that were detected were active at the time of sampling or if these were simply cysts that had persisted and whose DNA was extracted and amplified.

4.4 Microbial Processes

4.4.1 Photosynthesis and Primary Production

Photosynthetic microorganisms form the base of the food chain in most aquatic ecosystems and GSL appears to be no exception. The presence of abundant phototrophic plankton has been noted since the first scientific observations of GSL

Fig. 4.5 (**a**) Cross-sectional image of a microbialite hand sample from the south arm (SA) of Great Salt Lake (GSL) showing the layers of biomass on top of hardened gray carbonate. *Ephydra* larvae are visible in association with the top layers of the microbialite (on the left side), with casings also present on the surface. (**b**) *Artemia* in a laboratory aquarium at Montana State University with only microbialites present as potential food. Microbialites and *Artemia* have been sustained in our aquaria for >3 years by maintaining water levels (and salinity) with deionized water. (**c**) Schematic depicting a highly simplified food web in the SA of GSL, limited to microbialites, invertebrates, and primary producers, with individual components not to scale. Microbialite biomass, consisting of primary producers such as *Navicula* (algal diatom) and *Euhalothece* (Cyanobacteria), directly support *Ephydra* and *Artemia* secondary consumers. Additionally, the microbialite biomass can slough off and provide free-floating microbial detritus which can also support *Ephydra* or *Artemia*. *Ephydra* or *Artemia* biomass supports higher trophic levels including a diversity of shorebirds and waterbirds at GSL. Food web adapted from (Wurtsbaugh 2009), with images and clip art of birds [Gull from clipart-library.com, water bird (Phalarope) from National Audubon Society], *Artemia/Ephydra* (learn.genetics.utah.edu, Wurtsbaugh/ASLO, 2000), and *Cyanothece* microorganisms adapted for use. Image in part **a** adapted from Lindsay et al. (2017)

were made (Crisp et al. 1880; Tilden 1898), which date to the 1870s (Baxter 2018). The overall health of the GSL aquatic ecosystem is likely closely tied to the activity of phototrophic organisms (Belovsky et al. 2011; Thompson et al. 2012), including both plankton and periphyton primary producers. As such, any change to the attributes that define the niches of primary producers in GSL could influence ecosystem level functioning through what has been referred to as a trophic cascade (Paine 1980).

Two abundant photosynthetic microorganisms have been identified in plankton and periphyton microbial mats that form in association with microbialites in the SA of GSL. These include the algal diatom *N. salinicola* and the cyanobacterium *Euhalothece* [the latter is almost certainly the same cyanobacterium previously referred to as *Aphanothece* sp. (Wurtsbaugh 2009)]. Both organisms are oxygenic phototrophs, using light energy to oxidize water and to reduce carbon dioxide to organic carbon. In the case of *Euhalothece* (data not readily available for *N. salinicola*), these activities are adapted to function at the high salinities found in the SA of GSL, with the laboratory-determined optimal salinity for the growth of this genus at around 15–20% (Clavero et al. 2000; Garcia-Pichel et al. 1998). Photosynthate and biomass produced by these dominant organisms is available to support the diversity of heterotrophic secondary consumers that co-inhabit the microbial mat ecosystem (Fig. 4.6).

4.4.2 Heterotrophy and Secondary Production

A diversity of putative heterotrophs has been identified in microbial mats associated with microbialites sampled from the SA of GSL (Lindsay et al. 2017; Pace et al. 2016). The majority of the heterotrophs identified in SA microbialite mats are affiliated with aerobes, including the most abundant OTUs that are affiliated with *Marinimicrobium haloxylanilyticum* (Møller et al. 2010) within the order Alteromonadales and *Saccharospirillum salsuginis* (Chen et al. 2009) within the order Oceanospirillales (Lindsay et al. 2017). Both *M. haloxylanilyticum* and *S. salsuginis,* the former of which was isolated from GSL and the latter of which was isolated from a salt mine brine, can degrade a variety of simple and complex carbohydrates, perhaps including those produced by primary producers in the mat. Other putatively heterotrophic taxa comprising the most abundant community members of SA microbialites were also aerobes, including species within the orders Flavobacteriales, Cytophagales, and Bacillales (Van Trappen et al. 2004; Xu et al. 2015; Yaakop et al. 2015). Together, these five most abundant heterotrophic taxa comprise 40% of the SA microbialite bacterial community (Fig. 4.4b) (Lindsay et al. 2017), and it is likely that other aerobic heterotrophic species are present albeit at smaller relative abundances. Like *M. haloxylanilyticum* and *S. salsuginis*, these heterotrophic taxa are inferred to be supported by nutrients produced or excreted by phototrophs within the microbialite mat ecosystem. The presence of abundant organic carbon and oxygen (O_2) in productive SA microbialite mats likely allows for these aerobic heterotrophs to dominate the community, while other species such as sulfate (SO_4^{2-})-reducing bacteria (SRB), which do not typically tolerate high concentrations of O_2 (Muyzer and Stams 2008), are less abundant.

Intriguingly, despite abundant O_2 in the mats, radioisotopic data indicate the presence of SRB in SA microbialite mats (Pace et al. 2016). While SRBs have been shown to be a major component of benthic sediment-associated microbial communities in the SA of GSL (Boyd et al. 2017; Brandt et al. 2001), they tend to be found in relatively low abundance (~2–3% of total reads) in SA microbialite communities based on DNA- (Lindsay et al. 2017, 2019) and RNA-based analyses (Lindsay et al.,

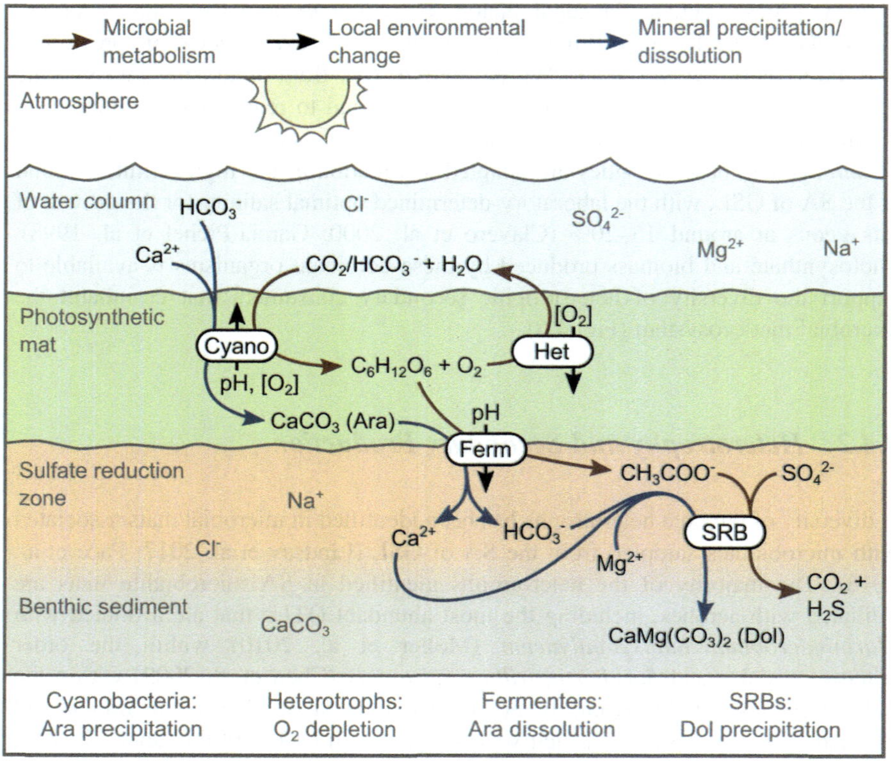

Fig. 4.6 Model for the formation of aragonite ($CaCO_3$) and protodolomite ($CaMg(CO_3)_2$) in mats and sediments in the south arm (SA) of Great Salt Lake (GSL). Benthic cyanobacterial mats, such as those covering the majority of submerged SA microbialites, generate a localized increase in alkalinity, promoting the precipitation of aragonite. Detrital biomass and photosynthate from these mats such as glucose ($C_6H_{12}O_6$) then can provide nutrients supporting other microorganisms such as anaerobic fermentative bacteria in deeper layers of the microbialite mat. Organic acids (including CH_3COO^-) produced by fermentation of photosynthetic biomass or photosynthate can lower local pH, which promotes the partial dissolution of aragonite, and also provides substrate in the form of organic acids to support heterotrophic sulfate (SO_4^{2-})-reducing bacteria (SRB) in underlying microbialite layers. The SRB then consume SO_4^{2-}, releasing free magnesium (Mg^{2+}), generate sulfide thereby reducing the energy required for dehydration of Mg^{2+}–water complexes, and generate alkalinity. Together, these activities can then allow for the formation of (proto) dolomite from the partial dissolution of aragonite. Abbreviations: Ara, aragonite, Dol, dolomite; Cyano, Cyanobacteria; Ferm, fermentative bacteria; Cl^-, chloride; Ca^{2+}, calcium; Na^+, sodium; HCO_3^-, bicarbonate; CO_2, carbon dioxide; H_2S, hydrogen sulfide. Figure adapted from Dunham et al. (2020)

unpublished). Nonetheless, it has been suggested that SRBs localized on and within microbialites subsist on and actively degrade organic acids generated during fermentation of phototrophic biomass or photosynthate (Pace et al. 2016). Indeed, the accumulation of sulfide in SA microbialite mats was more pronounced during dark hours when mats go anoxic due to heterotrophic respiration with O_2 (Pace et al.

2016). Thus, in simple terms, oxygenic phototrophs facilitate the activity of aerobic heterotrophs in SA microbialite mats by producing O_2 and organic carbon. Likewise, aerobic heterotrophs facilitate the activity of anaerobes such as fermenters and SRB through consumption of O_2.

4.4.3 A Role for Biology in Microbialite Formation?

Microbial mats are thought to play a key role in the formation of microbialites (Riding 2002, 2011), including those in GSL (Carozzi 1962; Chidsey et al. 2015; Lindsay et al. 2017; Pace et al. 2016) (Fig. 4.6). Extracellular polymeric substances (EPS) produced by microorganisms that inhabit/form the mats are thought to facilitate the trapping of detrital sand, ooids, and other material. These materials can be "cemented" together by carbonate that is likely precipitated through localized increases in pH. An increase in pH can be driven by any number of processes, but in photosynthetic mats, it is likely driven primarily by light-driven CO_2 uptake. A localized decrease in CO_2, in turn, can result in a shift in the inorganic carbon equilibrium [toward carbonate ion (CO_3^{2-})] and a rise in pH that favors deposition of CO_3^{2-}, typically as calcium carbonate [$CaCO_3$; (Merz 1992)]. Importantly, EPS produced by phototrophs is further suggested to serve as (*i*) a diffusion barrier that further allows for a localized pH increase and (*ii*) nucleation points for carbonate minerals to form (Riding 2011). Alternatively, others have suggested that decomposition of cyanobacterial EPS by microbial heterotrophs leads to the release of organic-bound calcium (Ca^{2+}) ions, making those Ca^{2+} ions more available for precipitation as $CaCO_3$ (Paerl et al. 2001).

In addition to a shift in pH and its effects on carbonate precipitation, shifts in alkalinity have been suggested to be a primary driver of carbonate mineral precipitation. Important biological processes implicated in the generation of localized increases in alkalinity include ammonification, denitrification, and SO_4^{2-} reduction, among others (Riding 2002, 2011). In GSL, the alkalinity generating process of heterotrophic SO_4^{2-} reduction, rather than a localized increase in pH due to CO_2 consumption by phototrophs, has been suggested to be the primary process responsible for formation and precipitation of $CaCO_3$ (as aragonite) in microbialite-associated mats (Pace et al. 2016). In this model, degradation of the low-molecular-weight organic carbon fraction of the extracellular organic material (EOM) produced largely by phototrophs contributes CO_2 that results in an increase in alkalinity due to an increase in the saturation state of waters with respect to aragonite. Secondly, the EOM serves as a nucleation point for aragonite precipitation. As aragonite progressively replaces the EOM during its degradation, the mat begins to lithify. The final stage of mat lithification was proposed to take place as fermentative bacteria drive the partial dissolution of precipitated aragonite at night, which allows for magnesium (Mg^{2+}) ions to replace Ca^{2+} ions in the process of aragonite dolomitization. However, while mineralogical analyses of microbialite-associated microbial mats revealed the presence of aragonite, as predicted (Pace et al. 2016), dolomite was not detected. Moreover, as described above, molecular characterizations of microbialite-associated mats in the SA of GSL have not shown abundant putative SRB or fermentative bacteria (Lindsay et al. 2017, 2019). These observations, combined with new data collected on biogeochemical processes taking place at the mat/sediment interface in GSL, led to the proposal of a slightly different model for formation of aragonite and (proto)dolomite in GSL microbialites (see Dunham et al. 2020).

In the newly proposed model (Dunham et al. 2020) (Fig. 4.6), the CO_2-consuming activity of phototrophs, which as mentioned above play a key role in carbonate precipitation in photosynthetic mats in other environments (Merz 1992), is proposed to generate a localized pH increase. This in turn promotes precipitation of aragonite on EPS surfaces within the mat. This slightly contrasts with the model proposed previously (Pace et al. 2016), which suggested that SRBs were the primary drivers of aragonite precipitation by promoting an increase in alkalinity. The model invoking an increase in pH as the primary driver of aragonite precipitation was based primarily on the detection of abundant aragonite in mats from across the SA of GSL (Dunham et al. 2020) and the absence of evidence for abundant SRB in the mats based on 16S rRNA gene (Lindsay et al. 2017, 2019) and metagenomic data (E.S. Boyd, *unpublished*). Rather, abundant transcripts of SRB were detected at the interface of the mats and benthic sediments and these decreased exponentially with depth into the sediment column. Also, this localized zone of SO_4^{2-} reduction was accompanied by high (500 μM) concentrations of sulfide and the highest rates of acetate oxidation, a common substrate for SRB (Widdel and Pfennig 1981). Finally, this zone was characterized by the detection of both aragonite and (proto)dolomite as abundant minerals. These collective observations led to the hypothesis that aragonite and photosynthetic biomass/photosynthate are deposited to the mat/benthic sediment interface and these stimulate anaerobic processes, including fermentation and SO_4^{2-} reduction (Dunham et al. 2020). Indeed, the carbon to nitrogen ratio of organic matter in sediments from this zone is low (~4) and is similar to that of fresh photosynthetic biomass associated with diverse Cyanobacteria (Geider and La Roche 2002; Kulasooriya et al. 1972).

Similar to the previously proposed model (Pace et al. 2016), the slightly refined model (Dunham et al. 2020) (Fig. 4.6) suggests that fermentation of organic matter drives a decrease in pH that results in partial dissolution of aragonite. The products of fermentation stimulate a zone of active SO_4^{2-} reduction that drives an increase in alkalinity. At the same time, a localized increase in sulfide is thought to promote its sorption to calcite (or aragonite) facies, thereby reducing the energy required for dehydration of Mg^{2+}–water complexes on those same surfaces and allowing for formation of authigenic (proto)dolomite (Lu et al. 2018; Zhang et al. 2012, 2013). It has been suggested that the consumption of SO_4^{2-} due to SRB activity can promote dolomitization since it disrupts the bonding between SO_4^{2-} and Mg^{2+}, potentially rendering Mg^{2+} available for replacement of Ca^{2+} in calcite or aragonite (Baker and Kastner 1981; Bontognali et al. 2010; Vasconcelos and McKenzie 1997; Warthmann et al. 2000). However, the concentration of SO_4^{2-} in sediment porewaters corresponding to the SO_4^{2-} reduction zone was similar to that of the overlying water column, suggesting that SO_4^{2-} depletion is unlikely to be responsible for (proto)dolomite formation in the SA of GSL. Regardless of whether it is SO_4^{2-} consumption or sulfide production that drives dolomitization of aragonite, the stark transition from aragonite as the dominant carbonate mineral in mat-associated biomass to a near equal mixture of aragonite and (proto)dolomite at the active zone of SO_4^{2-} reduction suggests a link between SRB activity and dolomitization of aragonite.

While the merits of both models (Dunham et al. 2020; Pace et al. 2016) for dolomite formation in the SA of GSL await further scrutiny through controlled experimentation, they do provide a plausible explanation for why microbialites in the SA and NA of GSL differ to such a large extent and why those in the NA have been suggested to be no longer actively forming (Lindsay et al. 2017). As mentioned above, the primary architects of GSL microbialites in the SA are likely to be the photosynthetic primary producers *Euhalothece* and *Navicula*, due both to production of EPS that can trap detrital grains and through their putative direct and/or indirect role in precipitating aragonite. However, photosynthesis is highly sensitive to salinity stress (Joint et al. 2002) and neither the cyanobacterium *Euhalothece* nor the eukaryotic alga *Navicula* were detected in microbialites in the NA (Lindsay et al. 2017). In fact, no abundant phototrophs were detected in association with the NA microbialites via molecular characterization. This likely decreases the extent by which aragonite can be produced either through (1) localized pH increases due to photosynthesis or by (2) SO_4^{2-} reduction since photosynthate is presumably needed to support robust heterotrophic SRB activity.

Similar to photosynthetic microorganisms, SRBs suggested to be important in mineral precipitation or alteration in SA-associated microbial mats (Dunham et al. 2020; Pace et al. 2016) are also inhibited by high salinity (Oren 2006a). In GSL sediments, rates of SO_4^{2-} reduction were significantly reduced in the extremely hypersaline NA when compared with the SA (32 nmol cm^{-3} day^{-1} in the NA, while rates in the SA range from 363 to 6131 nmol cm^{-3} day^{-1}). The SRBs responsible for this activity in both the SA and the NA had optimal growth at 10%, which is far lower than the salinity of the NA (Brandt et al. 2001). Consistent with these observations, the abundance of transcripts associated with putative SRB was found to be inversely correlated with salinity in sediments of GSL (Boyd et al. 2017).

If phototrophs and SRBs are inhibited by the high salinity of the NA and are potentially involved in carbonate mineral precipitation, then how did the microbialites that are present in the NA form? While the salinity of present-day NA water is saturated relative to sodium chloride, between 1900 and 1959 the salinity ranged from 20 to 27% and it is thought to have varied even more in the past due to climate-induced variation in the water level of the lake (Stephens 1990). Thus, it was not until construction of the railroad causeway in 1959 that the salinities of the two "arms" of GSL began to diverge due to localized input of as much as 95% of freshwater into the SA and a lack of mixing of waters between the SA and NA (Belovsky et al. 2011; Gwynn 1996; Naftz et al. 2011). Thus, prior to this time, it is possible that salinity conditions were permissible for the development of productive photosynthetic mats that could then drive the precipitation of aragonite.

The detection of (proto)dolomite in microbialites dated to ~2000–10,000 BP collected from paleo-shoreline deposits (Bouton et al. 2016) and sediments (Dunham et al. 2020) in the SA of GSL is significant since this mineral is not favored to form at low temperature and pressure, such as those present in GSL. This is despite waters being strongly oversaturated with respect to this mineral (Fairbridge 1957). The detection of (proto)dolomite in a modern-day sedimentary environment like GSL is consistent with its detection in other marginal hypersaline environments (Alderman 1958; Pace et al. 2016; Van Lith et al. 2002; Vasconcelos et al. 1995) and evaporitic environments that have no history of high temperature and pressure (e.g., Bontognali et al. 2010; Wells 1962). Together, these findings contribute to a growing body of literature pointing to the role of microorganisms in promoting the formation of authigenic (proto)dolomite, as described above and as reviewed recently (Petrash et al. 2017). Intriguingly, aragonite grains identified in sediments from the SA of GSL exhibited micrometer-sized euhedral (proto)dolomite crystals and these were absent on the faces of aragonite grains in a sediment core from the NA (Dunham et al. 2020). This suggests direct precipitation of (proto)dolomite. Moreover, if formation/precipitation of these micrometer-sized dolomite crystals indeed is shown to be microbially mediated, it is possible that such structures could serve as a biomarker for microbial activity in the rock record.

4.4.4 Contributions of Microbialites to the Greater Lake Ecosystem

Along with phytoplankton, microbialite-hosted periphyton biomass also supports several secondary consumers in GSL. These include both *Ephydra* (brine flies) and *Artemia* (brine shrimp) (Stephens 1990; Wurtsbaugh et al. 2011). In addition to supplying organic carbon and other nutrients such as fixed nitrogen, microbialite structures themselves are important support systems for *Ephydra* offspring. In this role, larvae utilize the carbonate structure to complete a key life cycle step which involves them physically attaching to submerged microbialites allowing them to grow a pupal case and transform into adults (Fig. 4.5c) (Collins 1979; Wurtsbaugh 2009). *Ephydra* and *Artemia* represent key food sources for birds (Wurtsbaugh 2009), including the ~10 million birds that visit GSL annually. This number includes both shorebirds and waterbirds that dive and feed on *Ephydra* and *Artemia* (Roberts 2013). As such, the food web of the SA of GSL is separated into pelagic and benthic compartments, with the microbialites categorized as part of the benthic compartment (Wurtsbaugh 2009). However, the microbial community associated with benthic microbialite-associated mats is almost identical to the surrounding phytoplankton community, which could indicate spatial crossover of communities (Lindsay et al. 2017). These compartmentalized food webs are also linked by apex consumers (bird species), which are the same for both the pelagic and benthic food web compartments (Belovsky et al. 2011; Wurtsbaugh 2009).

4.5 Microbialites in the Context of Natural and Anthropogenic Change

Over the past several hundred years, GSL has undergone short-term increases in salinity due to both natural (i.e., climate change) and anthropogenic reasons (i.e., river diversions, construction of railroad and motorway causeways). Collectively, these changes have driven changes in the lake environment, including increases in salinity of both the SA and NA of GSL (Baxter 2018; Baxter and Butler 2020; Null and Wurtsbaugh 2020; Stephens 1990). Here we briefly review the potential influence of changing salinity (and water level) on the GSL ecosystem.

4.5.1 Historic Changes in Great Salt Lake and Effects on Biology

Changes in the water level, salinity, and biology of GSL have been documented, in particular over the last 100 years (Stephens 1974, 1990; Wurtsbaugh et al. 2017). The shallow depth (average depth of ~5 m; Fig. 4.1b) and the terminal nature of GSL make it particularly susceptible to changes in water level and volume that, in turn, can impact salinities (Belovsky et al. 2011; Jones et al. 2009; Lindsay et al. 2019). This is especially true for the SA, as changes in riverine input (susceptible to changes in atmospheric and yearly snow conditions as well as anthropogenic water diversions) will affect this ecosystem to a greater extent than the NA since the NA does not receive any major input of freshwater via rivers (Chidsey et al. 2015; Wurtsbaugh et al. 2017). Changes in the salinity of GSL have been shown to impact the aquatic community, with the effects on *Artemia* and *Ephydra* fairly well documented (Stephens 1990). For example, an increase in lake elevation in the mid-1980s was accompanied by a decrease in salinity to 6%, with phytoplankton communities shifting from being dominated by halophilic microorganisms to those dominated by phototrophic microorganisms (Stephens 1990). This change was also associated with decreased abundances of *Artemia* due to cysts sinking to the bottom of the lake (Stephens 1990, 1998). Alternatively, it has been suggested to have been due to top-down control by a predaceous insect which apparently flourished when GSL had lower salinities (Wurtsbaugh 1992). These changes are contrasted with an increase in salinity between 1960 and 1963 that are not associated with regular seasonal changes. Here, the increase in salinity resulted in a decrease in the amount of phytoplankton and *Artemia* in the SA (Stephens 1998). Thus, changes in lake level and salinity have major effects on plankton and the higher trophic levels of the lake, including the abundance of *Artemia* and their cysts. Production, harvesting, and exporting *Artemia* cysts are key contributors to the regional economy, representing an estimated $57 million dollars as of 2012 (Belovsky et al. 2011; GSLAC 2012; Wotipka 2014). These studies estimate that ~35–45% of the *Artemia* cysts for use in aquaculture come from GSL.

4.5.2 Microbialite Response to Salinity Changes

Despite the substantial efforts made to document links between decreased lake level, increased salinity, changes in plankton abundance and composition, and a decline in the abundance of higher trophic levels such as *Artemia* in the GSL ecosystem, it was not until a recent study that a true mechanistic understanding for this link was uncovered. As mentioned previously, microbialites cover approximately 1000 km^2 (~20%) of the lake bottom of GSL (Baskin 2014; Eardley 1938) but contribute a disproportionately large amount of the total lake primary production via the activity of the phototrophs *Euhalothece* and *Navicula*, supporting the secondary consumers (Lindsay et al. 2017, 2019; Wurtsbaugh et al. 2011). Thus, a perturbation that negatively impacts the productivity of phototrophs in GSL could have consequences for the functioning and productivity of the entire ecosystem.

In a recent study, the ecological "tipping point" for the functioning of microbialite communities as it relates to changing salinity was investigated through a combination of molecular- and microcosm-based approaches (Lindsay et al. 2019). Photosynthetic mats that cover microbialite structures were sampled from near Stansbury Island in the southern portion of GSL. The salinity of the overlying water column was 15.6%. Microcosms were used to artificially increase or decrease the salinity and to study the effects on microbialite-associated microbial mat communities and the hatchability and survivability of *Artemia* cysts. The abundance of 16S rRNA gene templates (a proxy for total microbial biomass) was tracked for 7 weeks at salinities ranging from 8 to 30% salinity.

At a salinity (15%) corresponding closely to that measured in the SA when the mats were collected (15.6%), the abundance of 16S rRNA genes increased by 128% over 7 weeks (Fig. 4.7a). A decrease in salinity to 10 or 8% resulted in large overall increases in 16S rRNA gene abundances (537% and 161%, respectively), suggesting that the communities were stressed at even the *in situ* salinity at the time of sample collection (Lindsay et al. 2019). As expected, incubations of microbialites at elevated salinities of 20%, 25%, and 30% resulted in overall net decreases in 16S rRNA gene abundances of 43%, 97%, and 84%, respectively, over seven total weeks (Fig. 4.7a). The decrease in gene abundances at high salinities indicates the presence of maladapted populations (Lindsay et al. 2019).

In addition to a negative effect of increased salinity on the abundance of 16S rRNA gene templates in microcosm incubations, increased salinity also significantly impacted the composition of those communities as inferred by sequence analysis of those templates. The abundances of 16S rRNA gene templates attributable to primary producers *Euhalothece* and *Navicula* decreased markedly when incubated

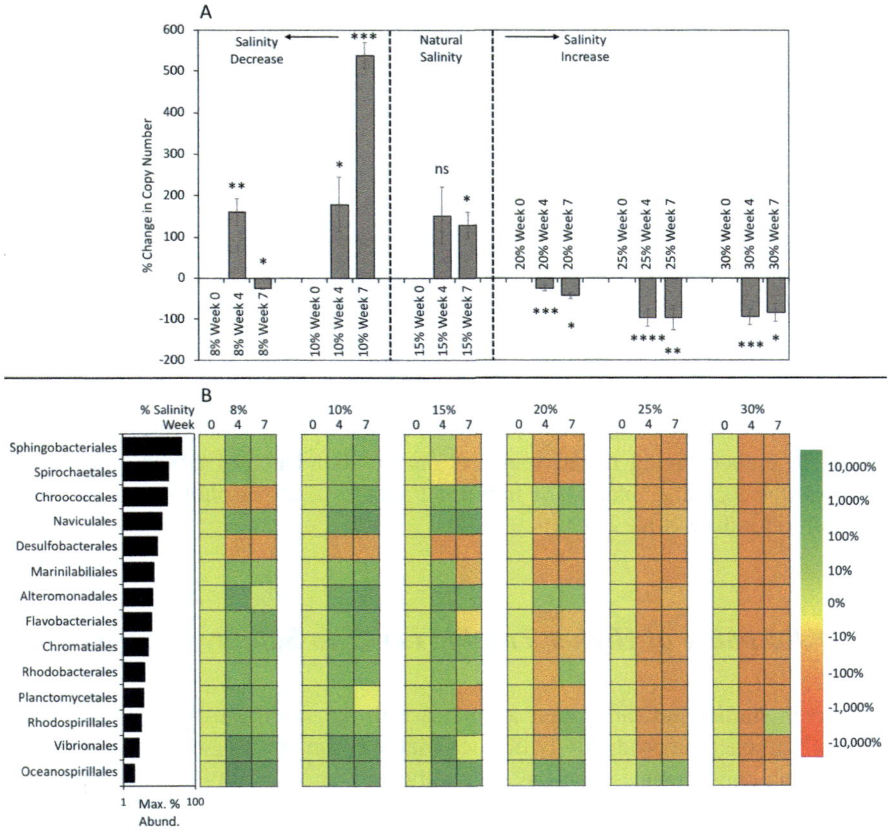

Fig. 4.7 (**a**) The change in the abundances of 16S rRNA gene templates in microbialite mat communities collected from microcosms that had been incubated at specified durations at natural salinity or at decreased or increased salinities. The abundances of templates at week 4 and at week 7 were normalized to abundances at week 0. Asterisks denote statistically significant differences ($*p < 0.1$; $**p < 0.01$; $***p < 0.001$; $****p < 0.0001$; ns, not significant with a $p > 0.1$) between measurements made at 0 weeks incubation and those made at 4 and 7 weeks incubation. (**b**) Taxonomic affiliation of 16S rRNA gene OTUs detected in microbialite communities from microcosms incubated at specified salinities after incubation for 0, 4, and 7 weeks. OTUs (grouped at order level) are sorted according to their maximum relative abundance. A rank abundance plot for the taxonomic bins is presented at the left of the figure. The abundances of sequences affiliated with orders as normalized to their abundances at week zero (for each salinity) are denoted by the color of each block, as denoted in the legend. Abbreviations: Max. % Abund., maximum percent abundance. Figure modified from Lindsay et al. (2019)

at higher salinities of 25 and 30% (Fig. 4.7b). In contrast, templates affiliated with *Euhalothece* exhibited increases when mats were incubated at 10 and 15% salinity for 7 weeks, and surprisingly were even shown to slightly increase when the salinity was increased to 20% (Fig. 4.7b) (Lindsay et al. 2019). This is consistent with optimal growth for *Euhalothece* species typically observed at salinities <15%, but with growth still possible at 20–24% salinity (Garcia-Pichel et al. 1998). The second most abundant phototroph *Navicula* exhibited increases at 8, 10, and 15% after both 4 and 7 weeks of incubation but decreased in abundance after 4 weeks of incubation at 20% salinity (Fig. 4.7b). These observations are consistent with the optimal salinity for the phototrophic species *Navicula* which is only ~5% (Clavero et al. 2000).

Following 7 weeks incubation at decreased, in situ (15%), and elevated salinity, the microbialite mats were collected and subjected to an analysis of CO_2 fixation activity when incubated in full sunlight. At lower salinities of 8 and 10%, the rates of DIC assimilation were not significantly different than the rate measured for microbialites incubated at the salinity (15%) corresponding closest to that of the SA of GSL at the time of microbialite sampling (15.6%) (Lindsay et al. 2019). However, at the elevated salinities of 20, 25, and 30%, the rates of DIC assimilation in microbialite-associated mats were all significantly lower than the rate measured at 15% salinity (Lindsay et al. 2019). Overall, this study points to increases in salinity to 20% or higher having negative effects on the abundance of biomass and their activities due to maladaptation of important species such as phototrophs to such conditions (Lindsay et al. 2019).

4.5.3 Response of Higher Trophic Levels to Salinity Changes

Microcosm studies indicating that an increase in the salinity of GSL is likely to negatively impact the productivity of microbialite communities (see above) suggest far-reaching consequences for the fecundity and survivability of *Artemia* and *Ephydra* that depend on the mats as a food source (Lindsay et al. 2019). GSL exhibits a relatively simple trophic organization, which allows for determination of the effect of environmental perturbations on individual species and trophic levels. Thus, a study was conducted to examine the effects of changing salinity on the hatch and survivability of *Artemia* in microcosms containing microbialite microbial mats (Lindsay et al. 2019). A previous study also examined the effects of salinity changes on *Ephydra*, copepod, and rotifer biomass abundances (Barnes and Wurtsbaugh 2015). Data from both studies indicated that salinities over ~20% did not allow for *Artemia* cysts to hatch (Barnes and Wurtsbaugh 2015; Lindsay et al. 2019). These observations are consistent with previous studies of other systems that indicate that *Artemia* populations cannot reproduce at high salinities due to osmotic stress, dessication, or low oxygen tension [dependent on temperature (Browne and Wanigasekera 2000)]. Additionally, salinities of 15% and greater were shown to significantly retard the time required for the first hatching of *Artemia* cysts (Lindsay et al. 2019).

In addition to these secondary consumers, many species of birds are likely affected by changes in biomass associated with trophic levels directly below them in the food web and at the level of primary producers. For example, eared grebe numbers were directly linked to the abundance of *Artemia* on a macro scale in GSL (Belovsky et al. 2011). Indeed, decreases in the abundance of *Artemia* at increased salinities were correlated with decreased numbers of eared grebes frequenting the lake. This finding, coupled with the above data indicating the negative effects of increased salinity on the abundance of microbialite-associated microbial biomass and composition, their activities, and *Artemia* hatch rate and survivability, suggests the interdependence of these trophic level compartments.

4.5.4 Potential Effects of Natural and Anthropogenically Driven Climate Change on the Great Salt Lake Food Web

Changes in the level and salinity of GSL due to changes in climate over the next several decades in the greater Salt Lake City, Utah area (Baxter and Butler 2020; Cook et al. 2015), could be further exacerbated by changes in water use. For example, stream diversion projects for agricultural uses, which have already resulted in a 40% reduction of freshwater flow into GSL and a decline in lake surface area by 50% since the mid-nineteenth century (Wurtsbaugh et al. 2017), are likely to continue and are thus expected to further shape the functioning and productivity of the GSL ecosystem. Given that microbialites are almost all located near (~5 km) the shoreline and at shallow depths in GSL, a substantial decrease in surface area may expose these structures and could result in a significant decrease in the primary production associated with mat communities. As discussed above, these periphyton communities contribute a disproportionate amount of the primary production to the lake and likely have done so for hundreds or thousands of years, if not longer (Bouton et al. 2016). Arguably, the most significant modern change to the GSL ecosystem occurred in the late 1950s when a railroad causeway was built across the lake, resulting in its separation into what are now termed the NA and SA (Stephens 1974). As previously mentioned, this change impacted movement of freshwater between the two "arms" of the lake, leading to the NA being an evaporative basin. The change in salinity that accompanied the construction of the causeway likely resulted in major shifts in the composition of the communities associated with microbialites (Lindsay et al. 2017), including the demise of the primary producers *Euhalothece* and *Navicula* as salinity surpassed the "tipping point" (~20%) for these taxa. It is likely that the demise of the primary producers in the NA, including those associated with microbialites, was accompanied shortly thereafter by a major decline in the *Artemia* numbers due to the combined effects of decreased food availability and increased salinity on their hatch rate and survivability.

The effects of increased salinity in GSL likely extend beyond the ecological compartments that have previously been studied. For example, in other saline lakes, increased salinities or the disappearance of lake surface area (i.e., the exposure of lake beds) have resulted in significant ecological, atmospheric, and economic impacts (Wurtsbaugh et al. 2017). The dessication of Lake Urmia in Iran (also a terminal lake) resulted in salinity excursions that now exceed 28.5%, leading to eradication of *Artemia* and consequently a loss of apex species such as flamingos and other birds (Lotfi 2012; Stone 2015; Wurtsbaugh et al. 2017). If GSL salinities are maintained at elevated concentrations over the thresholds tolerated by primary producers and secondary consumers [~20–22% salinity (Barnes and Wurtsbaugh 2015; Lindsay et al. 2019)], the downstream effects are likely to result in a cascading effect throughout the food web (Paine 1980; Ripple et al. 2016; Terborgh et al. 2006). For example, a decline in *Artemia* and *Ephydra* numbers is likely to negatively affect the nearly ten million migratory birds that use GSL as an important feeding ground annually (Roberts 2013; Vest and Conover 2011). Such a decline in *Artemia* and *Ephydra* numbers, which would undoubtedly be linked to a decline in the functioning of microbialite-associated primary producers, could negatively impact the economics of the *Artemia* cyst industry and tourism in the form of birding.

4.6 Concluding Remarks

GSL's microbialites lie at the intersection of geology and biology. The utility of GSL microbialites to inform on the potential for such structures to serve as biosignatures of microbial life in the rock record of early Earth and to sustain a healthy contemporary GSL ecosystem points to the need to further our understanding of these unique geomicrobial assemblages. The application of molecular approaches to better understand the taxonomic composition and function of the predominant members of these communities has led to the hypothesis that photosynthetic Cyanobacteria and algae direct the formation of microbialite structures largely through their role in binding/trapping sediment and creating localized increases in pH. This in turn is suggested to promote the precipitation of carbonate in association with EPS. Photosynthetic biomass and associated carbonate, when deposited into the deeper layers of the mat, promote the anaerobic microbial processes of fermentation and sulfate reduction that together are suggested to promote (proto)dolomite formation and lithification. Lithified mat structures are preserved locally in GSL rocks dated to >21,000 years ago (Bouton et al. 2016). The extensive and easily accessible salinity gradients offered by GSL provide a natural laboratory to evaluate the role of specific microorganisms or processes in the lithification of GSL mats and the influence of specific processes in promoting the formation of aragonite and (proto)dolomite as possible biosignatures of microbial activity in the rock record.

GSL is expected to be affected by both natural (i.e., climate change) and anthropogenic causes (i.e., river diversions, construction of railroad and motorway causeways) that are together expected to influence the input and movement of freshwater into the lake and thus water level and/or salinity. The extensiveness of microbialites in GSL that span spatial salinity gradients has allowed for researchers to begin to deduce the potential effect of temporal changes in salinity on the functioning of these highly productive assemblages. Results suggest a "tipping point" for the primary producers that drive microbialite mat formation at ~20% salinity. Lake salinities that rise above this are expected to be associated with significant and negative effects on several levels of ecosystem production. The negative effects of increased salinity on microbialite-associated photosynthetic production are expected to cascade through the highly simplified GSL ecosystem, reducing numbers of *Artemia* and *Ephydra* that could ultimately impact the lake's ability to support abundant shore and migratory birds. In turn, a decrease in *Artemia* and bird numbers could create economic uncertainty for industries that depend on these species, including the *Artemia* cyst industry and tourism.

Acknowledgements This work was supported by the Doyle W. Stephens scholarship from Friends of Great Salt Lake, the NASA Earth and Space Science Fellowship Program (NNX16AP51H) to M.R. Lindsay, and the NSF Graduate Research Fellowship Program to E.C. Dunham. E.S.B. acknowledges support from the NASA Astrobiology Institute (NNA15BB02A).

References

Alderman AR (1958) Aspects of carbonate sedimentation. J Geol Soc Aust 6:1–10

Allwood AC, Walter MR, Kamber BS, Marshall CP, Burch IW (2006) Stromatolite reef from the Early Archaean era of Australia. Nature 441:714–718. https://doi.org/10.1038/nature04764

Almeida-Dalmet S, Sikaroodi M, Gillevet P, Litchfield CD, Baxter BK (2015) Temporal study of the microbial diversity of the north arm of Great Salt Lake Utah, U.S. Microorganisms 3:310–326

Anderson RB, Naftz DL, Day-Lewis FD, Henderson RD, Rosenberry DO, Stolp BJ, Jewell P (2014) Quantity and quality of groundwater discharge in a hypersaline environment. J Hydrol 512:177–194

Antón J, Oren A, Benlloch R, Rodríguez-Valera F, Amann R, Rosselló-Mora R (2002) *Salinibacter ruber* gen. nov., sp. nov., a novel, extremely halophilic member of the Bacteria from saltern crystallizer ponds. Int J Syst Evol Microbiol 52:485–491

Baker PA, Kastner M (1981) Constraints on the formation of sedimentary dolomite. Science 213:214–216. https://doi.org/10.1126/science.213.4504.214

Barnes BD, Wurtsbaugh WA (2015) The effects of salinity on plankton and benthic communities in the Great Salt Lake, Utah, USA: a microcosm experiment. Can J Fish Aquat Sci 72:807–817

Baskin R (2014) Occurrence and spatial distribution of microbial bioherms in Great Salt Lake, Utah. The University of Utah

Baskin R, Wright V, Driscoll N, Kent G, Hepner C (2012) Microbialite bioherms in Great Salt Lake, Utah: influence of active tectonics and anthropogenic effects. In: American Association of Petroleum Geologists Hedburg conference, Houston, TX

Baskin R, Driscoll N, Wright V (2013) Controls on lacustrine microbialite distribution in Great Salt Lake, Utah. In: Microbial carbonates in space and time: implications for global exploration and production. Geological Society of America, Boulder, CO, pp 70–71

Bateson MM, Ward DM (1988) Photoexcretion and fate of glycolate in a hot spring cyanobacterial mat. Appl Environ Microbiol 54:1738–1743

Baxter BK (2018) Great Salt Lake microbiology: a historical perspective. Int Microbiol 21:79–95

Baxter BK, Butler JK (2020) Climate change and Great Salt Lake. In: Baxter BK, Butler JK (eds) Great Salt Lake biology: a terminal lake in a time of change. Springer, Cham

Belovsky GE, Stephens D, Perschon C, Birdsey P, Paul D, Naftz D, Baskin R, Larson C, Mellison C, Luft J, Mosley R (2011) The Great Salt Lake Ecosystem (Utah, USA): long term data and a structural equation approach. Ecosphere 2:1–40

Bontognali TRR, Vasconcelos C, Warthmann RJ, Bernasconi SM, Dupraz C, Strohmenger CJ, McKenzie JA (2010) Dolomite formation within microbial mats in the coastal sabkha of Abu Dhabi (United Arab Emirates). Sedimentology 57:824–844. https://doi.org/10.1111/j.1365-3091.2009.01121.x

Bosak T, Knoll AH, Petroff AP (2013) The meaning of stromatolites. Ann Rev Earth Planet Sci 41:21–44. https://doi.org/10.1146/annurev-earth-042711-105327

Bouton A, Vennin E, Boulle J, Pace A, Bourillot R, Thomazo C, Brayard A, Désaubliaux G, Goslar T, Yokoyama Y, Dupraz C (2016) Linking the distribution of microbial deposits from the Great Salt Lake (Utah, USA) to tectonic and climatic processes. Biogeosciences 13:5511–5526. https://doi.org/10.5194/bg-2015-647

Boyd ES, Hamilton TL, Swanson KD, Howells AE, Baxter BK, Meuser JE, Posewitz MC, Peters JW (2014) [FeFe]-hydrogenase abundance and diversity along a vertical redox gradient in Great Salt Lake, USA. Int J Mol Sci 15:21947–21966

Boyd ES, Yu R, Barkay T, Hamilton TL, Baxter BK, Naftz DL, Marvin-DiPasquale M (2017) Effect of salinity on mercury methylating benthic microbes and their activities in Great Salt Lake. Utah Sci Tot Environ 581:495–506

Braithwaite CJR, Zedef V (1994) Living hydromagnesite stromatolites from Turkey. Sed Geol 92:1–5. https://doi.org/10.1016/0037-0738(94)90051-5

Brandt KK, Vester F, Jensen AN, Ingvorsen K (2001) Sulfate reduction dynamics and enumeration of sulfate-reducing bacteria in hypersaline sediments of the Great Salt Lake (Utah, USA). Microb Ecol 41:1–11

Browne RA, Wanigasekera G (2000) Combined effects of salinity and temperature on survival and reproduction of five species of *Artemia*. J Exp Mar Biol Ecol 244:29–44

Burne RV, Moore LS (1987) Microbialites: organosedimentary deposits of benthic microbial communities. Palaios 2:241–254

Cannon JS, Cannon MA (eds) (2002) The Southern Pacific railroad trestle—past and present. Great Salt Lake, an overview of change. Utah Department of Natural Resources, Salt Lake City, UT

Carozzi AV (1962) Observations on algal biostromes in the Great Salt Lake, Utah. J Geol 70:246–252

Chen YG, Cui XL, Li QY, Wang YX, Tang SK, Liu ZX, Wen ML, Peng Q, Xu LH (2009) *Saccharospirillum salsuginis* sp. nov., a gammaproteobacterium from a subterranean brine. Int J Syst Evol Microbiol 59:1382–1386

Chidsey TCJ, Vanden Berg MD, Eby DE (2015) Petrography and characterization of microbial carbonates and associated facies from modern Great Salt Lake and Uinta Basin's Eocene Green River formation in Utah, USA. In: Bosence DWJ, Gibbons KA, Le Heron DP, Morgan WA, Pritchard T, Vining BA (eds) Microbial carbonates in space and time: implications for global exploration and production, vol 418. Geological Society, London

Clavero E, Hernández-Mariné M, Grimalt JO, Garcia-Pichel F (2000) Salinity tolerance of diatoms from thalassic hypersaline environments. J Phycol 36:1021–1034

Cohen AS, Talbot MR, Awramik SM, Dettman DL, Abell P (1997) Lake level and paleoenvironmental history of Lake Tanganyika, Africa, as inferred from late Holocene and modern stromatolites. Geol Soc Am Bull 109:444–460. https://doi.org/10.1130/0016-7606(1997)109<0444:Llapho>2.3.Co;2

Cohenour R, Thompson K (eds) (1966) Geologic setting of Great Salt Lake. Guidebook to the geology of Utah. Utah Geological Society, Salt Lake City, UT

Collins N (1979) Population ecology of *Ephydra cineria jones* (Diptera: Ephydridae), the only benthic metazoan of the Great Salt Lake, U.S.A. Hydrobiologia 69:99–112

Collins L, Jahnert R (2014) Stromatolite research in the Shark Bay World Heritage Area. Roy Soc West Aust 97:189–219

Cook BI, Ault TR, Smerdon JE (2015) Unprecedented 21st century drought risk in the American Southwest and Central Plains. Sci Adv 1

Crisp F, Bell FJ, Bennett AW, Hebb RG (1880) Transactions of the society, vol III, issue 2, Journal of the Royal Microscopical Society. Williams and Norgate, London

Dana GL, Lenz PH (1986) Effects of increasing salinity on an *Artemia* population from Mono Lake, California. Oecologia 68:428–436

Darley WM, Volcani BE (1969) Role of silicon in diatom metabolism: a silicon requirement for deoxyribonucleic acid synthesis in the diatom *Cylindrotheca fusiformis* Reimann and Lewin. Exp Cell Res 58:334–342. https://doi.org/10.1016/0014-4827(69)90514-X

Domagalski JL, Orem WH, Eugster HP (1989) Organic geochemistry and brine composition in Great Salt, Mono, and Walker Lakes. Geochim Cosmochim Acta 53:2857–2872. https://doi.org/10.1016/0016-7037(89)90163-4

Dunham EC, Fones EM, Fang Y, Lindsay MR, Steuer C, Fox N, Willis M, Walsh A, Colman DR, Baxter BK, Lageson D (2020) An ecological perspective on dolomite formation in Great Salt Lake, Utah. Front Earth Sci 8:24. https://doi.org/10.3389/feart.2020.00024

Dupraz C, Visscher PT (2005) Microbial lithification in marine stromatolites and hypersaline mats. Trends Microbiol 13:429–438

Eardley AJ (1938) Sediments of Great Salt Lake, Utah. AAPG Bull 22:1305–1411

Edgcomb VP, Bernhard JM, Summons RE, Orsi W, Beaudoin D, Visscher PT (2014) Active eukaryotes in microbialites from Highborne Cay, Bahamas, and Hamelin Pool (Shark Bay), Australia. ISME J 8:418–429. https://doi.org/10.1038/ismej.2013.130

Fairbridge RW (1957) The dolomite question. In: Blanc RJL, Breeding JG (eds) Regional aspects of carbonate deposition. SEPM Society for Sedimentary Geology, Tulsa, OK

Garcia-Pichel F, Nubel U, Muyzer G (1998) The phylogeny of unicellular, extremely halotolerant cyanobacteria. Arch Microbiol 169:469–482. https://doi.org/10.1007/s002030050599

Geider R, La Roche J (2002) Redfield revisited: variability of C:N:P in marine microalgae and its biochemical basis. Eur J Phycol 37:1–17. https://doi.org/10.1017/S0967026201003456

Gerard E, Menez B, Couradeau E, Moreira D, Benzerara K, Tavera R, Lopez-Garcia P (2013) Specific carbonate-microbe interactions in the modern microbialites of Lake Alchichica (Mexico). ISME J 7:1997–2009. https://doi.org/10.1038/ismej.2013.81

Great Salt Lake Advisory Council (2012) Economic significance of the Great Salt Lake to the State of Utah. Bioeconomics. Accessed 25 Sept 2018

Guiry MD (2019) *Navicula salinicola* Hustedt. National University of Ireland. www.algaebase.org. Accessed 30 Jan 2019

Gwynn JW (1996) Commonly asked questions about Utah's Great Salt Lake and Ancient Lake Bonneville, vol 39. Utah Geological Survey, Salt Lake City, UT

Halley RB (1976) Textural variation within Great Salt Lake algal mounds. Dev Sediment 20:435–445

Jahnert RJ, Collins LB (2011) Significance of subtidal microbial deposits in Shark Bay, Australia. Mar Geol 286:106–111. https://doi.org/10.1016/j.margeo.2011.05.006

Joint I, Henriksen P, Garde K, Riemann B (2002) Primary production, nutrient assimilation and microzooplankton grazing along a hypersaline gradient. FEMS Microb Ecol 39:245–257. https://doi.org/10.1111/j.1574-6941.2002.tb00927.x

Jones DL, Baxter BK (2017) DNA repair and photoprotection: mechanisms of overcoming environmental ultraviolet radiation exposure in halophilic Archaea. Front Microbiol 8:1–16

Jones BF, Naftz DL, Spencer RJ, Oviatt CG (2009) Geochemical evolution of Great Salt Lake, Utah, USA. Aquat Geochem 15:95–121

Keck W, Hassibe W (1979) The Great Salt Lake. U.S. Geological Survey, 25

Kemp BL, Tabish EM, Wolford AJ, Jones DL, Butler JK, Baxter BK (2018) The biogeography of Great Salt Lake halophilic archaea: testing the hypothesis of avian mechanical carriers. Diversity 10:124

Kendall MM, Boone DR (2006) The order Methanosarcinales. In: Dworkin M, Falkow S, Rosenberg E, Schleifer K-H, Stackebrandt E (eds) The prokaryotes, vol 3, A handbook on the biology of bacteria: ecophysiology and biochemistry. Springer, New York

Kulasooriya SA, Lang NJ, Fay P (1972) The heterocysts of blue-green algae. III. Differentiation and nitrogenase activity. Proc Roy Soc Lond B 181:199–209. https://doi.org/10.1098/rspb.1972.0046

Laval B, Cady SL, Pollack JC, McKay CP, Bird JS, Grotzinger JP, Ford DC, Bohm HR (2000) Modern freshwater microbialite analogues for ancient dendritic reef structures. Nature 407 (6804):626–629

Lindsay MR, Anderson C, Fox N, Scofield G, Allen J, Anderson E, Bueter L, Poudel S, Sutherland K, Munson-McGee JH, Van Nostrand JD (2017) Microbialite response to an anthropogenic salinity gradient in Great Salt Lake, Utah. Geobiology 15:131–145

Lindsay MR, Johnson RE, Baxter BK, Boyd ES (2019) Effects of salinity on microbialite-associated production in Great Salt Lake, Utah. Ecology 100:e02611

Logan BW (1961) Cryptozoon and associate stromatolites from the recent, Shark Bay, Western-Australia. J Geol 69:517–533

Lotfi A (2012) Lake Uromiyeh: a concise baseline report. Impel Review Initiative, Department of Environment, United Nations Development Program

Lu Y, Sun X, Xu H, KoniSHi H, Lin Z, Xu L, CHen T, Hao X, Lu H, PeCKmann J (2018) Formation of dolomite catalyzed by sulfate-driven anaerobic oxidation of methane: mineralogical and geochemical evidence from the northern South China Sea. Am Mineral 103:720–734. https://doi.org/10.2138/am-2018-6226

Merz MUE (1992) The biology of carbonate precipitation by cyanobacteria. Facies 26:81–101

Meuser JE, Baxter BK, Spear JR, Peters JW, Posewitz MC, Boyd ES (2013) Contrasting patterns of community assembly in the stratified water column of Great Salt Lake, Utah. Microb Ecol 66:268–280. https://doi.org/10.1007/s00248-013-0180-9

Mogany T, Swalaha FM, Allam M, Mtshali PS, Ismail A, Kurami S, Bux F (2018) Phenotypic and genotypic characterisation of an unique indigenous hypersaline unicellular cyanobacterium, *Euhalothece* sp. nov. Microbiol Res 211:47–56

Mohapatra GK, Johnson RA (1998) Localization of listric faults at thrust fault ramps beneath the Great Salt Lake Basin, Utah: evidence from seismic imaging and finite element modeling. J Geophys Res Sol Earth 103:10047–10063. https://doi.org/10.1029/98jb00023

Møller MF, Kjeldsen KU, Ingvorsen K (2010) *Marinimicrobium haloxylanilyticum* sp. nov., a new moderately halophilic, polysaccharide-degrading bacterium isolated from Great Salt Lake, Utah. Antonie van Leeuwenhoek 98:553–565

Muyzer G, Stams AJM (2008) The ecology and biotechnology of sulphate-reducing bacteria. Nat Rev Microbiol 6:441–454

Myshrall KL, Mobberley JM, Green SJ, Visscher PT, Havemann SA, Reid RP, Foster JS (2010) Biogeochemical cycling and microbial diversity in the thrombolitic microbialites of Highborne Cay, Bahamas. Geobiology 8:337–354. https://doi.org/10.1111/j.1472-4669.2010.00245.x

Naftz DL, Millero FJ, Jones BF, Green WR (2011) An equation of state for hypersaline water in Great Salt Lake, Utah, USA. Aquat Geochem 17:809–820

Null SE, Wurtsbaugh WA (2020) Water development, consumptive water uses, and Great Salt Lake. In: Baxter BK, Butler JK (eds) Great Salt Lake biology: a terminal lake in a time of change. Springer, Cham

Oren A (ed) (2006a) Life at high salt concentrations, vol 2. Prokaryotes. Springer, Berlin

Oren A (ed) 2006b The order Halobacteriales, vol 3. Prokaryotes. Springer, New York

Oren A, Ventosa A (2017) Haloferacales. In: Whitman WB et al (eds) Bergey's manual of systematics of archaea and bacteria. Wiley, Chichester

Oviatt CG, Thompson RS, Kaufman DS, Bright J, Forester RM (1999) Reinterpretation of the Burmester Core, Bonneville Basin, Utah. Quat Res 52:180–184. https://doi.org/10.1006/qres.1999.2058

Oviatt CG, Madsen DB, Miller DM, Thompson RS, McGeehin JP (2015) Early Holocene Great Salt Lake, USA. Quat Res 84:57–68. https://doi.org/10.1016/j.yqres.2015.05.001

Pace A, Bourillot R, Bouton A, Vennin E, Galaup S, Bundeleva I, Patrier P, Dupraz C, Thomazo C, Sansjofre P, Yokoyama Y (2016) Microbial and diagenetic steps leading to the mineralisation of Great Salt Lake microbialites. Sci Rep 6:31495

Paerl HW, Steppe TF, Reid RP (2001) Bacterially mediated precipitation in marine stromatolites. Environ Microbiol 3:123–130

Pages A, Welsh DT, Teasdale PR, Grice K, Vacher M, Bennett WW, Visscher PT (2014) Diel fluctuations in solute distributions and biogeochemical cycling in a hypersaline microbial mat from Shark Bay, WA. Mar Chem 167:102–112. https://doi.org/10.1016/j.marchem.2014.05.003

Paine RT (1980) Food webs: linkage, interaction strength and community infrastructure. J Anim Ecol 49:666–685

Parnell JJ, Rompato G, Latta IV LC, Pfrender ME, Van Nostrand JD, He Z, Zhou J, Andersen G, Champine P, Ganesan B, Weimer BC (2010) Functional biogeography as evidence of gene transfer in hypersaline microbial communities. PLoS One 5:e12919

Pedone VA, Folk RL (1996) Formation of aragonite cement by nannobacteria in the Great Salt Lake, Utah. Geology 24:763–765. https://doi.org/10.1130/0091-7613(1996)024<0763:Foacbn>2.3.Co;2

Pepe-Ranney C, Berelson WM, Corsetti FA, Treants M, Spear JR (2012) Cyanobacterial construction of hot spring siliceous stromatolites in Yellowstone National Park. Environ Microbiol 14:1182–1197. https://doi.org/10.1111/j.1462-2920.2012.02698.x

Petrash DA, Bialik OM, Bontognali TRR, Vasconcelos C, Roberts JA, McKenzie JA, Konhauser KO (2017) Microbially catalyzed dolomite formation: from near-surface to burial. Earth Sci Rev 171:558–582

Phyto'pedia (2012) *Navicula* spp. The University of British Columbia. https://www.eoas.ubc.ca/research/phytoplankton/diatoms/pennate/navicula/navicula_spp.html. Accessed 8 Sept 2018

Post FJ (1980) Biology of the north arm. In: Gwynn JW (ed) Great Salt Lake: a scientific, historical, and economic overview, Bulletin 116. Utah Geological and Mineral Survey, pp 313–322

Riding R (1991) Classification of microbial carbonates. In: Riding R (ed) Calcareous algae and stromatolites. Springer, Berlin, pp 21–51

Riding R (2002) Microbial carbonates: the geological record of calcified bacterial-algal mats and biofilms. Sedimentology 47:179–214

Riding R (2011) Microbialites, stromatolites, and thrombolites. In: Reitner J, Thiel V (eds) Encyclopedia of geobiology. Springer, Dordrecht, pp 635–654

Ripple WJ, Estes JA, Schmitz OJ, Constant V, Kaylor MJ, Lenz A, Motley JL, Self KE, Taylor DS, Wolf C (2016) What is a trophic cascade? Trends Ecol Evol 31:842–849

Roberts AJ (2013) Avian diets in a saline ecosystem: Great Salt Lake, Utah, USA. Hum Wildl Interact 7:158–168

Sei A, Fathepure BZ (2009) Biodegradation of BTEX at high salinity by an enrichment culture from hypersaline sediments of Rozel Point at Great Salt Lake. J Appl Microbiol 107:2001–2008

Smith VH (2007) Microbial diversity—productivity relationships in aquatic ecosystems. FEMS Microb Ecol 62:181–186

Stansbury H (1855) Exploration of the valley of the Great Salt Lake: including a reconnaissance of a new route through the Rocky Mountains. Lippincott, Gramabo, Philadelphia, PA

Stephens DW (1974) A summary of biological investigations concerning the Great Salt Lake, Utah (1861-1973). Great Basin Nat 34:221–229

Stephens DW (1990) Changes in lake levels, salinity and the biological community of Great Salt Lake (Utah, USA), 1847-1987. Hydrobiologia 197:139–146

Stephens DW (1998) Salinity-induced changes in the aquatic ecosystem of Great Salt Lake, Utah. In: Pitman J, Carroll A (eds) Modern and ancient lake systems, Utah Geological Survey Guidebook 26

Stone R (2015) Saving Iran's great salt lake. Science 349:1044–1047

Sugitani K, Mimura K, Takeuchi M, Lepot K, Ito S, Javaux E (2015) Early evolution of large microorganisms with cytological complexity revealed by microanalyses of 3.4 Ga organic-walled microfossils. Geobiology 13:507–521

Tazi L, Breakwell DP, Harker AR, Crandall KA (2014) Life in extreme environments: microbial diversity in Great Salt Lake, Utah. Extremophiles 18:525–535. https://doi.org/10.1007/s00792-014-0637-x

Terborgh J, Feeley K, Silman M, Nuñez P, Balukjian B (2006) Vegetation dynamics of predator-free land-bridge islands. J Ecol 94:253–263

Theisen CH, Sumner DY, Mackey TJ, Lim DS, Brady AL, Slater GF (2015) Carbonate fabrics in the modern microbialites of Pavilion Lake: two suites of microfabrics that reflect variation in microbial community morphology, growth habit, and lithification. Geobiology 13:357–372

Thompson RM, Brose U, Dunne JA, Hall RO Jr, Hladyz S, Kitching RL, Martinez ND, Rantala H, Romanuk TN, Stouffer DB, Tylianakis JM (2012) Food webs: reconciling the structure and function of biodiversity. Trends Ecol Evol 27:689–697

Tilden JE (1898) American algae. Century III:298

van Kranendonk MJ, Philippot P, Lepot K, Bodorkos S, Pirajno F (2008) Geological setting of Earth's oldest fossils in the ca. 3.5 Ga Dresser Formation, Pilbara Craton, Western Australia. Precamb Res 167:93–124

Van Lith Y, Vasconcelos C, Warthmann R, Martins JCF, McKenzie JA (2002) Bacterial sulfate reduction and salinity: two controls on dolomite precipitation in Lagoa Vermelha and Brejo do Espinho (Brazil). Hydrobiologia 489:35–49

Van Trappen S, Vandecandelaere I, Mergaert J, Swings J (2004) *Gillisia limnaea* gen. nov., sp. nov., a new member of the family *Flavobacteriaceae* isolated from a microbial mat in Lake Fryxell, Antarctica. Int J Syst Evol Microbiol 54:445–448

Vasconcelos C, McKenzie JA (1997) Microbial mediation of modern dolomite precipitation and diagenesis under anoxic conditions (Lagoa Vermelha, Rio de Janeiro, Brazil). J Sed Res 67:378–390. https://doi.org/10.1306/D4268577-2B26-11D7-8648000102C1865D

Vasconcelos C, McKenzie JA, Bernasconi S, Grujic D, Tiens AJ (1995) Microbial mediation as a possible mechanism for natural dolomite formation at low temperatures. Nature 377:220

Vest JL, Conover MR (2011) Food habits of wintering waterfowl on the GSL, Utah. Waterbirds 34:40–50

Walter MR, Buick R, Dunlop JSR (1980) Stromatolites 3,400-3,500 Myr old from the North Pole area, Western Australia. Nature 284:443–445

Warthmann R, van Lith Y, Vasconcelos C, McKenzie JA, Karpoff AM (2000) Bacterially induced dolomite precipitation in anoxic culture experiments. Geology 28:1091–1094. https://doi.org/10.1130/0091-7613(2000)28<1091:BIDPIA>2.0.CO;2

Wells AJ (1962) Recent dolomite in the Persian Gulf. Nature 194:274. https://doi.org/10.1038/194274a0

Widdel F, Pfennig N (1981) Studies on dissimilatory sulfate-reducing bacteria that decompose fatty acids. Arch Microbiol 129:395–400

Wotipka S (2014) Brine shrimp by the billions in the Great Salt Lake. High Country News. Accessed 25 Sept 2018

Wurtsbaugh WA (1992) Food-web modification by an invertebrate predator in the Great Salt Lake (USA). Oecologia 89:168–175

Wurtsbaugh WA (2009) Biostromes, brine flies, birds and the bioaccumulation of selenium in Great Salt Lake, Utah. Nat Res Environ 15:2

Wurtsbaugh WA, Gardberg J, Izdepski C (2011) Biostrome communities and mercury and selenium bioaccumulation in the Great Salt Lake (Utah, USA). Sci Tot Environ 409:4425–4434. https://doi.org/10.1016/j.scitotenv.2011.07.027

Wurtsbaugh WA, Miller C, Null SE, DeRose RJ, Wilcock P, Hahnenberger M, Howe F, Moore J (2017) Decline of the world's saline lakes. Nat Geosci 10:816–821

Xu Y, Zhang R, Li Q, Liu K, Jiao N (2015) *Marivirga lumbricoides* sp. nov., a marine bacterium isolated from the South China Sea. Int J Syst Evol Microbiol 65:452–456

Yaakop AS, Chan K, Ee R, Kahar UM, Kon WC, Goh KM (2015) Isolation of *Jeotgalibacillus malaysiensis* sp. nov. from a sandy beach, and emended description of the genus *Jeotgalibacillus*. Int J Syst Evol Microbiol 65:2215–2221

Zhang F, Xu H, Konishi H, Kemp JM, Roden EE, Shen Z (2012) Dissolved sulfide-catalyzed precipitation of disordered dolomite: implications for the formation mechanism of sedimentary dolomite. Geochim Cosmochim Acta 97:148–165

Zhang F, Yan C, Teng HH, Roden EE, Xu H (2013) In situ AFM observations of Ca–Mg carbonate crystallization catalyzed by dissolved sulfide: implications for sedimentary dolomite formation. Geochim Cosmochim Acta 105:44–55

Chapter 5
Unexpected Complexity at Salinity Saturation: Microbial Diversity of the North Arm of Great Salt Lake

Swati Almeida-Dalmet and Bonnie K. Baxter

Abstract The isolated north arm of Great Salt Lake, Utah, is a unique and complex environment with salinity at saturation, above 25% total salts. It is separated from the larger south arm, which experiences more freshwater input, due to a rock-filled causeway installed around 1960. Prior studies using both cultivation and molecular methods have shown that the microbial community of this part of the lake is diverse and dynamic, experiencing year-round fluctuations in salinity and temperature. The data emerging from our published studies and others have demonstrated the presence of microbial genera from all three domains of life, with the archaeal diversity being the greatest. When we cultivated approximately 50 isolates, the majority of these were genotyped as archaea, and only four cultivars belonged to the Domain bacteria. Thus, initial studies, reviewed herein, focused on understanding the diversity of the overrepresented archaea, using molecular, culture-independent methods to assess temporal diversity and significance of environmental parameters. Cultivation studies revealed details about how the stable members of the communities maintained their lifestyle using differential gene expression. But bacteria also live in this archaeal world, and they remain understudied in hypersaline systems. Therefore, we analyzed the bacterial isolates, genetically and biochemically, to reveal more information about the bacteria of the Great Salt Lake north arm. The genus *Salinibacter* was present throughout the year and mostly dominated the bacterial population. 16S rRNA gene sequencing of these bacterial cultivars demonstrated relationships to strains of *Salinibacter*, strains of *Halomonas,* and other uncultured deposited DNA sequences. To look at temporal diversity profiles of this bacterial minority, next-generation DNA sequencing (with semiconductor sequencing technology) was employed on DNA extracted from four water samples collected at different time points. The analysis showed that the majority of bacteria matched the genus *Salinibacter*, and the minority members of the microbial population were of the genera *Anaeromyxobacter*,

S. Almeida-Dalmet (✉)
Department of Environmental Science and Policy, George Mason University, Manassas, VA, USA

B. K. Baxter
Great Salt Lake Institute, Westminster College, Salt Lake City, UT, USA

B. K. Baxter, J. K. Butler (eds.), *Great Salt Lake Biology*,
https://doi.org/10.1007/978-3-030-40352-2_5

Perexilibacter, *Halomonas*, *Psychroflexus*, *Schlesneria*, *Pseudomonas*, *Roseovarius*, *Haliscomenobacter*, and *Vulgatibacter*. Here, we discuss methods for microbial diversity studies in hypersaline aquatic systems and review the work on the microbial diversity of the north arm. We give an overview of the predominant halophilic archaea, but we present a broader picture by including new data on the underrepresented bacterial component of this fascinating community that manages a lifestyle at salt saturation.

Keywords Halophiles · Great Salt Lake · Microbial diversity · Hypersaline

5.1 Introduction

5.1.1 The Setting

Great Salt Lake (GSL) receives water from precipitation and inflow of rivers, but water leaves this terminal lake only through evaporation, leaving behind salts and other minerals which tell the geochemical history of the basin (Jones et al. 2009). The construction of a solid railroad causeway around 1960 bisected the lake and isolated the north arm, restricting exchange and creating an artificial salinity gradient (Madison 1970; Cannon and Canon 2002; Baxter et al. 2005). Within 7 years, the north arm approached salt saturation (Greer 1971), currently fluctuating with temperature effects on solubility, between 24 and 34% sodium chloride (Almeida-Dalmet et al. 2015; Kemp et al. 2018). The brine in this part of the lake is colored pink by the carotenoid pigments of the resident microorganisms (Fig. 5.1) (Jones and Baxter 2017). Around 90% of the freshwater input enters the system through the

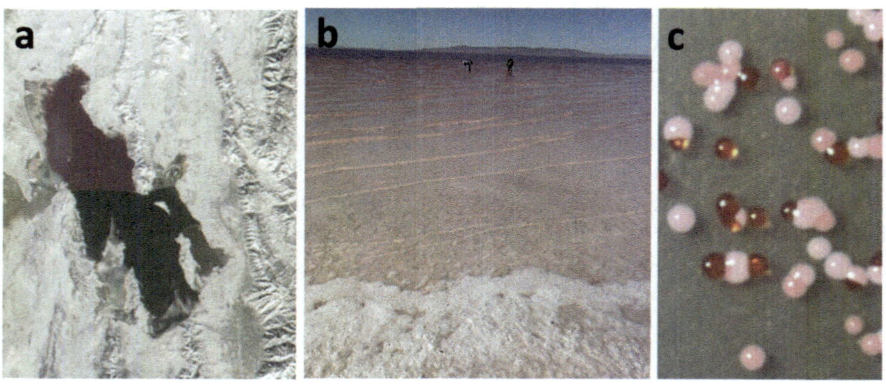

Fig. 5.1 The microbiota of the Great Salt Lake hypersaline north arm colors the water of the lake. (**a**) This pigmentation of the north arm, above the railroad causeway, can be seen from the International Space Station, image credit: NASA, 2001. (**b**) Pink waves at the shoreline near Rozel Point. (**c**) Carotenoid-containing microorganisms are grown on solid salt media after inoculating with the rosy brine

south arm, which maintains a lower salinity in comparison (Naftz et al. 2011). This lake is also situated at an elevation of more than 1200 m above sea level, which increases the exposure to ultraviolet (UV) light (Baxter et al. 2007; Jones and Baxter 2017). The temperature of the lake waters in winter can drop below freezing and in summer rise above 26 °C (Almeida-Dalmet et al. 2015). At this site, robust microbial communities persist, tolerating the UV irradiation and dramatic shifts in temperature conditions, in addition to the osmotic challenges of salt-saturated brine.

5.1.2 Microbial Communities

The microbiology of Great Salt Lake has been studied since the late nineteenth century (Baxter 2018; Baxter and Zalar 2019), but systematic studies looking at microbial communities have occurred only in recent years (e.g., Weimer et al. 2009; Parnell et al. 2011; Almeida-Dalmet 2011). These data present a complex picture of life at the microbial level in this lake, even in the saltiest parts, including members of all three domains: Bacteria, Archaea, and Eukaryota (Baxter and Zalar 2019). The hypersaline north arm is rich in archaea but has minority representation by bacteria (Fig. 5.2a, b), and even eukaryotes such as algae and fungi. There are also viruses that infect the halophilic bacteria and archaea, likely serving to control populations of particular species, mimicking predator–prey relationships (Baxter et al. 2011) (Fig. 5.2c).

To live in concentrated brine, "halophiles" must balance osmotically such that their cells do not shrivel up due to water loss. This is accomplished in part by the intracellular accumulation of osmotica, which balance against the salt on the outside

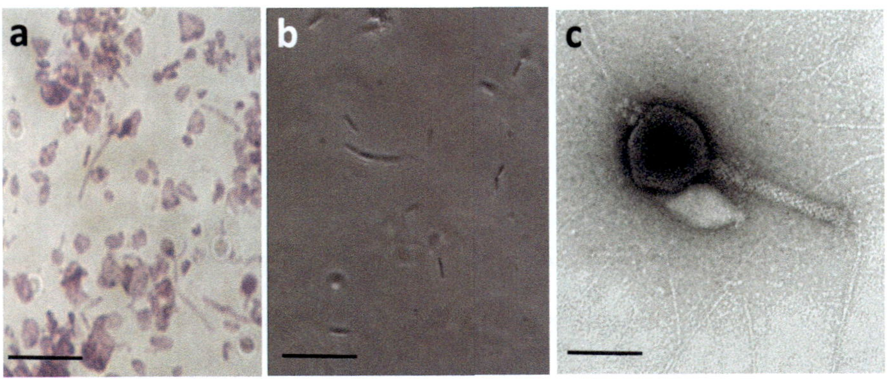

Fig. 5.2 Microscopy reveals the microbial diversity of the Great Salt Lake north arm. (**a**) Bacteria and archaea of various sizes and morphotypes compose the majority of the microbial community; scale bar is 4 μm. (**b**) A *Salinibacter* cultivar is characterized by long curved rods; scale bar is 4 μm. (**c**) Haloviruses shown are a head-tail morphotype and a fusiform, typical of hypersaline haloarchaeal viruses (Baxter et al. 2011); scale bar is 100 nm

of the cell membrane (Brown 1976). Halophilic archaea accumulate potassium ions and organic compatible solutes as their osmotica (Larsen 1967; Galinski and Trüper 1982; Galinski 1993, 1995; Oren 1999), which explains their success in salty environments like GSL. These extremophiles also have modifications in their proteins that help them function at high salt (Litchfield 1998). Eukaryotes such as *Dunaliella* species accumulate glycerol (Ben-Amotz et al. 1982). For bacteria to survive in such conditions, they also require osmophily strategies. Some bacteria, such as the *Salinibacter* genus, share some genes with haloarchaea and thus manage osmotic regulation in similar ways (Oren 2013). They also have some unusual sulfonolipids that are normally not found in bacteria. There are some bacteria, such as extremely halotolerant *Bacillus*, which can tolerate up to 25% salinity (Garabito et al. 1998). Some bacteria undergo dormancy (Lennon and Jones 2011) which protect them from extinction under severe environmental conditions, and this is similar to the ability of halophilic microorganisms to survive in salt crystals over geologic time (e.g., Vreeland et al. 2000).

The goal in much of our work is to determine the diversity of archaea and bacteria in the hypersaline brine, to identify and characterize new isolates, and to determine how these microorganisms adapt to extreme salinity. Two approaches were employed to this end in the work that we present below (Fig. 5.3): (1) a culture-dependent (cultivation) approach, which included isolating bacteria by culturing in the lab and studying their biochemical characteristics, and (2) a culture-independent (molecular) approach, which used cloning of environmental DNA and subsequent sequencing of the 16S rRNA gene of archaea and bacteria for identification. Also employed was a strategy involving next-generation sequencing (NGS) with semiconductor sequencing technology (Liu et al. 2012), a high-throughput sensitive methodology, which revealed a hidden diversity of the minority population of bacteria that were present in a robust representation of archaea.

Here, we describe the methods employed in a variety of studies. We will review our prior data as well as that of other researchers who have explored the hypersaline part of GSL. We have focused our north arm studies in the last decade on archaea,

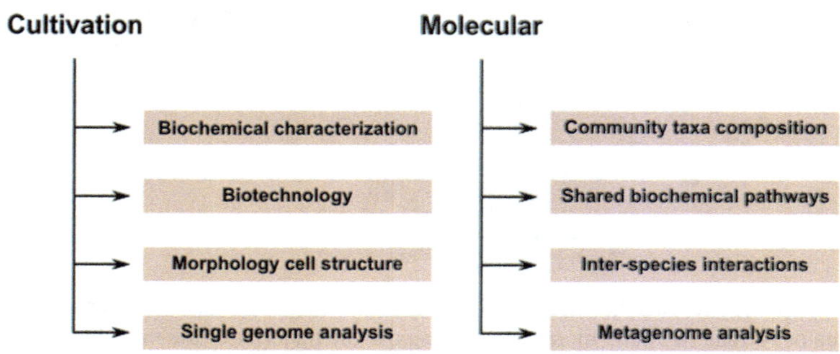

Fig. 5.3 A combination of cultivation and molecular (cultivation-independent) approaches are necessary to achieve a complete picture of the microbial diversity of any site. The information and applications that may be derived from cultivation are distinct from that of molecular studies

but herein we also present information on eukaryotes and data about the bacterial component of these assemblages to build a broader understanding of the entire community and its metabolic potential.

5.2 Microbiological Methods of Study in Hypersaline Systems

Building an understanding of the microbial consortia, the foundational level, of any ecosystem requires a combination of approaches (Fig. 5.3). Exploring the microbiota with cultivation, biochemical characterization, and microscopy is critical with the caveat that one loses information about those species that could not be isolated in the laboratory. However, the isolation of a species allows one to understand all of the features, cell structures, and genetic capacities of this isolate, including potential applications in biotechnology. Using molecular techniques to sequence the DNA present in a sample captures the whole community. For example, a metagenome study assessing all of the genes present in the location allows one to understand the genetic capacities of the entire consortia, including shared metabolisms. We describe below the application of these types of methodologies to hypersaline systems such as the GSL north arm.

5.2.1 Culture-Dependent (Cultivation) Approaches

Microbial identification by cultivation is done by the isolation and characterization of microorganisms from the natural environment. Such studies show that when both cultivation-independent and cultivation-dependent approaches were used to study diversity in a solar saltern, the diversity obtained by using the cultivation method was either the same (Burns et al. 2004) or sometimes even higher (Rodríguez-Valera et al. 1999) than the diversity obtained by using the molecular approach. In a direct approach comparing cultivars to those species not cultivated, 16S rRNA gene length heterogeneity fingerprinting was done on DNA prepared from washing plates which had colonies "too numerous to count," and on DNA from an uncultivated whole community of the same solar saltern (Milstead 2002). The result had very little overlap in the operational taxonomic units (OTU) observed, suggesting both approaches were required to answer the questions about community members.

These types of studies indicate that the cultivation approach is still important to uncover the diversity, and if we use the 16S rRNA gene sequencing approach alone, a significant fraction of the diversity will be overlooked (Donachie et al. 2007). New molecular techniques can be useful for phylogenetic identification of the microorganisms, but it is still necessary to grow the microorganisms to confirm their physiology (Fig. 5.3). Cultivation is also essential to isolate new biotechnologically important bacteria. Moreover, growing the bacteria in the laboratory is relatively the easiest and most inexpensive approach to the study of microorganisms and their relationship to environment. Recently, bacteria were isolated from saline and alkaline sodic soils from lakes in Mexico (Delgado-García et al. 2018). Principal component analysis showed strong correlation of isolated bacteria with their environment. For example, bacteria isolated from Cuatro Cienegas were associated with calcium and magnesium ions, and bacteria isolated from Sayula and San Marcos lakes were associated with sodium and bicarbonate ions (Delgado-García et al. 2018). Similar results were reported regarding the culturable diversity of unexplored meromictic and hypersaline Transylvanian lakes in Romania (Baricz et al. 2015). When the authors studied 191 isolates, they found that different genera were predominant in different lakes, despite the lakes being similar in location (central Romania) and habitat, due to ionic composition differences. Cultivation work was critical to these conclusions, and also allowed them to address the biogeography of similar strains, contemplating birds as mechanical carriers of halophilic microbial species.

Using culturing methods previously described (Almeida-Dalmet 2011, 2015), we isolated and identified dozens of microorganisms from GSL. The majority of our isolates belonged to the domain archaea; however, we have isolated a few halophilic bacteria. Interestingly, we have also collected data consistent with dispersal related to avian mechanical carriers of GSL microorganisms, which relied on culturing experiments from feathers of migrating birds (Kemp et al. 2018). The results of our cultivation work will be summarized below.

5.2.2 Culture-Independent (Molecular) Approaches

In most natural environments, microorganisms grow as a multispecies community in the form of aggregates (Amann et al. 1995). This theory suggests that traditional methods of cultivation result in the disruption of these community interactions and structure, and thus, many organisms cannot grow and form a colony on a plate. Also, total microscopic counts may be several times higher than the plate counts owing to the fact that a large number of microorganisms in the environment are unculturable on typically rich laboratory media (Amann et al. 1995; Hugenholtz et al. 1998) or the cells in the microscopic observation may be dead (Jennison 1936). Even when the microorganisms can be cultivated, their activities and physiologies in the laboratory may not necessarily reflect those in the environment when associated with an array of other microorganisms (Amann et al. 1995). Indeed, increasingly we understand that shared metabolites and interspecies signaling are hallmarks of a microbial community and not an exception. Utilizing culture-independent approaches in combination with culturing methods allows the researcher to collect phylogeny data and also make inferences regarding the lab versus the natural habitat.

Culture-independent methods such as DNA fingerprinting and DNA sequencing are useful to understand the genetic diversity of the majority of microorganisms, their population structure, and role in the ecosystem (Fig. 5.3) (Riesenfeld et al. 2004). PCR fingerprinting methods such as length heterogeneity PCR are used to compare relative abundances of bacteria present in different environments (Litchfield and Gillevet 2002; Litchfield et al. 2006). Each peak represents a common OTU, which can be at the level of family, genus, or species. Therefore, these methods are more informative when they are used in conjunction with DNA sequencing which allows individual members of the community to be identified.

A large number of studies on hypersaline environments have been done using traditional Sanger sequencing (Cytryn et al. 2000; Mesbah et al. 2007). With the advent of NGS, sequencing technology took an unprecedented leap. NGS is less expensive, faster, and produces tremendous number of reads. Since it does not require cloning and PCR, it avoids all biases associated with them (Schuster 2007). There are several platforms on which NGS is used, such as pyrosequencing, which detects release of pyrophosphate; sequencing by ligation; and semiconductor sequencing, which detects release of hydrogen ions as each of the sequencing product is constructed (Liu et al. 2012). One of the shortfalls of this technology is that it produces very short reads of DNA sequence; however, other new systems, such as Roche 454 that produces a read length 700 bp, deliver sequence data at the level of traditional Sanger sequencing (typically 1 kb) (Liu et al. 2012).

In hypersaline environments, NGS (454 pyrosequencing) was employed to study the metagenomics of hypersaline microbial mats of Shark Bay, Australia, a sequestered inlet where salinity is higher than the surrounding marine environment (Ruvindy et al. 2016). The data revealed that communities of the phyla Proteobacteria, Cyanobacteria, and Bacteroidetes dominated the shark bay mats and influenced each other with respect to shared capabilities. Proteobacteria deposited carbonate by sulfate reduction, while the photosynthetic Cyanobacteria provided nutrients and also produced extracellular polymeric substances (EPS) that help in deposition of carbonates and aggregation of sedimentary material. In addition, the Bacteroidetes produced alkaline conditions for deposition of calcium carbonate and they also degrade EPS mats and release nutrients. NGS has also been used to build a virtual microbial community structure present in soda lake brines (Vavourakis et al. 2016). The team constructed draft genomes of several novel bacteria and archaea from this environment to study their physiology, including shared genetic/metabolic capacities.

Metagenome studies endeavor to sequence all of the DNA in an environmental sample, not just the ribosomal RNA genes, which may help understand the system beyond the taxa present. These approaches can build a picture of metabolic capacities of microbial communities, a holistic view where shared metabolites may be at play. Hypersaline systems have been explored with this tool; a meta-analysis of metagenomes of high salt aqueous environments yielded important conclusions about dominant community members and their function (Ventosa et al. 2015). A north arm metagenome project was funded by the US Department of Energy, but results have not been published to date (Joint Genome Institute 2019).

5.2.3 Temporal Versus Spatial Approaches

The study of both temporal and spatial diversity is crucial to understanding the relationship of ecosystem and organisms (Bardgett et al. 2005). The temporal approach studies microbial diversity over timescales as seasonal changes can shift the community composition. On the other hand, the spatial approach checks the diversity at different depths or at different sites as nutrient availability changes with respect to inputs at different locations. In hypersaline waters, where haloclines can temporarily form due to density differences and weather conditions (Naftz et al. 2014), both temporal and spatial methods may be advised. Therefore, a one-time sampling or a one-layer survey of microbial communities might underestimate microbial diversity or miss important members of the microbial consortium. In addition, the study of the temporal and spatial patterns together is paramount to understand the processes that drive biogeography and the compartmentalization of biochemical capacities (Jones et al. 2012). For example, when the spatiotemporal microbial diversity in a tropical hypersaline lake was studied, it was found that photosynthetic bacteria were dominant in the surface layers of the lake whereas chemo-heterotrophs and anaerobes were dominant in the deeper layers of water (Hugoni et al. 2018). Similarly, the distribution of archaea in a hypersaline lake in Romania was affected by important factors such as salinity and oxygen, which varied over time (Andrei et al. 2017)

Spatial studies of hypersaline lakes may include studying various compartments such as microbial mats, the water column, and the sediment (Mesbah et al. 2007). Or it may focus on a single region, such as soils and sediments, but samples are taken along a transect (Hollister et al. 2010). A global survey of microbial communities found that sediment consortia are more phylogenetically diverse than other environments and salinity is a major driver of diversity (Lozupone and Knight 2007), suggesting much can be assessed by spatial approaches. Moreover, studies on sediments of hypersaline lakes lend insight on processes such as sulfate reduction and methane production (Henneke et al. 1997) and can demonstrate unknown facultative capabilities such as primarily aerobic haloarchaea which can produce energy under anaerobic conditions (Fernandez et al. 2016).

At GSL, spatial studies have sometimes focused on the anthropomorphic impact of the causeway construction (Fig. 5.1a) (Madison 1970), such as the effect of this salinity gradient on the microbial community of the north and south arm microbialites, carbonate rocks deposited by the actions of photosynthetic microorganisms (Lindsay et al. 2017). These structures in the hypersaline north arm are likely remnants from when the lake was a single body, prior to causeway construction, and they do not appear to be actively precipitating like the similar structures in the less saline south arm. Perhaps the north arm microorganisms now associated are inhabitants of the structure, but they were not the "architects" or "builders." Transect studies at GSL have also been undertaken. When the vertical column of a south arm site at GSL was studied for microbial community composition, little overlap of communities was observed as dissolved oxygen level decreased in the hypolimnion, and the halocline of the deep brine layer in that area formed discreet compartments (Meuser et al. 2013). One may assess spatial diversity also by stitching together various studies of the same region. For example, GSL sediments have been studied because they are rich in methanogens (Paterek and Smith 1985; Brandt et al. 2001; Baxter et al. 2005), and sulfate-reducing bacteria (Boyd et al. 2017), which relates to noted high methyl-mercury concentrations (Naftz et al. 2008, 2011, 2014). These metabolic activities are critical components of the sediment consortia, and spatial views support the notion that as salinity increases in the north arm, these activities are slow and less robust (Ward and Brock 1978)

Temporal studies are typically more challenging than spatial studies that look at various compartments as the field work component requires multiple trips to the work site, with longitudinal timelines that might stretch into years. But given impacts of seasonal parameters, these projects answer very important questions that snapshot studies cannot. In the north arm of GSL, Almeida-Dalmet and coworkers demonstrated stable community members versus those that were more transient (2015). Dissecting the data from this study inspired further work in the lab to look at genetic factors involved in the more stable microorganisms that could survive the stress of changing temperature and salinity (Almeida-Dalmet et al. 2018).

5.3 Microbial Diversity of the Great Salt Lake North Arm

GSL microbiological studies in the nineteenth and early twentieth centuries resulted in a number of isolations, including "pink bacteria" which were likely archaea, and reported hundreds of microorganisms per milliliter of brine (Baxter 2018). These rose-colored microorganisms were first reported by graduate students studying at the University of Utah (Daniels 1917; Frederick 1924; Kirkpatrick 1934). But this work was done before the installation of the causeway (Fig. 5.1a), and though the lake had periods of low elevation, it was never as concentrated in salts as the north arm is today.

Post was the first to really focus on cultivation in the north arm, where he isolated and described a number of halophilic bacteria and archaea (Post 1975, 1977, 1981). Later molecular studies suggest that the microbial communities in the north arm are composed predominantly of halophilic archaea with a minority contingent of bacteria (Weimer et al. 2009; Parnell et al. 2011; Tazi et al. 2014; Almeida-Dalmet et al. 2015). While we think of the north arm as rich in archaea, and this is likely true, one caveat of these molecular studies is that they are primarily based on 16S rRNA gene data and thus ignored eukaryotic community members.

Metabolic activities by microorganisms, in general, occur more slowly as the salinity increases (e.g., Ward and Brock 1978). At moderate salinities, nitrogen cycling in GSL south arm microcosms was stimulated by glutamate, but glutamate was degraded more slowly at higher salinities (Post and Stube 1988), especially in north arm microcosms (Stube et al. 1976). Conversion of glucose, glycerol, and acetate to CO_2 was much slower in the north arm than in the south arm (Fendrich and Schink 1988). Reduced metabolism at high salinity may suggest a restriction on microbial life, or alternatively, it may suggest that life must be very diverse and interconnected. Winogradsky columns from the north arm routinely demonstrate biofilm and microbial mat formation indicative of cell–cell interactions (Baxter 2018)

5.3.1 Eukaryotes

Eukaryotic microorganisms such as phytoplankton, fungi, and protists have been largely ignored in GSL until very recently. The south arm algae have received some attention as their population is critical to the health of the *Artemia* (brine shrimp), the cysts of which are harvested by industrial operations (Belovsky et al. 2011). The eukaryotic phytoplankton constituents of the north arm community, autotrophs who power the ecosystem with photosynthesis, are certainly less diverse than the less saline south arm, but a couple of species have been detected. *Dunaliella salina* has been observed in the north arm at saturated salinities above 300 g/L (Post 1977, 1980; Felix and Rushforth 1979). This carotenoid-rich alga may survive salt saturation and desiccation by becoming encysted; the round cyst-like cells (aplanospores) of *D. salina* increased in representation in north arm microcosms at lower temperatures and higher salinity (Post 1977). More recently, a molecular analysis indicated the presence of another chlorophyte, *Tetracystis*, in the north arm, which was associated with vestige microbialite carbonate structures (Lindsay et al. 2017).

A single fungus was described in 1977, isolated from a piece of wood soaking in the north arm water, and identified as a *Cladosporium* species (Cronin and Post 1977). However, recent investigations of the north arm water column and oolitic sand showed immense fungal diversity (Baxter and Zalar 2019). The authors reported the isolation of 32 fungal strains belonging to the phyla *Ascomycota* (genera *Acremonium, Alternaria, Aspergillus, Cladosporium, Coniochaeta, Neocamarosporium, Parengydontium, Penicillium, Stemphylium*) and *Basidiomycota* (genus *Wallemia*). Among the isolates, strains belonging to the genus *Cladosporium* (*Davidiellaceae, Capnodiales*) are the most numerous and several of them have not yet been identified to the species level, but presumably also represent some undescribed species.

It was noted as early as 1917 that other single-celled eukaryotes (that were not fungi or phototrophs), the chemoheterotrophic protists, were also present in GSL (Vorhies 1917). As salinity changes, some protists may enter an encysted phase, which allows them to tolerate high salinities, and then reemerge when the salinity decreases (Evans 1958; Post et al. 1983). Two ciliates were isolated from north arm samples, a *Uronema* species (Stube et al. 1976), and *Chilophyra* (*Prorodon*) *utahensis* (Pack 1919; Evans 1958; Flowers and Evans 1966). North arm hypersaline flagellates and amoeba were observed and described later in an aquarium experiment utilizing north arm sediment and water (Post 1977). To date, no molecular analyses have been undertaken to identify the protozoa community of GSL.

5.3.2 Archaea

Hypersaline environments such as GSL harbor predominantly extreme halophilic archaea (Fig. 5.1c). In order to survive harsh conditions such as high salt, low water activity, starvation, and radiation, these microorganisms have evolved unique strategies (reviewed in Stan-Lotter and Fendrihan 2015). *Haloquadratum*, a haloarchaeon, most commonly found in GSL north arm (Almeida-Dalmet et al. 2015), is square-shaped and extremely flat which increases its surface area for absorption of nutrients (Ventosa et al. 2015). Also, it produces mucoid substance called halomucin which forms a water cloud around the cell and protects it from low water activity and also from the attacks of phages. Another haloarcheaon, *Halobacterium*, transforms from rod shape to spherical shape to survive in low water activity. *Haloferax volcanii*, and perhaps other haloarchaea, uses polyploidy to obtain phosphate from its own DNA during the period of starvation (Zerulla et al. 2014). Due to the ability of GSL halophilic archaea to be sequestered in fluid inclusions in salt crystals over time, they may also be transported to other salty bodies of water on the feathers of birds (Kemp et al. 2018).

GSL is an unexplored reservoir of many uncultured and undescribed species; the majority of the 16S rRNA gene sequences in the GSL north arm were from uncultured taxa (Almeida-Dalmet et al. 2015). In general, the higher the salinity, the more archaeal genera are present relative to bacterial genera. These stable hypersaline north arm microorganisms also have a lower phylogenetic diversity relative to communities in the south arm (Parnell et al. 2009, 2010, 2011; Almeida-Dalmet et al. 2015).

The spatial microbial diversity of GSL was studied by Tazi and coworkers using 100 ml surface water samples from 20 different sites across a salinity gradient from the north arm at Rozel point and the south arm just off Antelope Island (Tazi et al. 2014). An analysis of their clone libraries showed that the majority of archaeal 16S rRNA gene sequences matched the following species: *Haloquadratum walsbyi* (17 clones), *Uncultured archaeon* (57 clones), and *Halonotius pteroides str.* (19 clones). Most of the known genera did not match with the clone sequences, and they form separate clusters in their phylogenetic analysis. Other researchers collected samples from similar sites, but they also included less saline Farmington Bay (Weimer et al. 2009). This study employed a phylogenetic array approach containing a total of 8741 probes for bacteria and archaea for hybridization. Haloarchaeal groups were only found at the Rozel point sampling site in the north arm. The phylochip method was also used to determine metabolic diversity and revealed genes for metal and nitrogen metabolism, in the highest numbers in microorganisms of the Rozel point.

Utilizing a temporal approach, cloning 16SrRNA gene sequences from the water column over seasons, we studied the microbial diversity at Rozel Point at different time intervals (Almeida-Dalmet et al. 2015) (Fig. 5.4). Similar to Tazi et al. (2014), the phylogenetic analysis of archaea showed that all known genera were clustered together. 16SrRNA genes matching uncultured archaea were dominant throughout

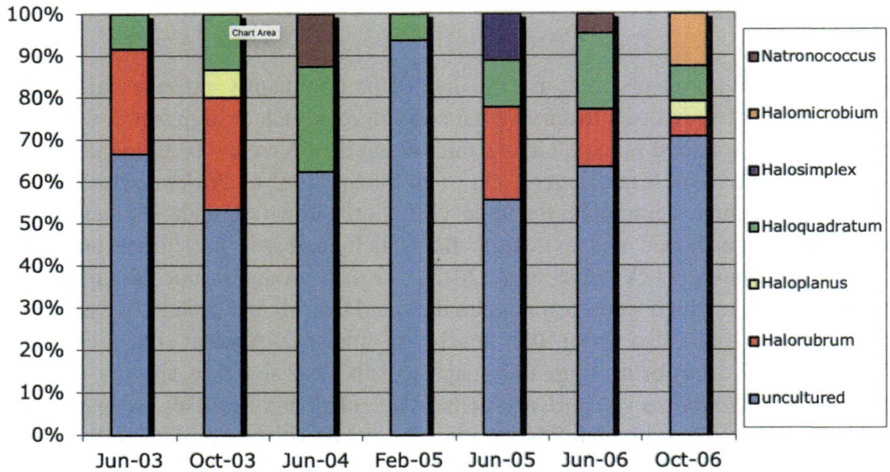

Fig. 5.4 Distribution of 16S rRNA gene phylotypes of cloned DNA sequences collected from brine at different time intervals at the north arm of Great Salt Lake [adapted from Almeida-Dalmet et al. (2015)]

the year and formed separate clades showing that they were very distantly related to known genera. Out of total 600 sequences of archaea, the majority (237) were related to *Haloquadratum* genus group and were present in all samples. The minor communities changed in each sampling period. Interestingly, there was no correlation observed in sampling dates and phylogenetic diversity. Canonical correspondence analysis of five environmental variables (temperature, salinity, pH, dissolved oxygen, and lake elevation) and relative abundance of seven samples indicated that these factors do not have strong influence on the community.

The stability of the GSL north arm microbiota shown by molecular means in the temporal study (Almeida-Dalmet et al. 2015) suggests that some members of these assemblages are able to respond to the changes in salinity and temperature that accompany the seasons in the high-altitude desert where they reside, which involves changes in gene regulation (Almeida-Dalmet et al. 2018).

We cultured microorganisms from the same north arm location, Rozel Point, and identified the isolates with thin-layer chromatography and 16S rRNA gene sequencing. The overwhelming majority (46 of 50 isolates) were archaea at all seasonal sampling times, including the genera *Haloarcula*, *Halobacterium*, *Halococcus*, and *Halorubrum* (Fig. 5.5).

5.3.3 Bacteria

In the cultivation experiments described above, the minority of Rozel Point cultivars were bacteria, including *Halomonas* and *Salinibacter* genera (Fig. 5.5). Studies on

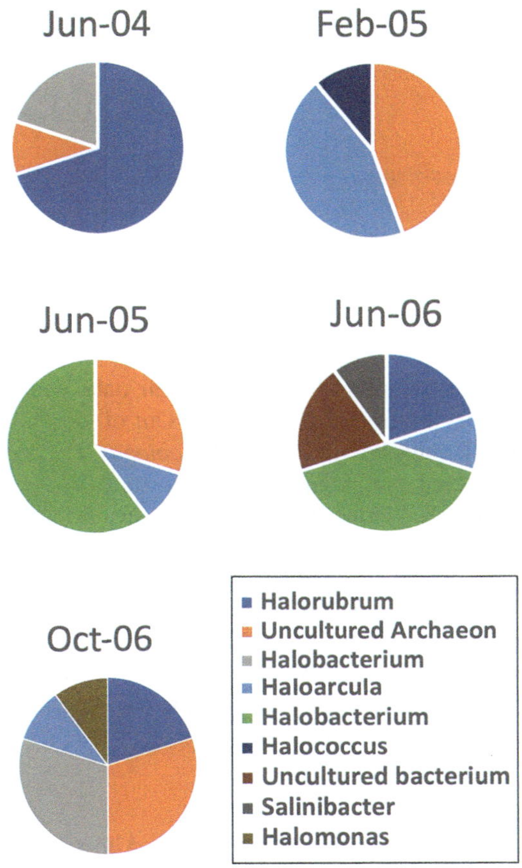

Fig. 5.5 Great Salt Lake cultivated isolates from a temporal study of the north arm. The quantitative cultivatable diversity of archaea and bacterial genera is shown for different time periods of seasonal sampling

isolation of bacteria from the GSL north arm are sparse, which suggests low bacterial diversity in this salt-saturated locale. Several bacteria have been isolated and identified previously from GSL, which show very little similarity to the known members of the archaeal *Halobacteriaceae* family (Wainø et al. 2000). These include a halophilic methanogen from the sediment samples of the south arm (Paterek and Smith 1985), several sulfate-reducing halophilic bacteria (Brandt et al. 2001; Kjeldsen et al. 2007), the obligatory anaerobe *Haloanaerobium praevalens* (Zeikus et al. 1983), *Halovibrio variabilis*, and *Pseudomonas halophila* (Fendrich 1988). There are a few north arm bacterial cultivars stored in culture banks, shown in Table 5.1 [adapted from Baxter and Zalar (2019)].

More comprehensive molecular studies suggest a bit more bacterial diversity. Tazi and coworkers found that many novel bacterial sequences, not found in other hypersaline environments, were present in GSL when they examined the samples that included south arm, lower salinity areas (Tazi et al. 2014). For bacteria, the Shannon–Weaver diversity index was 3.97 and the library coverage was 9% which indicates an abundant bacterial community, but the authors did not separate the

Table 5.1 Great Salt Lake bacterial isolates from north arm brine, available from culture banks. DSM is the Leibniz Institute DSMZ (Germany), and VKM is the All-Russian Collection of Microorganisms (Russia)

Strain	Cell Culture Bank Accession Number	References
Desulfohalobium utahense	DSM 17720 VKM B-2384	Jakobsen et al. (2006)
Desulfosalsimonas propionicica	DSM 17721 VKM B-2385	Kjeldsen et al. (2010)
Halomonas utahensis	DSM 3051	Fendrich (1988), Sorokin and Tindall (2006)

different sampling locations in their data, and thus it is hard to determine which bacteria are from the north arm. Out of their 212 bacterial clones, the majority of them belonged to the genera *Halomonas*, *Pseudomonas*, *Salicola*, and unidentified/uncultured strains. In our temporal study of the north arm, *Salinibacter* dominated the bacterial clone library; 98% sequences of the library belonged to this bacterial genus (Almeida-Dalmet 2011; Almeida-Dalmet et al. 2015). The remaining bacteria were in the *Halomonas* genus or previously unidentified/uncultured bacteria (Almeida-Dalmet 2011). Weimer and coworkers showed the presence of Gamma-proteobacteria and some Firmicutes at the three points of the north arm using phylochip method (Weimer et al. 2009). With sampling along a salinity gradient, this analysis showed that with increasing salinity, the number of bacterial genera decreased from 641 to 100.

5.3.3.1 Cultivation Methodology Allows Subsequent Characterization of Bacteria

Culturing GSL bacteria in the laboratory has allowed for further biochemical characterization and understanding of the lifestyles of particular species (Fig. 5.3). In our own cultivation experiments, we isolated only four bacterial strains among the larger numbers of archaea we cultivated. These bacteria most closely matched the families of *Halomonads* and *Bacteroides*, but two of the strains had a low homology (81–83%) with the 16S rRNA gene matches (NA10-65 and NA9-35) (Table 5.2). Since one *Halomonads* strain had been studied previously (Fendrich 1988; Sorokin and Tindall 2006), we chose to further analyze the three *Bacteroides* isolates (NA9-35, NA9-38, and NA9-83) to determine their morphological and physiological characteristics. Prior to this work, nothing was known about these bacterial members of the north arm community.

Once in isolation, the three *Bacteroides* cultures were studied for their optimal salt concentration. The range of tolerance differed for each isolate: 10–20% NaCl for both NA9-35 and NA9-83, and 15–20% NaCl for NA9-38. Although all three grew in 15–20% salt concentrations (Fig. 5.6), they could not be grown to a high-density culture in the lab, not even to a logarithmic growth phase. This suggests that in

Table 5.2 Cultured isolates from the Great Salt Lake north arm and 16S rRNA gene analysis for closest relatives

Isolated GSL Strain	Closest Matched Species (Genbank Accession Number)	% Similarity
NA10-65	Halomonas spp. (EU870508.1)	81
NA9-35	Salinibacter ruber (AF323500.1)	83
NA9-38	Uncultured Bacteroides (AM981372.1)	96
NA9-83	Uncultured Bacteroides (FN393447.1)	99

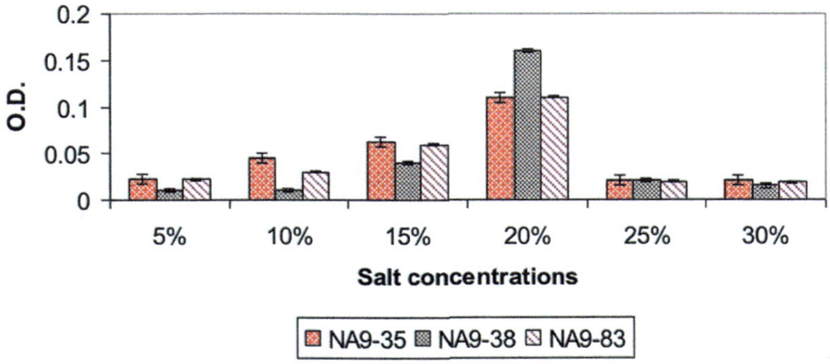

Fig. 5.6 Growth over a three-week period of Great Salt Lake north arm bacterial isolates at 37 °C at various salt concentrations in *Salinibacter* medium (Antón et al. 2002). O.D. represents the optical density of cultures at an absorbance of 600 nm. Error bars show ± standard error

natural north arm environment, where that salinity is above 25%, these bacteria likely remain dormant and in small numbers. We determined the optimal temperature for all of these bacterial cultures was 37 °C, but we saw distinctions for the toleration range. The temperatures tolerated for growth for both strains NA9-35 and NA9-83 were 37–46 °C. This was different for NA9-38, which tolerated temperatures in the 24–42 °C range.

When grown on solid media formulations, all three isolates shared morphological and biochemical characteristics. NA9-35, NA9-38, and NA9-83 formed orange colonies of 1–2 mm diameter that were opaque and slightly elevated with smooth consistency. All three strains were Gram negative and motile. In addition, all were negative for anaerobic growth in the presence of nitrate, reduction of nitrate to nitrite, formation of gas from nitrate, anaerobic growth in the presence of arginine, formation of indole, and hydrolysis of gelatin, starch, casein, and tween 80. The optimal $MgCl_2$ concentration for growth was 0.1 M, and optimum pH range was 8–9 for each isolate.

We did observe distinctions in nutritional requirements. Using BIOLOG (Hayward, CA) phenotyping plates, we studied the growth of these isolates on sole nitrogen and sole carbon sources. The bacterial strains NA9-35 and NA9-83 grew slowly in α-ketobutyric acid, maltose, and glyoxylic acid. In addition, NA9-35 used uridine as carbon source, whereas NA9-38 used alpha-D-lactose, sucrose, maltotriose, and acetoacetic acid as carbon source. The NA9-38 used more of the amino acids than

Table 5.3 Fatty acid profiles of Great Salt Lake north arm bacteria in % of whole-cell hydrolysates using MIDI

	Isolated GSL Strain			
Major components of fatty acids	NA9-35 (%)	NA9-38 (%)	NA9-83 (%)	NA10-65 (%)
C15:0 iso	29.98%	17.38%	26.24%	0
C16:1 w7c/16:1 w6c	40.03%	16.95%	32.49%	17.33%
C18:1 w7c	11.53%	20.26%	21.03%	28.17%
C18:0	0	12%	0	0
C16:0	0	0	0	29.87%

NA9-35 and NA9-83. The bacterial control, *S. ruber*, did not use any of the simple sugars as predicted by the phenotype of this species.

Both antibiotic sensitivity and fatty acid analysis can aid in the identification of bacteria. All three of our *Bacteroides* isolates, NA9-35, NA9-38, and NA9-83, were sensitive to novobiocin (25 μg), erythromycin (15 μg), ampicillin (10 μg), and chloramphenicol (30 μg). They were resistant to bacitracin (10 IU), anisomycin (25 μg), penicillin (2 U), rifampin (5 μg), and neomycin (30 μg). Regarding the fatty acid analysis, we observed that the whole cell fatty acid profiles of three bacterial strains contained the iso-branched C15:0, a characteristic component of the fatty acids of *Salinibacter* (Corcelli et al. 2004), and also included C16:1 w7c/16:1 w6c, and C18:1 w7c as major fatty acids (Table 5.3). For comparison, the C15:0 component was absent in our *Halomonads* isolate, NA10-65.

DNA sequencing of the 16S rRNA gene for the three isolates was used to determine the significant intraspecies differences. BLAST (Altschul et al. 1997) queries and the phylogenetic tree of bacterial isolates showed identical results. Known strains of *Salinibacter* and NA9-35 and NA9-83 clustered together forming one clade, while NA9-38 formed another cluster with an uncultured halophilic bacterium with a bootstrap value of 98 (Fig. 5.7). At least two of our cultivated species from the GSL north arm are related to the *Salinibacter* genus. The NA9-35 and NA9-83 were closest to *S. ruber* isolated from Alicante and Santa Pola, Spain, and also to a *Salinibacter* 5Sm6 strain, isolated from Maras saltern of Peruvian Andes, with a bootstrap value 84. The other isolate NA9-38 was closest to a previously uncultured halophilic bacterium isolated from a Tunisian saltern with a bootstrap value 99.

5.3.3.2 Non-cultivation Methodology Reveals the True Bacterial Diversity

In our previous work, amplifying and cloning 16S rRNA genes from the GSL north arm (Almeida-Dalmet et al. 2015), we used the traditional Sanger sequencing method for analysis. We did not observe significant diversity in bacterial sequences; 98% of the clones of the bacterial library belonged to *Salinibacter*-related species. Therefore, to reveal true diversity of bacteria, we used NGS semiconductor

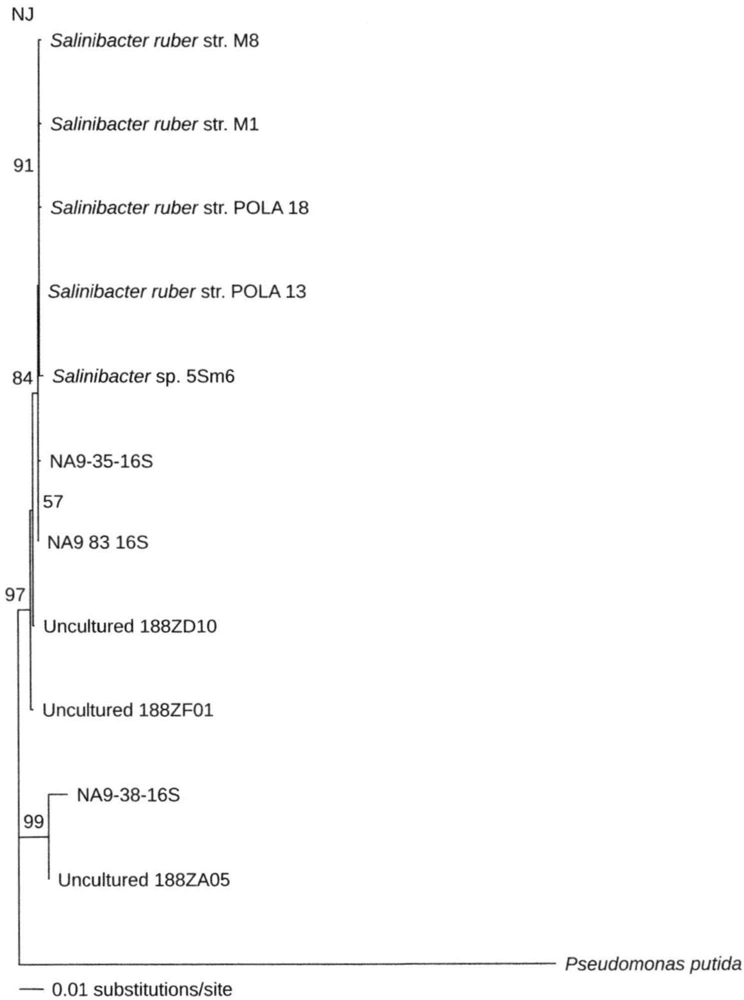

Fig. 5.7 Neighbor joining 16S rRNA gene phylogenetic tree of bacteria showing positions of Great Salt Lake cultivars, NA9-35, NA9-38, and NA9-83. Bootstrap values are at the node

sequencing (Ion Torrent) technology as it does not cause loss of genomic sequences, which is a bias of cloning and PCR-associated method (Metzker 2010).

We subjected few north arm samples to NGS technology which revealed the presence of *Schlesneria* of the family Planctomycetaceae and *Perexilibacter* in all five north arm samples. Although they are 8% of total bacterial population in our May 2004 sampling, their population reduced in subsequent samples. Other groups like *Halomonas* (~1%) and *Anaeromyxobacter* (<1%) were also present in all the samples. Some groups like *Ferrithrix* were very rare and occasionally were discovered, in only one or two samples. This study showed the presence of other bacteria

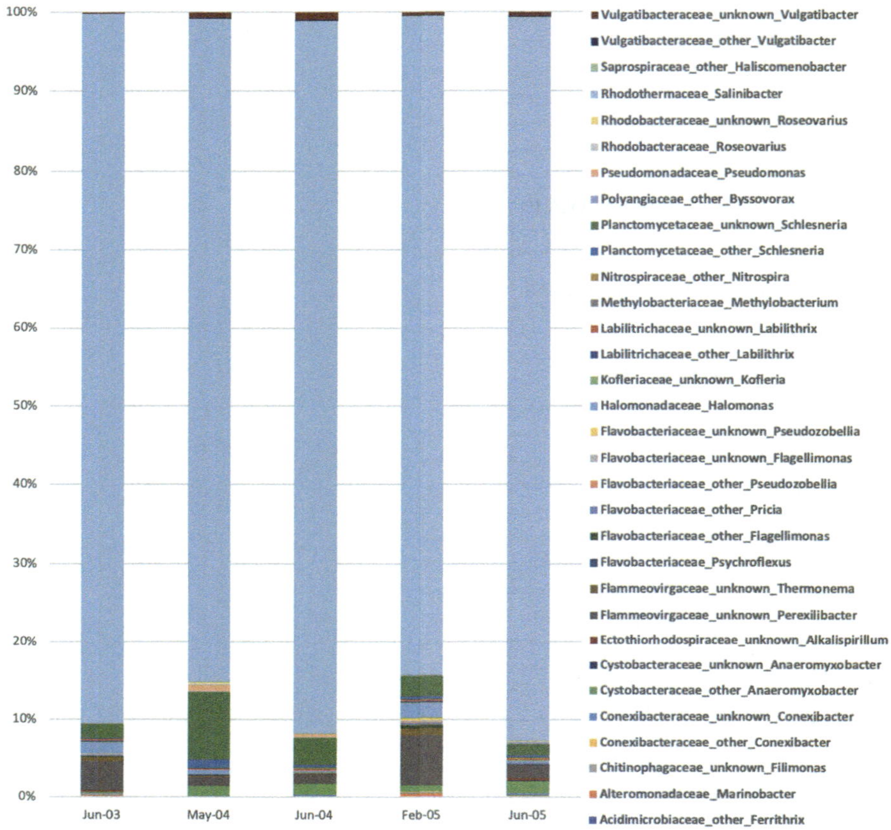

Fig. 5.8 Bacterial diversity of Great Salt Lake north arm brine obtained at different time intervals, using ion torrent next-generation multi-tag sequencing (Solano-Aguilar et al. 2018)

such as *Psychroflexus, Pseudomonas, Roseovarius, Haliscomenobacter*, and *Vulgatibacter* (Fig. 5.8).

5.4 Insights and Conclusions

5.4.1 Halophilic Archaea Dominate the Great Salt Lake North Arm

Through our studies (Almeida-Dalmet et al. 2015, 2018) and others (Parnell et al. 2009, 2010; 2011; Weimer et al. 2009; Tazi et al. 2014), it is clear that the salt-saturated north arm of GSL is dominated by halophilic archaea. We have explored the effect of environmental factors on archaeal population (Almeida-Dalmet et al. 2015) and the biogeography of these archaea, connected globally to other strains by avian mechanical carriers (Kemp et al. 2018); see below. It appears that salinity is the main driver that supports this microbial population and that gene regulation allows the stable members of the consortia to overcome the challenges of the fluctuating environment (Almeida-Dalmet et al. 2018). However, focusing only on the dominant population is problematic in understanding the consortia of this brine.

5.4.2 The Unexpected Reservoir of Bacteria in the Great Salt Lake North Arm, Especially *Salinibacter* Species

As reviewed above, members of the bacterial Domain do live in the GSL north arm waters. By molecular methods, we identified 80–90% of the bacterial population as related to the *Salinibacter* genus, and two of our four bacterial cultivars were related. Some of the other strains identified may not be living in their optimal conditions, and at salinity saturation they may be dormant until they find themselves near a spring with a localized reduction in salinity, but *Salinibacter* species seem to be thriving in the hypersaline north arm conditions of GSL.

Very few types of bacteria can grow at high salt concentrations, likely because they are outnumbered by halophilic archaea. *Salinibacter,* which belongs to *Cytophaga-Flavobacterium Bacteroides* group, is a notable exception. The members of this genus use KCl to provide osmotic balance and do not build high concentration of organic osmotic solute. The best characterized example is *Salinibacter ruber*, a red, aerobic, and extremely halophilic bacterium first isolated from crystallizer ponds in Alicante and Mallorca, Spain (Antón et al. 2002). Strains isolated from geographically different sites in Europe and from GSL (NA9-35 and NA9-83) showed more than 96% similarity (Fig. 5.7). Previous studies with genomic fingerprinting and fatty acid analyses of *Salinibacter* to determine the geographical differences did not produce any obvious trend (Peña et al. 2005). When a metabolomic study was undertaken to see the metabolic similarities among the *Salinibacter* isolates from different locations in the world, the strains were found to produce different metabolites which were responsible for their geographical differentiation (Antón et al. 2008). This suggests that GSL *Salinibacter* strains may be phylogenetically related to other strains but metabolically distinct.

5.4.3 Biogeography of Extremely Halophilic Bacteria

Studies on bacteria from other hypersaline environments around the world demonstrated the presence of more or less the same bacterial genera, which is a comment on the biogeography of extreme halophilic bacteria; how do the same species appear in salt-saturated locations? Bacteria found in hypersaline Lake Urmia of Iran were Gram-negative bacteria such as *Idiomarina*, *Saliocola*, *Halomonas, Pseudomonas*, and *Marinobacter* and Gram-positive bacteria such as *Bacillus* and *Halobacillus* (Vahed et al. 2011). Bacteria isolated from the Dead Sea were members of both Gram-positive bacteria such as *Arthrobacter* sp., *Kocuria erythromyxa*, and *Bacillus licheniformis* and Gram-negative bacteria such as *Salinivibrio costicola*, *Vibrio alginolyticus*, and *Chromohalobacter salexigens* (Jacob et al. 2017). These studies used culture-dependent methods to isolate and characterize the microorganisms. The metagenomics studies showed a completely different picture. The bacteria of the Dead Sea constituted 45% and archaea constituted 55% diversity. The bacteria belonged mostly to *Proteobacteria* and *Firmicutes* (Jacob et al. 2017). Moderately halophilic bacteria which include *Halomonas*, *Salinivibrio*, *Chromohalobacter* were also found in hypersaline environment. *Anaeromyxobacter* were isolated previously from the sands of hypersaline lakes of Kiritimati Island (Mohr et al. 2016). *Psychroflexus* found in hypersaline GSL possesses proteorhodopsins to produce energy from light. Abundance of this bacterium depends on high salinity. As the salinity increases, the proteorhodopsin activity increases (Feng et al. 2013). *Pseudomonas* has been previously found in hypersaline environments including GSL (Quesada et al. 1982; Fendrich 1988). *Roseovarius tolerans* has been isolated from Ekho Lake, East Antarctica, and a species of *Roseobacter pacificus* was isolated from deep sea sediment (Labrenz et al. 1999).

Since GSL is a terminal lake and it is not connected to other water sources, and given the similar species of bacteria that inhabit hypersaline locations around the world, it is possible that birds are mechanical carriers. Haloarchaea have been shown to survive on avian hosts (Yim et al. 2015; Brito-Echeverría et al. 2009), and this hypothesis was supported by a recently GSL biogeography study (Kemp et al. 2018). Given the longevity in desiccated salt crystals (e.g., Vreeland et al. 2000), halophilic bacteria certainly may also be transported from place to place by migratory birds.

5.4.4 Haloviruses as Predators in the North Arm System

Viruses (phages) are considered to be in predator–prey relationships with bacterial or archaeal species in many extreme environments such as hot springs or hypersaline lakes (Breitbart et al. 2004; Rodriguez-Valera et al. 2009; Rohwer et al. 2009; Clokie et al. 2011; Jończyk et al. 2011). The major composition of the microbial population in GSL, archaea and bacteria, are likely controlled by the endemic haloviruses (Post 1981; Baxter et al. 2011; Shen et al. 2012) which can shift the genera or species present based on specific virus–host interactions. The ratio of viruses to prokaryotes in the north arm brine is 100:1, making them significant as predators in this extreme ecosystem, which is limited in invertebrate grazers (Baxter et al. 2011). The diversity of GSL north arm viral morphologies include spherical, head-tail, fusiform (lemon-shaped), and filamentous structures (Fig. 5.2c). Viruses, though not classified as life per se, must be considered as significant community members in GSL.

5.4.5 *The Great Microbial Diversity at Salt Saturation*

The studies reviewed here show the breadth of diversity among species of GSL in the remote north arm of the lake. These consortia represent all three domains of life, and also the viruses that infect these cells. Recent work highlights the importance of communities versus individuals, their shared metabolites, and abilities to form higher order structures. The current approaches, utilizing a variety of tool sets for hypersaline systems, indicate the whole picture is visible only when we use this array of techniques. A comprehensive view of microbial diversity allows us to ask larger questions about how life evolves and persists in Earth's low water activity environs.

Acknowledgements The authors would like to thank Carol D. Litchfield for her mentorship, stimulation of this work, and her passion for life in hypersaline waters. Also, we appreciate Masoumeh Sikaroodi for technical assistance in ion torrent multi-tag sequencing of the halophilic bacteria.

References

Almeida-Dalmet S (2011) A study of the microbial diversity in the north arm of Great Salt Lake. Dissertation, George Mason University

Almeida-Dalmet S, Sikaroodi M, Gillevet PM, Litchfield CD, Baxter BK (2015) Temporal study of the microbial diversity of the north arm of Great Salt Lake, Utah, USA. Microorganisms 3:310–326

Almeida-Dalmet S, Litchfield C, Gillevet P, Baxter B (2018) Differential gene expression in response to salinity and temperature in a *Haloarcula* strain from Great Salt Lake, Utah. Genes 9:52

Altschul SF, Madden TL, Schäffer AA, Zhang J, Zhang Z, Miller W, Lipman DJ (1997) Gapped BLAST and PSI-BLAST: a new generation of protein database search programs. Nucleic Acids Res 25:3389–3402

Amann RI, Ludwig W, Schleifer K-H (1995) Phylogenetic identification and in situ detection of individual microbial cells without cultivation. Microbiol Rev 59:143–169

Andrei A-Ş, Baricz A, Păuşan M, Muntean V, Sicora CI, Alexe M, Rakosy-Tican E, Banciu HL (2017) Spatial distribution and molecular diversity of archaeal communities in the extreme hypersaline meromictic Brâncoveanu Lake (Transylvanian Basin, Romania). Geomicrobiol J 34:130–138

Antón J, Oren A, Benlloch S, Rodríguez-Valera F, Amann R, Rosselló-Mora R (2002) *Salinibacter ruber* gen. nov., sp. nov., a novel, extremely halophilic member of the Bacteria from saltern crystallizer ponds. Int J Syst Evol Microbiol 52:485–491

Antón J, Peña A, Santos F, Martínez-García M, Schmitt-Kopplin P, Rosselló-Mora R (2008) Distribution, abundance and diversity of the extremely halophilic bacterium *Salinibacter ruber*. Saline Syst 4:15

Bardgett RD, Bowman WD, Kaufmann R, Schmidt SK (2005) A temporal approach to linking above ground and below ground ecology. Trends Ecol Evol 20:634–641

Baricz A, Cristea A, Muntean V, Teodosiu G, Andrei A-Ş, Molnár I, Alexe M, Rakosy-Tican E, Banciu HL (2015) Culturable diversity of aerobic halophilic archaea (Fam. Halobacteriaceae) from hypersaline, meromictic Transylvanian lakes. Extremophiles 19:525–537

Baxter BK (2018) Great Salt Lake microbiology: a historical perspective. Int Microbiol:1–17

Baxter B, Zalar P (2019) Model ecosystems in extreme environments. In: Seckbach J, Rampelotto P (eds) The extremophiles of Great Salt Lake: complex microbiology in a dynamic hypersaline ecosystem. Elsevier, Amsterdam, pp 57–99

Baxter BK, Litchfield CD, Sowers K, Griffith JD, Dassarma PA, Dassarma S (2005) Microbial diversity of Great Salt Lake. In: Gunde-Cimerman N, Oren A, Plemennitas A (eds) Adaptation to life at high salt concentrations in Archaea, Bacteria, and Eukarya. Springer, Dordrecht, pp 9–25

Baxter BK, Eddington B, Riddle MR, Webster TN, Avery BJ (2007) Great Salt Lake halophilic microorganisms as models for astrobiology: evidence for desiccation tolerance and ultraviolet irradiation resistance. International Society for Optics and Photonics, pp 669415

Baxter BK, Mangalea MR, Willcox S, Sabet S, Nagoulat M-N, Griffith JD (2011) Haloviruses of Great Salt Lake: a model for understanding viral diversity. In: Ventosa A, Oren A, Ma Y (eds) Halophiles and hypersaline environments. Springer, Berlin, pp 173–190

Belovsky GE, Stephens D, Perschon C, Birdsey P, Paul D, Naftz D, Baskin R, Larson C, Mellison C, Luft J, Mosley R, Mahon H, Van Leeuwen J, Allen DV (2011) The Great Salt Lake Ecosystem (Utah, USA): long term data and a structural equation approach. Ecosphere 2: art33. https://doi.org/10.1890/ES10-00091.1

Ben-Amotz A, Sussman I, Avron M (1982) Glycerol production by Dunaliella. In: Mislin H, Bachofen R (eds) New trends in research and utilization of solar energy through biological systems. Springer, Basel, pp 55–58

Boyd ES, Yu R-Q, Barkay T, Hamilton TL, Baxter BK, Naftz DL, Marvin-DiPasquale M (2017) Effect of salinity on mercury methylating benthic microbes and their activities in Great Salt Lake, Utah. Sci Total Environ 581:495–506

Brandt K, Vester F, Jensen A, Ingvorsen K (2001) Sulfate reduction dynamics and enumeration of sulfate-reducing bacteria in hypersaline sediments of the Great Salt Lake (Utah, USA). Microb Ecol 41:1–11

Breitbart M, Wegley L, Leeds S, Schoenfeld T, Rohwer F (2004) Phage community dynamics in hot springs. Appl Environ Microbiol 70:1633. https://doi.org/10.1128/AEM.70.3.1633-1640.2004

Brito-Echeverría J, López-López A, Yarza P, Antón J, Rosselló-Móra R (2009) Occurrence of *Halococcus* spp. in the nostrils salt glands of the seabird *Calonectris diomedea*. Extremophiles 13:557–565

Brown AD (1976) Microbial water stress. Bacteriol Rev 40:803–846

Burns DG, Camakaris HM, Janssen PH, Dyall-Smith ML (2004) Combined use of cultivation-dependent and cultivation-independent methods indicates that members of most haloarchaeal groups in an Australian crystallizer pond are cultivable. Appl Environ Microbiol 70:5258. https://doi.org/10.1128/AEM.70.9.5258-5265.2004

Cannon JS, Canon MA (2002) The southern pacific railroad trestle—past and present. In: Gwynn JW (ed) Great Salt Lake, an overview of change. Special Publication of the Utah Department of Natural Resources, Salt Lake City, UT, pp 283–294

Clokie MR, Millard AD, Letarov AV, Heaphy S (2011) Phages in nature. Bacteriophage 1:31–45

Corcelli A, Lattanzio VM, Mascolo G, Babudri F, Oren A, Kates M (2004) Novel sulfonolipid in the extremely halophilic bacterium *Salinibacter ruber*. Appl Environ Microbiol 70:6678–6685

Cronin E, Post F (1977) Report of a dematiaceous hyphomycete from the Great Salt Lake, Utah. Mycologia 69:846–847

Cytryn E, Minz D, Oremland RS, Cohen Y (2000) Distribution and diversity of archaea corresponding to the limnological cycle of a hypersaline stratified lake (Solar Lake, Sinai, Egypt). Appl Environ Microbiol 66:3269–3276

Daniels L (1917) On the flora of great salt Lake. Am Nat 51:499–506

Delgado-García M, Contreras-Ramos SM, Rodríguez JA, Mateos-Díaz JC, Aguilar CN, Camacho-Ruíz RM (2018) Isolation of halophilic bacteria associated with saline and alkaline-sodic soils by culture dependent approach. Heliyon 4:e00954

Donachie SP, Foster JS, Brown MV (2007) Culture clash: challenging the dogma of microbial diversity. ISME J 1:97

Evans F (1958) Culture of protozoa from Great Salt Lake. J Protozool 5:13

Felix E, Rushforth S (1979) The algal flora of the Great-Salt-Lake, Utah, USA. Nova Hedwigia 31:163–195

Fendrich C (1988) *Halovibrio variabilis* gen. nov. sp. nov., *Pseudomonas halophila* sp. nov. and a new halophilic aerobic coccoid Eubacterium from Great Salt Lake, Utah, USA. Syst Appl Microbiol 11:36–43

Fendrich C, Schink B (1988) Degradation of glucose, glycerol and acetate by aerobic bacteria in surface water of Great Salt Lake, Utah, USA. Syst Appl Microbiol 11:94–96. https://doi.org/10.1016/S0723-2020(88)80054-7

Feng S, Powell SM, Wilson R, Bowman JP (2013) Light-stimulated growth of proteorhodopsin-bearing sea-ice psychrophile *Psychroflexus torquis* is salinity dependent. ISME J 7:2206

Fernandez AB, Rasuk MC, Visscher PT, Contreras M, Novoa F, Poire DG, Patterson MM, Ventosa A, Farias ME (2016) Microbial diversity in sediment ecosystems (evaporites domes, microbial mats, and crusts) of hypersaline Laguna Tebenquiche, Salar de Atacama, Chile. Front Microbiol 7:1284

Flowers S, Evans FR (1966) The flora and fauna of the Great Salt Lake region, Utah. In: Boyko H (ed) Salinity and aridity. Springer, Dordrecht, pp 367–393

Frederick E (1924) On the bacterial flora of Great Salt Lake and the viability of other microorganisms in Great Salt Lake water. Master's thesis, University of Utah

Galinski EA (1993) Compatible solutes of halophilic eubacteria: molecular principles, water-solute interaction, stress protection. Experientia 49:487–496

Galinski EA (1995) Osmoadaptation in bacteria. In: Poole R (ed) Advances in microbial physiology. Elsevier, London, pp 273–328

Galinski E, Trüper H (1982) Betaine, a compatible solute in the extremely halophilic phototrophic bacterium *Ectothiorhodospira halochloris*. FEMS Microbiol Lett 13:357–360

Garabito MJ, Márquez MC, Ventosa A (1998) Halotolerant Bacillus diversity in hypersaline environments. Can J Microbiol 44:95–102. https://doi.org/10.1139/w97-125

Greer DC (1971) Annals map supplement number fourteen: Great Salt Lake, Utah. Ann Assoc Am Geogr 61:214–215

Henneke E, Luther GW III, De Lange GJ, Hoefs J (1997) Sulphur speciation in anoxic hypersaline sediments from the eastern Mediterranean Sea. Geochim Cosmochim Acta 61:307–321

Hollister EB, Engledow AS, Hammett AJM, Provin TL, Wilkinson HH, Gentry TJ (2010) Shifts in microbial community structure along an ecological gradient of hypersaline soils and sediments. ISME J 4:829

Hugenholtz P, Goebel BM, Pace NR (1998) Impact of culture-independent studies on the emerging phylogenetic view of bacterial diversity. J Bacteriol 180:4765–4774

Hugoni M, Escalas A, Bernard C, Nicolas S, Jézéquel D, Vazzoler F, Sarazin G, Leboulanger C, Bouvy M, Got P (2018) Spatiotemporal variations in microbial diversity across the three domains of life in a tropical thalassohaline lake (Dziani Dzaha, Mayotte Island). Mol Ecol 27:4775–4786

Jacob JH, Hussein EI, Shakhatreh MAK, Cornelison CT (2017) Microbial community analysis of the hypersaline water of the Dead Sea using high-throughput amplicon sequencing. MicrobiologyOpen 6:e00500. https://doi.org/10.1002/mbo3.500

Jakobsen TF, Kjeldsen KU, Ingvorsen K (2006) *Desulfohalobium utahense* sp. nov., a moderately halophilic, sulfate-reducing bacterium isolated from Great Salt Lake. Int J Syst Evol Microbiol 56:2063–2069

Jennison M (1936) Relationships between plate counts and direct microscopic counts of *Escherichia coli* during the logarithmic growth period. J Bacteriol 33:461–477

Joint Genome Institute (2019) https://genome.jgi.doe.gov/portal/GSLNorRozelPoint_FD/GSLNorRozelPoint_FD.info.html, Accessed 12 Mar 2019

Jończyk E, Kłak M, Międzybrodzki R, Górski A (2011) The influence of external factors on bacteriophages. Folia Microbiol (Praha) 56:191–200

Jones DL, Baxter BK (2017) DNA repair and photoprotection: mechanisms of overcoming environmental ultraviolet radiation exposure in halophilic archaea. Front Microbiol 8:1882

Jones BF, Naftz DL, Spencer RJ, Oviatt CG (2009) Geochemical evolution of Great Salt Lake, Utah, USA. Aquat Geochem 15:95–121

Jones SE, Cadkin TA, Newton RJ, McMahon KD (2012) Spatial and temporal scales of aquatic bacterial beta diversity. Front Microbiol 3:318

Kemp B, Tabish E, Wolford A, Jones D, Butler J, Baxter B (2018) The biogeography of Great Salt Lake halophilic archaea: testing the hypothesis of avian mechanical carriers. Diversity 10:124

Kirkpatrick R (1934) The life of Great Salt Lake, with special reference to the algae. Master's thesis, University of Utah

Kjeldsen KU, Loy A, Jakobsen TF, Thomsen TR, Wagner M, Ingvorsen K (2007) Diversity of sulfate-reducing bacteria from an extreme hypersaline sediment, Great Salt Lake (Utah). FEMS Microbiol Ecol 60:287–298

Kjeldsen KU, Jakobsen TF, Glastrup J, Ingvorsen K (2010) Desulfosalsimonas propionicica gen. nov., sp. nov., a halophilic, sulfate-reducing member of the family Desulfobacteraceae isolated from a salt-lake sediment. Int J Syst Evol Microbiol 60:1060–1065

Labrenz M, Collins MD, Lawson PA, Tindall BJ, Schumann P, Hirsch P (1999) *Roseovarius tolerans* gen. nov., sp. nov., a budding bacterium with variable bacteriochlorophyll a production from hypersaline Ekho Lake. Int J Syst Evol Microbiol 49:137–147

Larsen H (1967) Biochemical aspects of extreme halophilism. In: Rose A, Wilkinson J (eds) Advances in microbial physiology. Elsevier, Scotland, pp 97–132

Lennon JT, Jones SE (2011) Microbial seed banks: the ecological and evolutionary implications of dormancy. Nat Rev Microbiol 9:119

Lindsay M, Anderson C, Fox N, Scofield G, Allen J, Anderson E, Bueter L, Poudel S, Sutherland K, Munson-McGee J (2017) Microbialite response to an anthropogenic salinity gradient in Great Salt Lake, Utah. Geobiology 15:131–145

Litchfield CD (1998) Survival strategies for microorganisms in hypersaline environments and their relevance to life on early Mars. Meteorit Planet Sci 33:813–819

Litchfield CD, Gillevet PM (2002) Microbial diversity and complexity in hypersaline environments: a preliminary assessment. J Ind Microbiol Biotechnol 28:48–55

Litchfield CD, Sikaroodi M, Gillevet PM (2006) Characterization of natural communities of halophilic microorganisms. Methods Microbiol 35:513–533

Liu L, Li Y, Li S, Hu N, He Y, Pong R, Lin D, Lu L, Law M (2012) Comparison of next-generation sequencing systems. Biomed Res Int 2012:251364

Lozupone CA, Knight R (2007) Global patterns in bacterial diversity. Proc Natl Acad Sci U S A 104:11436–11440

Madison RJ (1970) Effects of a causeway on the chemistry of the brine in Great Salt Lake, Utah. Utah Geological and Mineralogical Survey. Water-Resources Bulletin 14, 52 p

Mesbah NM, Abou-El-Ela SH, Wiegel J (2007) Novel and unexpected prokaryotic diversity in water and sediments of the alkaline, hypersaline lakes of the Wadi An Natrun, Egypt. Microb Ecol 54:598–617

Metzker ML (2010) Sequencing technologies—the next generation. Nat Rev Genet 11:31

Meuser JE, Baxter BK, Spear JR, Peters JW, Posewitz MC, Boyd ES (2013) Contrasting patterns of community assembly in the stratified water column of Great Salt Lake, Utah. Microb Ecol 66:268–280

Milstead C (2002) Characterization of bacteria isolated from a solar saltern, the Salina Margherita Di Savoia. Master's thesis, George Mason University

Mohr KI, Stechling M, Wink J, Wilharm E, Stadler M (2016) Comparison of myxobacterial diversity and evaluation of isolation success in two niches: Kiritimati Island and German compost. MicrobiologyOpen 5:268–278

Naftz D, Angeroth C, Kenney T, Waddell B, Darnall N, Silva S, Perschon C, Whitehead J (2008) Anthropogenic influences on the input and biogeochemical cycling of nutrients and mercury in Great Salt Lake, Utah, USA. Appl Geochem 23:1731–1744

Naftz DL, Millero FJ, Jones BF, Green WR (2011) An equation of state for hypersaline water in Great Salt Lake, Utah, USA. Aquat Geochem 17:809–820

Naftz DL, Carling GT, Angeroth C, Freeman M, Rowland R, Pazmiño E (2014) Density-stratified flow events in Great Salt Lake, Utah, USA: implications for mercury and salinity cycling. Aquat Geochem 20:547–571

Oren A (1999) Bioenergetic aspects of halophilism. Microbiol Mol Biol Rev 63:334–348

Oren A (2013) *Salinibacter*: an extremely halophilic bacterium with archaeal properties. FEMS Microbiol Lett 342:1–9

Pack DA (1919) Two ciliata of Great Salt Lake. Biol Bull 36:273–282

Parnell JJ, Crowl TA, Weimer BC, Pfrender ME (2009) Biodiversity in microbial communities: system scale patterns and mechanisms. Mol Ecol 18:1455–1462

Parnell JJ, Rompato G, Latta LC IV, Pfrender ME, Van Nostrand JD, He Z, Zhou J, Andersen G, Champine P, Ganesan B (2010) Functional biogeography as evidence of gene transfer in hypersaline microbial communities. PLoS One 5:e12919

Parnell JJ, Rompato G, Crowl TA, Weimer BC, Pfrender ME (2011) Phylogenetic distance in Great Salt Lake microbial communities. Aquat Microb Ecol 64:267–273

Paterek JR, Smith PH (1985) Isolation and characterization of a halophilic methanogen from Great Salt Lake. Appl Environ Microbiol 50:877–881

Peña A, Valens M, Santos F, Buczolits S, Antón J, Kämpfer P, Busse H-J, Amann R, Rossello-Mora R (2005) Intraspecific comparative analysis of the species *Salinibacter ruber*. Extremophiles 9:151–161

Post F (1975) Life in the Great Salt Lake, Utah. Science 36:43–47

Post FJ (1977) The microbial ecology of the Great Salt Lake. Microb Ecol 3:143–165

Post F (1980) Biology of the north arm. In: Gweynn J (ed) Great Salt Lake: a scientific, historical and economical overview. Geological and Mineral Survey, Bulletins, Salt Lake City, UT. pp 313–321

Post F (1981) Microbiology of the Great Salt Lake north arm. Hydrobiologia 81:59–69

Post FJ, Stube JC (1988) A microcosm study of nitrogen utilization in the Great Salt Lake, Utah. In: Melack JM (ed) Saline Lakes. Springer, Dordrecht, pp 89–100

Post FJ, Borowitzka LJ, Borowitzka MA, Mackay B, Moulton T (1983) The protozoa of a Western Australian hypersaline lagoon. Hydrobiologia 105:95–113. https://doi.org/10.1007/BF00025180

Quesada E, Ventosa A, Rodriguez-Valera F, Ramos-Cormenzana A (1982) Types and properties of some bacteria isolated from hypersaline soils. J Appl Bacteriol 53:155–161

Riesenfeld CS, Schloss PD, Handelsman J (2004) Metagenomics: genomic analysis of microbial communities. Annu Rev Genet 38:525–552

Rodríguez-Valera F, Acinas SG, Antón J, Oren A (1999) Contribution of molecular techniques to the study of microbial diversity in hypersaline environments. CRC Press, Boca Raton, FL, pp 27–38

Rodriguez-Valera F, Martin-Cuadrado A-B, Rodriguez-Brito B, Pašić L, Thingstad TF, Rohwer F, Mira A (2009) Explaining microbial population genomics through phage predation. Nat Rev Microbiol 7:828

Rohwer F, Prangishvili D, Lindell D (2009) Roles of viruses in the environment. Environ Microbiol 11:2771–2774

Ruvindy R, White RA III, Neilan BA, Burns BP (2016) Unravelling core microbial metabolisms in the hypersaline microbial mats of Shark Bay using high-throughput metagenomics. ISME J 10:183

Schuster SC (2007) Next-generation sequencing transforms today's biology. Nat Methods 5:16

Shen PS, Domek MJ, Sanz-García E, Makaju A, Taylor RM, Hoggan R, Culumber MD, Oberg CJ, Breakwell DP, Prince JT (2012) Sequence and structural characterization of Great Salt Lake bacteriophage CW02, a member of the T7-like supergroup. J Virol 86:7907–7917

Solano-Aguilar GI, Lakshman S, Jang S, Beshah E, Xie Y, Sikaroodi M, Gupta R, Vinyard B, Molokin A, Urban JF Jr, Gillevet P, Davis CD (2018) The effect of feeding cocoa powder and *Lactobacillus rhamnosus* on the composition and function of pig intestinal microbiome. Curr Dev Nutr 2(5). https://doi.org/10.1093/cdn/nzy011

Sorokin DY, Tindall B (2006) The status of the genus name *Halovibrio Fendrich* 1989 and the identity of the strains *Pseudomonas halophila* DSM 3050 and *Halomonas variabilis* DSM 3051. Request for an opinion. Int J Syst Evol Microbiol 56:487–489

Stan-Lotter H, Fendrihan S (2015) Halophilic archaea: life with desiccation, radiation and oligotrophy over geological times. Life 5(3):1487–1496. https://doi.org/10.3390/life5031487

Stube J, Post F, Procella D (1976) Nitrogen cycling in microcosms and application to the biology of the north arm of Great Salt Lake. Logan, UT, Utah Water Research Laboratory Report, Utah State University

Tazi L, Breakwell DP, Harker AR, Crandall KA (2014) Life in extreme environments: microbial diversity in Great Salt Lake, Utah. Extremophiles 18:525–535

Vahed SZ, Forouhandeh H, Hassanzadeh S, Klenk H-P, Hejazi MA, Hejazi MS (2011) Isolation and characterization of halophilic bacteria from Urmia Lake in Iran. Microbiology 80:834–841

Vavourakis CD, Ghai R, Rodriguez-Valera F, Sorokin DY, Tringe SG, Hugenholtz P, Muyzer G (2016) Metagenomic insights into the uncultured diversity and physiology of microbes in four hypersaline soda lake brines. Front Microbiol 7:211

Ventosa A, de la Haba RR, Sánchez-Porro C, Papke RT (2015) Microbial diversity of hypersaline environments: a metagenomic approach. Curr Opin Microbiol 25:80–87

Vorhies C (1917) Notes on the fauna of the Great Salt Lake. Am Nat 61:494–499

Vreeland RH, Rosenzweig WD, Powers DW (2000) Isolation of a 250 million-year-old halotolerant bacterium from a primary salt crystal. Nature 407:897

Wainø M, Tindall B, Ingvorsen K (2000) *Halorhabdus utahensis* gen. nov., sp. nov., an aerobic, extremely halophilic member of the Archaea from Great Salt Lake, Utah. Int J Syst Evol Microbiol 50:183–190

Ward DM, Brock TD (1978) Hydrocarbon biodegradation in hypersaline environments. Appl Environ Microbiol 35:353–359

Weimer BC, Rompato G, Parnell J, Gann R, Ganesan B, Navas C, Gonzalez M, Clavel M, Albee-Scott S (2009) Microbial biodiversity of Great Salt Lake, Utah. Nat Resour Environ Issues 15:3

Yim KJ, Kwon J, Cha I-T, Oh K-S, Song HS, Lee H-W, Rhee J-K, Song E-J, Rho JR, Seo ML (2015) Occurrence of viable, red-pigmented haloarchaea in the plumage of captive flamingoes. Sci Rep 5:16425

Zeikus J, Hegge P, Thompson T, Phelps T, Langworthy T (1983) Isolation and description of *Haloanaerobium praevalens* gen. nov. and sp. nov., an obligately anaerobic halophile common to Great Salt Lake sediments. Curr Microbiol 9:225–233

Zerulla K, Chimileski S, Näther D, Gophna U, Papke RT, Soppa J (2014) DNA as a phosphate storage polymer and the alternative advantages of polyploidy for growth or survival. PLoS One 9:e94819

Chapter 6
Invertebrates and Phytoplankton of Great Salt Lake: Is Salinity the Driving Factor?

Katherine L. Barrett and Gary E. Belovsky

Abstract Great Salt Lake (GSL) is a hypersaline terminal lake and has varied historically in salinity from 6 to 28%. Because the lake's salinity is much greater than in marine environments (~3.5%), salinity is often assumed to be the driving factor for GSL benthic and pelagic food webs. Certainly, many species cannot live in a hypersaline environment (e.g., fish), and the diversity of species capable of coping with hypersaline conditions is limited. However, the GSL's benthic and pelagic food webs are adapted to these extreme saline conditions, and their dynamics (primary and secondary production, species abundances, etc.) respond in a complex fashion to the interplay of salinity, temperature, and nutrient availability. Therefore, focusing solely on salinity is not appropriate. In this chapter, we first explore historically how GSL food webs have been reported to change and found salinity to have limited impact. We next demonstrate that in recent years (1994–2018) GSL food webs varied far less with salinity than might be expected, even though salinity varied by 8.2–17.5%, because temperatures and nutrient availability covaried with salinity and showed more impacts than salinity alone. Finally, we employ the observations on the interplay of salinity, temperature, and nutrients to project how future climatic changes in the GSL watershed will affect primary producers and consumers and impact GSL food webs. These future climatic changes will have profound effects on GSL food web dynamics.

Keywords Salinity · Nutrients · Temperature · Abiotic factors · Food webs · Climate change · *Artemia franciscana* · *Ephydra* spp. · Historical trends · Great Salt Lake food web

K. L. Barrett · G. E. Belovsky (✉)
Department of Biological Sciences, University of Notre Dame, Notre Dame, IN, USA
e-mail: gbelovsk@nd.edu

B. K. Baxter, J. K. Butler (eds.), *Great Salt Lake Biology*,
https://doi.org/10.1007/978-3-030-40352-2_6

6.1 Great Salt Lake as an Extreme Environment

Situated in the arid Great Basin of the United States, Great Salt Lake (GSL) is hypersaline (averaging ~4 times more saline than the ocean) and hypereutrophic, as it has no outlet and accumulates salt and nutrients from surrounding watershed inflows (Williams 1998). The GSL's terminal nature and shallow bathymetry make it sensitive to climate variability that creates a balance between evaporation (temperature) and inflows (precipitation), which impact salinity and nutrient concentrations (Tweed et al. 2011).

Saline lakes, like GSL, are considered harsh environments, and therefore contain a limited diversity of species because high salinity imposes osmoregulatory stress on organisms (Williams 1998). This has led to the idea that primary productivity, species richness, and diversity should decline with increasing salinity (Herbst and Blinn 1998; Herbst and Bradley 1989; Marcarelli et al. 2006; Smith et al. 2010). However, are saline lakes truly harsh environments to species that are adapted to the range of salinities encountered there?

Figure 6.1 presents a classic perception of the role of salinity in structuring biological communities in saline lakes. Fluctuations in salinity within the bounds of moderately saline and hypersaline lakes do not typically have a strong influence on production and diversity (Ben-Amotz and Avron 1983). However, in marine systems (top right corner of Fig. 6.1), salinity fluctuations that surpass the upper bounds of salinity can lead to changes in productivity, species richness, and diversity (Galat and Robinson 1983; Williams 2001).

Yet, in hypersaline lakes, species tend to be extremely halotolerant and their ability to osmoregulate at high salinities allows biota to survive across a greater range of salinities than organisms in moderately saline lakes (Javor 1989; Schapira et al. 2010; Velasco et al. 2006). As a result, organisms in high salinity environments (bottom left corner of Fig. 6.1) should be more impacted by other factors such as temperature, nutrients, and pH, rather than salinity. In other words, while these are

Fig. 6.1 Conceptualization of the role of salinity changes on lake biota [adapted from Williams (1998) and Herbst (2001)]

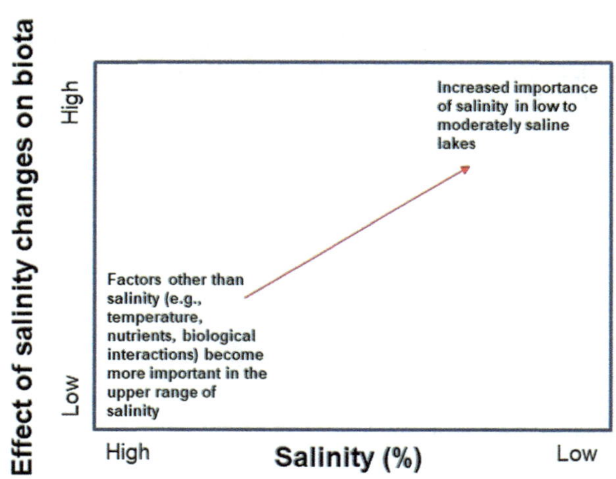

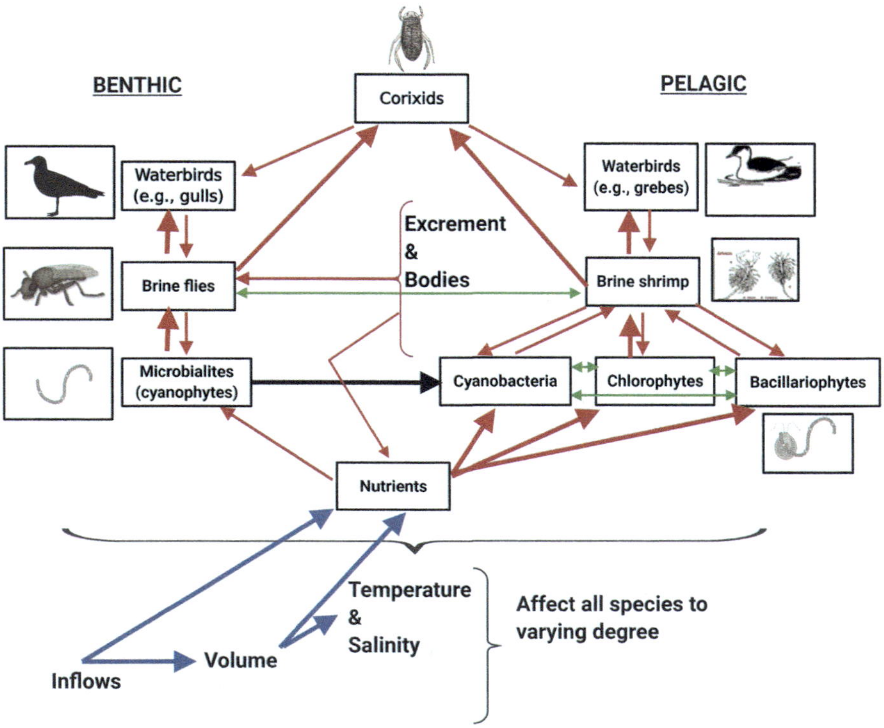

Fig. 6.2 Great Salt Lake south arm ecosystem. Based on 1994–2006 studies, interactions and their directions are depicted by arrows: blue are physical, red are consumption, and green are competitive effects, while black are disturbances [modified from Belovsky et al. (2011)]

harsh environments, they are not harsh to species adapted to these conditions (Hammer 1986; Sanders 1969; Slobodkin and Sanders 1969).

Considering the above scenario, GSL may not be an extreme environment to species adapted to live there. The species inhabiting GSL must be adapted to salinity fluctuations because the highly variable freshwater inflows impact lake volume, which directly affects salinities and nutrient concentrations (Fig. 6.2). As the GSL is terminal and has an endogenous pool of nutrients, increased inflows not only reduce salinity, but also nutrient concentrations (Belovsky et al. 2011; Naftz 2017). Due to shallow bathymetry, GSL has a high surface area-to-volume ratio, which contributes to water temperatures responding to variable air temperatures and water volumes (Arnow and Stephens 1990). Therefore, changes in water temperatures, salinities, and nutrient levels affect all components of the GSL food web to varying degrees.

The GSL food web is composed of benthic and pelagic food chains that respond to changes in temperature, nutrients, and salinity (see Fig. 6.2). Halotolerant Cyanobacteria, Chlorophytes (green algae), and diatoms (Bacillariophytes) comprise the phytoplankton, the primary resource base of brine shrimp (*Artemia franciscana*)

(Belovsky et al. 2011; Fig. 6.2). Benthic microbialites composed of Cyanobacteria and diatom biofilms (Lindsay et al. 2017) are the primary grazing habitat of brine fly larvae (*Ephydra* sp.) (Collins 1980; Fig. 6.2). Both brine flies and brine shrimp are abundant, as they are adapted to high and low salinities (Croghan 1958; Gajardo and Beardmore 2012; Riisgård et al. 2015).

Brine shrimp reproduce by both oviparity (dormant cysts) and ovoviviparity (live births) (Belovsky et al. 2011). During periods of abundant phytoplankton and warm water temperatures in the lake, live births predominate. By the fall, females switch to cyst production in response to lowered food availability and water temperatures. The cysts produced in the fall then overwinter and hatch the following spring to initiate the next year's population. This is important because brine shrimp are a primary food source for many species of waterbirds that use GSL as staging and nesting habitat. In particular, brine shrimp comprises ~90% of the diet of Eared Grebes (*Podiceps nigricollis*), while gulls feed mostly on brine flies (see Fig. 6.2). Another prey item of birds, but not a permanent resident of the core GSL food web, corixids are predatory insects that live and reproduce in the brackish and freshwater bays of GSL. When salinity is low (3–6%), corixids may move into shallow littoral areas of GSL and prey upon brine shrimp (Mellison 2000).

The GSL food web dynamics are a reflection of the responses of primary and secondary production, and species abundances to interactions of salinity, temperature, and nutrients. In the following sections, we review historical observations of GSL water volume, salinity, and biological communities, to determine if a relationship exists between salinity and biota. Then, we examine what current lake monitoring and lab studies reveal about the multiple factors influencing the GSL food web. Lastly, we apply current observations of relationships between biota and lake conditions to ask if we can project how the GSL food web will change in the future with climate change.

6.2 Are There Historical Patterns in Great Salt Lake Food Web Changes with Salinity?

Over the last two centuries, fluctuations in GSL salinity and lake volume have occurred due to precipitation patterns and anthropogenic activities (Stephens 1998a). The GSL's water levels (a proxy for lake volume) have been estimated since 1850 (Fig. 6.3); these historical records can be qualitatively related to reports on brine shrimp and phytoplankton abundances to see whether patterns emerge with salinity (Table 6.1; Fig. 6.3). Since salinity varies inversely with water levels, it is logical that early studies of the lake were aimed at correlating salinity changes with phytoplankton and brine shrimp. In this section, three time periods are examined: (1) the historical GSL prior to causeway construction (1847–1959); (2) post-causeway construction before a commercial brine shrimp harvesting industry (1959–1980); and (3) the emergence of a commercial brine shrimp harvesting industry (1981–1994).

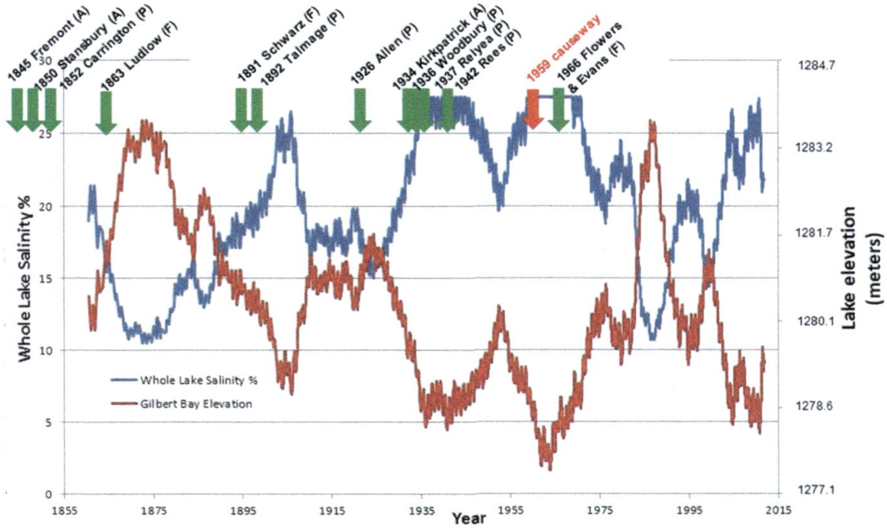

Fig. 6.3 Historical timeline of changes in water level and salinity. Note that after 1959 causeway construction, lake levels and salinity refer to that of the south arm (Gilbert Bay). Green downward pointing arrows indicate years in which relevant studies or observations provide qualitative information about brine shrimp abundance at that time. Brine shrimp abundance is categorized as follows: *A* abundant, *P* present, *F* few/absent. Records from Fremont, Carrington, Stansbury, and Ludlow found in Morgan (1947) and Rawley (1980)

6.2.1 The Historical Great Salt Lake Prior to Causeway Construction (1847–1959)

When Mormons settled in the Salt Lake Valley in 1847, little was known about aquatic life in the historical GSL, and it was common for early explorers to conclude that no living organism could thrive there (The Deseret News 1907). From the 1850s to 1959, GSL water levels fluctuated substantially, and salinity ranged between 13 and 25% (see Fig. 6.3). Early explorers of the lake reported a low diversity of algae comprising of the cyanobacteria *Coccochloris elebans* and *Oscillatoria* sp., five species of green algae, including *Dunaliella viridis*, and two diatom species (Flowers and Evans 1966; Packard 1879; Stephens 1974; Tilden 1898). The benthic component of the GSL was mentioned by Kirkpatrick (1934) and Flowers (1934), who reported that microbialites were covered with cyanobacteria biofilms. Brine fly larvae, which graze on the surface of microbialites, were reported to be abundant (Rushforth and Felix 1982; Schwarz 1891).

The yellow portion of Table 6.1 provides pre-causeway salinity ranges and phytoplankton taxa composition. Due to a common perception that most diatoms could only grow in freshwater (Patrick 1936), some debated as to whether diatoms were actively growing in the lake (Kirkpatrick 1934). However, Daines (1910, 1917)

Table 6.1 Overview of historical salinity, associated pelagic phytoplankton taxa, and brine shrimp qualitative abundance (Abundant, Present, Few, and NA if no data are available)

Time period	Salinity (%)	Algae present	Brine shrimp abundance	Source
1800s	13–23	5 species of cyanobacteria and green algae, some diatoms	Present	Tilden (1898)
1900–1959	20–27	Cyanobacteria, *Dunaliella viridis*, halophilic bacteria abundant; gelatinous cyanobacteria abundant on microbialites; few diatoms	Abundant	Kirkpatrick (1934), Carozzi (1962), Patrick (1936)
1960s-early 1970s	26–28	Halophilic bacteria and *Coccochloris elebans* abundant	Few	Arnow and Stephens (1990)
1977–1980	12–14	Diatoms abundant and cyanobacteria absent on microbialites,	NA; few surveys during this time	Collins (1980), Stephens (1990)
1980s–1994	12–14; 6	Cyanobacterium *Nodularia spumigena* and several diatoms present	Few	Arnow and Stephens (1990), Wurtsbaugh and Berry (1990)

Qualitative microbialite algae abundance and sources for each time period are also presented. The different colors indicate the three time periods examined: pre-causeway, early post-causeway; and brine shrimp industry

and Kirkpatrick (1934) reported that diatoms could survive and grow in GSL phytoplankton laboratory cultures between salinities of 7–17%, while the cyanobacterium *Coccochloris elebans* was most abundant between 22 and 26%. Thus, decreasing salinity would be predicted to foster diatom growth and limit production of *C. elebans* (Felix and Rushforth 1980).

Historical observations reported brine shrimp as either abundant, present, or few/absent (Kirkpatrick 1934; Morgan 1947; Stephens and Birdsey 2002; see Fig. 6.3). The earliest records of brine shrimp date back to the 1830s, when B. E. L. Bonneville noted high numbers of brine shrimp in GSL (Relyea 1937). Although several historical observations exist since then, no clear pattern emerges between shrimp abundance and the range of salinity recorded throughout historical accounts. For instance, during periods of high and low salinity, most investigators reported shrimp as present, but provided no indication of relative abundance (see Fig. 6.3).

Due to the qualitative nature of studies during this period, we cannot assess how the full range of salinities, temperatures, and nutrients affected biota. In fact, no studies during this time reported temperature and nutrients, which restricts understanding of GSL dynamics in relation to changing abiotic factors. No relationship appeared to emerge between salinity, phytoplankton, and shrimp, except perhaps diatoms, which appeared to be more abundant during periods of low salinity.

6.2.2 Post-causeway Construction Before a Commercial Brine Shrimp Harvesting Industry (1959–1980)

Beginning in 1959, causeways were constructed in GSL to allow vehicle and railroad traffic across the lake (Madison 1970; Cannon and Cannon 2002). The rock-filled railroad causeway between Promontory Point and Lakeside divided the GSL into two major bays, the north (Gunnison Bay) and south arm (Gilbert Bay) (Arnow and Stephens 1990; Stephens 1974, 1990, 1998a). The north arm, because it receives fewer freshwater inflows, became highly saline (26–28%), near saturation as a result, and exceeded salinity tolerances for the major components of the original GSL food web (Almeida-Dalmet et al. 2018; Baxter 2018; Brock 1975; Lindsay et al. 2017; Post 1977; Stephens 1990; see Fig. 6.2). Gilbert Bay receives 95% of freshwater inflows and the salinity range encountered here (9–17%) is more typical of historical conditions experienced on the original GSL (Belovsky et al. 2011).

Another causeway isolated Farmington Bay in the southeast portion of GSL. Farmington Bay receives freshwater inflows, which led to salinities below the range encountered in Gilbert Bay (see upper right portion of Fig. 6.1). As a result, a very different community comprised of stenohaline species that are adapted to lower salinities characterizes the food web of Farmington Bay. For example, the cyanobacterium *Nodularia spumigena*, which thrives at lower salinities, is an abundant component of the phytoplankton (Arnow and Stephens 1990; Marcarelli et al. 2006; Wurtsbaugh et al. 2012, 2015). Brine flies are absent, brine shrimp are present in low abundance, and other invertebrates more typical of freshwater, such as chironomids and corixids, are abundant in Farmington Bay (Gray 2012; Wurtsbaugh and Berry 1990). Brine shrimp populations in Farmington Bay may be less abundant than in Gilbert Bay because of predation by corixids (Wurtsbaugh 1992; Fig. 6.2).

Neither the north arm nor Farmington Bay is reflective of the original GSL conditions. Therefore, the remainder of this chapter will focus solely on food web dynamics in Gilbert Bay.

After construction of causeways, salinity of Gilbert Bay ranged between 12 and 28% (see orange portion of Table 6.1 and Fig. 6.3). Aside from Cuellar (1990) and Wirick (1972) who reported GSL brine shrimp natural history and Flowers and Evans (1966) who reported low brine shrimp abundance in GSL, there were no quantitative brine shrimp surveys. This restricts analysis of how fluctuating lake conditions impacted brine shrimp (Stephens 1998a; Table 6.1).

Little was also known about phytoplankton at this time, when salinities ranged between 12 and 14%. The one study that examined this in detail occurred in the 1970s, when Felix and Rushforth (1979) reported a few species of diatoms, four species of cyanobacteria, and seven species of green algae.

While few pelagic data are available, some benthic studies occurred. Carozzi (1962) reported that cyanobacteria were dominant on microbialites during periods of high salinity, while Collins (1980) found that diatoms tended to predominate at lower salinity. Using lab studies, Nemenz (1960) reported that GSL brine fly larvae were highly adapted to the high salt concentrations and fluctuations in salinity. These lab findings were in accord with Collins (1980), who also reported high densities of brine fly larvae on microbialites throughout his study. Compared to densities on microbialites, brine fly larval densities were low on sand and mud substrates, and Collins (1980) suggested microbialites were important grazing habitat for larvae. However, the lack of long-term brine fly larval dynamics prevents analysis on how their populations changed with varying salinity.

Despite the lack of detailed food web studies, considerable progress was made on understanding limnological characteristics of GSL. A deep brine layer (DBL) was first observed in Gilbert Bay in the mid-1960s as a result of the flow of heavy brines from the north arm through causeway culverts (Handy 1967). The DBL is a layer of very saline water that is denser than the saline water above it, creating a saline chemocline and establishing meromixis, a semipermanent stratified condition (Belovsky et al. 2011). The DBL was recognized as a potential nutrient sink in GSL, and work by Stephens and Gillespie (1976) documented its significance to whole-lake circulation of nutrients.

During this period, Stephens and Gillespie (1976) used lab studies to determine that *Dunaliella* sp. standing crop was primarily nitrogen (N)-limited. This was an important finding because *Dunaliella* sp. is a dominant component of the phytoplankton community and a prime food item for brine shrimp.

6.2.3 The Emergence of a Commercial Brine Shrimp Harvesting Industry (1981–1994)

By the 1980s, salinity in Gilbert Bay ranged from 12 to 14% (green portion of Table 6.1). During this time, 20 species of algae were identified, including 14 species of diatoms, which was a significant increase compared to previous observations of higher salinity and lower diatom diversity (Carozzi 1962; Felix and Rushforth 1980; Stephens 1998a). The cyanobacterium *Coccochloris elebans* was reported to be rare during this period, which was consistent with previous reports of this species thriving at high salinities (Stephens 1990).

By the late 1980s, brine shrimp cyst harvesting, which began in 1952 by the Sanders Brine Shrimp Company, emerged as a large-scale commercial industry

(Belovsky and Perschon 2019). From the 1980s to mid-1990s, commercial harvesters annually removed on average 70% of cysts produced in the fall.

As the commercial harvest of shrimp cysts was underway, brine shrimp were reported to be less abundant compared to previous surveys (Wurtsbaugh and Berry 1990; red portion of Table 6.1). Several authors attributed the decline in shrimp densities to low salinities (6%) that occurred in the mid-1980s (Arnow and Stephens 1990; Stephens 1990). It was believed that lower salinities caused cysts to sink to the bottom of the lake, where they would not be able to hatch in the freshwater lens. However, Wurtsbaugh and Berry (1990) suggested the lower salinities in Gilbert Bay did not directly impact shrimp abundance; rather, low salinities were within the tolerance range of corixids, which became common in Gilbert Bay and could prey on shrimp. However, corixids do not reproduce in GSL and are intolerant of salinities greater than 6%, and therefore, their impact on shrimp would have been limited to brackish areas (Mellison 2000).

6.2.4 Summary

This section examined three major time periods in recent GSL history: (1) the original GSL prior to causeways; (2) post-causeway years; and (3) the expanding brine shrimp industry in the 1980s.

The observed salinity ranges in GSL prior and post causeway are certainly within the range of the tolerances of species, as the major components of the GSL food web are broadly halotolerant species. Thus, fluctuations in salinity within the range normally experienced would not strongly influence dynamics. While many studies have inferred that the high salinities and variations should be driving patterns within the lake, brine shrimp, phytoplankton, and brine fly larvae did not appear to change over the range of historical salinities (Table 6.1, Fig. 6.3). Diatoms may be an exception, as their abundance appeared to change with salinity throughout pre- and post-causeway periods (Table 6.1), but even this relationship recorded in historical accounts is questionable (see below).

6.3 Current Conditions and Changing Paradigms (1994–Present)

The historical review of GSL neither provides a long-term ecological database to understand how the food web has changed over time or salinity's impact on it. This lack of information and the need to monitor the harvesting of brine shrimp led the State of Utah to develop the Great Salt Lake Ecosystem Program (GSLEP), as well as ancillary lab studies, beginning in 1994 (Belovsky et al. 2011). Analyses of these long-term data since 1994 show that salinity may not have as pronounced an effect on the GSL food web as previously thought (Belovsky et al. 2011). We examine below the results of these long-term data and additional studies to address the factors controlling the GSL food web.

We begin with the 1995–1997 diatom bloom as an example of how focusing only on salinity can be misleading. Based on historical records, diatoms are expected to be most abundant at low salinities. Indeed, centric and pennate diatoms dominated the phytoplankton assemblage throughout 1995–1997, when there were high inflows to the lake and salinities ranged from 13 to 15% (Stephens and Birdsey 2002; Belovsky et al. 2011). However, after 1997, even with reduced inflows, salinity continued to decline, and diatoms were replaced by Cyanobacteria and green algae, which increased in abundance (Belovsky et al. 2011; Stephens 1998b). This is counter to the argument that salinity alone determines diatom abundance. In this case, the diminishing watershed inflows after 1997 likely led to a reduction in silica, a limiting nutrient for diatoms but not limiting to Cyanobacteria or green algae. It is likely that changes in nutrients, not salinity, may have been the driving factor.

6.3.1 Great Salt Lake Dynamics: An Interplay of Salinity, Temperature, and Nutrients

Monthly monitoring by GSLEP and lab experiments of brine shrimp and environmental parameters in Gilbert Bay since 1994 has contributed to an improved understanding of nutrient, salinity, and temperature effects on primary producers, brine shrimp, and brine fly larvae (Belovsky et al. 2011). The effects of different abiotic factors on pelagic and benthic primary producers and consumers are discussed in the following sections.

6.3.2 Pelagic Dynamics: Controls on Phytoplankton

Nutrients, salinity, and temperature all affect phytoplankton dynamics to varying degrees (see Fig. 6.2). Both long-term monitoring and laboratory experiments show that phytoplankton dynamics (production, taxa composition, and richness) are influenced by bottom-up (N levels) and top-down consumer control (Belovsky et al. 2011; Larson and Belovsky 2013; Ogata et al. 2017).

Phytoplankton Production Belovsky et al. (2011) demonstrated that bottom-up control of phytoplankton production in GSL needs to be examined during times of the year when brine shrimp are absent. This is because shrimp are efficient grazers that can reduce phytoplankton to low levels even when N is abundant (Belovsky et al. 2011). When brine shrimp are absent from December until February, phytoplankton production is primarily determined by N concentrations, with smaller effects of salinity (Belovsky et al. 2011; Stephens and Gillespie 1976; see Fig. 6.4).

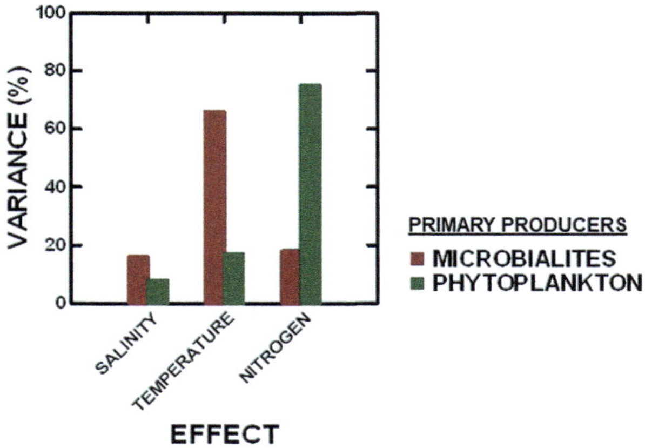

Fig. 6.4 Relative importance (% variance explained) of salinity, temperature, and nitrogen on phytoplankton and microbialite primary production in laboratory studies. Temperature and nitrogen account for ~70% and 80% of the variance in benthic and pelagic primary production, respectively. Salinity and nitrogen have comparable effects on benthic production, and salinity is more important in benthic than pelagic production [data from Belovsky et al. (2000)]

Lab studies further reinforced that N, not salinity, is the primary driver of GSL phytoplankton (chlorophyll-*a*), as it accounts for ~75% of the variance in production (Fig. 6.4). These studies also showed that production increases with temperature and N, and maximum production occurs at intermediate salinities (Belovsky et al. 1999). Although GSL is primarily N-limited, Ogata et al. (2017) reported that phytoplankton growth is enhanced with the addition of Phosphorus (P), which suggests P may modify phytoplankton growth responses to N. Furthermore, phytoplankton production is not significantly affected by salinity when high levels of N are available (Larson and Belovsky 2013). However, production declines at higher temperatures approaching 30 °C and at low salinities (Larson and Belovsky 2013).

Phytoplankton Taxa Composition More current studies also show that phytoplankton taxa composition changes in response to varying nutrients, temperatures, salinities, and competitive interactions among taxa (Larson and Belovsky 2013). Belovsky et al. (2011) reported that green algae, in particular *Dunaliella* sp., are more abundant in the lake at high salinities, low temperatures, high nutrient levels, and low diatom abundance. In the lab, Larson and Belovsky (2013) also found *Dunaliella* sp. abundance is greatest at high salinities and N levels, low temperatures, and low relative abundance of other taxa. These findings suggest that *Dunaliella* sp. may experience reduced competition in these conditions because other taxa may be intolerant of cold temperatures (Larson and Belovsky 2013).

Diatom abundance may also be influenced by competitive interactions, as lab and field studies show it increases as Chlorophyte and Cyanobacteria abundances decrease (Belovsky et al. 2011; Larson 2004). However, diatom responses in the lab to abiotic factors are variable, as they do not appear to be very sensitive to changing temperature, N, and salinity (Larson 2004).

Cyanobacteria abundance in the lake is negatively correlated with green algae and diatom abundance, likely a result of competitive interactions (Belovsky et al. 2011). Lab studies also showed that Cyanobacteria thrive at higher temperatures (Belovsky and Mellison 1998). In addition, lake observations showed that Cyanobacteria abundance is negatively correlated with N levels (Belovsky et al. 2011). This relationship may be explained by N-fixing Cyanobacteria taxa that thrive when nutrients are less abundant, which would lead to reduced competition by other phytoplankton taxa (Belovsky et al. 2011).

Phytoplankton Species Richness Larson and Belovsky (2013) found that phytoplankton species richness increases with N availability and decreases with rising salinity. However, salinity only has a strong negative effect on richness at 10 and 20 °C, but not at 30 °C, which suggests that the impact of salinity on phytoplankton richness changes with water temperature.

6.3.3 Pelagic Dynamics: Controls on Brine Shrimp Grazing, Survival, and Reproduction in the Great Salt Lake

Belovsky et al. (2011) demonstrated that the resource base of the GSL food web is strongly affected by N, temperature, and salinity. There is no direct effect of salinity on shrimp survival or reproduction.

Effects on Shrimp Grazing Sura et al. (2017) found that brine shrimp grazing rates increase with food concentration and temperature, and decrease with salinity. The authors also reported that brine shrimp grazing rates are higher when fed *Dunaliella* sp. compared to *Coccochloris*. However, changing salinities can modify shrimp grazing rates (Barnes and Wurtsbaugh 2015; Sura et al. 2017). At high salinities, the viscosity of water increases, leading to reduced grazing efficiency.

Effects on Survival Brine shrimp survival is strongly dependent on food availability and the taxa composition of their phytoplankton food sources. Lab studies show that food type and abundance account for ~50 and 45% of the variance in shrimp survival, respectively (Fig. 6.5). Brine shrimp in the lab survive better when fed *Dunaliella* sp. compared to *Coccochloris* or diatoms (Belovsky and Mellison 1998; Belovsky unpubl.). Analysis of long-term lake data also show that shrimp abundance is negatively impacted by Cyanobacteria, but not affected by green algae (Belovsky et al. 2011). Observed declines in shrimp abundance in GSL in 1996 were attributed to reduced phytoplankton abundance and shifts in taxa composition from predominantly *Dunaliella* sp. to diatoms (Belovsky and Mellison 1998).

In addition to food amount and type impacting survival, water temperature and food supply interact to influence shrimp survival in GSL (Belovsky et al. 1999, 2000). Shrimp survival is greatest at intermediate temperatures, increases with food level, and decreases at low and high temperatures (Belovsky et al. 2011; Belovsky and Mellison 1997).

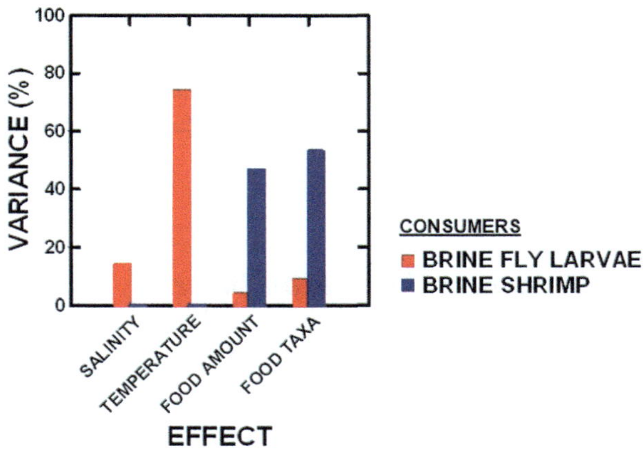

Fig. 6.5 Relative importance (% variance explained) of salinity, temperature, food amount, and food taxa on brine fly larval and shrimp survival in laboratory studies. Temperature has nearly 7 times more impact on fly survivorship than salinity. Shrimp are impacted by taxa composition and amount of food, and salinity and temperature have a relatively smaller impact on their survival [adapted from Belovsky and Larson (2001)]

Effects on Reproduction Belovsky and Larson (2001) found that temperature has the greatest impact on reproductive output of female shrimp reared in the lab, but the mode of reproduction (live births versus cysts) is strongly dependent on temperature and food abundance. In the lab, optimal reproduction occurs at intermediate temperatures and high food levels, and cyst production always declines as food availability increases (Belovsky 2002; Belovsky and Mellison 1998).

6.3.4 Benthic Dynamics: Controls on Benthic Primary Production

Compared to studies on phytoplankton, few researchers have studied the effects of N, temperature, and salinity on microbialite primary producer abundance. However, a recent lab study (Anderson et al. submitted) demonstrated that temperature and N have a greater impact than salinity on microbialite primary producer abundance (measured as chlorophyll-*a*). Anderson and coworkers showed that primary production increases with temperature and N levels, with salinity having the least impact (Fig. 6.4). Additionally, other studies that have examined a wider range of salinities at the same range of temperatures used by these authors reported the same pattern.

At present, the effects of changing temperatures, nutrients, and salinity on microbialite taxa composition is not fully understood. Lindsay et al. (2019) reported that the diatom *Navicula* and cyanobacterium *Euhalothece*, which are hypothesized to be the main primary producers on microbialites, were abundant in low to moderate salinity treatments and declined as salinities exceeded 20%, which is also in accord with Anderson et al. (submitted). At salinities higher than 20%, heterotrophic bacteria and halophiles become more abundant (Lindsay et al. 2019). The reduction in diatom abundance with rising salinities seems plausible given historical and current (1995–1997) observations that diatoms appeared more abundant during periods of lower salinity (Table 6.1).

6.3.5 Benthic Dynamics: Controls on Brine Fly Larvae

Similar to microbialites, brine fly larvae are most affected by temperature, which has nearly seven times more impact compared to salinity, with a smaller effect of food type and amount (Belovsky et al. 2011; see Fig. 6.5). Brine fly larvae survive better at higher temperatures, because they graze more efficiently than at colder temperatures. In the lab, brine fly larvae survive slightly better when fed diatoms compared to *Coccochloris* and chlorophytes (Belovsky and Barrett unpublished). In addition to grazing on benthic algae on microbialites, brine fly larvae consume a range of organic detrital materials (Belovsky et al. 2011), which may explain why the type of food has little impact on their survival (Fig. 6.5).

6.3.6 Summary of Current Food Web Dynamics

Clearly, salinity is not the primary driver of GSL food web dynamics, as other factors such as temperature and N can have a greater impact and modify how salinity influences consumer and producer dynamics. Isolating the effects of salinity and N on food web dynamics in the lake is not straightforward because these abiotic factors are linked through hydrology (see Fig. 6.2). The studies described in this section show why long-term data and laboratory experiments are important in determining the effects of these factors on GSL food web components. Furthermore, these studies support the idea that extreme environments are not extreme to species adapted to live there because salinity was shown to have a small effect on GSL food web dynamics.

The current studies show that GSL's benthic and pelagic components respond differently to abiotic and biotic factors (see Figs. 6.4 and 6.5). As Fig. 6.4 illustrates, temperature is the primary driver of microbialite primary production, while phytoplankton production is driven by N availability. In addition, lab studies suggest that benthic primary producers may be more sensitive to changes in salinity than phytoplankton (Barnes and Wurtsbaugh 2015). Perhaps diatoms, an important component of microbialites, may not be as well adapted as other taxa to the range of salinity experienced (Clavero et al. 2000; Herbst and Blinn 1998).

While brine shrimp survival is driven primarily by the availability and composition of their food, brine fly larvae are most impacted by temperature and salinity (see Fig. 6.5). Although larval survival is influenced least by type and amount of food, the type of food accounts for a slightly higher proportion of the variance, a trend also observed in brine shrimp. This suggests that changes in the taxa composition of their food sources may impact their survival.

These responses need to be acknowledged when considering how future climate changes will affect the GSL food web. Understanding the varied responses of benthic and pelagic organisms to lake conditions is also important because ongoing laboratory and field studies indicate these food chains may be linked (Barrett unpublished). Pelagic monitoring shows that shrimp overgraze phytoplankton over the summer, which may lead them to consume benthic algae on microbialites. Additionally, lab studies show that shrimp can survive on benthic algae (Barrett unpublished). Together, these observations suggest that microbialites may provide a critical summer food for shrimp (Barrett unpublished). Therefore, the impact of lake conditions on benthic producers could impact pelagic ecology in the GSL.

Studies in other saline lakes show that temperature and nutrients are driving factors structuring food webs (Afonina and Tashlykova 2018; Wen et al. 2005). To be clear, these observations are based on annual variability or longer time periods, not seasonal variability (Melack 1988). For instance, temperature and nutrients are a strong determinant of phytoplankton abundance and composition (Salm et al. 2009; Schagerl and Odour 2008). Temperature is an important factor that interacts with salinity to control benthic primary production and taxa composition (Chiu et al. 2005; Stenger-Kovács et al. 2014). Combined with the review of current GSL studies, these observations support the idea that salinity is not the only force at work in structuring food webs.

6.4 What Is the Future of Great Salt Lake?

Like all saline lakes (Wurtsbaugh et al. 2017), GSL is threatened by water diversions, effluents, and mineral extraction, as these anthropogenic activities affect salinity and N levels. Furthermore, water temperatures change with lake volume. All of these variables impact the food web in GSL. However, these activities, which are short-term policy decisions by society, must be framed in terms of the larger effect of anthropogenic climate change that is beyond short-term human control.

While the western US region is expected to become warmer and drier with periodic extreme droughts in the late twenty-first century as climate changes (Kunkel et al. 2013; Cook et al. 2015), the GSL watershed climate is a more local phenomenon. For the GSL watershed in Utah, Wyoming, and Idaho, we employ temperature and precipitation projections given by EPA (1998).

We acknowledge that recent climate models for the western United States provide updated projections on temperature and precipitation changes (Kunkel et al. 2013). However, these projections are not specific to Utah, and EPA (1998) provides seasonal ranges of projected temperature and precipitation changes for Utah. Additionally, we sought to be conservative in the projections for GSL, and note that the more recent seasonal projections for the western United States (Kunkel et al. 2013) are comparable to those given in EPA (1998). Therefore, we believe that the

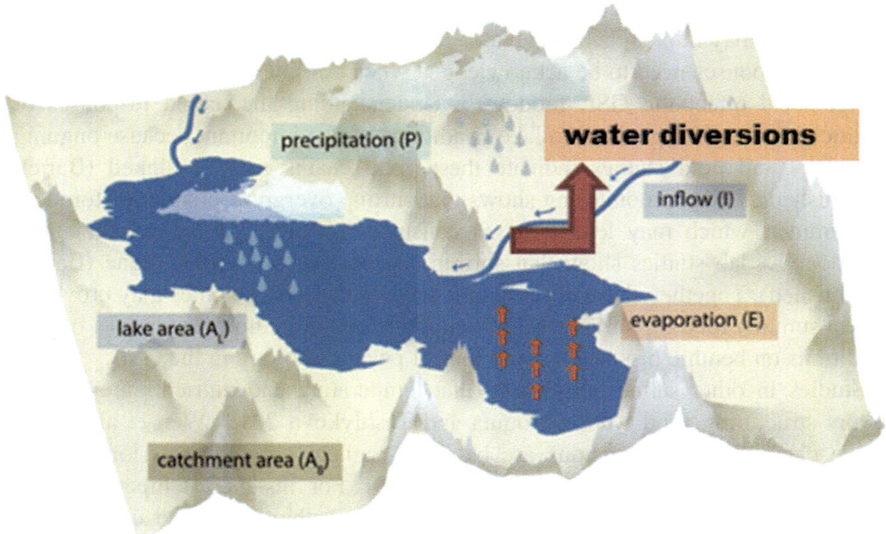

Fig. 6.6 Conceptual diagram of major processes influencing water levels in Great Salt Lake [modified from Huybers et al. (2016)]

temperature and precipitation projections are representative of potential future climate scenarios for the GSL watershed.

The projection for 2100 is for precipitation to increase by 13% in winter and decrease by 14% in summer, producing net greater precipitation as most precipitation occurs in winter (EPA 1998). By 2100, EPA (1998) predicted that temperature will increase on average by 5–6 °F in winter and summer and by 3–4 °F in spring and fall. The GSL watershed precipitation projections differ from the Western forecast due to higher temperatures and increased precipitation in winter and spring that promote more snowpack on the Wasatch Front surrounding GSL, which would contribute to increased freshwater inflows to the lake (Crosman and Horel 2009; Fig. 6.6). Given these projections and assuming no further water diversions, effluent addition, or mineral extraction, we ask how these climate changes may impact the GSL food web.

6.4.1 Projected Changes in Lake Conditions with Climate Change

Note that the 2100 climate projections are based on season (Table 6.2); therefore, our projected effects on the GSL food web must be seasonal. Furthermore, we compute the climate changes based on current seasonal average climate. These are conservative projections, as climate varies among years so that actual seasonal climate changes will be greater and less over years than the average values that we present.

Table 6.2 Projections of the Great Salt Lake ecosystem's components given anticipated climate changes in its watershed (EPA 1998)

Environment								
1994–2006 average	Temperature (C: degrees)		Inflow (I: proportion)		Salinity (S: ppt)		Max. Nitrogen (N: g/L)	
Winter	4				122		0.8	
Spring	16				117			
Summer	23				125			
Fall	9.1				131			
Future	Max	Min	Max	Min	Max	Min	Max	Min
Winter	9.58	5.13	1.7	1.2	114.26	75.8	0.7	0.46
Spring	19.73	16.96	1.2	1.05	109.39	72.5		
Summer	29	24.55	1.2	1.05	116.59	77.3		
Fall	12.4	9.63	1.5	1.1	122.86	81.5		
Primary producers								
Max. Phytoplankton (A: μg chl α/L)					Max. Microbialites (M: μg chl α/cm^2)			
1994–2006 average	64				54			
Future	Max		Min		Max		Min	
Winter	57.95 (Hi S/Lo T)		34.36 (Lo S/Hi T)					
Spring								
Summer					50.44 (Hi S/Lo T)		33.47 (Lo S/Hi T)	
Grazers								
	Avg. brine shrimp (BS: #/L)		Peak shrimp cysts (BC: #/L)		Spring cysts (SC: #/L)		Brine fly larvae (BF: #/m^2)	
1994–2006 average	7.4		90		36		4278.67	
Future	Max	Min	Max	Min	Max	Min	Max	Min
Winter					19.07 (HiS/Lo T)	4.22 (Lo S/Hi T)	6517.6 (Lo S/Hi T)	4668.83 (Hi S/Lo T)
Spring								
Summer	9.7 (Hi S/Hi T)	7.49 (LoS/Lo T)						
Fall			89.04 (Hi S/Hi T)	28.74 (Lo S/Lo T)				

(continued)

Table 6.2 (continued)

Predators		
Grebes (G: #/1000)		
1994–2006 average	1362.6	
Future	Max	Min
Fall	2554.3 (HiS/HiT)	1972.3 (Lo S/LoT)

Equations used in predictions	Source
Equation (1): S = annual I × current S; (r = 0.92, N = 132, P < 0.000005)	Belovsky et al. (2011)
Equation (2): N = annual I × current N; (r = 0.93, N = 10, P < 0.00008);	Belovsky et al. (2011)
Equation (3): $P = 0.16 \times S - 5.70 \times T + 0.12 \times T\text{^}2 - 0.18/N + 66.62$; ($r$ = 0.96, N = 11, P < 0.000001)	Larson and Belovsky (2013)
Equation (4): $M = 0.43 \times S - 0.000045 \times S\text{^}2 + 0.18 \times T - 0.0045 \times T\text{^}2 + 0.45 \times N - 1.18$; ($r$ = 0.77, N = 135, P < 0.0000001)	Anderson et al. (submitted)
Equation (5): BS = 0.044 × P + 0.61 × T − 9.72; (r = 0.91, N = 14, P < 0.01);	Belovsky et al. (2011)
Equation (6): PC = 4.12 × BS + 1.24 × S − 102.99; (r = 0.74, N = 16, P < 0.005)	Belovsky and Perschon (2019)
Equation (7): SC = PC × e^(−0.62–0.16 × T); (r = −0.69, N = 16, P < 0.05)	Belovsky et al. (2019)
Equation (8): BF = 7984.87–42.39 × S + 170.28 × BS; (r = 0.53, N = 47, P < 0.0008)	Belovsky et al. (2011), Barrett unpublished
Equation (9): G = 263.33 × BS; (r = 0.93, N = 17, P < 0.000001)	Belovsky and Perschon (2019)

The range of projections (max, min) and the associated projected environmental circumstances (temperature = T, salinity = S, and high = Hi, low = Lo) are presented. Current values are provided as the 1994–2006 average (Belovsky et al. 2011). If the current or projected values are seasonal, this also is presented. The equations employed to make projections and their source are listed as well

Because salinity and nutrient concentrations are currently observed to depend simply on lake volume, a function of inflows (Belovsky et al. 2011), future precipitation changes can be used to scale lake volume and thereby, the range of salinity and N concentrations. In addition, air temperature should directly affect lake water temperatures (Strong et al. 2014).

Annual surface water temperatures are predicted to increase on average by 33% (range: 11.7–56.2%) (Fig. 6.7, Table 6.2). With anticipated increases in inflows due to increased winter and spring precipitation (see Fig. 6.7), salinity will decrease on average by 22.2% (range: 6.4–38%) and N concentrations will decrease by 27.5% (range: 12.5–42.5%) (see Fig. 6.7 and Eqs. (1) and (2) in Table 6.2). Our projections for declining salinity are counter to most projections for saline lakes throughout the

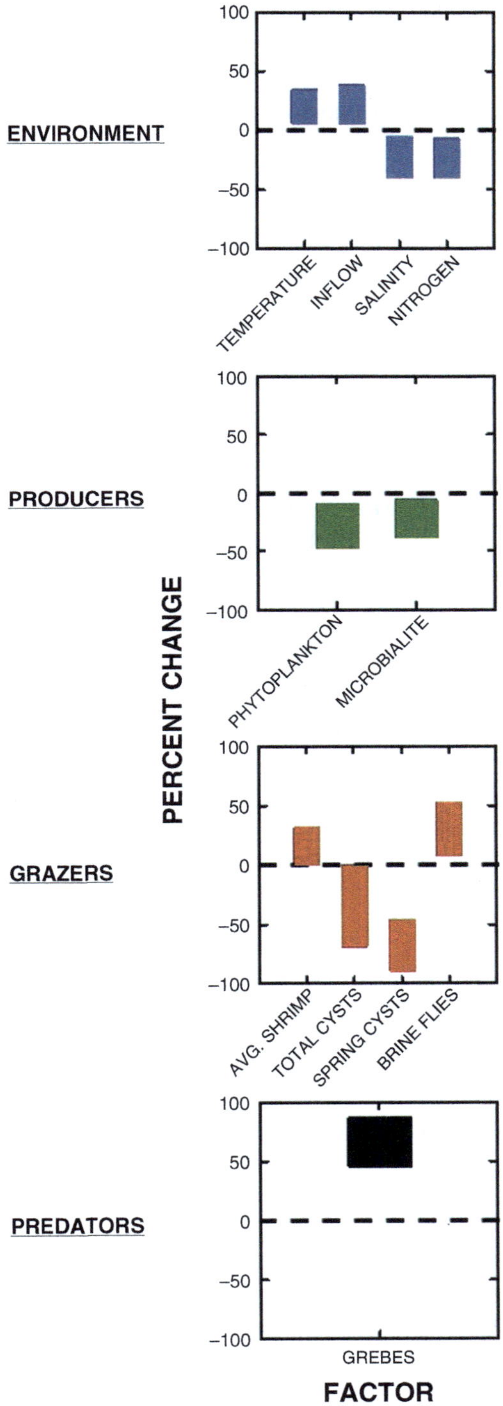

Fig. 6.7 Projected changes (percent change from current average conditions) of the abiotic conditions, primary producers, grazers, and predators in Great Salt Lake given climate change predictions

world based upon warmer conditions increasing evaporation and thereby salinity (Bedford 2009; Moore 2016; Williams 1993, 2002). However, the specific predictions for GSL suggest increased precipitation as temperature increases, which is distinct from the predictions of watershed conditions of other terminal basins.

6.4.2 Projected Changes in Pelagic and Benthic Primary Production

Given the projected changes in salinity, N, and temperature, changes in pelagic and benthic primary productivity were computed (Eqs. (3) and (4) in Table 6.2). Peak seasonal phytoplankton and microbialite primary producer abundance (measured as chlorophyll-*a*) are projected to decrease by an average of 27.3% (range: 9.5–46.3%) in winter and 22.3% (range: 6.6–38%) in summer, respectively, relative to current conditions (Fig. 6.7). The predicted declines are primarily due to reduced N concentrations as the higher inflows dilute nutrient pools. The increase in temperature further reduces phytoplankton production, but slightly increases benthic production (see Fig. 6.7). The reduced salinity has a much smaller impact than N and temperature. Finally, the seasons providing peak production (winter for the pelagic and summer for the benthic primary producers) do not change from current observations.

These projections do not consider a possible further reduction in N availability due to the DBL (deep brine layer). Dead organisms can sink into the DBL where the N sequestered in these organisms is unavailable until the lake infrequently mixes. Mixing of the deep and upper layers becomes less frequent as the upper layer becomes less saline (Belovsky et al. 2011; Naftz 2017); therefore, the projection for reduced future salinity may reduce mixing frequency and hold N longer in the DBL, where it is inaccessible to primary producers.

6.4.3 Projected Changes in Brine Shrimp, Brine Fly Larvae, and Eared Grebes

Given the projected changes in salinity, peak pelagic primary production and temperature, changes in brine shrimp abundance were computed (Eq. (5) in Table 6.2). Average brine shrimp abundance will increase in abundance by about 16.2% (range: 1.2–31.1%). Peak phytoplankton abundance is the limiting factor for brine shrimp in GSL, and the extent to which phytoplankton limits shrimp depends on the rate at which brine shrimp can harvest. Harvesting rate at a given phytoplankton abundance increases with temperature and decreases with salinity (Riisgård et al. 2015; Sura et al. 2017). Therefore, the projected decrease in phytoplankton abundance is countered by more efficient harvesting, as in 2100 GSL is predicted to be warmer and less saline.

While average brine shrimp abundance is projected to increase, peak brine shrimp cyst production in fall is predicted to decrease by 34.6% on average (range: 1.1–68.1%) (Eq. (6) in Table 6.2, Fig. 6.7). In addition, the cysts surviving through winter to start the brine shrimp population will decrease on average by 67.6% (range: 47–88.3%) (Eq. (7) in Table 6.2, Fig. 6.7). Several factors account for the predicted declines in Fall and Spring cyst abundances: (1) warmer conditions and more efficient harvesting of the phytoplankton in Summer and Fall lead the shrimp to reproduce more by live births than cysts (Belovsky et al. 2011), and (2) warmer conditions in winter lead to cysts breaking diapause when water temperatures are too cold for the nauplii to feed (~5 °C) (Belovsky et al. 2019). This decline becomes exacerbated when fewer spring cysts lead to fewer nauplii, causing the population to take longer to deplete phytoplankton, which leads female shrimp to shift from producing live young to cysts (Belovsky et al. 2011). These projected cyst dynamics will not provide the cyst numbers needed to permit a sustainable commercial harvesting of cysts (Belovsky and Perschon 2019).

Average brine fly larval abundance in summer is predicted to increase on average by 30.7% (range: 9.1–52.3%) (Eq. (8) in Table 6.2, Fig. 6.7). Larval responses to projected changes are similar to brine shrimp; their abundance will be positively affected by lower salinity and warmer water temperatures, because they feed more efficiently under these conditions.

Brine fly larvae are omnivorous, grazing on microbialite production, as well as a range of organic materials. Their density increases as the density of brine shrimp increases because shrimp produce fecal pellets and carcasses which settle on benthic surfaces that brine fly larvae graze. This may explain why the increase for brine fly larvae is greater than for brine shrimp (Fig. 6.7).

Eared Grebe abundance on GSL is projected to increase on average by 66% (range: 44.7–87.5%) (Eq. (9) in Table 6.2), as they are highly dependent on brine shrimp abundance. This is not to say that the overall Eared Grebe population increases, but the seasonal migrating population on GSL would increase in this model.

6.4.4 *Summary of Projections*

Based on the predicted climate changes in the GSL watershed, the food web will respond to increased precipitation and temperature because these factors alter inflows, water temperature, N, and salinity (Meng 2019). The projections reported above are best estimates given current knowledge from climatology and GSL ecology. By applying current knowledge of how the various GSL food web components respond to abiotic factors, we have presented how factors interact in complex ways to produce novel responses to climate change. Our projections also emphasize how temperature and nutrients impact the GSL food web more strongly than salinity.

The responses of lake conditions, primary producers, and consumers are based on field and laboratory studies in which the experimental conditions were representative of the range of conditions typically experienced in the lake. However, our projections, while based on quantitative regressions, are qualitative in nature and need to be interpreted with caution. We did not include in our projections how benthic and pelagic taxa composition may change under future predictions. Indeed, changes in taxa composition could exacerbate or dampen the responses reported here. Further studies on the pelagic and benthic food web components will strengthen future projections on how GSL will respond to climate change.

6.5 Concluding Remarks

After examining historical observations and employing future climate projections to current lake conditions, we may finally have an answer to the question posed at the beginning of this chapter: is salinity the driving factor in GSL?

Prior to current laboratory studies and long-term monitoring in GSL, people viewed salinity as the dominant driver of food web dynamics, but a review of historical and current studies shows that this is not the case. Rather, salinity has a minor impact and interacts with temperature and N levels to influence biota in indirect ways. Furthermore, the core communities in the GSL food web are adapted to tolerate the range of salinities historically experienced here (see Fig. 6.1).

By first exploring historical observations of salinity, phytoplankton, and brine shrimp abundance, one can see how changes in salinity were thought to be the primary influence on food web dynamics. Although some taxa, such as diatoms, appeared to decline with rising salinity, other factors not accounted for in historical surveys were also likely at work. Indeed, the review of current studies shows that since GSL is a hypersaline system, other factors become more important in structuring its food web (Fig. 6.1).

The projections presented in Sect. 6.4 indicate that GSL salinity and N concentrations may decline as a result of climate changes in precipitation and temperature (Fig. 6.7). In turn, we predicted declines in primary production and brine shrimp cysts, and increases in invertebrate and Eared Grebe abundance. The different responses of the GSL ecosystem to climate change will have direct effects on not only the benthic and pelagic food chains, but also on birds that depend on GSL for staging, foraging, and nesting habitat. Furthermore, other anthropogenic impacts such as water diversions and mineral extractions may exacerbate climate impacts and modify lake conditions. In this section we consider the conservation, economic, and management implications of GSL food web dynamics and our projections of how these may change in the future, as well as the immediate stressors surrounding GSL.

6.5.1 Conservation Implications

GSL is an important oasis in the arid Great Basin. Each year, millions of migratory birds and shorebirds use the lake and surrounding bays for staging and nesting habitat (Aldrich and Paul 2002). Among these are Eared Grebes that feed on abundant brine shrimp and brine flies, as well as the world's largest staging population of Wilson's Phalaropes (Belovsky et al. 2011; Frank and Conover 2017; Jellison et al. 2008). Because many bird species, including species of conservation concern, depend on GSL, it is a designated site on the Western Hemisphere Shorebird Reserve Network (USGS 2013). Under the above projections, Eared Grebes may be positively affected by climate change, but further work is needed to predict how other bird species abundances will be affected.

6.5.2 Implications for the Brine Shrimp Harvesting Industry

In addition to their ecological and conservation value for birds, brine shrimp cysts are harvested annually from GSL for their use in aquaculture, representing an estimated economic output of $56.7 million each year (Edwards and Null 2019). Long-term monitoring of GSL ecology since 1994 culminated in an adaptive management strategy to ensure enough cysts remained in the lake to sustain the brine shrimp population in the future (Belovsky and Perschon 2019). The qualitative projections in Sect. 6.4 present a potentially upsetting scenario for the brine shrimp cyst industry (Fig. 6.7). A future decline in cyst production would not only have negative impacts on Utah's economy, but it could lead aquaculture industries to begin exploiting resources in other saline environments. Hopefully, the projections highlight the need for continued adaptive management that considers the interplay of factors affecting brine shrimp dynamics.

6.5.3 Potential Impacts of Water Diversions, Mineral Extraction, and Effluents

Because most saline lakes are located in arid to semiarid regions and are terminal, the effects of climate change will be amplified compared to freshwater systems (Williams 2002). However, GSL is also affected by more immediate policies of water diversions, mineral extraction industries, and wastewater effluents (Jellison et al. 2008). These activities were not considered in Sect. 6.4 projections and will undoubtedly add further complexity to future changes in the GSL ecosystem.

Worldwide, water diversions are a major anthropogenic activity threatening saline lakes, including GSL (Bedford 2009; Shadkam et al. 2016). To be clear, salinity is an important determinant of lake productivity and species abundances when water diversions increase salinity beyond the tolerance limits of organisms (Williams 1998). Indeed, several saline lakes have experienced reductions in biological diversity and productivity with increasing salinity due to water diversions, e.g., the Aral Sea, Uzbekistan/Kazakhstan (Bedford 2005); Lake Urmia, Iran, (AghaKouchak et al. 2015); Mono Lake, California, USA (Ficklin et al. 2013); Lake Abert, Oregon, USA (Moore 2016); Lake Clifton, Australia (Smith et al. 2010); and Lake Corangamite, Australia (Tweed et al. 2011).

As the watersheds surrounding GSL become more developed, significant agricultural, stormwater, and wastewater discharges will impact the lake. These effluents contain possible contaminants such as excessive nutrients and heavy metals that could negatively impact GSL food web (UDEQ 2012). Excessive nutrients could alter nutrient limitation in GSL and affect phytoplankton dynamics by potentially changing taxa composition. Mineral extraction industries, such as magnesium and potassium, also generate potentially toxic byproducts that may leak into the lake, adding further stress on GSL (Bedford 2009; Larsen 2016). Projected increases in precipitation could make extracting minerals difficult and lead industries to intensify evaporation efforts which could harm the GSL ecosystem (Bedford 2009).

6.5.4 *Final Remarks*

Over the course of recorded history, GSL has experienced changes in water volume due to climate and human activities. The early studies of GSL were short term and narrow in scope, which prevented understanding how the food web changed with lake conditions. Current studies have been critical in establishing how the interplay of factors impacts the food web in various ways, and ongoing studies will deepen our understanding of these dynamics.

Climate change and the immediate threats described above will undoubtedly affect the GSL food web by altering abiotic and biotic factors in the lake. By forecasting changes to the GSL ecosystem, we hope to encourage stakeholders and managers to consider the interplay among multiple abiotic and biotic factors, not just salinity, and their effects on the GSL food web. Going forward, perhaps appropriate management strategies can be developed to preserve existing food web productivity, conservation values, and economic activities.

References

Afonina EY, Tashlykova NA (2018) Plankton community and the relationship with the environment in saline lakes of Onon-Torey plain, Northeastern Mongolia. Saudi J Biol Sci 25:399–408

AghaKouchak A, Norouzi H, Madani K, Mirchi A, Azarderakhsh M, Nazemi A, Nasrollahi N, Farahmand A, Mehran A, Hasanzadeh E (2015) Aral Sea syndrome desiccates Lake Urmia: call for action. J Great Lakes Res 41:307–311

Aldrich TW, Paul DS (2002) Avian ecology of Great Salt Lake. In: Gwynn JW (ed) Great Salt Lake: an overview of change, DNR Special Publication. Utah Geological Survey, Salt Lake City, UT, pp 343–385

Allen WC (1926) Some interesting animal communities of northern Utah. Sci Monthly 23:481–495

Almeida-Dalmet S, Litchfield CD, Gillevet P, Baxter BK (2018) Differential gene expression in response to salinity and temperature in a haloarcula strain from Great Salt Lake, Utah. Genes 9:52

Arnow T, Stephens DW (1990) Hydrologic characteristics of the Great Salt Lake, Utah. United States Geological Survey, Water-Supply Paper 2332

Barnes BD, Wurtsbaugh WA (2015) The effects of salinity on plankton and benthic communities in the Great Salt Lake, Utah, USA: a microcosm experiment. Can J Fish Aquat Sci 72:807–817

Baxter BK (2018) Great Salt Lake microbiology: a historical perspective. Int Microbiol 21 (3):79–95

Bedford D (2005) Utah's Great Salt Lake: a complex environmental-societal system. Geogr Rev 95:73–96

Bedford D (2009) The Great Salt Lake America's Aral Sea? Environment 51:8–19

Belovsky GE (2002) Brine shrimp population dynamics and sustainable harvesting in the Great Salt Lake, Utah. 2002 Progress report to the Utah Department of Wildlife Resources, Salt Lake City, Utah

Belovsky GE, Larson C (2001) Brine shrimp population dynamics and sustainable harvesting in the Great Salt Lake, Utah. 2001 Progress report to the Utah Department of Wildlife Resources, Salt Lake City, Utah

Belovsky GE, Mellison C (1997) Brine shrimp population dynamics and sustainable harvesting in the Great Salt Lake. 1997 Progress Report to the Utah Division of Wildlife Resources, Salt Lake City, Utah

Belovsky GE, Mellison C (1998) Brine shrimp population dynamics and sustainable harvesting in the Great Salt Lake, Utah. 1998 Progress report to the Utah Department of Wildlife Resources, Salt Lake City, Utah

Belovsky GE, Perschon WC (2019) A management case study for a new commercial fishery: brine shrimp harvesting in Great Salt Lake, Utah, USA. Ecol Appl 29:e01864

Belovsky GE, Kilham S, Larson C, Mellison C (1999) Brine shrimp population dynamics and sustainable harvesting in the Great Salt Lake, Utah. 1999 Progress report to the Utah Department of Wildlife Resources, Salt Lake City, Utah

Belovsky GE, Larson C, Mellison C (2000) Brine shrimp population dynamics and sustainable harvesting in the Great Salt Lake, Utah. 2000 Progress report to the Utah Department of Wildlife Resources, Salt Lake City, Utah

Belovsky GE, Stephens D, Perschon C, Birdsey P, Paul D, Naftz D, Baskin R, Larson C, Mellison C, Luft J, Mosley R, Mahon H, Van Leeuwen J, Allen DV (2011) The Great Salt Lake Ecosystem (Utah, USA): long term data and a structural equation approach. Ecosphere 2:40

Belovsky GE, Perschon C, Larson C, Mellison C, Slade J, Mahon H, Appiah-Madson H, Luft J, Mosley R, Neill J, Stone K, Kijowski A, Van Leeuwen J (2019) Overwinter survival of crustacean diapausing cysts: brine shrimp (Artemia franciscana) in Great Salt Lake, Utah. Limnol Oceanogr 64(6):2538–2549

Ben-Amotz A, Avron M (1983) Accumulation of metabolites by halotolerant algae and its industrial potential. Annu Rev Microbiol 37:95–119

Brock TD (1975) Salinity and the ecology of *Dunaliella* from Great Salt Lake. J Gen Microbiol 89:285–292

Cannon JS, Cannon MA (2002) The southern Pacific railroad trestle – past and present. In: Gwynn JW (ed) Great Salt Lake: an overview of change. Special Publication of the Utah Department of Natural Resources, Salt Lake City, UT, pp 283–294

Carozzi AV (1962) Observations of algal biostromes in the Great Salt Lake. J Geol 70:246–252

Chiu JM, Thiyagarajan YV, Tsoi MMY, Qian PY (2005) Qualitative and quantitative changes in marine biofilms as a function of temperature and salinity in summer and winter. Biofilms 2:183–195

Clavero E, Hernádez-Mariné M, Grimalt JO, Garcia-Pichel F (2000) Salinity tolerance of diatoms from thalassic hypersaline environments. J Phycol 36:1021–1034

Collins N (1980) Population ecology of Ephydra cinerea Jones (Diptera: Ephydridae), the only benthic metazoan of the Great Salt Lake, U.S.A. Hydrobiologia 68:99–11

Cook BI, Ault TR, Smerdon JE (2015) Unprecedented twenty-first century drought risk in the american southwest and central plains. Sci Adv 1(1):e1400082. https://doi.org/10.1126/sciadv.1400082

Croghan P (1958) The osmotic and ionic regulation of Artemia salina (L.). J Exp Biol 35:219–233

Crosman ET, Horel JD (2009) MODIS-derived surface temperature of the Great Salt Lake. Remote Sens Environ 113:73–81

Cuellar O (1990) Ecology of brine shrimp from Great Salt Lake, Utah, U.S.A. (Branchiopoda, Anostraca). Crustaceana 59:25–34

Daines LL (1910) On the flora of Great Salt Lake. Am Nat 51:499–506

Daines LL (1917) Physiological experiments on some algae of Great Salt Lake, Thesis, University of Utah, Salt Lake City, UT

Edwards EC, Null SE (2019) The cost of addressing saline lake level decline and the potential for water conservation markets. Sci Total Environ 651:435–442. https://doi.org/10.1016/j.scitotenv.2018.09.006

Environmental Protection Agency, United States (1998) Climate change and Utah, Elusive Documents, Paper 56. https://digitalcommons.usu.edu/elusive_docs/56

Felix EA, Rushforth SR (1979) The algal flora of the Great Salt Lake, Utah, USA. Nova Hedwigia 31:163–195

Felix EA, Rushforth SR (1980) Biology of the south arm of the Great Salt Lake, Utah. In: Gwynn JW (ed) Great Salt Lake: a scientific, historical and economic overview. Utah Geological and Mineral Survey Bulletin 116, Salt Lake City, UT, pp 305–312

Ficklin DL, Stewart IT, Maurer EP (2013) Effects of projected climate change on the hydrology in the Mono Lake Basin, California. Climate Change 116:111–131

Flowers S (1934) Vegetation of the Great Salt Lake Region. Bot Gaz 95(3):353–418

Flowers S, Evans FK (1966) The flora and fauna of the Great Salt Lake region, Utah. In: Boyko H (ed) Salinity and aridity New approaches to old problems. Dr. W. Junk Publishers, The Hague

Frank MG, Conover MR (2017) Weather and prey availability affect the timing of fall migration of Eared Grebes (*Podiceps nigricollis*) from Great Salt Lake. Wilson J Ornithol 129:98–111

Gajardo GM, Beardmore JA (2012) The brine shrimp *Artemia*: adapted to critical life conditions. Front Physiol 3:1–8

Galat DL, Robinson R (1983) Predicted effects of increasing salinity on the crustacean zooplankton community of Pyramid Lake, Nevada. Hydrobiologia 105:115–131

Gray LJ (2012) Report to Utah Department of Environmental Quality, Salt Lake City, UT, USA. https://deq.utah.gov/locations/G/greatsaltlake/willard-spur/docs/2013/11Nov/Macroinvertebrates2011_LJGray.pdf

Hammer UT (1986) Saline lake ecosystems of the world. Monographiae Biologicae 59. Dordrecht, Dr. W. Junk Publishers

Handy A (1967) Distinctive brines in Great Salt Lake, Utah. US Geol Surv Prof Paper 575-B:225–227

Herbst DB (2001) Gradients of salinity stress, environmental stability and water chemistry as a templet for defining habitat types and physiological strategies in inland salt waters. Hydrobiologia 466:209–219

Herbst DB, Blinn DW (1998) Experimental mesocosm studies of salinity effects on the benthic algal community of a saline lake. J Phycol 34:772–778

Herbst DB, Bradley TJ (1989) Salinity and nutrient limitations on growth of benthic algae from two alkaline salt lakes of the western great basin (USA). J Phycol 25:673–678

Huybers K, Rupper S, Roe GH (2016) Response of closed basin lakes to interannual climate variability. Clim Dyn 46:3709–3723

Javor B (1989) Hypersaline environments. Springer, Berlin

Jellison R, Williams WD, Timms B, Alcocer J, Aladin NV (2008) Salt lakes: values, threats and future. In: Polunin NVC (ed) Aquatic ecosystems. Cambridge University Press, Cambridge

Kirkpatrick R (1934) The life of Great Salt Lake, with special reference to the algae. Thesis. Salt Lake City, UT: University of Utah

Kunkel KE, Stevens LE, Stevens SE, Sun L, Janssen E, Wuebbles D, Redmond KT, Dobson JG (2013) Regional climate trends and scenarios for the U.S. National Climate Assessment, NOAA Technical Report NESDIS 142-5, Part 5 Climate of the Southwest U.S.

Larsen L (2016) Mineral extraction on Great Salt Lake has local, national and global impact Standard Examiner. https://www.standard.net/news/environment/mineral-extraction-on-great-salt-lake-has-local-national-and/article_875d73e9-9100-54a5-a0e9-371e012c3945.html

Larson CA (2004) Experimental examination of the factors affecting growth and species of composition of phytoplankton from Great Salt Lake, Utah, Thesis, Utah State University, Logan, UT

Larson CA, Belovsky GE (2013) Salinity and nutrients influence species richness and evenness of phytoplankton communities in microcosm experiments from Great Salt Lake, Utah, USA. J Plankton Res 35:1154–1166

Lindsay MR, Anderson C, Fox N, Scofield G, Allen J, Anderson E, Bueter L, Poudel S, Sutherland K, Munson-McGee JH, Van Nostrand JD, Zhou J, Spear JR, Baxter BK, Lageson DR, Boyd ES (2017) Microbialite response to an anthropogenic salinity gradient in Great Salt Lake, Utah. Geobiology 2016:1–15. https://doi.org/10.1111/gbi.12201

Lindsay MR, Johnston RE, Baxter BK, Boyd ES (2019) Effects of salinity on microbialite-associated production in Great Salt Lake, Utah. Ecology 100(3):e02611

Madison RJ (1970) Effects of a causeway on the chemistry of the brine in Great Salt Lake. Utah Geological Survey Water-Resources Bulletin 14, Utah, 52 p

Marcarelli AM, Wurtsbaugh WA, Griset O (2006) Salinity controls phytoplankton response to nutrient enrichment in the Great Salt Lake, Utah, USA. Can J Fish Aquat Sci 63:2236–2248

Melack JM (1988) Primary producer dynamics associated with evaporative concentration in a shallow, equatorial soda lake (Lake Elmenteita, Kenya). Hydrobiologia 158:1–14

Mellison SC (2000) Functional response of a water boatman (Trichocorixa verticalis) and environmental conditions that affect its distribution in the Great Salt Lake, Utah, USA. Thesis, Utah State University, Logan, UT

Meng Q (2019) Climate change and extreme weather drive the declines of saline lakes: a showcase of the Great Salt Lake. Climate 7:1–12

Moore JN (2016) Recent desiccation of western great basin saline lakes: Lessons from Lake Abert, Oregon, U.S.A. Sci Total Environ 554–555:142–154. https://doi.org/10.1016/j.scitotenv.2016.02.161

Morgan DL (1947) The Great Salt Lake. University of Nebraska Press

Naftz D (2017) Inputs and internal cycling of nitrogen to a causeway influenced, hypersaline lake, Great Salt Lake, Utah, USA. Aquat Geochem 23:199–216

Nemenz H (1960) On the osmotic regulation of the larvae of Ephydra cinerea. J Insect Physiol 4:38–44

Ogata EM, Wurtsbaugh WA, Smith TN, Durham SL (2017) Bioassay analysis of nutrient and Artemia franciscana effects on trophic interactions in the Great Salt Lake, USA. Hydrobiologia 788:1–16

Packard AS Jr (1879) The sea-weeds of Great Salt Lake. Am Nat 13:701–703

Patrick R (1936) Some diatoms of Great Salt Lake. Bull Torrey Bot Club 63:157–166

Post FJ (1977) The microbial ecology of Great Salt Lake. Microb Ecol 3:143–165

Rawley EV (1980) Wildlife of the GSL. In: Gwynn JW (ed) Great Salt Lake: a scientific, historical and economic overview. Utah Geological and Mineral Survey Bulletin 116, Salt Lake City, UT, pp 287–312

Rees DM (1942) Animals living in the Great Salt Lake. Mineral Soc Utah News Bull 3:57–58
Relyea GM (1937) The brine shrimp of Great Salt Lake. Am Nat 71:612–616
Riisgård HU, Zalacáin D, Jeune N, Wiersma JB, Lüskow F, Pleissner D (2015) Adaptation of the brine shrimp *Artemia salina* (Branchipoda: Anostraca) to filter-feeding: effects of body size and temperature on filtration and respiration rates. J Crustac Biol 35:650–658
Rushforth SR, Felix EA (1982) Biotic adjustments to changing salinities in the Great Salt Lake, Utah, USA. Microb Ecol 8:157–161
Salm CR, Saros JE, Martin CS, Erickson JM (2009) Patterns of seasonal phytoplankton distribution in prairie saline lakes of the northern Great Plains (USA). Saline Syst 5:1–13
Sanders HL (1969) Benthic marine diversity and the stability-time hypothesis. Brookhaven Symp Biol 22:71–80
Schagerl M, Odour SO (2008) Phytoplankton community relationship to environmental variables in three Kenyan Rift Valley saline-alkaline lakes. Mar Freshw Res 59:125–136
Schapira M, Buscot MJ, Pollet T, Leterme SC, Seuront L (2010) Distribution of picophytoplankton communities from brackish to hypersaline waters in a South Australian coastal lagoon. Saline Syst 6:1–15
Schwarz EA (1891) Preliminary remarks on the insect fauna of the Great Salt Lake. Utah Can Entomol 23:235–241
Shadkam S, Ludwig F, Oel P, Kirmit C, Kabat P (2016) Impacts of climate change and water resources development on the declining inflow into Iran's Urmia Lake. J Great Lakes Res 42:942–952
Slobodkin LB, Sanders HL (1969) On the contribution of environmental predictability to species diversity. Brookhaven Symp Biol 22:82–93
Smith MD, Goater SE, Reichwaldt ES, Knott B, Ghadouani A (2010) Effects of recent increases in salinity and nutrient concentrations on the microbialite community of Lake Clifton (Western Australia): are the thrombolites at risk? Hydrobiologia 649:207–216
Stenger-Kovács E, Lengyel K, Buczkó FM, Tóth LO, Crossetti A, Pellinger Z, Doma Z, Padisák J (2014) Vanishing world: alkaline, saline lakes in Central Europe and their diatom assemblages. Inland Waters 4:383–396
Stephens DW (1974) A summary of biological investigations concerning the Great Salt Lake, Utah, 1861–1973. Great Basin Nat 34:221–229
Stephens DW (1990) Changes in lake levels, salinity and the biological community of Great Salt Lake (Utah, USA), 1847–1987. Hydrobiologia 197:139–146. https://doi.org/10.1007/BF00026946
Stephens DW (1998a) Salinity-induced changes in the aquatic ecosystem of Great Salt Lake, Utah. U.S. Geological Survey, Salt Lake City, Utah: Utah Geological Association Guidebook 26
Stephens DW (1998b) Brine shrimp ecology in the Great Salt Lake, Utah Progress Report to the Utah Division of Wildlife Resources Salt Lake City, Utah
Stephens DW, Birdsey PW (2002) Population dynamics of the brine shrimp, *Artemia franciscana*, in Great Salt Lake and regulations of commercial shrimp harvest. In: Gwynn, JW (ed) Great Salt Lake a scientific, historical, and economic overview. Utah Geological and Mineral Survey Bulletin 116, Salt Lake City, UT, pp 327–336
Stephens DW, Gillespie DM (1976) Phytoplankton production in the Great Salt Lake, Utah, and a laboratory study of algal response to enrichment. Limnol Oceanogr 21:74–87
Strong C, Kochanski AK, Crosman ET (2014) A slab model of the Great Salt Lake for regional climate simulation. J Adv Model Earth Syst 6:602–615
Sura S, Herlihy N, Mahon H, Belovsky G (2017) Environmental impacts on grazing of different brine shrimp (Artemia franciscana) life stages. Hydrobiologia 792:97–104
The Deseret News (1907) Weary path trodden by intrepid band, to the Shores of America's Dead Sea, pp 4–5, July 27, 1907
Tilden J (1898) American Algae. Century III:298
Tweed S, Grace M, Leblanc M, Cartwright I, Smithyman D (2011) The individual response of saline lakes to a severe drought. Sci Total Environ 409:3919–3933

USGS (2013) https://ut.water.usgs.gov/greatsaltlake/birds/
Utah Department of Environmental Quality/Division of Water Quality (2012) A Great Salt Lake Water Quality Strategy Salt Lake City, Utah, p 38
Velasco J, Hernández J, Gutiérrez C, Abellán P, Sánchez D, Ruiz M (2006) Response of biotic communities to salinity changes in a Mediterranean hypersaline system. Saline Syst 2:1–15
Wen Z, Mian-ping Z, Xian-zhong X, Xi-Fang L, Gan-lin G, Zhi-hui H (2005) Biological and ecological features of saline lakes in northern Tibet, China. Hydrobiologia 541:189–203
Williams WD (1993) The worldwide occurrence and limnological significance of falling water-levels in large, permanent saline lakes. Hydrobiologia 267:291–306
Williams WD (1998) Salinity as a determinant of the structure of biological communities in salt lakes. Verhandlungen der Internationalen Vereinigung für Theoretische und Angewandte Limnologie 25:980–983
Williams WD (2001) Anthropogenic salinization of inland waters. Hydrobiologia 466:329–337
Williams WD (2002) Environmental threats to salt lakes and the likely status of inland saline ecosystems in 2025. Environ Conserv 29:154–167
Wirick CD (1972) Dunaliella-Artemia plankton community of the Great Salt Lake, Utah, Thesis, University of Utah, Salt Lake City, UT
Woodbury AM (1936) Animal relationships of Great Salt Lake. Ecology 17:1–8
Wurtsbaugh WA (1992) Food-Web modification by an invertebrate predator in the Great Salt Lake (USA). Oecologia 89:168–175
Wurtsbaugh WA, Berry TS (1990) Cascading effects of decreased salinity on the plankton, chemistry, and physics of the Great Salt Lake (Utah). Can J Fish Aquat Sci 47:100–109
Wurtsbaugh WA, Marcarelli AM, Boyer GL (2012) Final report to the Utah Division of Water Quality Salt Lake City, UT. https://digitalcommons.usu.edu/cgi/viewcontent.cgi?article=1548&context=wats_facpub
Wurtsbaugh WA, Richards C, Hodson J, Rasmussen C, Winter C (2015) Report to the Utah Division of Water Quality, Salt Lake City, UT, USA. http://digitalcommons.usu.edu/wats_facpub/879/
Wurtsbaugh WA, Miller C, Null SE, DeRose RJ, Wilcock P, Hahnenberger M, Howe F, Moore J (2017) Decline of the world's saline lakes. Nat Geosci 10:816–823

Chapter 7
Great Salt Lake *Artemia*: Ecosystem Functions and Services with a Global Reach

Brad Marden, Phil Brown, and Thomas Bosteels

Abstract The anostracan crustacean *Artemia franciscana* is the most abundant zooplankter in Great Salt Lake (GSL) and generally the only zooplankton in the largest bay (Gilbert Bay) of this hypersaline system. Colloquially referred to as brine shrimp, *Artemia* are crucially important organisms in GSL and provide numerous ecosystem services including the control of eutrophication in this naturally productive lake, an abundant energy supply to a large avian population along hemispheric flyways, and critical support of global aquaculture through the large-scale commercial harvest of the resting eggs (cysts) for use as live feed in shrimp and finfish production across the world. This chapter examines the GSL *Artemia* population and its management from multiple angles. The successful adaptive management of the *Artemia* resource is discussed as a model of cooperative public and private research. An extensive body of research on the biochemistry and physiology of diapause and quiescence among *Artemia* cysts is reviewed. Population structure and patterns of GSL *Artemia* are examined across annual and multi-decadal timescales using large datasets of public and private research programs. Population level responses to spatial and temporal fluctuations in salinity are evaluated. Top-down and bottom-up controls on the *Artemia* population are reviewed, including the influence of salinity stratification (meromixis) on nutrient distribution within the lake and new molecular evidence of benthic linkages to the *Artemia* population via microbialites. Finally, we provide an assessment of threats to the GSL *Artemia* population and a summary of management structures and initiatives in place to mitigate them.

Keywords Great Salt Lake · *Artemia* · Meromixis · Salinity · Nutrients · Dormancy · Ecosystem · Adaptive management · Brine shrimp harvest · Sustainability

---

B. Marden · P. Brown · T. Bosteels (✉)
Great Salt Lake Brine Shrimp Cooperative, Ogden, UT, USA
e-mail: thomas@gsla.us

B. K. Baxter, J. K. Butler (eds.), *Great Salt Lake Biology*,
https://doi.org/10.1007/978-3-030-40352-2_7

7.1 Introduction

Great Salt Lake (GSL) differs from most other saline lakes in that it is a system of interconnected bays separated by causeways. The bays differ substantially in salinity, water residence time, community structure, and nutrient dynamics, yet all interact and contribute to the integrity of the GSL ecosystem. This chapter focuses on the south arm of the lake, Gilbert Bay, and its resident *Artemia franciscana* (brine shrimp) population. Other bays are considered in terms of their relationship to *Artemia* via nutrient loading and cycling, primary production, biotic assemblages, *Artemia* predators, and inflow. *Artemia* cysts have been documented in GSL sediments (and prior lakes including the Pleistocene Lake Bonneville) and there have been estimates of cysts dating back 600,000 years ago or longer (Eardley and Gvosdetsky 1960). *Artemia* were present throughout GSL prior to the construction of a railroad causeway in 1959, and have been mainly confined to Gilbert Bay since 1962 (Sturm et al. 1980). The salinity of Gilbert Bay generally remains in an intermediate range suitable for *Artemia*, whereas other bays are either too saline or too dilute to be optimal.

Throughout this chapter, we will be primarily using the genus name *Artemia* to refer to the aquatic invertebrate that is the main subject of this chapter. In more commonplace terminology, *Artemia* are often referred to as "brine shrimp," a term we will restrict to discussion of management of the resource and the history of the "brine shrimp industry." When commenting on scientific studies in which the species of *Artemia* was determined specifically, or is known with confidence, then the species name is provided as in: *Artemia franciscana* or *A. franciscana*. Furthermore, the use of the genus name only to reference this organism is widely used and accepted in the scientific literature and therefore has been adopted for use in this chapter.

Artemia are extraordinary organisms in possession of unique biochemical products and pathways, cellular functions, and evolved behavioral responses which confer a resiliency to extreme conditions unparalleled in other animal life forms (Gajardo and Beardmore 2012). The specialized reproductive ability to produce live young, "nauplii," or encysted embryos, "cysts," for overwintering gives this arthropod a flexible life cycle and the ability to survive desiccation over time (Browne and Lenz 1991) (Fig. 7.1). The resident *Artemia* population in GSL performs fundamental ecological roles through its exponential capacity to transfer energy, carbon, and nutrients through the ecosystem. It also provides a wealth of ecosystem services for the surrounding human population and supports critical protein (shrimp and fish) production across the globe.

An intensification of scientific research over the past two decades has illuminated some of the interrelationships between the GSL biota and the resident *Artemia* population. The information below represents a distillation of scientific investigations completed on GSL *Artemia* and the GSL ecosystem. Published scientific reports and previously unpublished research are provided with the intention of presenting a reasonably comprehensive view of this keystone species.

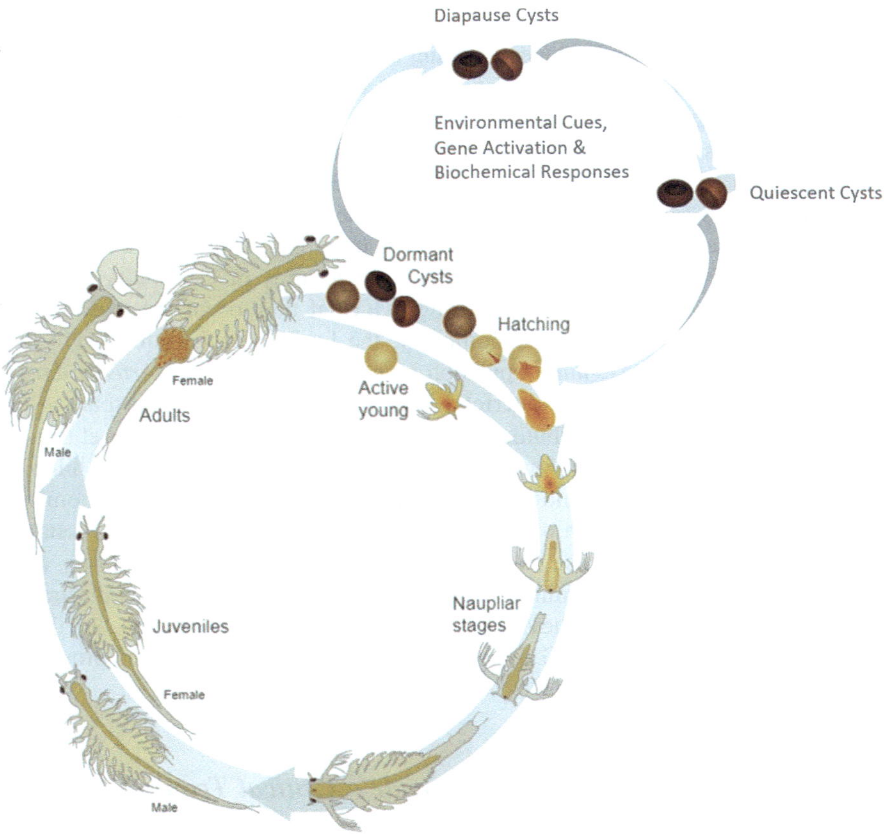

Fig. 7.1 Life cycle of Great Salt Lake *Artemia*. Gravid females produce young via two pathways: ovoviviparous (live free swimming nauplii) or oviparous (dormant encysted gastrula stage embryos). Dormant cysts are initially in diapause following which environmental conditions and endogenous changes cause the embryo to enter into a quiescent state which is then primed for hatching given suitable conditions. Upon hatching nauplii proceed through multiple instar stages until reaching a mature adult stage. Modified image is courtesy of https://learn.genetics.utah.edu/content/gsl/artemia/

7.2 Ecosystem Functions and Services

7.2.1 A Dependable Artemia Population Provides Ecosystem Services and Economic Benefits

Ecosystem services of GSL *Artemia* are defined as those features and attributes of the *Artemia* population that contribute to human well-being. These services can be divided into provisioning, regulating, supporting, and cultural categories (Millennium Ecosystem Assessment 2005; Zadereev 2018). With respect to *A. franciscana: Provisioning* includes the production of cysts for the multi-billion dollar international aquaculture industry which provides critical protein to meet growing global demand; *Regulating* includes reducing algal blooms which enhances evaporation in salt ponds to the benefit of the salt extraction industries; *Supporting* includes sustaining avian species desirable for tourism, bird watching, and hunting; *Cultural* includes environmental education, scientific research, and maintaining water clarity in a way that enhances recreational uses. Management of GSL elevation, volume, and salinity that is suitable for a robust *Artemia* population serves multiple ecosystem functions and services by providing a flow of carbon and nutrients through the food web that sustains a broad array of GSL biota and that prevents trophic bottlenecks from occurring, which if they occur are undesirable from a human use perspective. Economically, the businesses that are directly or indirectly related to the *Artemia* population include the commercial *Artemia* business, salt extraction industries, waterfowl hunting, bird watching, and tourism to GSL. These are all major contributors to the $1.32 billion in economic benefits that GSL brings annually for Utah with each contributing $56.7 million, $1.131 billion, and $135.8 million (hunting and tourism combined), respectively (Bioeconomics, Inc. 2012). Additionally, much of the funding for fundamental ecological research on GSL has come from license fees and royalties from the *Artemia* industry that contribute approximately $3.6 million annually to the State of Utah. This illustrates the financial linkages between the *Artemia* population and human benefits, and these values can be diminished if the GSL volume and salinity are not managed properly.

7.2.2 Great Salt Lake Artemia Provide a Critical Trophic Linkage Supporting Millions of Birds

Gilbert Bay and its *Artemia* population connect to a larger hemispheric ecosystem by supporting several species of waterfowl and shorebirds that utilize the bay in large numbers during critical life stages. GSL was designated in 1991 as a Western Hemisphere Shorebird Reserve site due to its importance for shorebirds (WHSRN 2019). Up to 4.5 million eared grebes stage annually in Gilbert Bay while feeding primarily on *Artemia* in preparation for migratory flight (Roberts 2013a). *Artemia* adults and cysts provide high energy food for northern shovelers and green-wing teal during food-limited winter months when adjacent wetlands have frozen (Roberts 2013a, b). Wilson's and red-necked phalaropes (Frank 2016) visit the lake by the hundreds of thousands to feed on the abundant brine fly *Ephydra* in Gilbert Bay (Paul and Manning 2002). The ecological importance extends to the outlying habitat of connected shoreline, wadable shallows, mudflat, and wetland habitats which support an expansive variety and abundance of waterfowl and shorebird species. Over 860,000 people visit GSL state parks and marinas each year with recreation and bird watching generating a net economic value of $26.3 million per year and annual waterfowl hunting revenues producing an additional $9.6 million for the economy of Northern Utah (Bioeconomics Inc. 2012).

7.2.3 *Artemia* Grazing Controls Eutrophication

One of the combined ecosystem functions and services that *Artemia* delivers is the functional capacity to transfer carbon, nutrients, and energy from primary producers to the food web while simultaneously cleaning the water in a way that makes it more visually appealing for human visitors to the lake. The enormous grazing capacity of the GSL *Artemia* population converts hypereutrophic conditions (secchi depth <0.5 m) into clear-water conditions (secchi depth >5 m), preferred for human usage of the lake (Smith et al. 1995; Riera et al. 2001; Belovsky et al. 2011; GSLBSC unpublished). The cycle begins in late fall and early winter when GSL experiences periods of rapid algal growth that result in average winter chlorophyll-*a* levels around 40–50 μg/L, but that can be well in excess of 80 μg/L, and that would classify the waterbody as hypereutrophic (Stephens and Gillespie 1976; Belovsky et al. 2011; Wurtsbaugh et al. 2012). These eutrophic conditions are a source of substantial concern in most lakes, especially freshwater systems, in which high chlorophyll levels are caused by cyanobacteria and where they lead to anoxia, cyanotoxin production, and significant declines in desirable taxa (Edmondson et al. 1956; Anderson et al. 2002; Smith et al. 2006; Heisler et al. 2008; Smith and Schindler 2009). However, concerns about adverse consequences of eutrophic conditions in GSL are unwarranted. Hypereutrophic status in Gilbert Bay does not represent an adverse condition and in fact it represents a "normal" cyclical characteristic of biological productivity within this system. High algal abundance in late winter and early spring is rapidly grazed by the burgeoning *Artemia* population, transforming the bay from hypereutrophic to pseudo-oligotrophic (<1 μg Chl-a/L) generally in 2–3 weeks (Fig. 7.2) (Stephens and Gillespie 1976; Melack and Jellison 1998; Belovsky et al. 2011). This boom-and-bust cycle is characteristic of other endorheic lakes and is necessary to support the robust exponential growth of the *Artemia* population. The grazing capacity of *Artemia* is also highly favorable for the salt extraction industry as it not only facilitates evaporation rates but also enhances the quality and timing of salt crystallization, thus supporting a billion-dollar industry of GSL (Sorgeloos and Tackaert 1991; Davis and Giordano 1995; Davis 2000).

7.2.4 *Artemia* Provide the Foundation for Global Aquaculture Production

Artemia have been used, as a living food or "live feed" for larger animals, in aquaculture since the 1930s (Bengtson et al. 1991), and GSL *A. franciscana* were first introduced for aquaculture production in the 1950s by the Sanders Brine Shrimp Company (Sturm et al. 1980). Since that time *A. franciscana*, and other live feeds, have become a foundation for the global aquaculture industry. That industry provides food and high-quality protein for a growing human population on a massive—and increasingly important—scale. All future predictions about the global need for protein include aquaculture as the major provider (Bostock et al. 2010) and fisheries

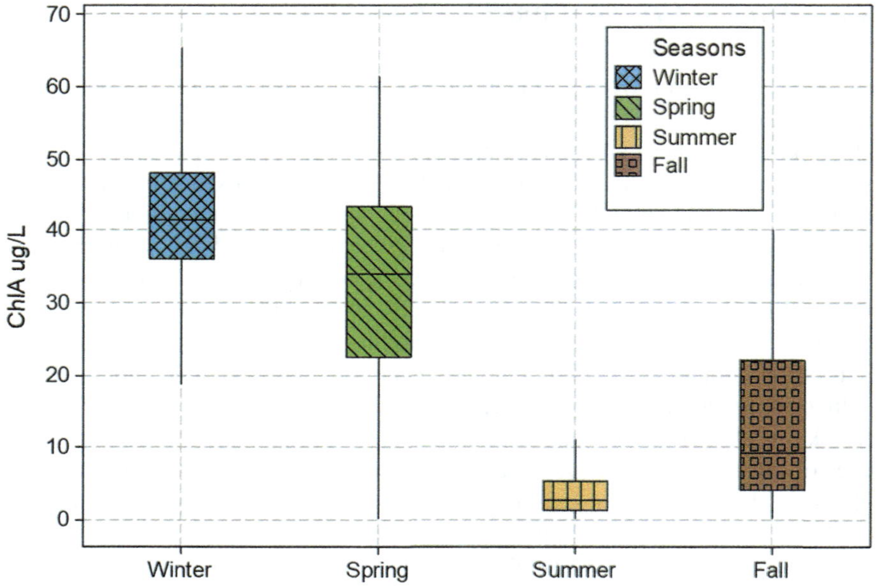

Fig. 7.2 Great Salt Lake epilimnetic chlorophyll-*a* levels (μg/L) from 2012 to 2017. Chlorophyll-*a* levels are severely depleted during summer due to excessive grazing pressure by *Artemia*. In fall this constraint on algal growth subsides and chlorophyll-*a* levels rebound. Peak concentrations approach hypereutrophic status, but do not indicate any adverse condition on Gilbert Bay. In fact, high winter chlorophyll-*a* levels are associated with more robust *Artemia* population dynamics during the summer

are undergoing a transition that resembles extensive aquaculture or is a hybridization of fisheries and aquaculture rather than traditional capture fisheries (Anderson et al. 2002; Klinger et al. 2013). The dominant role of aquaculture is in evidence in the 2014 fish consumption statistics that showed aquaculture production exceeding the global fish capture level for human consumption (FAO 2018a, b). In 2016, the aquaculture industry produced 80 million tons of fish (FAO 2018a), and this is projected to increase to 102 million tons by 2025. The total fisheries and aquaculture production in 2016 was estimated at USD $362 billion, of which USD $232 billion was from aquaculture production. The aquaculture industry has experienced robust growth since the 1980s, whereas capture fisheries have remained relatively flat or declined over the same period. By most estimates this trend will continue as 90% of fish stocks are depleted or overexploited (Moffitt and Cajas-Cano 2014; FAO 2018a). In order to meet the global demand of an estimated 9 billion people for protein and to reach projected levels of production, the aquaculture industry must steadily increase capacity while ensuring adequate livelihood and sustainability practices that minimize the dependence upon wild fish stocks for feed production (Godfray et al. 2010; Martins et al. 2010; Dittmann et al. 2017). That goal cannot be achieved without dependable live feeds, most notably *Artemia* (Naylor et al. 2000; Conceição et al. 2010a). In short, *Artemia* plays an essential role in meeting the world's nutritional needs.

The eminence of GSL *A. franciscana* as a foundational food for the global aquaculture industry is due to the selective foraging behaviors and specific feeding requirements of certain finfish and crustaceans. Essentially, there are larval and post-larval stages that are intolerant of artificial feeds or they develop poorly when fed substitutes for live feed (Bengtson 2003; Conceição et al. 2010a). Though live feeds are typically more expensive, their additional cost is offset by improved outcomes in production. There are several reasons that living food is required: their movement stimulates foraging behavior among larvae who otherwise will not forage for settled or clumped feed products; they are easily ingested and are highly digestible; their nutrient profile can be further enhanced and tailored to the specific dietary and physiological needs of larvae; and they do not settle in tanks and serve as substrate for undesirable bacterial growth such as *Vibrio* spp. (Sorgeloos 1980; Bengtson et al. 1991; Bengtson 2003; Bahabadi et al. 2018). The combination of these features of live feeds has been a major challenge of artificial feed producers to replicate (Conceição et al. 2010b). An incentive for *Artemia* replacement diets was the perceived unreliability of the *Artemia* supply; however, this concern has diminished greatly because the advancement of science-based management, and the effective exploitation of a diversity of saline lakes worldwide, has increased the reliability of the supply of *Artemia* over the past few decades (Lavens and Sorgeloos 2000; Van Stappen 2008). The value of *Artemia* to support global food production has been recognized officially by the FAO. In response to its importance for providing critical protein for the world's needs, both the EU and FAO have dedicated multiple resources to research, training, communication, education, and conservation efforts to ensure the sustainable harvest and production of *Artemia* (FAO 2018a, b).

The three primary options for live feeds are *Artemia*, rotifers, and copepods (Dhert et al. 2001; Conceição et al. 2010a, b). *Artemia* has the most convenient handling and usage requirements of all live feeds and has therefore been used most extensively by aquaculture producers (Léger et al. 1987; Bengtson et al. 1991; Lavens and Sorgeloos 1996; Sorgeloos et al. 2001). Furthermore, *Artemia* nauplii and meta-nauplii, which are non-selective filter feeders, can be enriched with a variety of nutrients, because they accumulate fine particulate matter that is sufficiently small (i.e., <26 microns), with the preferential size range for meta-nauplii between 4 and 8 microns and an optimum of 16 microns for a variety of age classes (Makridis and Vadstein 1999; Fernández 2001). Through this mechanism the dietary profile of *Artemia* can be further improved and altered to suit the specific physiological requirements for fish larval production (Evjemo et al. 1997; Watters et al. 2012). Omega 3 highly unsaturated fatty acids (HUFA) which are required for growth and development (Watanabe 1993; Furuita et al. 1996; Coutteau and Sorgeloos 1997; Conceição et al. 2010a; Viciano et al. 2013; Khudyi et al. 2017) and other nutrients, such as immune boosters, vitamins, and microalgae (e.g., *Chlorella salina*, *Chaetoceros calcitrans*, and *Nannochloropsis salina*), can also be delivered to fish larvae via *Artemia* and the results have shown positive outcomes (Sorgeloos et al. 1991; Ako et al. 1994; Chakraborty et al. 2007; Roo et al. 2019).

Although complete replacement of *Artemia* by artificial feeds across the spectrum of aquaculture products has been attempted, it has met with limited success. One of the most beneficial feeding regimes is the combination of *Artemia* with artificial feeds (Aristizábal and Suárez 2006; Curnow et al. 2006; Mai et al. 2009). This dual feeding program has repeatedly demonstrated positive outcomes in growth, maturation, survival, and immunocompetency and will likely be an ongoing approach in feeding regimes (Beck and Bengston 1978; Léger et al. 1986; Conceição et al. 2010a). Another factor that has increased the demand for live feeds is the introduction of novel species into aquaculture production. Whereas it is already extremely difficult to successfully transition well-established aquaculture production to entirely artificial feeds, this task is even more challenging when introducing new species, especially marine taxa, into commercial production (Houde 1994). The demonstrable need for live feeds indicates that *Artemia* will remain a preferred dietary component of the aquaculture industry well into the foreseeable future. This nexus between increasing demand for live feeds for the aquaculture industry and the trend of greater stability and dependability of the GSL *Artemia* supply represents a fortuitous convergence of supply and demand for a natural resource.

7.3 Resource Management

7.3.1 Management of the Artemia Resource

The brine shrimp industry began in 1952 and started with the collection of adult brine shrimp for the tropical fish market. The industry soon transitioned to collecting the dormant encysted embryos (brine shrimp cysts) which could be dried, stored, and later hatched into free-swimming larvae that were highly suitable as a dietary component for fish and crustaceans. From 1952 until 1992, there was no limitation on the annual harvest of cysts, which were collected, dried, and stored for market. Harvesters were required to pay a royalty to the state and had some restrictions on specific regions of the lake or salt ponds where harvesting was not allowed, but were not otherwise restrained (Stephens and Birdsey 2002). In 1992, a permit system was established and the management of the GSL *Artemia* resource underwent a substantial transition to a resource-based and access limited system. This was necessary as commercial interest expanded substantially in the late 1980s and early 1990s along with major technological advances in harvesting methods and efficiencies (Kuehn 2002). In the mid-1990s, the brine shrimp industry expressed concern to the Utah Division of Wildlife Resources (UDWR) about the potential for overharvest. Initially, resource managers from the UDWR resisted calls for regulation and opined that the resource would not be overharvested because it would not be economically feasible for the industry to harvest scarce cysts (Stephens and Birdsey 2002). Ultimately, however, the agency took into consideration the urging of the industry and the rapid expansion of the number of harvesters, and they recognized the need for a more rigorous management system. In 1996, the UDWR imposed a moratorium on the number of permits allowed and undertook a concerted effort to establish a science-based adaptive management approach for the brine shrimp resource.

With the support of the Utah *Artemia* Association (UAA), steps were taken to introduce a resource-based management method for the harvesting of cysts. The primary objective for the UAA was the consolidation of expertise within the brine shrimp industry to leverage skills and knowledge for the development of a sustainable management program for the GSL *Artemia* resource (Don Leonard personnel communication). The *Artemia* industry already had extensive experience in *Artemia* research both on GSL and on saline lakes throughout the world. Since the formation of the UAA, the industry has worked cooperatively with the UDWR throughout the process on its adaptive approach. The UDWR followed by organizing a Technical Advisory Group (TAG) composed of resource managers, environmental advocates, scientists, and *Artemia* industry members to establish research priorities and to review the current ecological information about GSL and the *Artemia* population. The TAG continues to meet quarterly, and its current role is to advise the UDWR through examination of the science that supports management decisions and to reassess management goals and outcomes. The TAG membership undergoes periodic changes, but always provides a broad swath of scientific experts from government, academic, and private sectors to inform and help shape the adaptive management process.

Under the advice of the TAG, the UDWR established a resource management policy that initially was based on an allowable harvest above the required remaining tonnage of cysts that needed to remain in the lake. This method was modified to the number of cysts per liter in spring necessary to repopulate the lake. The initial minimal cyst count was based on 1 year of data (1995–1996 harvest season), estimates of survival and viability, and a safety factor. UDWR refined the model in subsequent years to a hypothesized density-dependent empirical relationship between post-harvest spring cyst densities and maximum cyst densities the following autumn (Belovsky and Perschon 2019). The model is a simplified Ricker curve (Ricker 1954, cited by Belovsky and Perschon 2019) that results in strong coefficients of correlation when a piecewise regression is used to describe the relationship in lieu of Ricker's logistic function (Belovsky and Perschon 2019). Implicit to the management model is that the season closure is contingent upon the number of cysts, rather than tonnage, which need to remain in Gilbert Bay post-harvest to overwinter and reestablish the subsequent population. Autumn cyst densities and raw harvest have increased since 1999, coincident with the period of improved predictive power from this adaptive management model and Gilbert Bay salinities remaining within the important 100–160 g/L salinity range.

Notwithstanding the expansive research programs of the UDWR and reevaluation of the management model that takes place on an annual basis, the initial criterion of a spring cyst count of 21 cysts/L has remained relatively intact over the years (Stephens and Birdsey 2002; Belovsky and Perschon 2019). As a result, 21 cysts/L continues to be the regulatory standard. This model, which focuses not only on the resource, but also on the entire GSL ecosystem, and involves informed stakeholders committed to the long-term sustainability of the resource, has proven successful and has resulted in a dependable and stable source of *Artemia* cysts in spite of substantial fluctuations in climatic conditions. In this regard, it is a good example of a functional, cooperative, and adaptive system of resource management.

7.3.2 Management of Salinity for Artemia

The railroad causeway which bisects the lake into Gilbert Bay and Gunnison Bay and renders approximately 40% of the combined bays' volume largely salt-saturated and unsuitable for *Artemia* and benthic microbialites (Lindsay et al. 2017) may, paradoxically, provide an invaluable tool for managing Gilbert Bay salinity in the future. The large salt load sequestered in dissolved and precipitated form within Gunnison Bay (Loving et al. 2000) has kept Gilbert Bay salinities within a range supportive of *Artemia* despite the long-term anthropogenic reduction in water supply and lake volume (Wurtsbaugh et al. 2017). However, the isolation of this salt has caused Gilbert Bay to become progressively more dilute so that the salinity at a given elevation has decreased over time (Fig. 7.3). Such dilution of Gilbert Bay runs the risk of decreasing below a salinity level that maintains primary ecosystem functions as occurred in the mid-1980s and late 1990s. The 2016 causeway bridge and breach were therefore designed with an adaptive management berm that can be modified to restrict or increase salt flow between Gunnison and Gilbert bays (HDR Engineering, Inc. 2015). The restricted connectivity between the bays allows Gunnison Bay's saturated brine to be delivered gradually and in limited quantities to Gilbert Bay. The

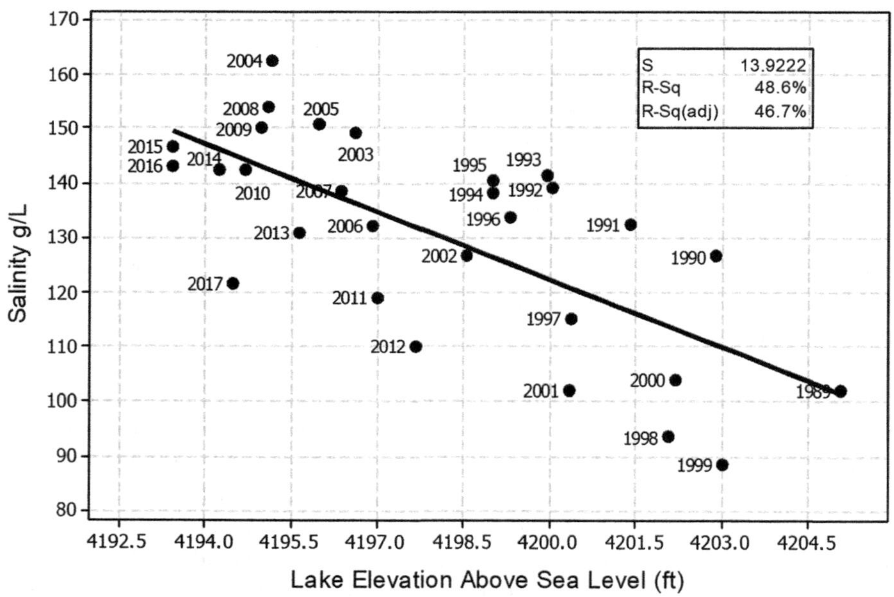

Fig. 7.3 Regression plot of year and salinity of Gilbert Bay, Great Salt Lake. There is a progressive trend of dilution in which at a given lake elevation the salinity is lower than previous years

railroad causeway and resulting salinity stratification in Gilbert Bay thereby act as a critical buffer, protecting Gilbert Bay from both the Gunnison Bay salt load and the dilution of Gilbert Bay to ecologically harmful low salinities due to river inflows entering the bay without a salt supply to counterbalance them. The management berm provides a rare and critical option for maintaining salinities supportive of Gilbert Bay ecological functions against the influences of chronic reduced water supply and acute high natural runoff.

7.4 *Artemia* Population Biology

7.4.1 Cyst Dormancy

Artemia are distributed throughout the world in saline environments where they have evolved to withstand extreme environmental conditions and, due to these adaptations, are able to achieve competitive success within a particular biotope (Herbst 2001; Gajardo et al. 2002; Van Stappen 2002; Gajardo and Beardmore 2012). *Artemia* have no demonstrable mechanism of defense, nor of superior foraging capabilities, relative to other organisms that compete for survival in a similar environmental niche. *Artemia* therefore rely upon their ability to endure severe environments as their strategy of competitive success. One of their unique capacities is the ability to produce a dormant encysted form to withstand harsh conditions such as temperature extremes (−273 °C to +90 °C), hydrogen sulfide-reducing environments, oxidation, desiccation (<0.0069 g H_2O/g tissue), UV radiation, and years of anoxia (Skoultchi and Morowitz 1964; Clegg 1978; van Stappen 1996, 2002; Willsie and Clegg 2001; Clegg and Trotman 2002; MacRae 2016). While these represent extremes beyond those encountered under most natural settings, severe environmental conditions tolerated by *A. franciscana* are unsuitable at best, and more typically lethal, for most other invertebrates and vertebrates. Paramount among *Artemia* adaptations is the ability to maintain homeostasis in spite of immersion in hyperosmotic solutions. This is achieved through phenomenal osmoregulatory processes that keep hemolymph ionic composition within tolerable limits. In this regard, *Artemia* nauplii, and all other larval stages, as well as the adults are the single most proficient osmoregulatory organisms in the entire animal kingdom (Clegg and Trotman 2002). Similarly, the dormant cysts are the single most resistant eukaryotic life form known to environmental extremes (Clegg and Trotman 2002; Gajardo and Beardmore 2012). These and other functional traits of *A. franciscana* have allowed them to survive for 400 million years, instilling a certain degree of evolutionary clout among earth's organisms (Gajardo et al. 2002; Dattilo et al. 2005; Grimaldi 2010).

Because of their unique capacities and commercial interest, *A. franciscana* embryos have been studied extensively, and the research has led to a broad understanding of the ability of their cells to reduce metabolism and to protect critical protein and membrane structure, which allows cysts to withstand punishing environmental conditions with intact cells, tissues, and preserved protein functions (Lenz

1987; Lavens and Sorgeloos 1996; Clegg 1997, 2007; Clegg and Trotman 2002). Early recognition of the unique capability of GSL *A. franciscana* encysted embryos to withstand desiccation, heat, and anoxia without compromising viability was elucidated in a number of studies (Clegg et al. 1996; Clegg 1997). Molecular chaperones, for example heat shock proteins (HSPs), which preserve protein structure and also stabilize phospholipid bilayer membranes such as liposomes and cell membranes, were identified and their multifaceted functions detailed (Miller and McLennan 1988; Liang et al. 1997; Willsie and Clegg 2001; Sun et al. 2004; Qiu and MacRae 2008; King and MacRae 2012; Clegg 2011; Hand et al. 2018). Late-embryogenesis abundant (LEA) proteins, associated with desiccation resistance in plants, were also discovered, notable as the first report of LEAs in an animal (Hand et al. 2011a; Moore et al. 2016). Another molecule, the disaccharide trehalose, which stabilizes membranes and proteins during desiccation has been characterized in *Artemia* (Clegg 1965; Hand and Carpenter 1986; Clegg and Jackson 1992a; Crowe 2004). These molecules and a host of biochemical pathways combine to support the complex transition from dormancy to development and then to emergence (Warner and Finamore 1967; Hand and Carpenter 1986; Drinkwater and Crowe 1987; Hand and Gnaiger 1988; Clegg and Jackson 1992b; Clegg et al. 1994, 1996; Hand 1998; Hand et al. 2001, 2011b; Qiu et al. 2004; MacRae 2010; Radzikowski 2013; King et al. 2014; MacRae 2016). Technical breakthroughs, like the ability to isolate and characterize *Artemia* mitochondria (Kwast and Hand 1993; Hand and Menze 2008), led to the understanding that *A. franciscana* are uniquely fortified by the preexistence of trehalose and LEA proteins embedded in mitochondrial matrices and signaling mechanisms of cell death.

The molecular research on *A. franciscana* facilitated a cascade of advancements and broader applications in the understanding of mitochondrial activity, desiccation resistance, and stress tolerance, including direct benefits for humans through advanced methods for storing and stabilizing stem cells and tissue preparations (Crowe and Crowe 2000; Guo et al. 2000; Puhlev et al. 2000; Eroglu et al. 2000; Chen et al. 2001; Shirakashi et al. 2002; Acker et al. 2004; Buchanan 2004; Collins and Clegg 2004; Crowe 2004; Han and Bischof 2004; Ma et al. 2005; Hand et al. 2011a, b; Li et al. 2012). *A. franciscana* has also become a standardized toxicity testing organism and has demonstrated enormous value as a predictive model for toxicological impacts on biological systems (Sorgeloos et al. 1978; Vanhaecke et al. 1981; Nunes et al. 2006; Hisem et al. 2011). The merits of research using GSL *A. franciscana* as a means of understanding complex processes and interactions in biological systems continue to be a promising avenue through which some of the most pressing contemporary issues of ecology, evolution, biology, and human health and pathology can be studied and elucidated (Codd et al. 1999; Abatzopoulos et al. 2003; Kappas et al. 2004; Tanguay et al. 2004; Baxevanis et al. 2006; Gajardo and Beardmore 2012).

Artemia are survivors and their extraordinary ability to thrive in hostile environments is achieved through competitive advantages over other taxa. One of the advantages *A. franciscana* possesses is the ability to alter reproductive mode between live free-swimming offspring (ovoviviparous reproduction) or alternatively to produce a dormant gastrula stage embryo encysted in a durable chitinous shell (Lavens and Sorgeloos 1987; van Stappen 1996; Clegg 2001; Clegg and Trotman 2002; Criel and MacRae 2002). Environmental cues that trigger the developmental and biochemical transition to the production of encysted embryos in diapause have been revealed through laboratory or controlled outdoor pond experiments and include temperature (heat shock), pH alteration, food deprivation, and oxygen stress (Drinkwater and Clegg 1991; Lavens and Sorgeloos 1996; Clegg and Trotman 2002; Radzikowski 2013; King et al. 2014). These cues forewarn the reproductive individual of potentially adverse conditions, thus triggering a "bet-hedging" response that instills a long-term assurance that offspring will have a greater chance of survival and propagation of their genome (Hand and Podrabsky 2000; Clegg and Trotman 2002; Hand et al. 2011a, b; Radzikowski 2013; MacRae 2015). Although we understand that *Artemia's* reproductive shift from ovoviviparous to oviparous reproduction is controlled by environmental factors that impact gene regulation and biochemical pathways, the key sequence of environmental cues that lead to the production of dormant cysts, and later deactivation of this process, remains unclear.

This reproductive shift in GSL *A. franciscana* correlates with a decline in food availability (as measured by the algal community size and composition as well as chlorophyll-*a* levels), increases in water temperature (below 30 °C), and reduced dissolved oxygen levels (Stephens and Gillespie 1976; Belovsky et al. 2011). However, given similar environmental cues, the responses vary both within the population and between years. The diversity of responses is notable and may signify genetic variability within the species that features divergent biochemical pathways (GSLBSC unpublished). However, studies of the diversity of cellular responses among field-collected GSL *A. franciscana* corresponding to changing conditions have not been completed. It would be especially important to examine spatially restricted *Artemia* conspecifics under conditions of environmental stress to unravel the complex interplay of cellular responses at the individual and population level.

Once the dominant shift to oviparity has happened, the encysted gastrula that is produced differs substantially in its developmental pathway from that of the ovoviviparous offspring. The oviparous gastrula is initially metabolically active, yet upon liberation from the female and release into a hypersaline solution, it enters into a state of diapause characterized by suppression of both metabolism and development (Clegg et al. 1996; Clegg and Trotman 2002; Patil et al. 2012; MacRae 2015; Hand et al. 2018). HSPs such as Asp26, Asp21, and Artemin are synthesized, which serves to protect critical cellular structures (Liang et al. 1997; Clegg et al. 1999; Criel and MacRae 2002). One of these small heat shock proteins (ArHsp21) develops exclusively in diapause-destined embryos and reaches a maximal level just prior to release of encysted embryos (Qiu and MacRae 2008). Metabolic depression follows as a defense mechanism that enhances survivability and individual fitness given unsuitable environmental conditions. Trehalose concentrations are at their highest during diapause, and upon inactivation and the onset of development trehalose is converted into glycogen and glycerol (MacRae 2016). The trehalose concentration in overwintering *A. franciscana* cysts from GSL was examined to determine if there is a temporal aspect of the concentration that corresponds to

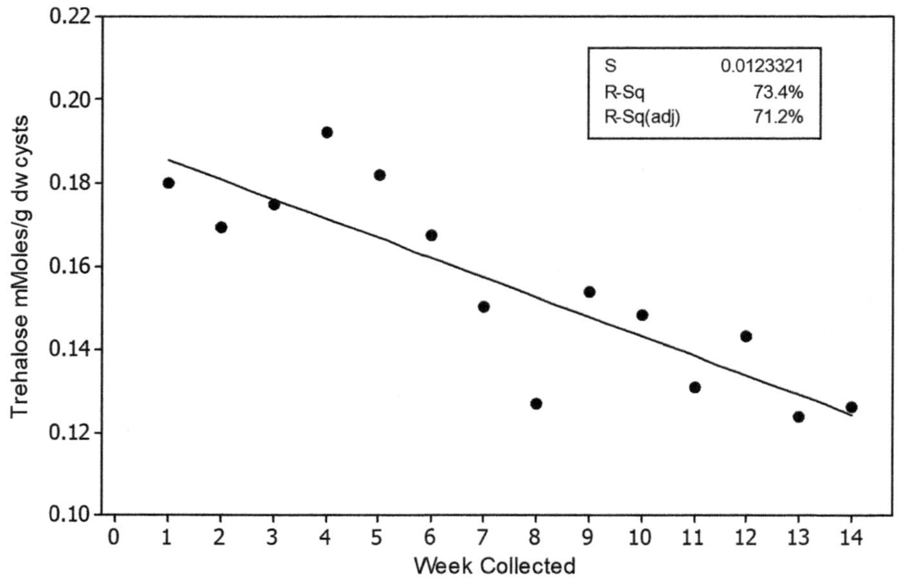

Fig. 7.4 Trehalose concentration in *A. franciscana* cysts collected from Gilbert Bay, Great Salt Lake. The collection period began in the first week of October 2008 (week 1) and continued weekly through the end of December (week 14). There was a significant and linear decline in the concentration in the cysts that coincided with deactivation of diapause

diapause deactivation. This test was conducted systematically during one brine shrimp cyst harvest season (October 1, 2008, to January 31, 2009) and showed a declining concentration in cysts that in fact corresponds to the pattern of diapause deactivation and quiescence initiation (Fig. 7.4).

Although many other taxa exhibit the ability to depress metabolic activity or development and to protect cells from environmental stressor damage, *A. franciscana* stands out as the most profound in its ability to elicit almost complete metabolic cessation (Clegg et al. 1999; Reynolds and Hand 2004; Hand et al. 2018). Cysts that remain in either dormancy or quiescence are able to minimize metabolic activity and preserve energy-storing molecules and protein structure necessary for prolonged survival and to be able to meet the caloric demands of emergence once environmental conditions are more favorable in spring. The diapause state responds to endogenous changes and will prevent the embryo from initiating differentiation and development even in the presence of suitable environmental conditions until diapause has been deactivated (Clegg 2011; MacRae 2016). The complexity of the transition from diapause to normal metabolism is illustrated by the involvement of an astonishing array of 324 differentially expressed genes (Chen et al. 2009). A quiescent embryo transitioning to metabolic reactivation will alter gene expression and reenter the process of development and cellular differentiation (Clegg 2011). The ability to remain in endogenously regulated diapause, or alternatively to deactivate diapause yet enter into a state of exogenously regulated quiescence, confers an advantage to overwintering GSL *Artemia* cysts. This is particularly advantageous for GSL *Artemia* because GSL is located in a temperate zone with four distinct seasons in which winter ambient temperature can reach minus 20 °C and GSL water temperature is typically minus 2 °C to plus 2 °C (GSLBSC unpublished). As the lake warms in the spring to between 5 °C and 10 °C quiescence is halted among sufficiently hydrated embryos and metabolism and development proceed (Belovsky et al. 2011; GSLBSC unpublished). The dynamic interaction of *A. franciscana* with its biotope is of importance for understanding adaptive capacities and the phylogeny of this particular species (Clegg et al. 2001; Gajardo and Beardmore 2012; Asem et al. 2018). Additionally, the diversity of hydrochemical conditions among the various embayments of GSL presents a unique opportunity for studying the life cycle and genetic vs. epigenetic responses to environmental changes on a population scale. More detailed research on *A. franciscana* and its response to environmental stressors within its natural setting can continue to expand our understanding of gene expression, cellular responses, and the role of phylogeny in predicting resiliency to perturbation.

In our research on GSL *Artemia,* we record data on population age structure, population size, and reproduction on a biweekly basis year-round. Through these records we are able to track embryo viability throughout the year and in so doing maintain a chronology of reproductive mode, embryo viability, and dormancy characteristics of cysts. Our work suggests that females typically shift to oviparity during hot summer months when food limitation, salinity, and temperature are elevated. These putative stressors apparently trigger the shift in reproductive mode, and freshly deposited cysts begin to noticeably accumulate in the water column. Dormancy, as indicated by a lack of hatching, is the dominant condition of summer and fall cysts in GSL. The typical pattern that is observed is that upon release from the ovisac cysts go through a process of diapause activation and metabolic cessation; hence they exhibit low hatching success under standardized "optimal" conditions (Sorgeloos 1980; Clegg et al. 1996; Sorgeloos et al. 2001). Through late fall, diapause is systematically deactivated and the quiescent embryos are metabolically prepared for differentiation, growth, and development given the necessary environmental cues such as rehydration, increased photoperiod, and warming temperature that occur in spring (Lenz 1987). Our field results (Fig. 7.5) documenting the early termination of diapause (November) call into question the utility of cold temperature as a predominant cue for diapause termination in this population. With *A. franciscana* from the San Francisco Bay population, Drinkwater and Crowe (1987) reported that exposure to 7.5 °C requires 30–90 days to promote diapause termination. For GSL *A. franciscana* cysts, cold may be more important for suppressing development and metabolism during quiescence after diapause has been terminated. Further elucidation is certainly needed to fully understand the transition from diapause to quiescence among the GSL *Artemia* population and to explore the predominant role that quiescence serves in protecting critical proteins during prolonged periods of overwintering on GSL.

Embryo viability tests of *A. franciscana* taken directly from the lake, and hatching tests conducted usually within 24 h, indicate that diapause deactivating

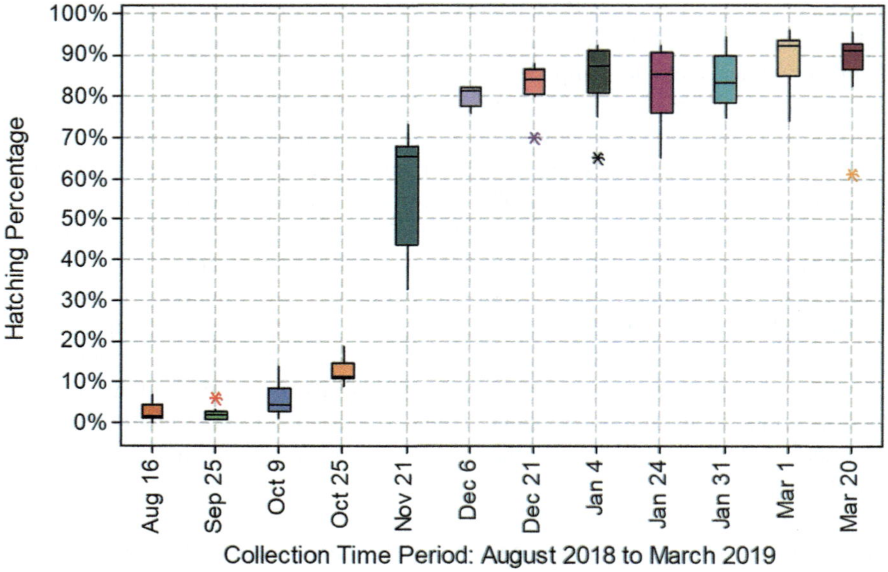

Fig. 7.5 Hatching results for overwintering *Artemia* cysts during the fall of 2018 to spring 2019. Cysts were collected from multiple sites (9 sites) in Gilbert Bay, Great Salt Lake, and from the entire water column at each site. Hatching tests were generally done within 24 h after collection and were carried out at 30 °C in 35 g/L hatching solution (Instant Ocean®)

environmental cues have exerted their influence by November and hatching increases with a sigmoidal pattern from summer through winter and into early spring months (Fig. 7.5). In general, beginning in October and progressing through December diapause has been deactivated, or is in the process of being deactivated at various rates, among the individuals within the population, and the resting embryo is in a state of quiescence primed to respond to favorable conditions. Water column cyst hatching in July is usually quite low (i.e., <10%) indicative of freshly deposited cysts in diapause. As diapause is inactivated and quiescence is established in late fall the maximum viability of cysts generally increases to 80–90%, though there have been some years in which hatching peaked well below this range. The viability of cysts remains relatively stable through February and March and by April early evidence of hatching begins. As the overwintering embryos hatch, the hatching results for remaining cysts decline as recalcitrant embryos are either in a state of quiescence or "nonviability" due to cellular/tissue damage (Warner and Finamore 1967; Lenz and Browne 1991). Spring hatchability of water column cysts is an important statistic for management purposes because overwintering survivability and spring viability of *Artemia* embryos are incorporated in management strategies that endeavor to regulate toward an "optimal" spring abundance of cysts (Stephens and Birdsey 2002; Belovsky and Perschon 2019).

The brine shrimp industry harvests cysts during October through January, during the time period when they are still in dormancy. The harvesting companies collect, clean, and store the cysts and then manage conditions in a manner that serves to deactivate diapause while maintaining the cysts in a state of quiescence. The quiescent cysts are subsequently dried, stored, and then shipped to aquaculture centers around the world. As a result of the proprietary handling and storage techniques, end users are able to realize reliable high hatchability of cysts that emerge into free-swimming nauplii which can then be enriched and fed to a wide variety of farm-raised crustaceans and finfish.

7.4.2 Two Decades of Artemia Population Dynamics in Great Salt Lake

Research on the Anostrocan crustacean genus *Artemia* has been extensive and varied over the past century, but the majority has been divorced from the natural systems in which *Artemia* operate. The organism itself has captured the attention of physiologists, geneticists, and aquaculture specialists intrigued by its halotolerance (e.g., Croghan 1958), desiccation resistance (e.g., Clegg 2005), diapause (e.g., Hand 1998), rapid evolution and adaptation (e.g., Browne and Bowen 1991), and essential role in aquaculture (e.g., Sorgeloos et al. 2001). However, this wealth of research is related to natural *Artemia* populations only tangentially, and field studies on *Artemia* populations have lagged behind those of freshwater and marine zooplankton. The limited effort available to document *Artemia* populations has therefore been spread thinly across their numerous and variable habitats, resulting primarily in short-term studies (e.g., Van Stappen et al. 2001; Torrentera and Dodson 2004) and cursory surveys or descriptions of occurrence (e.g., Abatzopolous et al. 2006a). Mono Lake (California, USA) has been a striking exception, with year-round *A. monica* densities and phytoplankton chlorophyll-*a* measurements across three decades (Melack et al. 2017), and nutrient information from multiple years (Jellison et al. 1993; Jellison and Melack 2001), demonstrating distinct annual patterns in the endemic *Artemia* population and strong bottom-up effects of lake stratification and nutrient cycling between years.

Despite similar geography, continental climate, and basic food web structure, information on the endemic GSL *A. franciscana* population was limited to a few short-term studies over the prior century (reviewed in Stephens 1974), and informative surveys of population change during a rare lake flooding phase (Stephens 1990; Wurtsbaugh and Berry 1990) until the late 1990s. A catalyst was needed to focus research attention on this population, and the rapid growth in commercial harvest of *Artemia* cysts from GSL during that decade provided it. The scant data on basic GSL *Artemia* population patterns was concerning because even data-rich and actively managed fisheries were collapsing from erroneous stock assessment methods (Myers et al. 1997). Neither the *Artemia* harvest companies nor the UDWR were interested in fostering a local analogue of the collapsed Atlantic cod fishery, so the pace of GSL *Artemia* research increased.

The resulting monitoring programs from the UDWR and the *Artemia* cyst harvest industry (Great Salt Lake Brine Shrimp Cooperative—GSLBSC) have generated a near-continuous and ongoing *Artemia* population dataset since 1994, with UDWR data used in a comprehensive ecosystem level analysis by Belovsky et al. (2011). The parallel state and private research efforts chronicle long-term *Artemia* demographics and ecosystem components such as nutrient concentrations, water temperatures and salinities, and phytoplankton abundance and composition. Here, we broadly examine the annual patterns of GSL's *Artemia franciscana* population over the past 25 years, and discuss the resiliency and stability of this population in the face of large interannual environmental changes typical of endorheic saline lakes.

7.4.2.1 Summary of Methods and Dataset

The presented data is the product of GSLBSC's intensive and ongoing year-round *Artemia* population and habitat monitoring program conducted from August 2007 to December 2019. Sampling occurs at 9 sites across Gilbert Bay, typically at monthly to bimonthly (twice a month) intervals as weather allows. Data collection consists of vertical tows for *Artemia* demographics and epilimnetic measures of oxygen concentration, salinity, temperature, and chlorophyll-*a*. Nitrogen (N) and phosphorous (P) samples were taken at monthly to bimonthly intervals from 2012 to 2018 and analyzed by IEH Laboratories (Seattle, WA). We have found good general agreement in demographic densities and trends between the two programs during collaborative meetings. For the purposes of this chapter, we provide monthly averages of the 13-year 2007–2019 GSLBSC dataset to describe the patterns of the *Artemia* population throughout a typical annual cycle, and bimonthly data for representative years to illustrate key short-term responses in greater resolution. The *Artemia* population patterns from our 2007–2019 dataset are compared to the 1994–2006 UDWR monitoring program results to cover more than two decades. Over this 25-year monitoring period, the *Artemia* population exhibits dynamic and responsive annual trends and remarkable resistance over a broad range of hypersaline conditions (e.g., 90 g/L to 180 g/L), and rapid recovery from rare periods of extreme salinity (e.g., >180 g/L).

7.4.2.2 Annual Patterns

The GSL *Artemia* population follows a distinct annual pattern most years (Fig. 7.6). Nauplii typically begin hatching from overwintering cysts in March, though this can occur as early as mid-February. Maturation is slowed by the cool ($\bar{X} = 7.7$ °C) water temperatures. Nauplii transition to juvenile and adult life stages occurs in April, and the first signs of ovoviviparous reproduction from the F1 generation are observed in this month as well. The population advances rapidly in May as temperatures warm and the adult F1 and new F2 generations typically encounter an abundant phytoplankton food base that had been flourishing during the winter. *Artemia* can mature

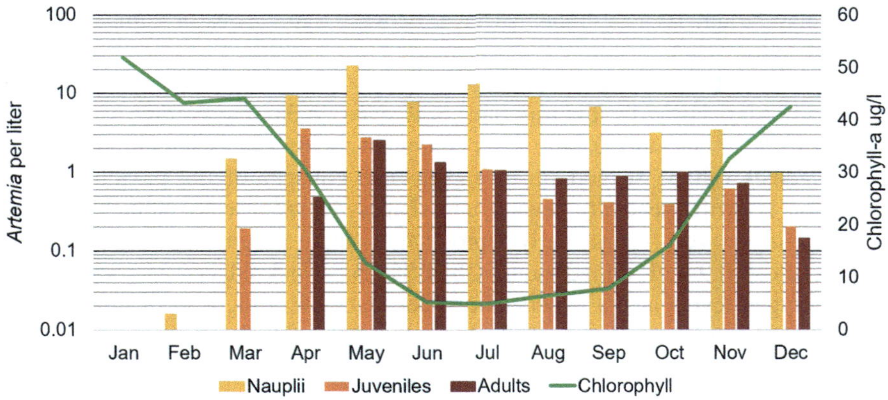

Fig. 7.6 Average monthly densities of Great Salt Lake *Artemia* age classes and chlorophyll-*a* from 2007 to 2019. Note logarithmic scale for *Artemia*

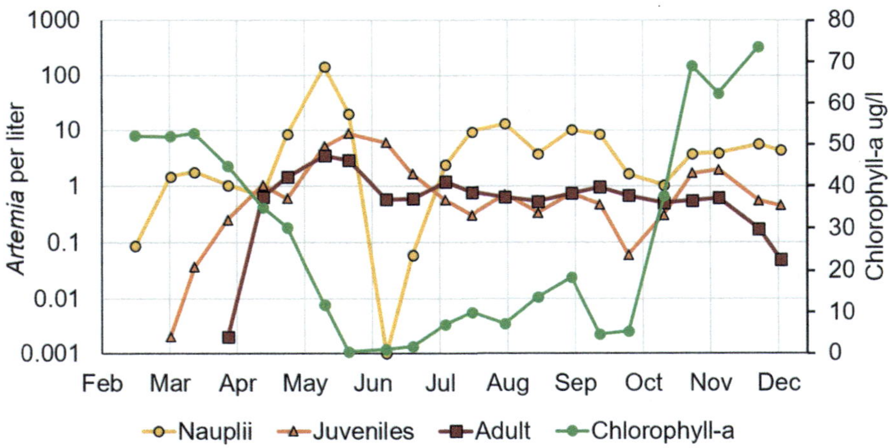

Fig. 7.7 Great Salt Lake *Artemia* population age classes and chlorophyll-*a* throughout an example year of bimonthly Great Salt Lake Brine Shrimp Cooperative field data

in as little as 2 weeks under favorable food and temperature conditions (Wurtsbaugh and Gliwicz 2001), and ovoviviparous clutch sizes of the first generation are the largest of the annual cycle. The rapid reproduction and maturation can lead to overlapping generations and simultaneous annual peaks for nauplii, juvenile, and adult densities in May or early June (Fig. 7.7). This rapid rise, transition, and fall of age classes can be problematic for bimonthly sampling and for monthly averages built from these measurements. Bimonthly intervals can miss these short-lived peaks and their true timing and magnitude can therefore be masked and blunted. Given that, it may be unwise to estimate survival between age classes or compare peak densities across years.

The peaking *Artemia* population exerts enormous grazing pressure on the phytoplankton in May. Using the filtering rates of Reeve (1963a), the average May adult density of 2.5 per liter can clear the water column in about 11 days, but filtration rates on peak dates can be several times higher. The collapse of the phytoplankton population is swift, generally severe, and shapes *Artemia* demographics for the remainder of the year—particularly reproductive output and mode. Chlorophyll-*a* declines to 0.1–0.9 μg/L on individual GSLBSC dates, and reduced photosynthesis lowers oxygen levels in the hypersaline water to below 3 mg/L even in daylight hours. The *Artemia* population declines to 0.3–0.9 adults/L in response to the exhausted food supply. The remnants of the phytoplankton are freed from intense grazing pressure and utilize the *Artemia*-excreted ammonia to rebuild. Chlorophyll-*a* and *Artemia* densities oscillate in response to each other throughout the following summer months, with neither achieving early summer densities and with fluctuations being somewhat less severe (Fig. 7.7). The *Artemia* reproductive mode shifts sharply from the ovoviviparity strongly dominant in spring to an oviparous majority for the remainder of the growth season. Abundance of oviparous females and clutch sizes increase alongside chlorophyll-*a* in September and October. Cyst densities within the water column increase accordingly. Winter temperatures are lethal to the free-swimming *Artemia*, and the population declines through November and December, leaving only the diapausing cysts to overwinter.

The results of the 1994–2006 UDWR monitoring program published by Belovsky et al. (2011) differ little from our subsequent 13-year 2007–2019 dataset in regards to the monthly *Artemia* population patterns within the annual cycle. The months of peak nauplii, juvenile, and adult age classes are the same, and the general stabilization of the population densities through late summer and fall remains. However, the magnitudes of the average annual juvenile and adult peaks are higher in the UDWR dataset. Similarly, the annual phytoplankton chlorophyll-*a* trend matched ours, but UDWR annual maximum chlorophyll-*a* was somewhat elevated, likely the result of several annual peaks in the late 1990s that exceeded 100 μg/L and one year that exceeded 200 μg/L—values far higher than the 78 μg/L maximum that we have observed on any single sampling date. The UDWR dataset included a period of higher lake elevations at the end of 1990s, which diluted Gilbert Bay salinities for several consecutive years to below the 100 g/L minimum we observed between 2007 and 2019. Belovsky et al. (2011) described an inverse correlation between maximum annual chlorophyll-*a* and salinity which reflects their higher average annual chlorophyll-*a* maximum, but this correlation is not present across the somewhat narrower salinity ranges of our 13-year dataset. These points of disagreement are minor in comparison to the larger annual patterns of *Artemia* life stage initiation, growth, and maturation which are sufficiently robust to persist across two individual research programs spanning a 25-year period.

7.4.2.3 Reproductive Output

The midseason phytoplankton and *Artemia* decline acts as a distinct trigger for oviparity, but the precise factor driving this reproductive shift has not been fully resolved. *Artemia* oviparity is generally understood to be a response to unfavorable conditions such as high salinity, low oxygen levels, and low food availability (Clegg and Trotman 2002), with stand-alone effects from oxygen stress (Sorgeloos et al. 1975), photoperiod, and temperature (Nambu 2004; Wang et al. 2017). These factors are all co-occurring within a typical GSL annual cycle and thus not readily teased apart. Separating the effect of food levels and oxygen is particularly difficult, as the inverse relationship between phytoplankton density and oxygen concentration is causative, and oxygen levels in GSL rarely exceed the oviparity-triggering 4 mg/L threshold identified in Sorgeloos et al. (1975) from July to October (Fig. 7.8).

More certain is the interannual trend of increasing *Artemia* cyst production and greater stability from 1994 to 2018. Water column cyst densities peak annually between September and December in these datasets. This is an important indicator of reproductive output because it integrates the population's continuous cyst production that is otherwise only represented as bimonthly snapshots of reproductive female densities and clutch sizes. By this measure, the reproductive output of the population has increased. The average of the highest monthly autumn cyst density between 2007 and 2019 is 138.1 per liter, which is notably higher than the prior 12 years chronicled in Belovsky et al. (2011). The estimated annual cyst harvest from the lake exhibits a flat trend over this same period, and a far greater stability in the last 12 years than the 12 years prior (Fig. 7.9).

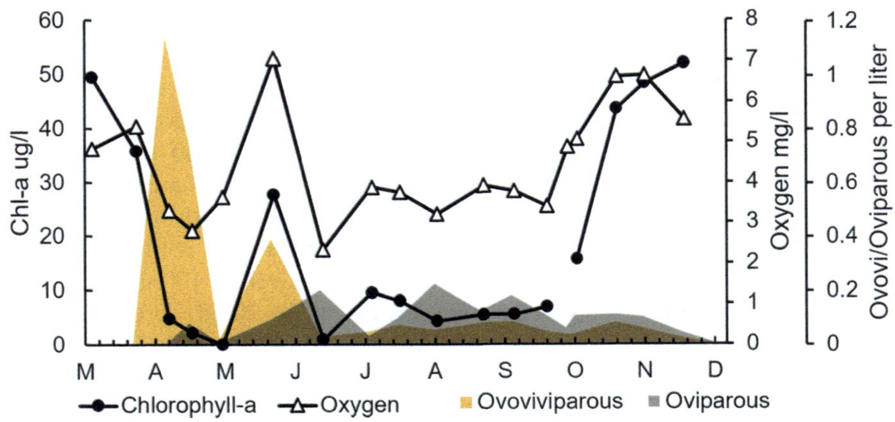

Fig. 7.8 Great Salt Lake Brine Shrimp Cooperative bimonthly 2017 data demonstrating the annual switch to oviparity, and its inverse correlation with the tandem changes in chlorophyll-*a* and oxygen concentrations

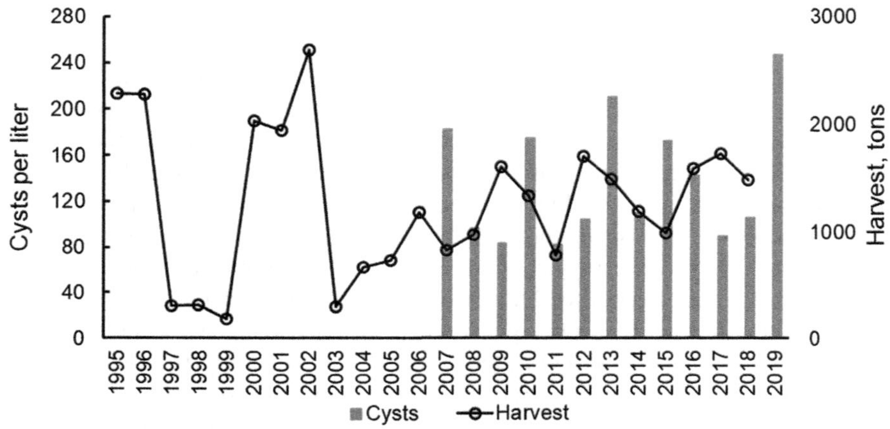

Fig. 7.9 Peak annual *Artemia* water column cyst densities from 2007 to 2019 and estimated cyst dry harvest weight from 1995 to 2018 from Great Salt Lake

7.4.2.4 Salinity Influences on *Artemia* Population Dynamics

GSL undergoes large annual and interannual fluctuations in volume and salinity expected in an endorheic lake governed by the balance of variable precipitation and evaporative losses. Salinity is a dominant physiological stressor both allowing *Artemia* to exist in the lake free of aquatic predators and limiting its individual and community growth through increased osmoregulatory exertion (Croghan 1958). The salinity range between predatory exclusion and physiological intolerance is broad and varies between *Artemia* species and populations. In a classic example of the intermediate salinity hypothesis (Herbst 1988, 2001), the lower salinity limit of GSL *A. franciscana* is determined not by the 3 g/L physiological tolerance of the genus (Vanhaecke et al. 1984), but by the upper salinity tolerance of the predatory invertebrate *Trichocorixa verticalis* which has been observed to severely limit or remove *Artemia* in periods and habitats of relatively low (<60 g/L) salinity (Wurtsbaugh and Berry 1990; Marden and Richards 2017). The functional lower salinity threshold is even higher—100 g/L—as population growth of the related *T. reticulata* in *A. franciscana*-producing evaporation ponds was not limited until salinities approached this concentration (Herbst 2006). Even at 100 g/L, caution should be exercised for GSL. Salinities typically decline by 20 g/L during a single runoff season, potentially diluting Gilbert Bay to the range in which *T. verticalis* presence is possible and unfavorable shifts in phytoplankton composition have been observed in prior years. Thus, we consider the low end of the optimal salinity range to be 120 g/L.

The upper salinity limit is physiological, and falls somewhere between the maximum salinity observed by the 25-year Gilbert Bay dataset (~160 g/L) and salt saturation in the largely *Artemia*-free Gunnison Bay. *Artemia* maintain hypotonic hemolymph that becomes metabolically expensive as the salinity gradient increases

(Croghan 1958; Van Den Branden et al. 1980; Triantaphyllidis et al. 1995). While physiological tolerances will vary between species and populations, evidence from the literature converges on salinities above 150–170 g/L leading to adverse responses in survival (Dana and Lenz 1986; Wear and Haslett 1986; Triantaphyllidis et al. 1995) and reproductive success (Dana and Lenz 1986; Browne and Wanigasekara 2000; Abatzopoulos et al. 2003). This salinity range is supported by observed drops in GSL *Artemia* density above 160 g/L in microcosm tests and the limited field data available in this upper range from both Gilbert and Gunnison bays (Figs. 7.10 and 7.11) (GSLBSC, unpublished), and is consistent with the precipitous decline in survival of a parthenogenetic *Artemia* clone at 160 g/L under laboratory conditions (Abatzopoulos et al. 2003). This evidence leads us to consider the upper salinity boundary for a healthy GSL *Artemia* population to be approximately 160 g/L. Thus, the GSL *Artemia* salinity optimum of 120–160 g/L is bracketed by physiological stress on the high end and complicating ecological interactions on the low end, aligning strongly with Herbst's intermediate salinity concept.

Rare climatic events have caused GSL salinity to fall well outside the intermediate range twice over the last 50 years, leading to detrimental impacts on the *Artemia* population which nonetheless recovered quickly when salinities returned to normal. Declining lake volume in the early 1960s drove salinities to near saturation at 260–280 g/L (Stephens 1998). No *Artemia* data accompany this salinity record, but this was likely disruptive to the GSL *Artemia* population because Barnes and Wurtsbaugh (2015) observed very poor survival of GSL *Artemia* above 225 g/L in a laboratory trial and we have found no evidence in the literature of an *Artemia* population performing well at these salinities. Two decades later, record precipitation in 1983 and 1984 filled GSL to a peak elevation last occurring a century earlier.

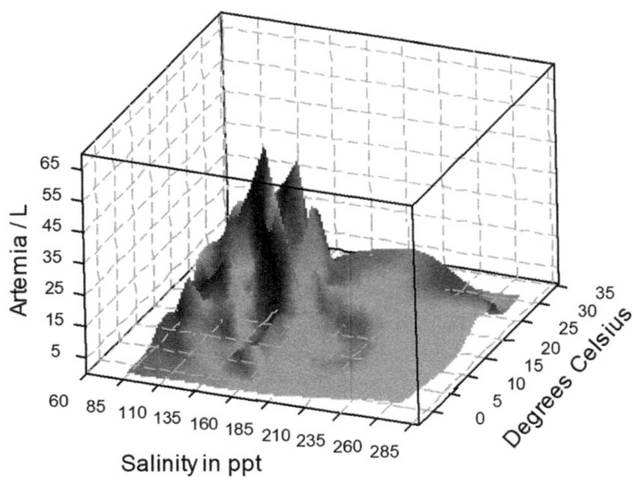

Fig. 7.10 *Artemia* population size in response to salinity and temperature in Gilbert Bay and Gunnison Bay. Results are from 1994 to 2017

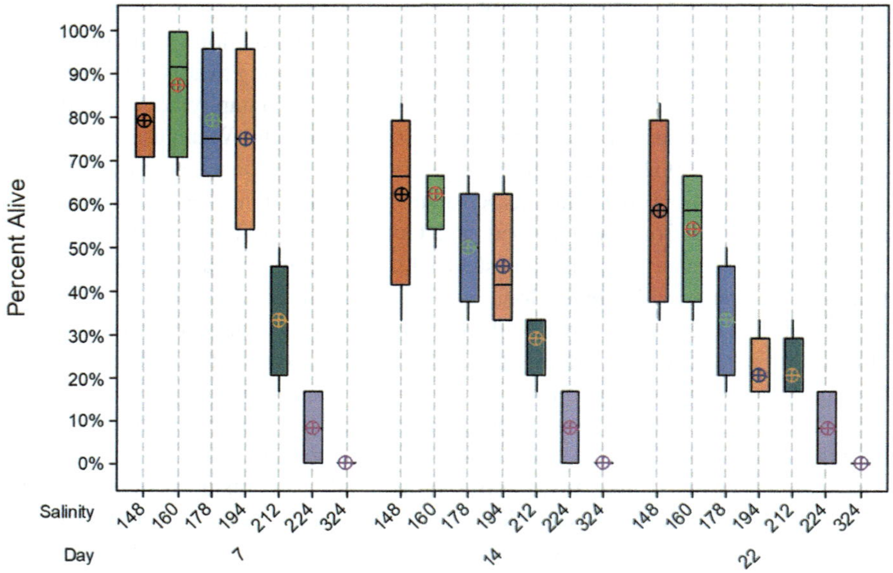

Fig. 7.11 Adult *Artemia* survival over a 3-week time period when Gilbert Bay water (148 g/L) was mixed with Gunnison Bay water (324 g/L). Significant declines in survival occurred on all dates of the experiment with 0% survival in the Gunnison Bay water. Boxes depict the interquartile range and 95% confidence intervals. Significant declines in survival relative to the control group (Gilbert Bay) occurred after 3 weeks when salinity was in excess of 178 g/L

Salinities declined to 50–90 g/L in Gilbert Bay, allowing competing zooplankton and predatory *Trichocorixa* to reduce the *Artemia* population to very low densities from 1985 to 1987 (Wurtsbaugh and Berry 1990). Despite this, the *Artemia* population recovered to dominance once salinities returned to a more favorable range above 100 g/L (D. Stephens, cited in Wurtsbaugh and Berry 1990). The unique structure of GSL may have contributed to this resilience, as Gunnison Bay provided a more saline refuge for the *Artemia* population during the flood years.

A less extreme salinity decline in the late 1990s may have adversely affected the *Artemia* population through the secondary influence of phytoplankton community assemblage shifts. Rising lake levels beginning in 1995 reduced Gilbert Bay salinities to as low as 76 g/L by 1999, a period of generally increasing adult *Artemia* densities (Belovsky et al. 2011) but very poor cyst production measured as water column densities and abundance of reproducing females—so poor in fact that the cyst harvest could not be opened in Gilbert Bay in 1999. The salinity decline over this period did not permit the establishment of competing or predatory zooplankton, but the phytoplankton assemblage shifted markedly from dominance by the chlorophyte *Dunaliella viridis* to alternating periods of centric and pennate diatom dominance in 1996 and 1997 (Stephens 1998), and then to co-dominance of diatoms and cyanobacteria (Belovsky et al. 2011) or diatoms and chlorophytes (GSLBSC unpublished).

The effect of the phytoplankton community shifts is unproven in this case but may have been substantial. *Dunaliella* is a nutritional food source for *Artemia*, but the suitability of diatoms varies. The small centric diatom *Chaetoceros* sp. is an effective diet even for gape-limited early instar nauplii (Sorgeloos and Kulasekarapandian 1984; Sanchez-Saavedra and Voltolina 1995), but this genus is rare in the hypersaline Gilbert Bay of GSL (Felix and Rushforth 1979; GSLBSC unpublished) where larger pennate diatoms are the more common form and may be too large for nauplii to consume (Stephens 1998). While it is possible that the shift to diatoms in GSL exerted a strong negative effect on *Artemia* reproduction in those years, without taxonomic resolution beyond the phylum (Belovsky et al. 2011) or class (GSLBSC unpublished) levels, no definitive statement can be made.

The assessment is complicated further by a co-occurring deviation in spring cyst densities from the current management target. The density-dependent management model of Belovsky and Perschon (2019) predicts poor fall cyst production at the low spring cyst densities of 1996, 1998, and 1999 and at the excessively high spring cyst densities of 1997. The combination of unfavorable spring cyst densities, phytoplankton community shifts, and declining salinities therefore provided overlapping potential stressors on the *Artemia* population that cannot be separated with available data. Regardless of cause, *Artemia* reproductive output recovered in 2000 and 2001 as salinities returned to the moderate range, *Dunaliella* regained dominance in the phytoplankton community, and evolving harvest management kept spring cyst densities closer to the putative optimum.

The relative stability of the GSL *Artemia* population across the past 25 years of fluctuating and extreme habitat conditions is remarkable, but perhaps expected. The salinities and temperatures common in GSL present enormous physiological challenges for any animal species, but because GSL is a large and permanent lake, they have consistently fallen within the broad range to which the endemic *Artemia* population has adapted. On the rare occasion when climatic factors cause the salinities to exceed this range, the detrimental effects persist only until salinities return to the moderate 120–160 g/L salinity bounds. The adaptive causeway berm now provides a management tool for keeping Gilbert Bay within this range.

7.4.3 Bottom-Up and Top-Down Effects on Artemia Population Cycles: The Dominant Role of Nutrients

Studies that examined top-down (i.e., predation) or bottom-up control (i.e., limited resources) on biological production in Gilbert Bay have arrived at similar conclusions—that both forces are inherent in the system with multiple abiotic and biotic factors influencing the relative dominance of the controls. Feedback loops and interactions among the top-down or bottom-up controls can result in trophic cascades in the system (Williams et al. 1990; Wurtsbaugh and Berry 1990; Wurtsbaugh 1992; Gruner and Taylor 2006; Gruner et al. 2008; Okun et al. 2008; Herbst et al.

2013a, b) or it can alter nutrient cycling pathways (Scherer et al. 1995). It is also recognized that the main factor controlling the relative strength of top-down or bottom-up controls is often not realized until the thoroughness of investigations is increased (Gruner 2004). Over most salinity conditions in Gilbert Bay bottom-up forces predominate in the simple food web; nutrients limit primary production, primary production limits zooplankton population growth rate and size, and zooplankton abundance affects avian fitness, reproductive success, and survivability during migration (Belovsky et al. 2011; Vest and Conover 2011). Predation, though important, is weakly coupled to population size of secondary and tertiary consumers and becomes uncoupled at higher salinity (>100–110 g/L) (Herbst 2006). Although conclusions about either top-down or bottom-up controls are defensible, it is also clear that there are always multiple simultaneous factors that are controlling biological production and population size and cycles (Hunter and Price 1992).

7.4.3.1 Bottom-Up Control

In early GSL investigations (Reeve 1963b; Wirick 1972; Porcella and Holman 1972), nutrient limitation was considered the main factor dictating biological production. Stephens and Gillespie (1972, 1976) identified self-shading and nutrients as the greatest influence on primary production and zooplankton abundance. In their 1976 investigation, Stephens and Gillespie specifically identified inorganic nitrogen (N) as the limiting factor controlling the chlorophyte population size. Others (Montague et al. 1982) modeled major feedback dynamics in GSL and concluded the alga species, *Dunaliella viridis*, is limited in spring by self-shading and in summer and fall by *Artemia* grazing. *Artemia* were determined to be food limited under the conditions of their study. Belovsky et al. (2011) documented inorganic N:P below the Redfield ratio of 16, considered typical for optimally growing phytoplankton (Redfield 1958), in all surveyed years from 1994 to 2006, and concluded that Gilbert Bay is constrained more by N than salinity and that 82% of variation in the *Artemia* population size was explained by phytoplankton abundance. Marcarelli et al. (2006) demonstrated a fivefold increase in algal growth over a wide range of salinities following N enrichment. Similar results have been found for other saline lakes. Jellison and Melack (2001) observed up to a 17-fold increase in chlorophyll-*a* in the first 48 h following NH_4-N enrichment of Mono Lake water, and N-limitation was observed in a wide variety of Canadian saline lakes (Evans and Prepas 1996). The consistency of findings of N-limitation is interpreted as being a result of substantial stores of legacy P in the sediments of GSL from decades of P loading (Wurtsbaugh et al. 2012) amid insufficiencies in bioavailable N.

N-limitation in the various bays of GSL is generally limited to salinities above 70 g/L (Wurtsbaugh 1988; Wurtsbaugh and Gliwicz 2001; Wurtsbaugh and Marcarelli 2004; Marcarelli et al. 2006). Below this salinity, N_2-fixing heterocystic cyanobacteria can be freed from the severe salinity constraints (Rai and Abraham 1993; McCulley 2014) and N-limitation can then be mitigated through N fixation (Marden et al. 2013). Indeed, shifts from N-limitation to P-limitation have been observed along salinity gradients within GSL embayments (Wurtsbaugh et al. 2004; Marcarelli et al. 2006; Marden et al. 2013; McCulley and Wurtsbaugh 2014).

Alternating limitations or dual non-additive co-limitation can occur given particular circumstances of nutrient availability and assimilation. In fact, evidence of co-limitation of N and P has been observed by Ogata et al. (2017) in microcosm experiments using Gilbert Bay water. In their experiments, they found that dual N and P treatments were 14-fold higher in chlorophyll-*a* than controls after 10 days. We conducted microcosm studies of algal and *Artemia* response following a single enrichment of GSL epilimnetic water with N (2.5 mg/L), P (1.0 mg/L), or N and P and found similar results. N enrichment increased primary production by 2.81 times, whereas P treatment was not statistically different from control. Dual N and P addition resulted in a pronounced synergistic increase in primary production that was 8.18 times greater than controls, 2.91 times above N treatment, and 8.36 times higher than P treatment (Fig. 7.12). Mixing of monimolimnetic water (DBL) with epilimnetic water produced similar results to N treatment. N and P co-limitation has received more scrutiny recently in terms of harmful algal bloom (HAB) management and may be a more relevant concern than has been previously appreciated, even in saline water bodies such as GSL (Paerl et al. 2008, 2016; Allgeier et al. 2011).

Additionally, there remains a role of elements such as P, Si, Fe, and Mb (Evans and Prepas 1997). Iron and Mb are necessary for cyanobacteria growth and their absence will reduce the possibility of N_2 fixation in N-limited systems (Wurtsbaugh and Horne 1983; Blomqvist et al. 2004), including GSL and other lakes within its

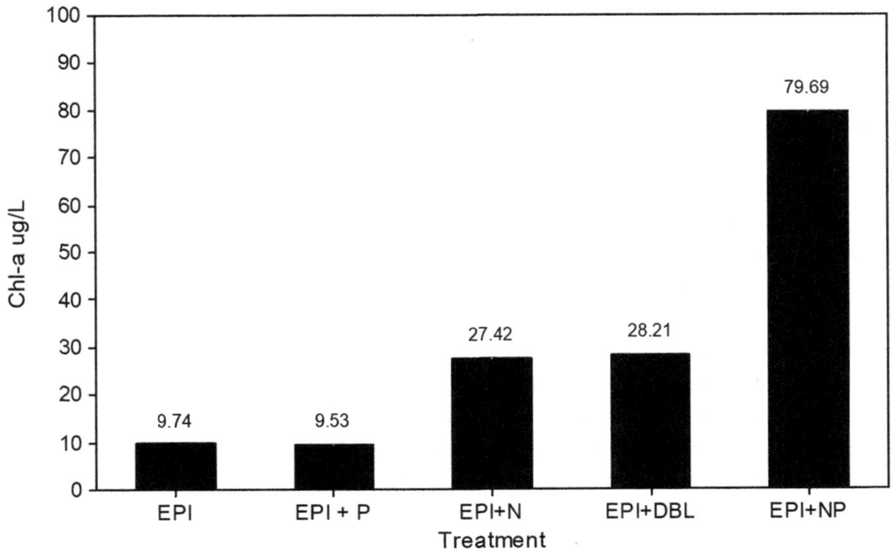

Fig. 7.12 Chlorophyll-*a* results after a 10-day incubation of Gilbert Bay, Great Salt Lake, epilimnetic water with various treatments of nitrogen (N), phosphorous (P), or combined nitrogen and phosphorous (NP). Concentrations used were 1.0 mg/L P or 2.5 mg/L N. Light conditions were 18/6 and temperature was maintained at 25 °C

watershed (Wurtsbaugh 1988). Multiple other studies have confirmed the importance of iron and molybdenum for cyanobacteria production and utilization of atmospheric N_2 in freshwater and marine systems (Paerl et al. 1987; Rueter and Petersen 1987; Marino et al. 1990). Despite this, the role of these elements in Gilbert Bay will be of reduced importance as long as high salinities restrict heterocystic cyanobacteria, and the role of fixed N will be limited to inflow from less saline bays.

Silicon (Si) availability can also directly affect algal community structure and taxa-specific abundance in GSL when combined with suitable salinities (Gilpin et al. 2004). Silicon may have contributed to the aforementioned *Artemia* population decline from 1995 to 1999. One interpretation of this population decline was available Si (6 mg/L) combining with declining salinities (below 80 g/L) to usher in a shift in the relative abundance of large (i.e., >30 μm) pennate diatoms (Stephens and Birdsey 2002). The authors hypothesized that the first generation of emergent *Artemia* in spring rapidly decimated the phytoplankton of suitable size (i.e., <30 microns), leaving non-ingestible large diatoms to flourish and the *Artemia* population to decline due to poor nutrition (evidenced by observations of black spots on the animals) in spite of chlorophyll-*a* levels normally indicative of high food availability. Within this 5-year time period (e.g., in 1996), shifts to smaller diatom cell sizes corresponded with increased *Artemia* population size and reproductive output. This example illustrates the community level transitions that can occur as a result of multiple trophic level interactions in response to alterations in salinity, nutrient composition, and bioavailability.

The paucity of detailed long-term accounting of nutrient budgets, fluxes, cycling, and loading for GSL is a persistent concern. Whereas the USGS has done periodic assessments of loading, losses, cycling, and mass balance of N in GSL, there remains a pronounced need for greater scrutiny and scope of research. Naftz (2017) provides statistics on loading, losses, and mass balance of Gilbert Bay by USGS from 2010 to 2014 and showed a range of 1.90–5.56 thousand MT annual N gain. Reported N loss was limited to the annual 0.96–1.51 MT export from Gilbert Bay into Gunnison Bay, but other loss pathways such as ammonia volatilization and denitrification need to be included in future work. The estimated 0.56–0.78 thousand MT annual atmospheric loading of N into Gilbert Bay was more than the input of some of the main tributaries during the same time, but it was calculated by proxy from the National Atmospheric Deposition Program (NADP) rather than direct measurement. These estimates differ somewhat from those of Belovsky et al. (2011), who used NADP data to estimate that annual input accounted for less than 1.5% of DIN—"a minor annual input." The USGS program provided valuable N loading statistics over multiple years using a robust combination of nutrient concentration samples and flow gauges from the tributaries. The comparison of mass balance to loading and loss estimates suggests that internal cycling of N is the primary mechanism providing bioavailable N for biological growth. The average annual mass of dissolved N in the epilimnion (31.2–33.1 thousand MT) and monimolimnion (2.0–2.9 thousand MT) dwarfs the loading estimates. Belovsky et al. (2011) concur with this interpretation and suggest that interannual variation is a function of dilution of an endogenous pool of nutrients.

Notwithstanding the magnitude of internal cycling of nutrients, riverine inputs of N into Gilbert Bay remain extremely important as they contribute an average of 9.6% (and up to 16.8% during high-runoff years), relative to the total mass of dissolved N in Gilbert Bay. These inputs of N are essential to offset the annual losses from the system. Any prolonged reduction in N contributions could negatively affect the balance of gains and losses and lead to further reductions in biological production. Nutrient investigations have been valuable, yet greater detail is needed on loss mechanisms, nutrient fluxes across the chemocline and sediment–water interface, nutrient retention, remineralization, and denitrification. Site-specific atmospheric sampling would fortify loading estimates and would benefit the interpretation of nutrient balance in GSL. Documentation of temporal cycles in bioavailable forms and relevant nutrient ratios combined with mass assessments would afford a more coherent interpretation of biological responses to nutrients within the system.

Artemia provide an essential role in the internal cycling of nutrients with consumption, digestion, and excretion facilitating the processing, assimilation, conversion, and release of nutrients, especially nitrogenous compounds. Fecal production associated with large numbers of *Artemia* is a major factor in the release of bioavailable forms of N that have the capacity to stimulate algal growth and have been reported to be one of the most essential factors in cycles of algal depletion and recovery (Vitousek and Howarth 1991; Jellison and Melack 1993). Figure 7.13 denotes the relationship between GSL *Artemia* population size and ammonia in co-located water. The released bioavailable N benefits algae and provides a positive return for *Artemia* via recovery of their food source and denotes the vital role that internal cycling of nutrients serves to promote biological integrity of the GSL ecosystem.

Nutrient limitation in Gilbert Bay is reflected in lower productivity of this lake compared to other saline lakes. Although GSL is often referred to as having high biological productivity (Belovsky et al. 2011), the few rigorous studies on GSL primary productivity indicate that GSL is well below most other terminal saline lakes. Primary production in Gilbert Bay of 140 g C m^2/year (Stephens and Gillespie 1976) is well below that of related systems such as Mono Lake, which typically exhibits 269–641 g C m^2/year but can exceed 1000 g C m^2/year (Jellison and Melack 1993; Wurtsbaugh and Gliwicz 2001). Other hypersaline lakes have higher productivity than GSL: Werowrap, Australia (435 g C m^2/year); the Alviso salterns of California (700 g C m^2/year); Borax Lake, California (386 g C m^2/year); saline lakes in Saskatchewan, Canada, (19 g C m^2/year to 2908 g C m^2/year) (Haynes and Hammer 1978; Hammer et al. 1983; Hammer 1981). In low carbonate lakes of Northwestern Ontario, Schindler and Holmgren (1971) calculated an annualized primary production of 65.3–402.6 g C m^2/year. Gilbert Bay is at the lower end of the biological production spectrum in comparison, and microcosm studies by others as well as our own have supported the higher primary production capacities of Gilbert Bay given increased input of N or N and P in combination (Ogata et al. 2017).

Similarly, *Artemia* population size and reproductive output in Gilbert Bay remains decidedly low in comparison to other *Artemia* biotopes. Gilbert Bay is

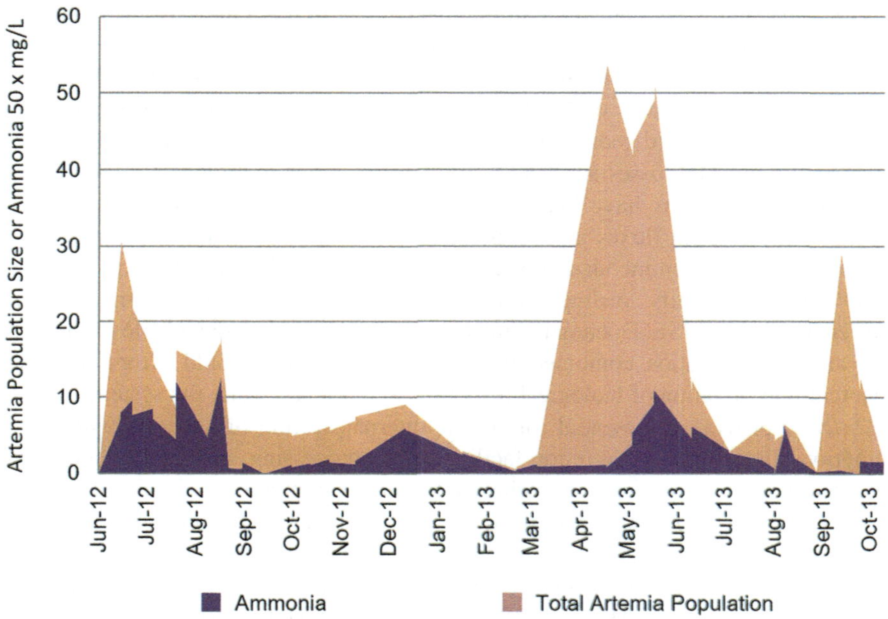

Fig. 7.13 Temporal pattern of ammonia concentration and *Artemia* population in the epilimnion during 2012 to 2013. Ammonia levels roughly followed a pattern of increasing concentration along with *Artemia* population size. *Artemia* are substantial "mineralizers" of nitrogen via digestion and excretion of ammonia-rich fecal plumes. Additionally the grazing pressure of *Artemia* reduces the uptake of ammonia by phytoplankton

lower in cyst production than Mono Lake, which generates 2,400,000–5,100,000 cysts/m^2 in contrast to 650,000 cysts/m^2 for Gilbert Bay (Dana et al. 1990; Wurtsbaugh and Gliwicz 2001). Wurtsbaugh and Gliwicz (2001) attribute the lower biological production to reduced primary production found in GSL relative to Mono Lake. These authors commented that clutch size among Mono Lake *Artemia* was 2-fold to 2.5-fold higher than GSL *A. franciscana*. The number of adult *Artemia* in Gilbert Bay during summer is between 1 and 3 adult/L (Fig. 7.14) compared to 6–8 adult/L in Mono Lake (Conte et al. 1988; Stephens and Birdsey 2002). These reproductive measures consistently reflect lower per capita reproductive output by GSL *A. franciscana* among the years of comparison.

The standing stock of *Artemia* cysts—a figure often calculated for *Artemia* harvesting management purposes—is also quite low in Gilbert Bay compared to saline lakes in Russia and Central Asia. Van Stappen (2008) in his survey of the global distribution of *Artemia* cites a variety of lakes that are known to have *Artemia* populations sufficiently large and dependable for commercial exploitation. Systematic studies by the State Research and Production Center of Fish Economy "Gosrybcenter," Tyumen, Siberia, Russia (2005), of many of these lakes have shown that only one commercially viable lake is less productive than GSL for *Artemia* cysts on a per area/volume basis: the Aral Sea of Kazakhstan and

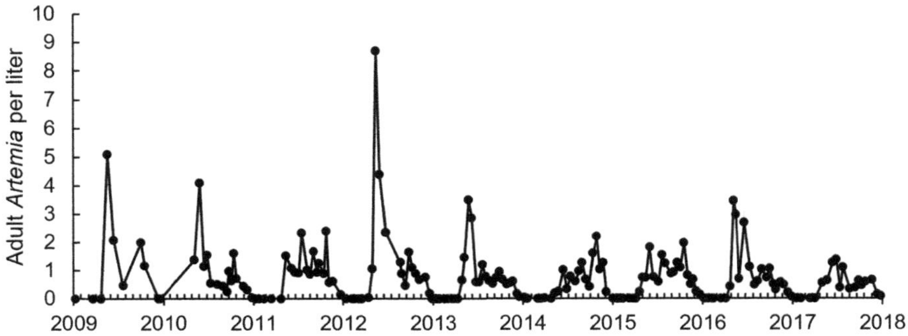

Fig. 7.14 Great Salt Lake Gilbert Bay adult *Artemia* population abundance over the time period 2009 to 2018 (data is from Great Salt Lake Brine Shrimp Cooperative research program)

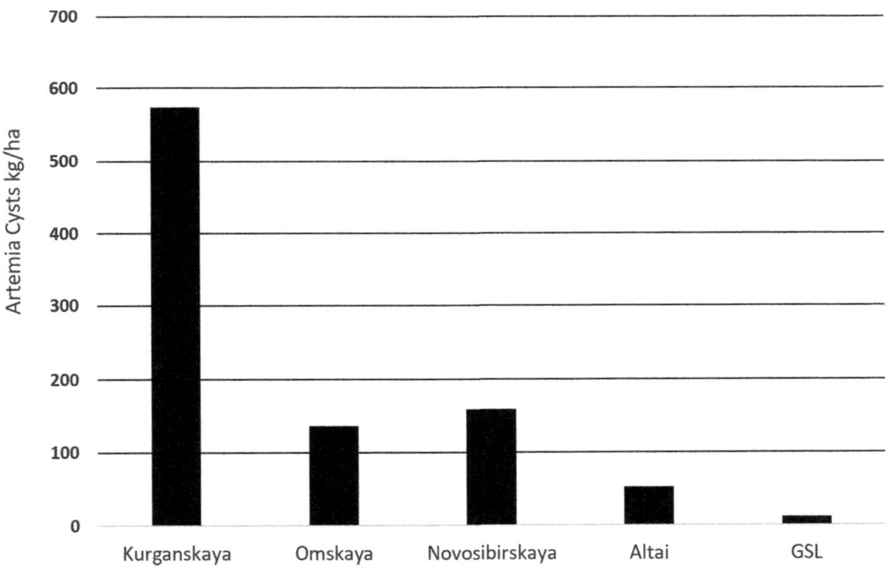

Fig. 7.15 Quantity of *Artemia* cysts produced per hectare of various saline lakes by region in Russia. Values are from 2000 to 2004. The lakes shown are the subject of multi-year systematic research studies. The results from these investigations are used to establish commercial harvesting quotas for each of the lakes and are therefore subject to strict scientific peer review and governmental oversight. Russian lakes included in the analysis ranged from 670 to 8330 ha

Uzbekistan (Fig. 7.15) (Marden et al. 2012; Litvinenko et al. 2015). Insofar as the Aral Sea is widely recognized as one of the greatest ecological calamities of the past century, any biological similarities between it and GSL come with a disquieting degree of additional concern (Micklin 2010; Marden et al. 2012; Micklin et al. 2016).

7.4.3.2 Top-Down Control

The relative dominance of either a bottom-up or top-down effect is a function of salinity—when the salinity of Gilbert Bay declines into a range (i.e., <90 g/L) that affords interspecific predation by the aquatic invertebrate *Trichocorixa verticalis* (corixid), there is greater influence of top-down control of the *Artemia* population (Wurtsbaugh and Berry 1990). Predation on *Artemia* and other zooplankton by aquatic invertebrates has been well documented in the scientific literature (Wurtsbaugh and Berry 1990; Wurtsbaugh 1992; Simonis 2013a, b; Céspedes et al. 2017) Top-down control of *Artemia* is also exerted by waterfowl which, in contrast to aquatic invertebrate predation, occurs over a broad range of salinity and has been shown to cause a pronounced impact on *Artemia* population size in saline lakes (Cooper et al. 1984; Mahoney and Jehl 1985; Jehl et al. 2002; Jehl 2007). Avian predation on *Artemia* occurs via large congregations of surface feeding and diving waterbirds such as eared grebes (*Podiceps nigricollis*), northern shoveler (*Anas clypeata*), Wilson's phalarope (*Phalaropus tricolor*), red-necked phalarope (*Phalaropus lobatus*), common goldeneye (*Bucephala clangula*), California gulls (*Larus californicus*), and green-winged teal (*Anas crecca*) (Paul and Manning 2002; Vest and Conover 2011; Wunder et al. 2012; Roberts and Conover 2013; Roberts 2013b). Along the margins of GSL shorebirds such as American avocets (*Recurvirostra americana*) and black-necked stilts (*Himantopus mexicanus*) forage for aquatic invertebrates (Caudell 2001; Roberts 2013b). The combined predation pressure by birds on GSL *Artemia* is substantial and is one of the important factors influencing *Artemia* population size and site-specific age structure.

Among avian species encountered in GSL, eared grebes are the most influential predator of *Artemia*, which comprise 81% of grebe diet on GSL (Roberts and Conover 2013) and 80–90% on Mono Lake (Cooper et al. 1984). In multiple studies (Caudell 2001; Caudell and Conover 2006a, b; Roberts and Conover 2013), calculations of consumption rates showed that each of the 1.5 million average eared grebes on GSL needs to consume 26,500–29,600 adult *Artemia* per day to meet their caloric demand for pre-migratory weight gain. Grebe predation can account for 55–84% of the Mono Lake brine shrimp population decline in fall (Cooper et al. 1984). Similarly, Varo et al. (2011) found that in the Odiel marshes black-necked grebes (also *Podiceps nigricollis*) could consume 0.2–2.0% of the brine shrimp population per day. In spite of this enormous grazing pressure, Belovsky et al. (2011) did not identify a significant population level impact on *Artemia*, and this was attributed to the size of GSL relative to population of grebes (Roberts and Conover 2014). However, an estimate based on the nutrient requirements calculated by Roberts and Conover (2014), average fall abundance of adult *Artemia* (over a 5-year period from 2012 to 2016), and the variety of reports on eared grebe population size on Gilbert Bay suggests predation by grebes can deplete between 9.72 and 19.33% of the adult *Artemia* population. We have also observed spatial differences in *Artemia* age structure and adult abundance during times of peak grebe abundance (October to December) in which sites associated with high grebe numbers are severely depleted of adult *Artemia* (Fig. 7.16).

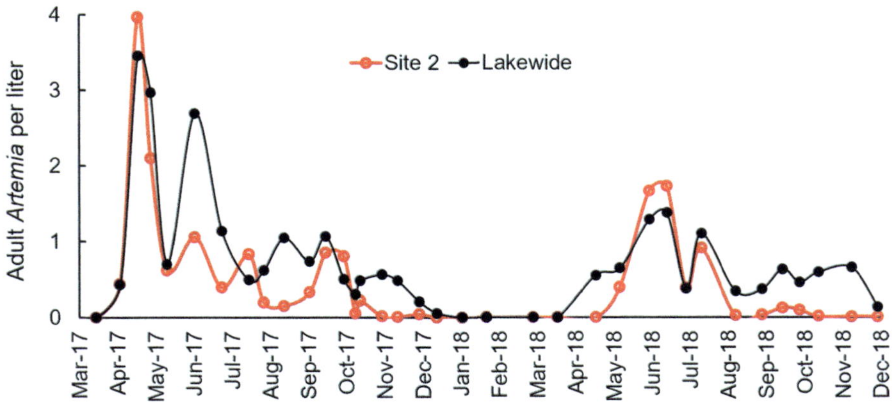

Fig. 7.16 Age structure of sites in Gilbert Bay during peak grebe presence. Over the 2 year period the lake-wide average of adults from August through December was 0.513 Adult/L, whereas the Site #2 abundance was 0.169 Adult/L. Site 2 is in the area of the lake with consistently high counts of eared grebes. Over the 2-year period, site 2 was significantly lower in adult abundance than other sites ($T = 3.95$; $P = 0.000$; DF = 34)

The impact of other bird species is comparatively small. While the diet of overwintering green-winged teal and northern shovelers can be dominated by *Artemia* cysts and adults (Vest and Conover 2011), the predatory effect of these two species of surface feeding ducks is mainly on dormant cysts and not the remaining live population of shrimp. The estimated cyst removal is <1% of the combined avian and commercial removal of cysts (Roberts and Conover 2014), and the consumption of cysts by birds has not been shown to affect the ability of overwintering cysts to repopulate the lake (Roberts and Conover 2014).

Predation by aquatic invertebrates on *Artemia* does occur in GSL, but is a relevant concern only at lower salinity levels than are typically experienced in Gilbert Bay (Wurtsbaugh 1992; Van de Meutter et al. 2010). Studies on Gilbert Bay during periods of substantial volume gain and salinity dilution (such as in the late 1980s) did show verifiable top-down control of *Artemia* by *T. verticalis*. Wurtsbaugh and Berry (1990) reported that top-down control of *Artemia* in the pelagic zone of Gilbert Bay occurred when the salinity declined from 100 g/L to about 50 g/L. In Farmington Bay, which has a salinity gradient that ranges from essentially brackish water to 50–70 g/L, multiple studies have demonstrated population level impacts on zooplankton by corixids. Demonstrable and significant depletion of cladocerans, copepods, rotifers, and *Artemia* occurred in Farmington Bay when corixid abundance approached 1 adult/L (Marden et al. 2013; Marden and Richards 2017), and the declines were not statistically attributable to cyanobacteria (Fig. 7.17). Temperature is also influential, as corixid eggs hatch at approximately 20 °C and population expansion occurs (Kelts 1979). This temperature influence coincides with the arrival and dominance pattern of corixids in Farmington Bay reported by Marden and Richards (2017).

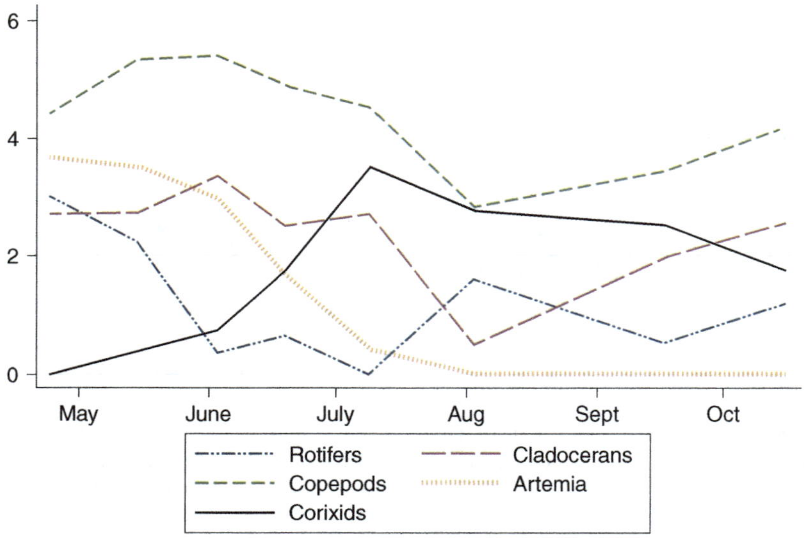

Fig. 7.17 Average zooplankton abundance (count per liter) contrasted with *Trichocorixa verticalis* (corixids) population development in Farmington Bay during the summer of 2014. The decline in *Artemia* was not statistically attributable to other factors such as competition, salinity, temperature, dissolved oxygen, or cyanotoxins. Figure with permission from Farmington Bay/Jordan River Water Quality Council 2019)

All studies of corixid predation on *Artemia* in GSL show that salinity was lower than 90 g/L in order for the corixid population to gain sufficient size and residency. Although 90 g/L appears to constitute the upper limit of corixid viability, there is evidence of adaptations to higher salinity among local corixid populations and climatic factors leading to an expanded distribution (Herbst 2006; Guareschi et al. 2013). In our own observations, substantial numbers of adult corixids were documented in littoral zones of Gilbert Bay in 2018 where the salinity was in excess of 120 g/L, thus emphasizing the need for management of Gilbert Bay that maintains salinity above 120 g/L to provide a buffer against a severe decline in salinity during high-runoff years into a level tolerated by corixids.

Harvesting of *Artemia* cysts by the brine shrimp industry represents a form of top-down control on the population and was a concern of resource managers and GSL stakeholders (Stephens and Birdsey 2002; Conover and Caudell 2009; Belovsky and Perschon 2019). Harvest tonnage over the period of 1985–2019 began with an initial harvest of 135 MT wet weight (ww) in 1985 and a peak harvest of 15,932 MTww in 2018. This increase is attributable to increased harvesting participation and changing harvest techniques that gathered more material like sand and other debris along with the cysts which are collectively reported as raw or “wet” product. The industry has harvested an average of 61% of the standing stock of cysts in Gilbert Bay, yet there have been no negative effects on subsequent growth seasons (Belovsky et al. 2011). The impact of harvesting on the *Artemia* population and potential consequences for birds, especially eared grebes, have been evaluated by multiple research teams (Belovsky et al. 2011; Vest and Conover 2011; Conover and Caudell 2009). In all cases investigations indicated that there was no evidence of adverse impacts on birds utilizing the GSL ecosystem, nor on the *Artemia* population, by the commercial harvest. Investigators have cautioned that there may be deficits in knowledge of species other than eared grebes and that other waterbirds need to be monitored during the cyst harvest. Nonetheless, over two decades of harvest results indicate that implementation of a resource-based limit on harvesting imposed in 1996 appears to have improved the dependability of the resource for birds, for the interannual stability of the *Artemia* population, and for the harvesting industry by ensuring that the number of viable cysts in spring is near the optima for population growth and fall cyst abundance (Belovsky and Perschon 2019). The weight of evidence suggests that the State of Utah's adaptive management strategy is benefitting the GSL ecosystem.

7.4.4 Benthic-Pelagic Linkages in Artemia Diet

Researchers on GSL have long suspected an *Artemia*-benthic link for carbon flow and nutrient cycling. Benthic and pelagic systems are now viewed as an integrated whole (Vadeboncoeur et al. 2002), and the GSL benthos may be a particularly productive component. Organo-sedimentary microbialites (Burne and Moore 1987) occupy approximately one-third of the benthic surface area (Baskin 2014) and are colonized by a diverse periphyton of archaea, bacteria, and microbial eukaryotes (Lindsay et al. 2017) responsible for a substantial portion of the lake's primary production (Lindsay et al. 2019; Wurtsbaugh 2007). Here, we present molecular evidence of microbialite taxa within the water column and its presence in *Artemia* diets.

Samples of *Artemia* adults and lake water were collected from five sites above microbialites (2 m average depth) and five nearby pelagic sites above silt substrate (6 m average depth) in October 2018. Microbialite periphyton samples were obtained with a small benthic dredge. Water samples were filtered onto syringe filters with a 0.45 μm pore size and immediately desiccated. *Artemia* and microbialite samples were immediately flash-frozen in liquid nitrogen. Composition of photosynthesizing microorganisms was analyzed for all samples by DNA metabarcoding using the universal primers for the plastid 23S rRNA gene (Sherwood and Presting 2007) by the lab of Jonah Ventures LLC (Boulder, Colorado). The 23S rRNA gene allows the detection of both eukaryotic algae and cyanobacteria and has high universality across taxonomic groups (Sherwood and Presting 2007; Steven et al. 2012; Sherwood et al. 2016). Isolated 23SrRNA gene sequences were grouped into operational taxonomic units (OTUs) using a 97% similarity threshold and assigned the closest taxonomic level available from reference sequences.

Metabarcoding results show strongly dissimilar photosynthetic communities between the microbialites and the pelagic water column, but also evidence of microbialite taxa entrainment into the water column of the shallow sites (Fig. 7.18). The pelagic phytoplankton was dominated by genes of the chlorophyte *Dunaliella*, with a secondary presence of unclassified diatoms and the genera *Navicula* and *Nitzschia*. *Dactylococcopsis* cyanobacteria and *Thalassiosira, Chaetoceros,* and *Cyclotella* diatom genes were present in small proportion. In contrast, microbialite periphyton was strongly dominated by *Dactylococcopsis* sequences in addition to

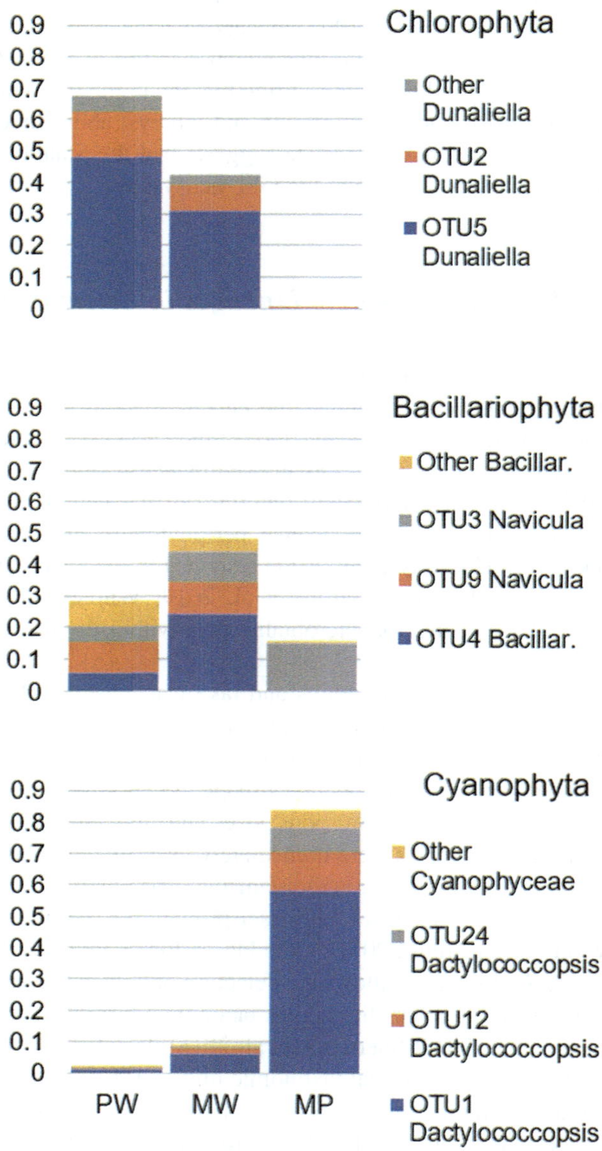

Fig. 7.18 Photosynthesizing microbial community of the water column at pelagic sites (PW), water column at microbialite sites (MW), and microbialite periphyton (MP), separated by guild and presented as the average proportion of total community 23S reads of the 5 replicates at each site type

one of the two *Navicula* OTUs common in the pelagic water. The phytoplankton assemblage above the microbialites was similar to the pelagic samples, but with higher proportions of the dominant microbialite *Dactylococcopsis* and *Navicula* OTUs.

Artemia gut content composition differed from the surrounding water column in all samples, and there is strong evidence that *Artemia* consumed microbialite taxa (Fig. 7.19). *Dunaliella* sequences were present in all *Artemia* samples but are underrepresented relative to the water column. Instead, a more diverse assemblage of sequences was present, with high variability between the five replicate sites. *Chlamydomonadales* was the most abundant in pelagic *Artemia*. *Dactylococcopsis* was the most abundant in *Artemia* from microbialite sites. Notably absent in the diet were the dominant diatom OTUs, suggesting selective grazing. Diatom cells are considerably larger than *Dunaliella* in GSL by direct microscopy (GSLBSC unpublished), and *Artemia* do exhibit size selectivity in dietary preferences (Makridis and Vadstein 1999).

The dominant view of the *Artemia* diet in GSL is that *Dunaliella sp.* is the primary component because it is more suitable to the gape limitations of the nauplii (Dobbeleir et al. 1980; Stephens 1998), is often the dominant taxa in GSL (Stephens 1998), and has served as an effective feed for GSL *Artemia* in culture (Wurtsbaugh and Gliwicz 2001). Our observed patterns of algal composition in *Artemia* diets, most notably *Dactylococcopsis*, may therefore be consequential. *Dactylococcopsis* is a halophilic genus of the *Chroococcales* order documented in GSL microbialites using 16S rRNA gene sequence comparisons (Lindsay et al. 2017). Walsby et al. (1983) determined that *Dactylococcopsis* could be digested by *Artemia*, which grazed them with sufficient intensity to clear the water column of a small hypersaline

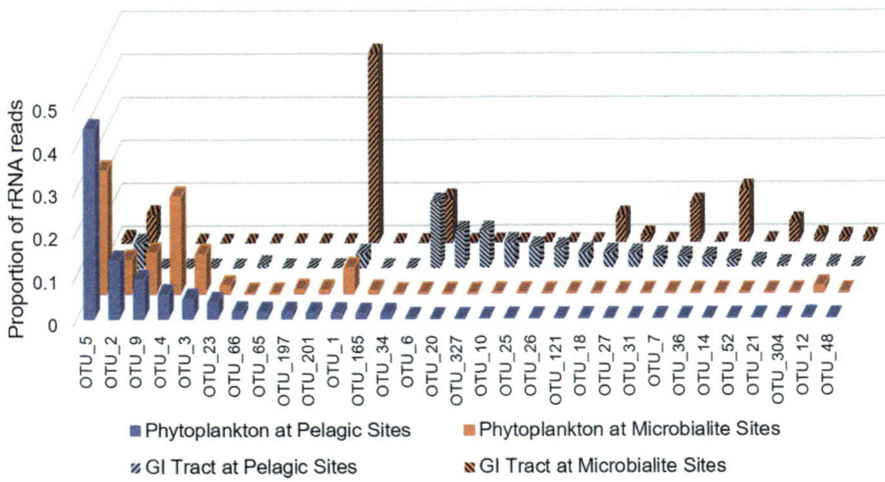

Fig. 7.19 *Artemia* gastrointestinal tract composition (striped columns) from animals collected in the water column of pelagic sites and microbialite sites compared to the phytoplankton community at those sites (solid columns)

Sinai Peninsula lake, suggesting this taxa may be a suitable and important food source for GSL *Artemia*. The presence of *Dactylococcopsis* and the OTU3 *Navicula* in the GSL water column may also indicate a larger microbialite contribution to *Artemia* diets than is revealed here by plastid primers alone. Microbialites are a diverse community dominated by archaea and bacteria taxa (Lindsay et al. 2017) that are likely to be entrained into the water column alongside *Dactylococcopsis*, but are undetectable with the plastid gene used here. Because *Artemia* can consume non-photosynthetic bacteria (Makridis et al. 2000) and may derive nutrition from them (Toi et al. 2013), the picture of *Artemia* diet in GSL is now complicated not only by the presence of microbialite photosynthetic plankton, but also the possibility of microbialite archaeal and bacterial contributions that could further strengthen the benthic-pelagic link in GSL.

Additional work is required before quantitative estimates of *Artemia* selective grazing or microbialite contribution can be made. Aggregate gut contents do not necessarily reflect the ratio of food taxa ingested. Phytoplankton cells consumed by zooplankton can experience rapid (Nejstgaard et al. 2008; Simonelli et al. 2009; Durbin et al. 2012) and differential (Porter 1973) digestion, and therefore the DNA assemblage at the beginning of the digestive tract will likely differ from that at the end. Quantitative dietary assessment of whole-gut contents is therefore problematic before even considering the effects of gene copy number differences within and between phytoplankton taxa (Nejstgaard et al. 2008; Conroy et al. 2017) and the effect of preservation method (Simonelli et al. 2009; Durbin et al. 2012). Nevertheless, our work underscores a linkage between the planktonic *Artemia* and the benthos through consumption of microbialite taxa in the shallow waters of the lake.

7.4.5 Lake Stratification Impacts: Meromixis and Monomixis

The *Artemia*-producing main arm of GSL, Gilbert Bay, has been strongly meromictic most years since 1965 (Madison 1970; Loving et al. 2000; Beisner et al. 2009). Meromixis is a long-term density stratification in the water column driven by chemical gradients that persist through a lake's mixing cycle (Hutchinson 1957). In the case of GSL, meromixis exists as a strong salinity gradient resulting from a 1959 railroad causeway bisecting the main body of the lake (Madison 1970) into a north arm (Gunnison Bay) and a south arm (Gilbert Bay). Gunnison Bay has limited hydrologic connection to Gilbert Bay, despite the original inclusion of small culverts (Madison 1970) and a larger opening in the western shallows in 1984 (Loving et al. 2000). A salinity differential exists between Gilbert and Gunnison bays because the former receives all three inflowing rivers and the latter became the de facto terminal basin in which evaporative loss increased salinity to saturation levels. Density-driven bidirectional flow through the culverts and causeway material established the monimolimnion in Gilbert Bay (Madison 1970). Meromictic lakes are typically deep or protected from wind mixing by local geography (Dodson 2005), but the continual supply of heavy brine and the enormous density difference between layers allows the monimolimnion to exist in Gilbert Bay despite the 55 km fetch and 8 m maximum current depth.

The persistence of meromixis in GSL has implications for the food web of Gilbert Bay that we have only begun to understand in recent years. The underlying concept of an anoxic reducing strata accumulating nutrients has been broadly recognized in GSL (Stephens and Gillespie 1976; Belovsky et al. 2011; Naftz 2017), but recent data collections more fully demonstrate the scale of this sequestration while raising questions and concerns about the bottom-up effects of meromixis on the pelagic food web, and the potential N loss from this nutrient-limited system if the anthropogenic meromixis is modified in the future. GSLBSC collected water column profiles, *Artemia* demographics, and N and P data from 2013 through 2018 to provide a continual 6-year record of GSL stratification. The period of record includes a rare 33-month span of complete or nearly complete monomixis and subsequent return to meromixis that allows some comparison in nutrient distribution and *Artemia* population response.

The monimolimnion is very resistant to complete mixing despite the thinness of this layer and the shallowness of GSL. Stratification persists in every monthly to bimonthly water column profile during periods of meromixis (Fig. 7.20), despite strong wind forcing events. For example, a storm front with average wind speeds of approximately 13 m/s and gusts of 28 m/s occurred between vertical profiles taken

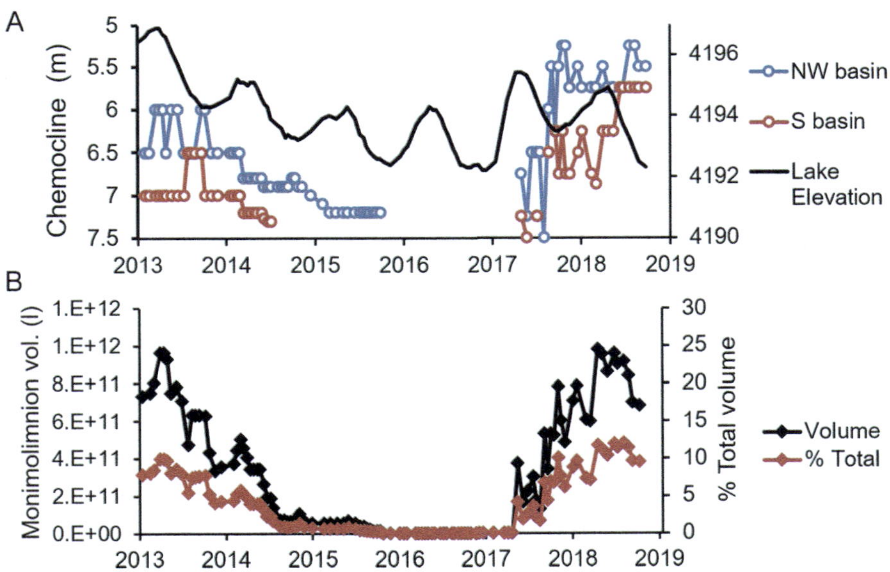

Fig. 7.20 (**a**) Depth of chemocline in 0.5 m intervals from Great Salt Lake Brine Shrimp Cooperative vertical profiles at two Gilbert Bay sites, and United States Geological Survey water surface elevations (Site 10010000) during the 2013–2018 period of alternating meromixis and monomixis. (**b**) Monimolimnion in absolute volume and proportion of total Gilbert Bay, estimated using chemocline depths, United States Geological Survey surface elevation, and bathymetry from Baskin (2005)

on October 20 and 23 of 2017, yet the chemocline depth and salinity remained unchanged. Wind forcing and seasonal temperature equilibria are therefore unlikely to affect stratification, but changes in the causeway bisecting the lake can be very influential. Closure of the causeway culverts in late 2012 and 2013 greatly reduced heavy brine supply from Gunnison Bay (Naftz 2017). Chemocline depths increased and monimolimnion volumes declined from late 2013 until disappearance in November 2015 (Fig. 7.20). Gilbert Bay became monomictic for the first time since the mid-1990s (Loving et al. 2000), and remained so until a new causeway breach was opened in December 2016. The monimolimnion did not return until the unidirectional forcing of Gilbert Bay water into Gunnison Bay from spring runoff had subsided. We first detected a chemocline in May of 2017 at a routine monitoring site 8 km from the causeway breach, and in October 2017 at a southerly site 55 km distant. The monimolimnion has since regained a proportional volume roughly equal to that before the culvert closures (Fig. 7.20).

The resistance to full mixing in Gilbert Bay has important implications for nutrient sequestration and cycling. The anoxic monimolimnion is rich in ammonia and P remineralized from organic material settled from the overlying oxic water layers (Fig. 7.21). Throughout 2013, the monimolimnion accounted for an average of 7.5% of total Gilbert Bay volume and covered an estimated 37% of the benthic surface area according to the bathymetric data of Baskin (2005). We estimate the monimolimnion contained only 12.6% of total Kjeldahl nitrogen load but 75.2% of the ammonia in that year (Fig. 7.22). Phosphorous was partitioned similarly, with 22.5% of total P in the monimolimnion (Fig. 7.22) and 75.5% of the soluble-reactive phosphorous.

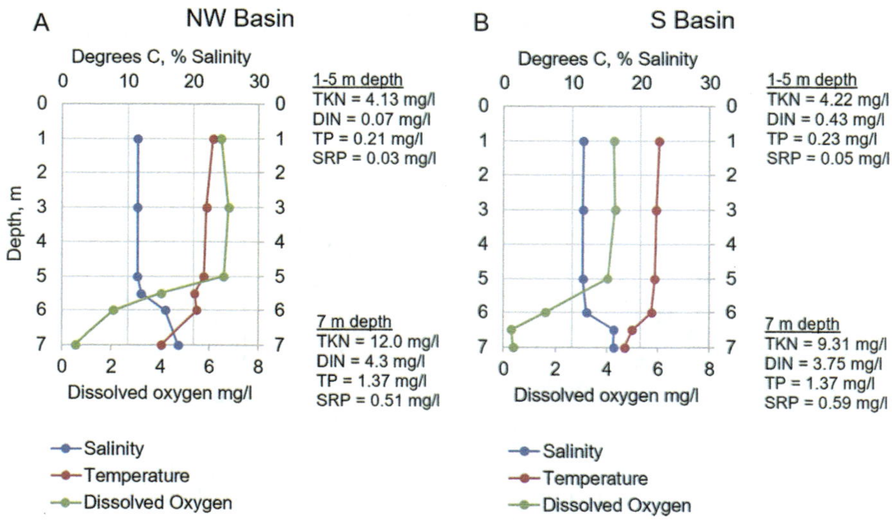

Fig. 7.21 Vertical profiles of salinity, oxygen, and temperature representing typical meromixis in the (**a**) northwest and (**b**) southern basins of Gilbert Bay of Great Salt Lake, with associated mixolimnetic and monimolimnetic nitrogen and phosphorous concentrations on June7, 2018

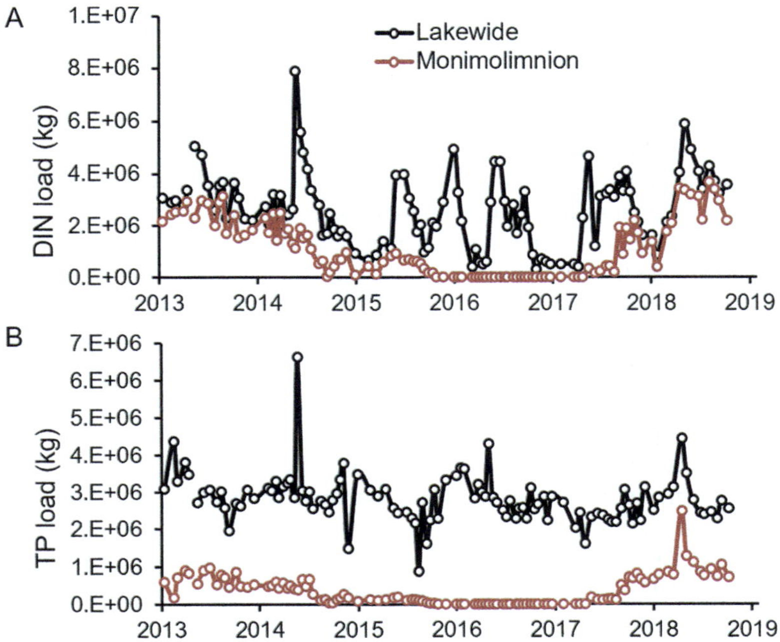

Fig. 7.22 (**a**) Dissolved inorganic nitrogen load (ammonia/ammonium + nitrate + nitrite) and (**b**) total phosphorous load within all of Gilbert Bay and monimolimnion only, estimated from average concentration within the 1–5 m mixolimnion of 9 sites and within 0.5–1.0 m of the benthos at 3 deep sites. Nitrogen and phosphorous analyses conducted by IEH Analytical Laboratories, Seattle, WA. Note summer peaks in dissolved inorganic nitrogen (DIN) driven by *Artemia*-excreted ammonia

Sequestration of limiting nutrients is common within the hypolimnia of lakes seasonally stratified by temperature, but the regular turnover cycle of these lakes results in a more predictable release of nutrients back into the epilimnion. The residence time of critical nutrients beneath the GSL chemocline, however, is almost entirely unknown. The relatively constant monimolimnetic volume despite nearly continual saline inflows suggests that gradual mixing at the chemocline releases nutrients and salts into the epilimnion through chemocline erosion (Boehrer et al. 2014; von Rohden et al. 2009). Evidence of this has been observed in GSL by Beisner et al. (2009), who measured short-term (12–24 h) breakdowns in stratification at a single site due to high sustained winds and probable internal seiches that tilted the monimolimnion at this location rather than mixing it into the epilimnion. Such wind-driven internal turbulence has resulted in measurable nutrient transport into the upper water layers of meromictic Mono Lake (MacIntyre and Jellison 2001). However, the areal distribution and mass of nutrients and salts reintroduced to the mixolimnion from these events is unquantified for GSL. At best, we have a cursory estimate of a two- to three-year residence time based upon the dates of culvert closure and the final disappearance of stratification.

The timing and magnitude of this N and P release may be important. Primary production in Gilbert Bay is generally N limited (Barnes and Wurtsbaugh 2015; Stephens and Gillespie 1976), but excreted ammonia from periods of dense *Artemia* populations can drive the observed dissolved inorganic nitrogen (DIN):TP ratio (Ptacnik et al. 2010; Morris and Lewis 1988) toward co-limitation with P. Because algal community nutrient demand is influenced by factors with a seasonal occurrence (grazing pressure, temperature, light limitation), the timing of nutrient release from the monimolimnion may determine whether they are readily assimilated into the biota or susceptible to exiting the system via mechanisms such as ammonia volatilization (e.g., Murphy and Brownlee 1981), which itself is partly governed by seasonal factors of temperature, salinity, and wind events (Johnson and Bell 2008).

The release of the entire labile monimolimnetic N and P pool during the 2015–2017 GSL monomixis resulted in elevated DIN, but progressively smaller effects in basic pelagic food chain responses. DIN peaks were notably higher following the loss of the monimolimnion across the extensive south basin of Gilbert Bay in the summer of 2014 (Fig. 7.23) and continuing through the monomixis of late 2015, 2016, and the first half of 2017. The return of meromixis across both basins of Gilbert Bay by late 2017 appears to have resumed sequestration of inorganic N and muted the 2018 DIN peak. Phytoplankton response to the elevated DIN during monomixis resulted in higher annual chlorophyll-a peaks, yet this did not translate into detectable increases in average annual *Artemia* abundances or reproductive output, whether measured as cyst densities in the water column (Fig. 7.23), or as the product of oviparous female abundances and clutch sizes.

The absence of a detectable increase in *Artemia* does not preclude substantial long-term effects of meromixis. The inherent inefficiency in energy transfer between trophic levels (Lindeman 1942) suggests the potential gain in *Artemia* densities and reproduction in a given year from the notable increases in inorganic N and chlorophyll-a may be small enough to be masked by stochastic interannual population variations and measurement errors. Additional DIN and primary production may result in incremental improvements to *Artemia* production that require more than a handful of monomictic years to detect. Although large increases in chlorophyll-a and *Artemia* densities were detected in Mono Lake the year after the breakdown of meromixis (Jellison et al. 1993), important differences exist between the two systems. The onset of holomixis in Mono Lake was sudden, whereas the decline of the GSL monimolimnion and its nutrient release occurred over 2 years. The monimolimnetic pool of limiting DIN in GSL was also proportionally smaller.

Long-term effects of meromixis in GSL are of far greater concern. The standing pool of nutrients within the lake is approximately an order of magnitude larger than annual inputs (Naftz 2017), and therefore alterations to internal cycling or nutrient loss from GSL as a result of changing meromixis will accumulate over multiple years. The rate of N loss through ammonia volatilization and denitrification is a prominent deficiency in our understanding of the nutrient balance of Gilbert Bay. Jellison et al. (1993) estimated rapid volatilization of the released ammonia, and while the rate from GSL is likely to be less due to the lower pH (Emerson et al. 1975), small annual rates can become meaningful at decadal timescales. Similarly,

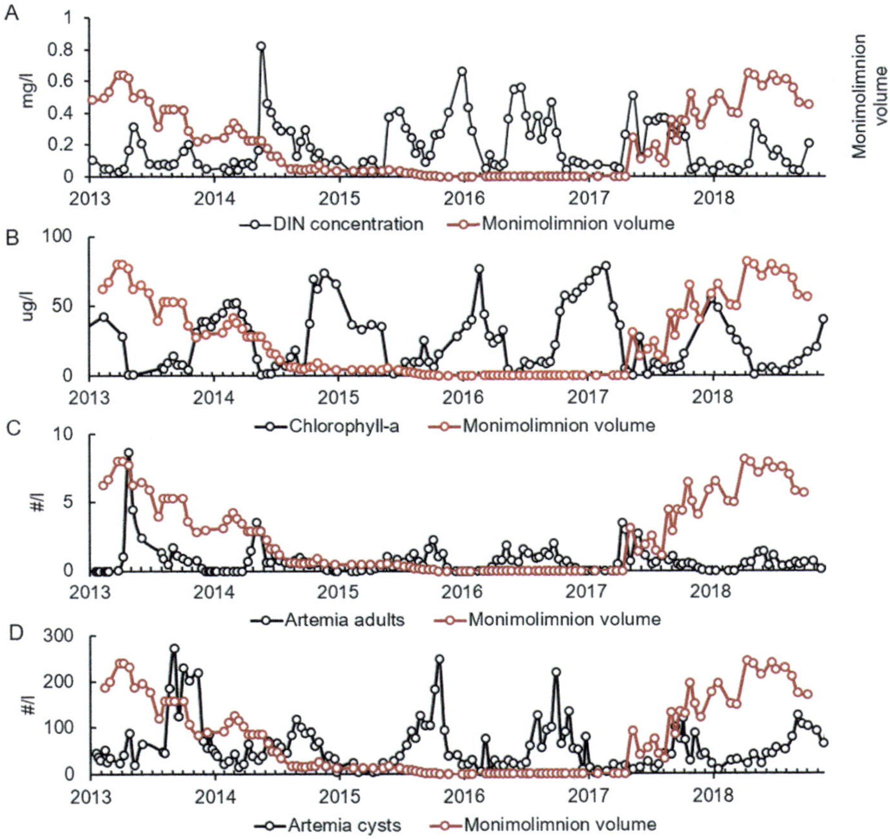

Fig. 7.23 The transition to monomixis from 2014 to 2015 resulted in (**a**) increased dissolved inorganic nitrogen (DIN) and (**b**) chlorophyll-*a* concentrations in the mixolimnion, but muted and inconsistent responses in (**c**) *Artemia* densities and (**d**) oviparous reproductive output

denitrification has been hypothesized from GSL mesocosm trials (Barnes and Wurtsbaugh 2015), and Gilbert Bay as a whole (Stephens and Gillespie 1976), but no direct measurement or detailed estimate has been given for this pathway despite the presence of denitrifying bacteria in the lake (Fendrich 1988). It is therefore unclear if and to what extent meromixis promotes or hinders primary productivity and nutrient retention in GSL. The monimolimnion is a nutrient remineralization and storage site and some proportion is transferred to the mixolimnion over time, yet it also provides conditions suitable for denitrification (Stephens and Gillespie 1976), isolates the complex and system-specific nutrient cycling processes of the sediments (Forsberg 1989), and may cause a net loss of nutrients to sediments capped almost permanently by the chemocline.

As with N and P, the monimolimnion also temporarily sequesters a portion of the annual production of *Artemia* cysts. Cyst buoyancy is not uniform in a given year (Abatzopolous et al. 2006b; Hajirostamloo 2008; GSLBSC unpublished), and an unknown fraction sink into the chemocline where, shielded from turbulent mixing, they can accumulate throughout the fall and winter. Direct field sampling and hatching assessments of these cysts in 2014 and 2015 demonstrate a viability similar to epilimnetic cysts, and a similar distinct annual hatchability trend (Fig. 7.24), strongly suggesting the majority of deep cysts were produced during the same annual cycle as the epilimnetic cysts. Epilimnetic and chemocline cysts both follow the seasonal pattern of peaking in autumn and declining by spring (Fig. 7.24), indicating proportionally little accumulation and carry over year-to-year. The ultimate fate of these cysts is unproven. Some settle into the benthos and have been found in GSL sediment cores (Clegg and Jackson 1997), while the remainder may become entrained into the mixolimnion where salinities are more conducive to hatching.

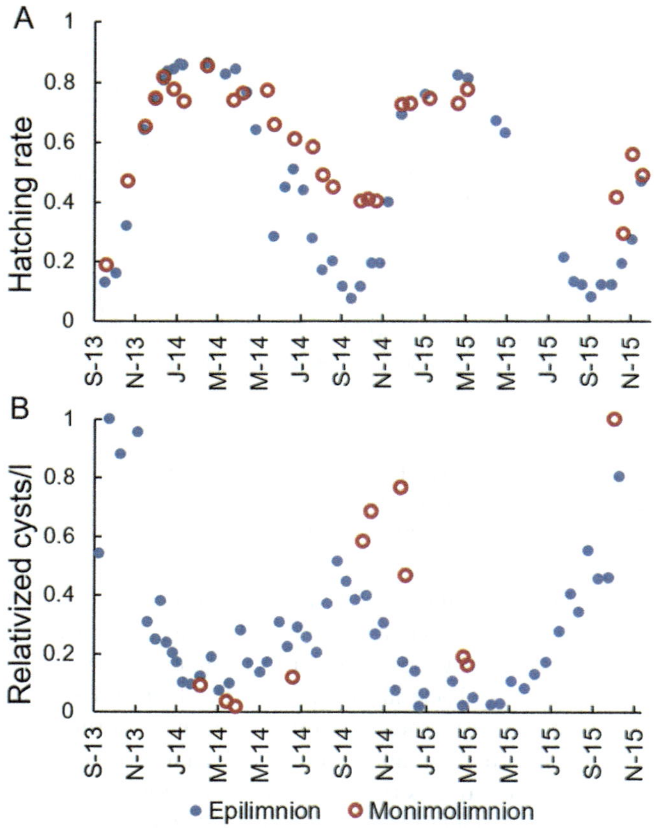

Fig. 7.24 (**a**) The average hatching rate of cysts collected by Great Salt Lake Brine Shrimp Cooperative within the chemocline and monimolimnion of two routine sampling sites compared to lake-wide average epilimnion hatching rates, October 2013 through November 2015 (**b**) Average monimolimnetic cyst abundance, relativized for clarity, from surveys of 12–17 deep water sites follows the annual trend of epilimnetic cyst abundance determined from routine biweekly *Artemia* population surveys

The seasonal timing of this viable cyst pool reentering the epilimnion may prove relevant for the cyst harvest management model in which spring and fall abundances are important (Belovsky and Perschon 2019). Such uncertainties are inherent in the anthropogenic meromixis of GSL, and the impacts on nutrient and *Artemia* cyst cycling and retention are not fully understood and cannot be controlled for. However, the moderating effect of the causeway on Gilbert Bay salinities and the modification of salt flow afforded by the adaptive management berm demonstrates a positive attribute to this unique lake stratification.

7.5 Threats to the Great Salt Lake Ecosystem and *Artemia*

Ecosystems are perpetually at risk—this is a given reality. The question is how well an ecosystem recovers from perturbations, what risk factors lead to alterations, and what management actions can be taken to mitigate these risks. GSL and the resident *Artemia* population face a number of risks, and we provide a cursory discussion of those we view as particularly crucial to GSL. Some risks have been mitigated by adaptive management and some by the inherent qualities of GSL itself, while other threats to the *Artemia* population and GSL ecosystem still remain and will require future action.

Some of the most important risk factors for GSL and the *Artemia* population have either been successfully mitigated by cooperative adaptive management or there are mechanisms in place to do so. Overharvesting of the *Artemia* population was a concern in the mid-1990s due to the rapid expansion of the brine shrimp industry. Overharvesting of the *Artemia* population could have led to consequential trophic cascades in the ecosystem. However, cooperative management of the harvest averted this and led to positive outcomes for the ecosystem and involved stakeholders. Severe shifts in Gilbert Bay salinity from natural hydrological events, climatic changes, and anthropogenic water use remain a current and future concern, but the recent adaptive management berm at the railroad causeway opening to Gunnison Bay, which regulates bidirectional flows between Gunnison Bay and Gilbert Bay, is a crucial tool that allows management agencies the ability to moderate salinity changes.

A third important and persistent risk is the reduction in Gilbert Bay volume through tributary water withdrawals and climatic change. A smaller lake volume brings the specific challenges of reduced *Artemia* habitat and diminished system-wide productivity, and the exposure of lakebed sediments. The additional nutrient assimilation capacity of the Gilbert Bay ecosystem strongly suggests that habitat loss can be offset to some degree by additional per-volume production if nutrient supply to the lake is not curtailed. The risks posed by exposed lakebed sediments, however, will require a far more proactive effort to avoid. The world's largest and most productive saline lakes and freshwater lakes in desert regions have met with disastrous depletions in size and volume. The Aral Sea, Lake Urmia, Dead Sea, Owens Lake, Pyramid Lake, Walker

Lake, Mono Lake, and Salton Sea (Vörösmarty et al. 2000; Micklin 2007; Herbst and Prather 2014; Zeinoddini et al. 2015; Moore 2016; Fazel et al. 2018; Wine et al. 2019) provide stark warnings about the vulnerability of saline lakes to human intervention and the astronomical economic and human health costs that can result from their decline. The economic damage from the decline of Lake Urmia and the irreversible damage to fertile soil through salt deposition is well into the billions of dollars (Weiss 2018; Salimi et al. 2019). More locally, the dewatered Owens Lake bed in California is the single largest PM10 hazard in the entire country and it is predicted that the city of Los Angeles will have spent $3.6 billion dollars over 25 years just for dust mitigation (Saint-Armand et al. 1986; Tyson 1997; LADWP 2015; Gutrich et al. 2016) that does not restore the functions and services of Owens Lake. This threat is looming for GSL, but it is not too late—effective management practices for scarce water resources are being explored and if effectively implemented can ensure the lake and its functions are sustained.

Contaminants are another concern and the most often cited concerns about pollutants in GSL pertain to anthropogenically sourced metals such as mercury, cadmium, chromium, zinc, copper, and lead (Sarabia et al. 2002; Brix et al. 2006; Naftz et al. 2009) and non-metals or metalloids such as selenium and arsenic (Lemly 1993, 1997, 2004; Canton and Van Derveer 1997; Skorupa 1998; Brix et al. 2003, 2004; Wurtsbaugh et al. 2011). Although these can be found in GSL biota, sediments, and water samples, most of the metals have been decreasing in GSL sediments since the 1950s (Wurtsbaugh 2012, 2014), reflecting the efficacy of federal and state laws mandating pollution discharge reductions and controls for wastewater.

The State of Utah Division of Water Quality has developed a Great Salt Lake Water Quality Strategy (Utah DEQ 2014) that details assessment methods for developing aquatic life criteria for priority pollutants. The priority pollutant list includes over 100 priority pollutants and each pollutant will be further evaluated and ranked based on species presence in GSL, known presence of pollutant in GSL or in discharges to GSL, and then using existing toxicity data, in combination with GSL concentrations, to establish prioritization. Among the pollutants on this list selenium is the only pollutant for which there are currently defined GSL site-specific water quality criteria (Ohlendorf et al. 2009). The water quality criteria established for selenium in GSL are a precedent as it is the first time in the United States that such criteria have been established on the basis of a wildlife standard rather than a human health standard. Another priority pollutant is mercury and early studies (Naftz et al. 2008; Peterson and Gustin 2008; Conover and Vest 2009; Naftz et al. 2009; Vest et al. 2009) caused speculation about trophic transfer of Hg throughout the GSL food web, yet more current studies have found Hg was only weakly coupled with epilimnetic organisms and demonstrated limited transfer between the monimolimnion and higher trophic level biota (Valdes et al. 2017). Our own data shows minimal transfer of Hg from adult *Artemia* to the cyst (Fig. 7.25a, b). Thus, contaminants, like other risks, remain but can be mitigated through careful attention and effective action.

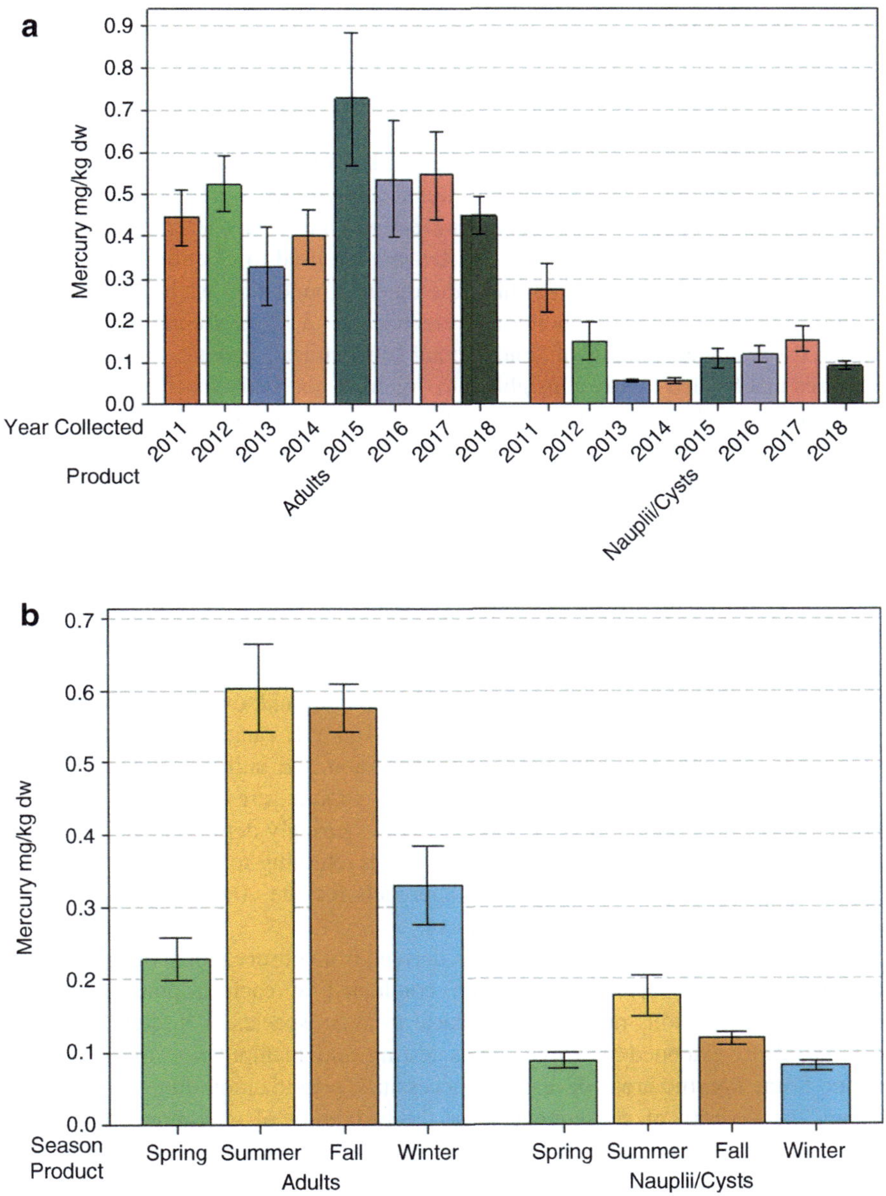

Fig. 7.25 Mercury (Hg) in Great Salt Lake *Artemia* tissue. Contaminants were routinely measured a minimum of four times per year, and often more frequently. Samples were taken seasonally (multiple times per season) from nine locations over an 8-year period. Adult age classes were separated from the cysts/nauplii fraction and total Hg levels were determined. The total number of samples analyzed for adult *Artemia* was: spring $n = 83$, summer $n = 96$, fall $n = 109$, winter $n = 17$, and for the nauplii/cyst age classes the numbers were: spring $n = 103$, summer $n = 98$, fall $n = 128$, winter $n = 78$. 95% confidence error bars are indicated. There were no long-term trends of increasing or decreasing concentration in *Artemia* tissue, though there was an indication of a transient increase in Hg during 2015 which coincided with the dissolution of the monimolimnion into the epilimnion

7.6 Conclusions

Extraordinary is an insufficient term to describe *Artemia franciscana*, an arthropod well adapted to extreme conditions where they serve pivotal roles in the GSL ecosystem. The reach of their influence extends well beyond the boundaries of GSL. *A. franciscana* provides the nutritional support for millions of migratory waterbirds that overwinter thousands of miles away in Mexico, Central and South America. Furthermore, the cysts of GSL *Artemia* are used to meet the needs of the world's demand for protein by forming the dietary foundation of the global aquaculture industry. Detailed laboratory research on *A. franciscana* has greatly expanded our understanding of fundamental biochemical, genetic, and metabolic processes that has led to novel insights into organism survival capacities via protein preservation, metabolic reductions, and developmental limitations. *Artemia* continue to be used for toxicological assessments and for testing of evolutionary and phylogenetic relationships. In short, *A. franciscana*, either directly or indirectly, is interconnected in ways that are beneficial for both humans and the environment across the globe.

Long-term *Artemia* population dynamics have been documented and analyzed revealing a remarkable response to large gradients in food availability and abiotic factors—most notably salinity. The *Artemia* population exhibits profound, yet normal and fundamentally predictable, seasonal swings in population size and reproductive response to seasonal changes in the phytoplankton population. Across years, oscillations in salinity beyond the 120–160 g/L range can influence the population indirectly through invertebrate predation and shifts in phytoplankton community composition, and directly through osmotic stress. Despite this, the two-decade record of the GSL *Artemia* population strongly demonstrates resilience across this salinity range and rapid recovery upon returning to it. Additionally, the cooperative and adaptive management approach for the *Artemia* resource is a successful example of resource management.

Most of the research on GSL *Artemia* is derived from observational or monitoring programs. Excellent studies have been conducted at each trophic level and documented interactions range from archaeal to avian species. Yet, there needs to be more effort extended into hypothesis testing and manipulation of influential factors. Some specific areas of research necessitate prioritization due to the rate of change, the severity of consequences, or the proximity of alterations. Since the system is driven primarily by bottom-up influences, research should focus on nutrient effects that are vectored throughout the food web. Nutrient assessments of the lake coupled with field and laboratory mesocosm/microcosm studies to elucidate biotic responses to nutrient alterations are in need. A more extensive assessment of the mass balance, cycling, loading, and losses of nutrients in GSL is required. Modeling of GSL and its biota is in its infancy, with the exception of the structural equation model developed by Belovsky et al. (2011). Similarly, research on the vast assemblages of microbes and their relationship to planktoners is just beginning to be understood and opens more questions than are solved (Meuser et al. 2013). As research progresses and data are shared among research teams and resource managers, the integration of field, laboratory, and modeling should coalesce and greatly facilitate our understanding of this surprisingly complex system.

GSL finds itself at a figurative crossroads. As a resilient ecosystem it has been able to absorb a number of stressors while still providing robust ecosystem functions and services, and the existing adaptive management of the lake system has assisted here. Yet limits do exist on the ability of the ecosystem and its adaptive management to handle anthropogenic alteration of the lake and its watershed, and these pressures will continue into the foreseeable future. Fortunately, an engaged and informed group of stakeholders concerned about GSL has been actively working on management options for the lake since the State of Utah gained ownership rights of the GSL lakebed in 1975 (Dewsnup and Jensen 1980). This continual effort by stakeholders to manage the resource in a prudent and sustainable manner still has many challenges to confront, but there are also a number of effective steps already taken and implemented that function to ensure the long-term integrity of the GSL ecosystem. If successful, the GSL ecosystem and its resident *Artemia* population will continue to provide ecosystem functions and services that are necessary and desirable well into the future.

References

Abatzopolous TJ, Agh N, Van Stappen G, Rouhani R et al (2006a) *Artemia* sites in Iran. J Mar Biol Asses UK 86:299–307

Abatzopolous TJ, Baxevanis AD, Triantaphyllidis GV et al (2006b) Quality evaluation of *Artemia urmiana* Günther (Urmia Lake, Iran) with special emphasis on its particular cyst characteristics (International Study on *Artemia* LXIX). Aquaculture 254:442–454

Abatzopoulos TJ, El-Bermawi N, Vasdekis C et al (2003) Effects of salinity and temperature on reproductive and life span characteristics of clonal *Artemia*. Hydrobiologia 492:191–199

Acker JP, Fowler A, Lauman B et al (2004) Survival of desiccated mammalian cells: beneficial effects of isotonic media. Cell Preserv Technol 1(2):129–140

Ako H, Tamaru CS, Bass P, Lee CS (1994) Enhancing the resistance of physical stress in larvae of *Mugil cephalus* by the feeding of enriched *Artemia* nauplii. Aquaculture 122:81–90

Allgeier JE, Rosemond AD, Layman CA (2011) The frequency and magnitude of non-additive responses to multiple nutrient enrichment. J Appl Ecol 48(1):96–101

Anderson DM, Glibert PM, Burkholder JM (2002) Harmful algal blooms and eutrophication: nutrient sources, composition, and consequences. Estuaries 25(4):704–726

Aristizábal EO, Suárez J (2006) Efficiency of co-feeding red porgy (*Pagrus pagrus L.*) larvae with live and compound diet. Rev Biol Mar Oceanogr 41(2):203–208

Asem A, Wang P, Sun SC (2018) Comparative phylogenetic perspectives on the evolutionary relationships in the brine shrimp *Artemia leach*, 1819 (Crustacea: Anostraca) based on secondary structure of ITS1 gene. J Genet Res 4(2):72–84

Bahabadi MN, Mozanzadeh MT, Agh N et al (2018) Enriched *Artemia* with L-lysine and DL-methionine on growth performance, stress resistance, and fatty acid profile of *Litopenaeus vannamei* postlarvae. J Appl Aquac 30(4):325–336. https://doi.org/10.1080/10454438.2018.1484838

Barnes BD, Wurtsbaugh WA (2015) The effects of salinity on plankton and benthic communities in the Great Salt Lake, Utah, USA: a microcosm experiment. Can J Fish Aquat Sci 72(6):807–817

Baskin RL (2005) Calculation of area and volume for the south part of Great Salt Lake, Utah U.S. Geological Survey Open-File Report 2005–1327

Baskin RL (2014) Occurrence and spatial distribution of microbial bioherms in Great Salt Lake, Utah. Dissertation, University of Utah

Baxevanis AD, Kappas I, Abatzopoulos TJ (2006) Molecular phylogenetics and asexuality in the brine shrimp *Artemia*. Mol Phylogenet Evol 40(3):724–738

Beck AB, Bengston DA (1978) Review of studies evaluating effects of live and artificial diets on survival and growth of the marine atherinid fish *Menidia menidia* the Atlantic Silverside. In: FAO, Rome (Italy). Fisheries Department. Symposium on Finfish Nutrition and Feed Technology, Hamburg

Beisner K, Naftz DL, Johnson WP et al (2009) Selenium and trace element mobility affected by periodic displacement of stratification in the Great Salt Lake, Utah. Sci Total Environ 407 (19):5263–5273

Belovsky GE, Perschon WC (2019) A management case study for a new commercial fishery: brine shrimp harvesting in Great Salt Lake, Utah, USA. Ecol Appl 29(3):e01864

Belovsky GE, Stephens D, Perschon C et al (2011) The Great Salt Lake ecosystem (Utah, USA): long term data and a structural equation approach. Ecosphere 2(3):1–40

Bengtson DA (2003) Status of marine aquaculture in relation to live prey: past, present and future. In: Støttrup JG, McEvoy LA (eds) Live feeds in marine aquaculture. Blackwell Science Ltd, Oxford, pp 1–16

Bengtson DA, Léger P, Sorgeloos P (1991) Use of *Artemia* as a food source for aquaculture. In: Brown RA, Sorgeloos P, Trotman CNA (eds) *Artemia* Biology, vol 11. CRC Press, Boca Raton, FL, pp 255–285

Bioeconomics, Inc. (2012) Economic significance of the Great Salt Lake to the State of Utah. Report to the State of Utah, Great Salt Lake Advisory Council. https://documents.deq.utah.gov/water-quality/standards-technical-services/great-salt-lake-advisory-council/Activities/DWQ-2012-006864.pdf. Accessed 14 May 2019

Blomqvist S, Gunnars A, Elmgren R (2004) Why the limiting nutrient differs between temperate coastal seas and freshwater lakes: a matter of salt. Limnol Oceanogr 49(6):2236–2241

Boehrer B, Kiwel U, Rahn K et al (2014) Chemocline erosion and its conservation by freshwater introduction to meromictic salt lakes. Limnologica 44:81–89

Bostock J, McAndrew B, Richards R et al (2010) Aquaculture: global status and trends. Philos Trans R Soc B Biol Sci 365(1554):2897–2912

Brix KV, Cardwell RD, Adams WJ (2003) Chronic toxicity of arsenic to the Great Salt Lake brine shrimp, *Artemia franciscana*. Ecotoxicol Environ Saf 54(2):169–175

Brix KV, Deforest DK, Cardwell RD et al (2004) Derivation of a chronic site-specific water quality standard for selenium in the Great Salt Lake, Utah, USA. Environ Toxicol Chem 23(3):606–612

Brix KV, Gerdes RM, Adams WJ et al (2006) Effects of copper, cadmium, and zinc on the hatching success of brine shrimp (*Artemia franciscana*). Arch Environ Contam Toxicol 51(4):580–583

Browne RA, Bowen ST (1991) Taxonomy and population genetics of *Artemia*. In: Browne RA, Sorgeloos P, Trotman CNA (eds) *Artemia* Biology. CRC, Boca Raton, p 384

Browne RA, Lenz PH (1991) Ecology of *Artemia*. In: Browne RA, Sorgeloos P, Trotman CNA (eds) *Artemia* Biology. CRC, Boca Raton, FL, pp 237–254

Browne RA, Wanigasekara G (2000) Combined effects of salinity and temperature on survival and reproduction of five species of *Artemia*. J Exp Mar Biol Ecol 267:107–119

Buchanan SS (2004) Cryopreservation of stem cells using trehalose: evaluation of the method using a human hematopoietic cell line. Stem Cells Dev 13(3):295–305

Burne RV, Moore L (1987) Microbialites; organosedimentary deposits of benthic microbial communities. PALAIOS 2:241–254

Canton SP, Van Derveer WD (1997) Selenium toxicity to aquatic life: an argument for sediment-based water quality criteria. Environ Toxicol Chem 16(6):1255–1259

Caudell JN (2001) Biology of Eared Grebes *(Podiceps nigricollis)* on the Great Salt Lake, Utah. Dissertation, Utah State University

Caudell JN, Conover MR (2006a) Energy content and digestibility of brine shrimp (*Artemia franciscana*) and other prey items of eared grebes (*Podiceps nigricollis*) on the Great Salt Lake, Utah. Biol Conserv 130(2):251–254

Caudell JN, Conover MR (2006b) Behavioral and physiological responses of Eared Grebes (*Podiceps nigricollis*) to variations in brine shrimp (*Artemia franciscana*) densities. West North Am Nat 66(1):12–23

Céspedes V, Sánchez MI, Green AJ (2017) Predator–prey interactions between native brine shrimp *Artemia parthenogenetica* and the alien boatman *Trichocorixa verticalis*: influence of salinity, predator sex, and size, abundance and parasitic status of prey. Peer J 5:e3554

Chakraborty RD, Chakraborty K, Radhakrishnan EV (2007) Variation in fatty acid composition of *Artemia salina* nauplii enriched with microalgae and baker's yeast for use in larviculture. J Agric Food Chem 55(10):4043–4051

Chen T, Acker JP, Eroglu A et al (2001) Beneficial effect of intracellular trehalose on the membrane integrity of dried mammalian cells. Cryobiology 43(2):168–181

Chen WH, Ge X, Wang W et al (2009) A gene catalogue for post-diapause development of an anhydrobiotic arthropod *Artemia franciscana*. BMC Genomics 10(1):52

Clegg JS (1965) The origin of trehalose and its significance during the formation of encysted dormant embryos of *Artemia salina*. Comp Biochem Physiol 14(1):135–143

Clegg JS (1978) Hydration-dependent metabolic transitions and the state of cellular water in *Artemia* cysts. In: Dry biological systems. Academic Press, New York, pp 117–153

Clegg JS (1997) Embryos of *Artemia franciscana* survive four years of continuous anoxia: the case for complete metabolic rate depression. J Exp Biol 200(3):467–475

Clegg JS (2001) Cryptobiosis—a peculiar state of biological organization. Comp Biochem Physiol B Biochem Mol Biol 128(4):613–624

Clegg JS (2005) Desiccation tolerance in encysted embryos of the animal extremophile, *Artemia*. Integr Comp Biol 45(5):715–724

Clegg JS (2007) Protein stability in *Artemia* embryos during prolonged anoxia. Biol Bull 212 (1):74–81

Clegg JS (2011) Stress-related proteins compared in diapause and in activated, anoxic encysted embryos of the animal extremophile, *Artemia* franciscana. J Insect Physiol 57(5):660–664

Clegg JS, Jackson SA (1992a) Aerobic heat shock activates trehalose synthesis in embryos of *Artemia franciscana*. FEBS Lett 303(1):45–47

Clegg JS, Jackson SA (1992b) The metabolic status of quiescent and diapause embryos of *Artemia franciscana* (*Kellog*). Special issues in the advancement of limnology. Evol Ecol Asp Arch Hydrobiol Crustac Diapause 52:425–439

Clegg JS, Jackson SA (1997) Significance of cyst fragments of *Artemia* sp. recovered from a 27,000 year old core taken under the Great Salt Lake, Utah, USA. Int J Salt Lake Res 6:207–216

Clegg JS, Trotman CNA (2002) Physiological and biochemical aspects of *Artemia* ecology. In: Abatzopoulos TJ, Beadmore JA, Clegg JS, Sorgeloos P (eds) *Artemia*: basic and applied biology, vol 1. Kluwer Academic, London, pp 128–170

Clegg JS, Jackson SA, Warner AH (1994) Extensive intracellular translocations of a major protein accompany anoxia in embryos of *Artemia franciscana*. Exp Cell Res 212(1):77–83

Clegg JS, Drinkwater LE, Sorgeloos P (1996) The metabolic status of diapause embryos of *Artemia franciscana* (SFB). Physiol Zool 69(1):49–66

Clegg JS, Willsie JK, Jackson SA (1999) Adaptive significance of a small heat shock/α-crystallin protein (p26) in encysted embryos of the brine shrimp, *Artemia franciscana*. Am Zool 39 (6):836–847

Clegg JS, Van Hoa N, Sorgeloos P (2001) Thermal tolerance and heat shock proteins in encysted embryos of *Artemia* from widely different thermal habitats. In: Melack JM, Jellison R, Herbst DB (eds) Saline lakes. Developments in hydrobiology, vol 162. Springer, Dordrecht

Codd GA, Bell SG, Kaya K et al (1999) Cyanobacterial toxins, exposure routes and human health. Eur J Phycol 34(4):405–415

Collins CH, Clegg JS (2004) A small heat-shock protein, p26, from the crustacean *Artemia* protects mammalian cells (Cos-1) against oxidative damage. Cell Biol Int 28(6):449–455
Conceição LE, Yúfera M, Makridis P et al (2010a) Live feeds for early stages of fish rearing. Aquac Res 41(5):613–640
Conceição LE, Aragão C, Richard N et al (2010b) Novel methodologies in marine fish larval nutrition. Fish Physiol Biochem 36(1):1–16
Conover MR, Caudell JN (2009) Energy budgets for eared grebes on the Great Salt Lake and implications for harvest of brine shrimp. J Wildl Manag 73(7):1134–1139
Conover MR, Vest JL (2009) Selenium and mercury concentrations in California gulls breeding on the Great Salt Lake, Utah, USA. Environ Toxicol Chem 28(2):324–329
Conroy BJ, Steinberg DK, Song B et al (2017) Mesozooplankton graze on cyanobacteria in the Amazon River plume and Western Tropical North Atlantic. Front Microbiol 8:1436
Conte FP, Jellison RS, Starrett GL (1988) Nearshore and pelagic abundances of *Artemia monica* in Mono Lake, California. Hydrobiologia 158(1):173–181
Cooper SC, Winkler DW, Lenz PH (1984) The effect of grebe predation on a brine shrimp population. J Anim Ecol 53:51–64
Coutteau P, Sorgeloos P (1997) Manipulation of dietary lipids, fatty acids and vitamins in zooplankton cultures. Freshw Biol 38(3):501–512
Criel GR, MacRae TH (2002) Reproductive biology of *Artemia*. In: *Artemia*: basic and applied biology. Springer, Dordrecht, pp 39–128
Croghan PC (1958) The osmotic and ionic regulation of *Artemia salina* (L.). J Exp Biol 35:219–233
Crowe JH (2004) Stabilization of cells in the dry state by trehalose. RedNova. http://www.rednova.com/modules/news/tool.php?tool=print&id=63450.html. Accessed 13 Nov 2004
Crowe JH, Crowe LM (2000) Preservation of mammalian cells-learning nature's tricks. Nat Biotechnol 18:145–146
Curnow J, Kin J, Bosmans J et al (2006) The effect of reduced *Artemia* and rotifer use facilitated by a new microdiet in the rearing of barramundi *Lates calcarifer* (BLOCH) larvae. Aquaculture 257 (1–4):204–213
Dana GL, Lenz PH (1986) Effects of increasing salinity on an *Artemia* population from Mono Lake, California. Oecologia 68:428–436
Dana GL, Jellison R, Melack JM (1990) *Artemia monica* cyst production and recruitment in mono Lake, California, USA. In: Saline lakes. Springer, Dordrecht, pp 233–243
Dattilo AM, Bracchini L, Carlini L et al (2005) Estimate of the effects of ultraviolet radiation on the mortality of *Artemia franciscana* in naupliar and adult stages. Int J Biometeorol 49(6):388–395
Davis JS (2000) Structure, function, and management of the biological system for seasonal solar saltworks. Global NEST J 2(3):217–226
Davis JS, Giordano M (1995) Biological and physical events involved in the origin, effects, and control of organic matter in solar saltworks. Int J Salt Lake Res 4(4):335–347
Dewsnup RL, Jensen DW (1980) Legal battle over ownership of the Great Salt Lake. In: Gwynn JW (ed) Great salt Lake: a scientific, historical, and economic overview, vol 116. Utah Geological and Mineral Survey, Salt Lake City, pp 15–18
Dhert P, Rombaut G, Suantika G et al (2001) Advancement of rotifer culture and manipulation techniques in Europe. Aquaculture 200(1–2):129–146
Dittmann KK, Rasmussen BB, Castex M et al (2017) The aquaculture microbiome at the centre of business creation. Microb Biotechnol 10(6):1279
Dobbeleir JN, Bossuyt AE, Bruggeman E, Sorgeloos P (1980) New aspects of the use of inert diets for high density culturing of brine shrimp. In: Persoone G, Sorgeloos P, Roels O, Jaspers E (eds) The brine shrimp Artemia, vol 3, Ecology, culturing, use in aquaculture, Universa Press, Wetteren, p 456
Dodson SI (2005) Introduction to limnology. McGraw Hill, New York, p 400
Drinkwater LE, Clegg JS (1991) Experimental biology of cyst diapause. In: Browne RA, Sorgeloos P, Trotman CNA (eds) *Artemia* biology. CRC, Boca Raton, FL, pp 93–115

Drinkwater LE, Crowe JH (1987) Regulation of embryonic diapause in *Artemia*: environmental and physiological signals. J Exp Zool 241:297–307
Durbin EG, Casas MC, Rynearson TA (2012) Copepod feeding and digestion rates using prey DNA and qPCR. J Plankton Res 34(1):72–82
Eardley AJ, Gvosdetsky V (1960) Analysis of Pleistocene core from Great Salt Lake, Utah. Geol Soc Am Bull 71:1323–1344
Edmondson WT, Anderson GC, Peterson DR (1956) Artificial eutrophication of Lake Washington. Limnol Oceanogr 1(1):47–53
Emerson K, Russo RC, Lund RE, Thurston RV (1975) Aqueous ammonia equilibrium calculations: effect of pH and temperature. J Fish Res Board Can 32:2379–2383
Eroglu A, Russo MJ, Bieganski R et al (2000) Intercellular trehalose improves the survival of cryopreserved mammalian cells. Nat Biotechnol 18(2):163–167
Evans JC, Prepas EE (1996) Potential effects of climate change on ion chemistry and phytoplankton communities in prairie saline lakes. Limnol Oceanogr 41(5):1063–1076
Evans JC, Prepas EE (1997) Relative importance of iron and molybdenum in restricting phytoplankton biomass in high phosphorus saline lakes. Limnol Oceanogr 42(3):461–472
Evjemo JO, Coutteau P, Olsen Y, Sorgeloos P (1997) The stability of docosahexaenoic acid in two *Artemia* species following enrichment and subsequent starvation. Aquaculture 155 (1–4):135–148
FAO (2018a) The state of world fisheries and aquaculture 2018. Food and Agriculture Organization of the United Nations. http://www.fao.org/3/i9540en/I9540EN.pdf. Accessed 6 Mar 2019
FAO (2018b) The state of world fisheries and aquaculture 2018. Meeting the sustainable development goals. Rome. License: CC BY-NC-SA 3.0 IGO
Fazel N, Berndtsson R, Uvo CB et al (2018) Regionalization of precipitation characteristics in Iran's Lake Urmia basin. Theor Appl Climatol 132(1–2):363–373
Felix EA, Rushforth SR (1979) The algal flora of the Great Salt Lake, Utah, USA. Nova Hedwigia 31:163–195
Fendrich C (1988) *Halovibrio variabilis* gen. nov. sp. nov., *Pseudomonas halophila* sp. nov. and new halophilic aerobic coccoid *Eubacterium* from Great Salt Lake, Utah, USA. Syst Appl Microbiol 11:36–43
Fernández RG (2001) *Artemia* bioencapsulation I. Effect of particle sizes on the filtering behavior of *Artemia franciscana*. J Crustac Biol 21(2):435–442
Forsberg C (1989) Importance of sediments in understanding nutrient cycling in lakes. Hydrobiologia 176(1):263–277
Frank MG (2016) Migratory Waterbird ecology at a critical staging area, Great Salt Lake, Utah. Dissertation, Utah State University
Furuita H, Takeuchi T, Toyota M et al (1996) EPA and DHA requirements in early juvenile red sea bream using HUFA enriched *Artemia* nauplii. Fish Sci 62(2):246–251
Gajardo GM, Beardmore JA (2012) The brine shrimp *Artemia*: adapted to critical life conditions. Front Physiol 3:185
Gajardo G, Abatzopoulos TJ, Kappas I et al (2002) Evolution and speciation. In: Abatzopoulos TJ, Beardmore JA, Clegg JS, Sorgeloos P (eds) *Artemia:* basic and applied biology. Springer, Dordrecht, pp 225–250
Gilpin LC, Davidson K, Roberts E (2004) The influence of changes in nitrogen: silicon ratios on diatom growth dynamics. J Sea Res 51(1):21–35
Godfray HCJ, Beddington JR, Crute IR, Haddad L, Lawrence D, Muir JF, Pretty J, Robinson S, Thomas SM, Toulmin C (2010) Food security: the challenge of feeding 9 billion people. Science 327(5967):812–818
Grimaldi DA (2010) 400 million years on six legs: on the origin and early evolution of Hexapoda. Arthropod Struct Dev 39(2–3):191–203
Gruner DS (2004) Attenuation of top-down and bottom-up forces in a complex terrestrial community. Ecology 85(11):3010–3022

Gruner DS, Taylor AD (2006) Richness and species composition of arboreal arthropods affected by nutrients and predators: a press experiment. Oecologia 147(4):714–724
Gruner DS, Smith JE, Seabloom EW et al (2008) A cross-system synthesis of consumer and nutrient resource control on producer biomass. Ecol Lett 11(7):740–755
Guareschi S, Coccia C, Sánchez-Fernández D et al (2013) How far could the alien boatman *Trichocorixa verticalis* spread? Worldwide estimation of its current and future potential distribution. PLoS One 8(3):e59757
Guo N, Puhlev I, Brown DR et al (2000) Trehalose expression confers desiccation tolerance on human cells. Nat Biotechnol 18:168–171
Gutrich JJ, Gigliello K, Gardner KV, Elmore AJ (2016) Economic returns of groundwater management sustaining an ecosystem service of dust suppression by alkali meadow in Owens Valley, California. Ecol Econ 121:1–11
Hajirostamloo M (2008) Differences of shell structure in cysts of *Artemia* from various depth of Urmia Lake (Iran). Res J Biol Sci 3(6):648–653
Hammer UT (1981) Primary production in saline lakes. In: Salt lakes. Springer, Dordrecht, pp 47–57
Hammer UT, Shamess J, Haynes RC (1983) The distribution and abundance of algae in saline lakes of Saskatchewan, Canada. Hydrobiologia 105(1):1–26
Han B, Bischof JC (2004) Engineering challenges in tissue preservation. Cell Preserv Technol 2 (2):91–112
Hand SC (1998) Quiescence in *Artemia franciscana* embryos: reversible arrest of metabolism and gene expression at low oxygen levels. J Exp Biol 201:1233–1242
Hand SC, Carpenter JF (1986) pH-induced metabolic transitions in *Artemia* embryos mediated by a novel hysteretic trehalase. Science 232:1535–1537
Hand SC, Gnaiger E (1988) Anaerobic dormancy quantified in *Artemia* embryos: a calorimetric test of the control mechanism. Science 239:1425–1427
Hand SC, Menze MA (2008) Mitochondria in energy-limited states: mechanisms that blunt the signaling of cell death. J Exp Biol 211(12):1829–1840
Hand SC, Podrabsky JE (2000) Bioenergetics of diapause and quiescence in aquatic animals. Thermochim Acta 349:31–42
Hand SC, Podrabsky JE, Eads BD et al (2001) Interrupted development in aquatic organisms: ecological context and physiological mechanisms. In: Atkinson D, Thorndyke M (eds) Environment and animal development. Genes, life histories and plasticity. BIOS Scientific Publishers, Oxford, pp 219–234
Hand SC, Menze MA, Toner M et al (2011a) LEA proteins during water stress: not just for plants anymore. Annu Rev Physiol 73:115–134
Hand SC, Menze MA, Borcar A et al (2011b) Metabolic restructuring during energy-limited states: insights from *Artemia franciscana* embryos and other animals. J Insect Physiol 57(5):584–594
Hand SC, Moore DS, Patil Y (2018) Challenges during diapause and anhydrobiosis: mitochondrial bioenergetics and desiccation tolerance. IUBMB Life 70(12):1251–1259
Haynes RC, Hammer UT (1978) The saline lakes of Saskatchewan IV. Primary production by phytoplankton in selected saline ecosystems. Internationale Revue der Gesamten Hydrobiologie und Hydrographie 63(3):337–351
HDR Engineering, Inc. (2015) Proposed compensatory mitigation and monitoring plan. Union Pacific Railroad Great Salt Lake causeway culvert closure and bridge construction project, SPK-2011-00755. https://deq.utah.gov/legacy/destinations/g/great-salt-lake/railroad-causeway/index.htm. Accessed 20 May 2019
Heisler J, Glibert PM, Burkholder JM et al (2008) Eutrophication and harmful algal blooms: a scientific consensus. Harmful Algae 8(1):3–13
Herbst DB (1988) Comparative population ecology of *Ephydra hians* Say (Diptera: Ephydridae) at Mono Lake (California) and Abert Lake (Oregon). In: Melack JM (ed) Saline lakes. Springer, Dordrecht, pp 145–166

Herbst DB (2001) Gradients of salinity stress, environmental stability and water chemistry as a templet for defining habitat types and physiological strategies in inland salt waters. In: Melack JM, Jellison R, Herbst DB (eds) Saline lakes. Springer, Dordrecht, pp 209–219
Herbst DB (2006) Salinity controls on trophic interactions among invertebrates and algae of solar evaporation ponds in the Mojave Desert and relation to shorebird foraging and selenium risk. Wetlands 26(2):475–485
Herbst DB, Prather M (2014) Owens Lake: from dustbowl to mosaic of salt water habitats. Lakeline 34:34–38
Herbst DB, Medhurst RB, Roberts SW et al (2013a) Substratum associations and depth distribution of benthic invertebrates in saline Walker Lake, Nevada, USA. Hydrobiologia 700(1):61–72
Herbst DB, Roberts SW, Medhurst RB (2013b) Defining salinity limits on the survival and growth of benthic insects for the conservation management of saline Walker Lake, Nevada, USA. J Insect Conserv 17(5):877–883
Hisem D, Hrouzek P, Tomek P et al (2011) Cyanobacterial cytotoxicity versus toxicity to brine shrimp *Artemia salina*. Toxicon 57(1):76–83
Houde ED (1994) Differences between marine and freshwater fish larvae: implications for recruitment. ICES J Mar Sci 51(1):91–97
Hunter MD, Price PW (1992) Playing chutes and ladders: heterogeneity and the relative roles of bottom-up and top-down forces in natural communities. Ecology 73(3):724–732
Hutchinson GE (1957) A treatise on limnology, geology, physics, and chemistry, vol 1. Wiley, New York
Jehl JR (2007) Why do eared grebes leave hypersaline lakes in autumn? Waterbirds 30(1):112–116
Jehl JR, Boyd WS, Paul DS et al (2002) Massive collapse and rapid rebound: population dynamics of eared grebes (*Podiceps nigricollis*) during an ENSO event. The Auk 119(4):1162–1166
Jellison R, Melack JM (1993) Algal photosynthetic activity and its response to meromixis in hypersaline Mono Lake, California. Limnol Oceanogr 38(4):818–837
Jellison R, Melack JM (2001) Nitrogen limitation and particulate elemental ratios of seston in hypersaline mono Lake, California, USA. Hydrobiologia 466:1–12
Jellison R, Miller LG, Melack JM et al (1993) Meromixis in hypersaliine Mono Lake, California. 2. Nitrogen fluxes. Limnol Oceanogr 38(5):1020–1039
Johnson MT, Bell TG (2008) Coupling between dimethylsulfide emissions and the ocean–atmosphere exchange of ammonia. Environ Chem 5:259–267
Kappas I, Abatzopoulos TJ, Van Hoa N et al (2004) Genetic and reproductive differentiation of *Artemia franciscana* in a new environment. Mar Biol 146(1):103–117
Kelts LJ (1979) Ecology of a tidal marsh corixid, *Trichocorixa verticalis* (Insecta, Hemiptera). Hydrobiologia 64(1):37–57
Khudyi O, Khuda L, Kushniryk O et al (2017) An effectiveness of *Artemia* nauplii enrichment with polyunsaturated fatty acids using a supplement easy DHA Selco. Acta Biol Univ Daugavp 17 (2):169–183
King AM, MacRae TH (2012) The small heat shock protein p26 aids development of encysting *Artemia* embryos, prevents spontaneous diapause termination and protects against stress. PLoS One 7(8):e43723. https://doi.org/10.1371/journal.pone.0043723
King AM, Toxopeus J, MacRae TH (2014) Artemin, a diapause-specific chaperone, contributes to the stress tolerance of *Artemia franciscana* cysts and influences their release from females. J Exp Biol 217(10):1719–1724
Klinger DH, Turnipseed M, Anderson JL et al (2013) Moving beyond the fished or farmed dichotomy. Mar Policy 38:369–374
Kuehn D (2002) The brine shrimp industry in Utah. In: Gwynn WA (ed) Great Salt Lake, an overview of change. DNR Special Publication, Utah Department of Natural Resources, Utah Geological Survey, Salt Lake City, pp 259–264
Kwast KE, Hand SC (1993) Regulatory features of protein synthesis in isolated mitochondria from *Artemia* embryos. Am J Phys 265(6):R1238–R1246

LADWP (2015) LADWP annual Owens Valley report. Available via Inyowater.org. http://www.inyowater.org/wp-content/uploads/2012/12/2015-Combined-Files.pdf. Accessed 21 Jun 2019
Lavens P, Sorgeloos P (1987) The cryptobiotic state of *Artemia* cysts, its diapause deactivation and hatching: a review. In: Sorgeloos P, Bengtson DA, Decleir W, Jaspers E (eds) *Artemia* research and its applications, vol. E. Ecology, culturing, use in aquaculture. Universa Press, Wetteren, pp 27–63
Lavens P, Sorgeloos, P (eds) (1996) Manual on the production and use of live food for aquaculture (No. 361). Food and Agriculture Organization (FAO), Rome, pp 79–195
Lavens P, Sorgeloos P (2000) The history, present status and prospects of the availability of *Artemia* cysts for aquaculture. Aquaculture 181(3–4):397–403
Léger P, Bengtson DA, Simpson KL, Sorgeloos P (1986) The use and nutritional value of *Artemia* as a food source. Oceanogr Mar Biol Annu Rev 24:521–623
Léger P, Bengtson DA, Sorgeloos P et al (1987) The nutritional value of *Artemia*: a review. In: Artemia research and its applications, vol 3, Ecology, culturing, use in aquaculture. Universa Press, Wetteren, pp 357–372
Lemly AD (1993) Guidelines for evaluating selenium data from aquatic monitoring and assessment studies. Environ Monit Assess 28(1):83–100
Lemly AD (1997) Ecosystem recovery following selenium contamination in a freshwater reservoir. Ecotoxicol Environ Saf 36(3):275–281
Lemly AD (2004) Aquatic selenium pollution is a global environmental safety issue. Ecotoxicol Environ Saf 59(1):44–56
Lenz PH (1987) Ecological studies on *Artemia*: a review. In: Sorgeloos P, Bengtson DA, Decleir W, Jaspers E (eds) *Artemia* research and its applications, vol 3. Ecology, culturing, use in aquaculture. Universa Press, Wetteren, pp 5–18
Lenz PH, Browne RA (1991) Ecology of *Artemia*. In: Browne RA, Sorgeloos P, Trotman CAN (eds) *Artemia* biology. CRC, Boca Raton, FL, pp 237–253
Li S, Chakraborty N, Borcar A et al (2012) Late embryogenesis abundant proteins protect human hepatoma cells during acute desiccation. Proc Natl Acad Sci U S A 109(51):20859–20864
Liang P, Amons R, Clegg JS, MacRae TH (1997) Molecular characterization of a small heat shock/alpha-crystallin protein in encysted *Artemia* embryos. J Biol Chem 272(30):19051–19058
Lindeman RL (1942) The trophic-dynamic aspect of ecology. Ecology 23(4):300–417
Lindsay MR, Anderson C, Fox N et al (2017) Microbialite response to an anthropogenic salinity gradient in Great Salt Lake, Utah. Geobiology 15(1):131–145
Lindsay MR, Johnston RE, Baxter BK, Boyd ES (2019) Effects of salinity on microbialite-associated production in Great Salt Lake, Utah. Ecology 100(3):e02611
Litvinenko LI, Litvinenko AI, Boiko EG et al (2015) *Artemia* cyst production in Russia. Chin J Oceanol Limnol 33(6):1436–1450
Loving BL, Waddell KM, Miller CW (2000) Water and salt balance of Great Salt Lake, Utah, and simulation of water and salt movement through the causeway, 1987-98. US Geological Survey, Water-Resources Investigations Report 00-4221
Ma X, Jamil K, MacRae TH et al (2005) A small stress protein acts synergistically with trehalose to confer desiccation tolerance on mammalian cells. Cryobiology 51(1):15–28
MacIntyre S, Jellison R (2001) Nutrient fluxes from upwelling and enhanced turbulence at the top of the pycnocline in Mono Lake, California. In: Melack JM, Jellison R, Herbst DB (eds) Saline lakes. Developments in hydrobiology, vol 162. Springer, Dordrecht, pp 13–29
MacRae TH (2010) Gene expression, metabolic regulation and stress tolerance during diapause. Cell Mol Life Sci 67(14):2405–2424
MacRae TH (2015) Small heat shock proteins and diapause in the crustacean, *Artemia franciscana*. In: Tanguay RM (ed) The big book on small heat shock proteins. Springer, Cham, pp 563–578
MacRae TH (2016) Stress tolerance during diapause and quiescence of the brine shrimp, *Artemia*. Cell Stress Chaperones 21(1):9–18

Madison RJ (1970) Effects of a causeway on the chemistry of the brine in Great Salt Lake, Utah (No. 14). Utah Geological and Mineralogical Survey. Available via USGS. https://pubs.er.usgs.gov/publication/70179728. Accessed on 21 Jun 2019

Mahoney SA, Jehl JR (1985) Adaptations of migratory shorebirds to highly saline and alkaline lakes: Wilson's Phalarope and American Avocet. Condor 87(4):520–527

Mai MG, Engrola S, Morais S et al (2009) Co-feeding of live feed and inert diet from first-feeding affects *Artemia* lipid digestibility and retention in Senegalese sole (*Solea senegalensis*) larvae. Aquaculture 296(3–4):284–291

Makridis P, Vadstein O (1999) Food size selectivity of *Artemia franciscana* at three developmental stages. J Plankton Res 21(11):2191–2201

Makridis P, Fjellheim AJ, Skjermo J, Vadstein O (2000) Control of the bacterial flora of *Brachionus plicatilis* and *Artemia franciscana* by incubation in bacterial suspensions. Aquaculture 185:207–218

Marcarelli AM, Wurtsbaugh WA, Griset O (2006) Salinity controls phytoplankton response to nutrient enrichment in the Great Salt Lake, Utah, USA. Can J Fish Aquat Sci 63(10):2236–2248

Marden B, Richards D (2017) Multi-year investigations of complex interactions between cyanobacteria blooms and the food web in Farmington Bay, Great Salt Lake, Utah: a progress report of scientific findings. Wasatch Front Water Quality Council. Available via Researchgate. https://www.researchgate.net/publication/315694208_Mulit-year_Investigations_of_Complex_Interactions_Between_Cyanobacteria_Blooms_and_the_Food_Web_in_Farmington_Bay_Great_Salt_Lake_Utah. Accessed 21 Jun 2019

Marden B, Van Stappen G, Musaev A et al (2012) Assessment of the production potential of an emerging *Artemia* population in the Aral Sea, Uzbekistan. J Mar Syst 92(1):42–52

Marden B, Miller T, Richards D (2013) Factors influencing cyanobacteria blooms in Farmington Bay, Great Salt Lake, Utah. Progress report of scientific findings. Wasatch Front Water Quality Council. Available via Researchgate. https://www.researchgate.net/publication/305488678_Factors_Influencing_Cyanobacteria_Blooms_in_Farmington_Bya_Great_Salt_Lake_Utah. Accessed 21 Jun 2019

Marino R, Howarth RW, Shamess J, Prepas E (1990) Molybdenum and sulfate as controls on the abundance of nitrogen-fixing cyanobacteria in saline lakes in Alberta. Limnol Oceanogr 35 (2):245–259

Martins CIM, Eding EH, Verdegem MC et al (2010) New developments in recirculating aquaculture systems in Europe: a perspective on environmental sustainability. Aquac Eng 43(3):83–93

McCulley EB (2014) Factors affecting the toxic cyanobacteria *Nodularia spumigena* in Farmington Bay of Great Salt Lake, Utah. Available via Utah State University. https://digitalcommons.usu.edu/etd/4014/. Accessed 21 June 2019

McCulley EB, Wurtsbaugh W (2014) Factors affecting the spatial and temporal variability of *Nodularia* blooms in Farmington Bay, Great Salt Lake, Utah. Available via Utah State University. https://digitalcommons.usu.edu/runoff/2014/2014Abstracts/55/. Accessed 21 Jun 2019

Melack JM, Jellison R (1998) Limnological conditions in Mono Lake: contrasting monomixis and meromixis in the 1990s. Hydrobiologia 384:21–39

Melack JM, Jellison R, MacIntyre S et al (2017) Mono Lake: plankton dynamics over three decades of meromixis and monomixis. In: Gulati RD, Zadereev ES, Degermendzhi AG (eds) Ecology of meromictic lakes. Springer International Publishing, Cham, pp 325–352

Meuser JE, Baxter BK, Spear JR et al (2013) Contrasting patterns of community assembly in the stratified water column of Great Salt Lake, Utah. Microb Ecol 66(2):268–280

Micklin P (2007) The Aral Sea Disaster. Annu Rev Earth Planet Sci 35:47–72

Micklin P (2010) The past, present, and future Aral Sea. Lakes Reserv Res Manag 15(3):193–213

Micklin P, Aladin NV, Plotnikov I (eds) (2016) Aral Sea. Springer, Berlin

Millennium Ecosystem Assessment (2005) Ecosystems and human well-being: synthesis. Island Press, Washington, DC

Miller D, McLennan AG (1988) The heat shock response of the cryptobiotic brine shrimp *Artemia*—II. Heat shock proteins. J Therm Biol 13(3):125–134

Moffitt CM, Cajas-Cano L (2014) Blue growth: the 2014 FAO state of world fisheries and aquaculture. Fisheries 39(11):552–553

Montague CL, Fey WR, Gillespie DM (1982) A causal hypothesis explaining predator-prey dynamics in Great Salt Lake, Utah. Ecol Model 17(3):243–270

Moore JN (2016) Recent desiccation of western Great Basin saline lakes: lessons from Lake Abert, Oregon, USA. Sci Total Environ 554:142–154

Moore DS, Hansen R, Hand SC (2016) Liposomes with diverse compositions are protected during desiccation by LEA proteins from *Artemia franciscana* and trehalose. Biochim Biophys Acta Biomembr 1858(1):104–115

Morris D, Lewis W (1988) Phytoplankton nutrient limitation in Colorado mountain lakes. Freshw Biol 20:315–327

Murphy TP, Brownlee BG (1981) Ammonia volatilization in a hypereutrophic Prairie lake. Can J Fish Aquat Sci 38:1035–1039

Myers RA, Hutchings JA, Barrowman NJ (1997) Why do fish stocks collapse? The example of cod in Atlantic Canada. Ecol Appl 7(1):91–106

Naftz D (2017) Inputs and internal cycling of nitrogen to a causeway influenced, hypersaline lake, Great Salt Lake, Utah, USA. Aquat Geochem 23(3):199–216

Naftz D, Angeroth C, Kenney T et al (2008) Anthropogenic influences on the input and biogeochemical cycling of nutrients and mercury in Great Salt Lake, Utah, USA. Appl Geochem 23 (6):1731–1744

Naftz D, Fuller C, Cederberg J et al (2009) Mercury inputs to Great Salt Lake, Utah: reconnaissance-phase results. Available via Digitalcommons USU. https://digitalcommons.usu.edu/nrei/vol15/iss1/5/. Accessed 21 Jun 2019

Nambu Z (2004) Influence of photoperiod and temperature on reproductive mode in the brine shrimp, *Artemia franciscana*. J Exp Zool 301A:542–546

Naylor RL, Goldburg RJ, Primavera JH et al (2000) Effect of aquaculture on world fish supplies. Nature 405(6790):1017

Nejstgaard JC, Frischer ME, Simonelli P et al (2008) Quantitative PCR to estimate copepod feeding. Mar Biol 153:565–577

Nunes BS, Carvalho FD, Guilhermino LM et al (2006) Use of the genus *Artemia* in ecotoxicity testing. Environ Pollut 144(2):453–462

Ogata EM, Wurtsbaugh WA, Smith TN et al (2017) Bioassay analysis of nutrient and *Artemia franciscana* effects on trophic interactions in the Great Salt Lake, USA. Hydrobiologia 788 (1):1–16

Ohlendorf HM, DenBleyker J, Moellmer WO, Miller T (2009) Development of a site-specific standard for selenium in open waters of Great Salt Lake, Utah. Natural Resources and Environmental Issues 15, Article 4. https://digitalcommons.usu.edu/nrei/vol15/iss1/4. Accessed 21 Jun 2019

Okun N, Brasil J, Attayde JL et al (2008) Omnivory does not prevent trophic cascades in pelagic food webs. Freshw Biol 53(1):129–138

Paerl HW, Crocker KM, Prufert LE (1987) Limitation of N_2 fixation in coastal marine waters: relative importance of molybdenum, iron, phosphorus, and organic matter availability. Limnol Oceanogr 32(3):525–536

Paerl HW, Joyner JJ, Joyner AR et al (2008) Co-occurrence of dinoflagellate and cyanobacterial harmful algal blooms in Southwest Florida coastal waters: dual nutrient (N and P) input controls. Mar Ecol Prog Ser 371:143–153

Paerl HW, Scott TJ, McCarthy MJ et al (2016) It takes two to tango: when and where dual nutrient (N & P) reductions are needed to protect lakes and downstream ecosystems. Environ Sci Technol 50(20):10805–10813

Patil YN, Marden B, Brand MD et al (2012) Metabolic downregulation and inhibition of carbohydrate catabolism during diapause in embryos of *Artemia franciscana*. Physiol Biochem Zool 86 (1):106–118

Paul DS, Manning AE (2002) Great Salt Lake Waterbird survey five-year report (1997–2001). Publication Number 08-38. Utah Division of Wildlife Resources, Salt Lake City

Peterson C, Gustin M (2008) Mercury in the air, water and biota at the Great Salt Lake (Utah, USA). Sci Total Environ 405(1–3):255–268

Porcella DB, Holman JA (1972) Nutrients, algal growth, and culture of brine shrimp in the southern Great Salt Lake. The Great Salt Lake and Utah's water resources. Utah Water Research Laboratory, Logan, UT, pp 142–155

Porter KG (1973) Selective grazing and differential digestion of algae by zooplankton. Nature 244:179–180

Ptacnik R, Andersen T, Tamminen T (2010) Performance of the Redfield ratio and a family of nutrient limitation indicators as thresholds for phytoplankton N vs. P limitation. Ecosystems 13:1201–1214

Puhlev GN, Brown DR, Mansbridge J et al (2000) Trehalose expression confers desiccation tolerance on human cells. Nat Biotechnol 18(2):168–171

Qiu Z, MacRae TH (2008) ArHsp21, a developmentally regulated small heat-shock protein synthesized in diapausing embryos of *Artemia franciscana*. Biochem J 411(3):605–611

Qiu Z, Viner RI, MacRae TH et al (2004) A small heat shock protein from *Artemia franciscana* is phosphorylated at serine 50. Biochim Biophys Acta 1700(1):75–83

Radzikowski J (2013) Resistance of dormant stages of planktonic invertebrates to adverse environmental conditions. J Plankton Res 35(4):707–723

Rai AK, Abraham G (1993) Salinity tolerance and growth analysis of the cyanobacterium *Anabaena doliolum*. Bull Environ Contam Toxicol 51(5):724–731

Redfield AC (1958) The biological control of the chemical factors in the environment. Am Sci 46:205–221

Reeve MR (1963a) The filter feeding of *Artemia*: I. In pure cultures of plant cells. J Exp Biol 40:195–205

Reeve MR (1963b) Growth efficiency in *Artemia* under laboratory conditions. Biol Bull 125 (1):133–145

Reynolds JA, Hand SC (2004) Differences in isolated mitochondria are insufficient to account for respiratory depression during diapause in *Artemia franciscana* embryos. Physiol Biochem Zool 77(3):366–377

Ricker WE (1954) Stock and recruitment. J Fish Res Board Can 11:559–623

Riera J, Voss PR et al (2001) Nature, society and history in two contrasting landscapes in Wisconsin, USA: interactions between lakes and humans during the twentieth century. Land Use Policy 18(1):41–51

Roberts AJ (2013a) Winter waterbird ecology on the Great Salt Lake, Utah, and interactions with commercial harvest of brine shrimp cysts. Dissertation, Utah State University. https://digitalcommons.usu.edu/etd/2042

Roberts AJ (2013b) Avian diets in a saline ecosystem: Great Salt Lake, Utah, USA. Hum Wildl Interact 7(1):15

Roberts AJ, Conover MR (2013) Eared grebe diet on Great Salt Lake, Utah, and competition with the commercial harvest of brine shrimp cysts. J Wildl Manag 77(7):1380–1385

Roberts AJ, Conover MR (2014) Diet and body mass of ducks in the presence of commercial harvest of brine shrimp cysts in the Great Salt Lake, Utah. J Wildl Manag 78(7):1197–1205

Roo J, Hernandez-Cruz C, Mesa-Rodriguez M et al (2019) Effect of increasing n-3 HUFA content in enriched *Artemia* on growth, survival and skeleton anomalies occurrence of greater amberjack *Seriola dumerili* larvae. Aquaculture 1:651–659

Rueter JG, Petersen RR (1987) Micronutrient effects on cyanobacterial growth and physiology. N Z J Mar Freshw Res 21(3):435–445

Saint-Armand P, Mathews LA, Gaines C et al (1986) Dust storms from Owens and Mono valleys, California (No. NWC-TP-6731). Available via Naval Weapons Center. https://apps.dtic.mil/dtic/tr/fulltext/u2/a226243.pdf. Accessed 21 Jun 2019

Salimi J, Maknoon R, Meijerink S (2019) Designing institutions for watershed management: a case study of the Urmia Lake Restoration National Committee. Water Alt 12(2):2

Sanchez-Saavedra MP, Voltolina D (1995) The effect of different light quality on the food value of the diatom *Chaetocerous* sp. for *Artemia franciscan* Kellogg. Riv Ital Acquacolt 20:135–138

Sarabia R, Del Ramo J, Varo I et al (2002) Comparing the acute response to cadmium toxicity of nauplii from different populations of *Artemia*. Environ Toxicol Chem 21(2):437–444

Scherer NM, Gibbons HL, Stoops KB et al (1995) Phosphorus loading of an urban lake by bird droppings. Lake and Reservoir Manage 11(4):317–327

Schindler DW, Holmgren SK (1971) Primary production and phytoplankton in the Experimental Lakes Area, northwestern Ontario, and other low-carbonate waters, and a liquid scintillation method for determining 14C activity in photosynthesis. J Fish Board Canada 28(2):189–201

Sherwood AR, Presting GG (2007) Universal primers amplify a 23S rDNA plastid market in eukaryotic algae and cyanobacteria. J Phycol 43:605–608

Sherwood AR, Dittbern M, Johnston E et al (2016) A metabarcoding comparison of windward and leeward airborne algal diversity across the Ko'olau mountain range on the island of O'ahu, Hawai'i. J Phycol 53:437. https://doi.org/10.1111/jpy.12502

Shirakashi R, Kostner CM, Muller KJ et al (2002) Intracellular delivery of trehalose into mammalian cells by electropermeabilization. J Membr Biol 189:45–54

Simonelli P, Troedsson C, Nejstgaard JC et al (2009) Evaluation of DNA extraction and handling procedures for PCR-based copepod feeding studies. J Plankton Res 31(12):1465–1474

Simonis JL (2013a) Predator ontogeny determines trophic cascade strength in freshwater rock pools. Ecosphere 4(5):art62

Simonis JL (2013b) Prey (*Moina macrocopa*) population density drives emigration rate of its predator (*Trichocorixa verticalis*) in a rock-pool metacommunity. Hydrobiologia 715(1):19–27

Skorupa JP (1998) Selenium poisoning of fish and wildlife in nature: lessons from twelve real-world examples. In: Frankenberger WT Jr, Engberg RA (eds) Environmental chemisty of selenium, vol 64. Marcel Dekker, New York, pp 315–354

Skoultchi AI, Morowitz HJ (1964) Information storage and survival of biological systems at temperatures near absolute zero. Yale J Biol Med 37(2):158

Smith VH, Schindler DW (2009) Eutrophication science: where do we go from here? Trends Ecol Evol 24(4):201–207

Smith DG, Croker GF et al (1995) Human perception of water appearance: 2. Colour judgment, and the influence of perceptual set on perceived water suitability for use. N Z J Mar Freshw Res 29 (1):45–50

Smith VH, Joye SB, Howarth RW (2006) Eutrophication of freshwater and marine ecosystems. Limnol Oceanogr 51(1part2):351–355

Sorgeloos P (1980) The use of the brine shrimp *Artemia* in aquaculture. In: Persoone G, Sorgeloos P, Roels O, Jaspers E (eds) The brine shrimp *Artemia*, vol 3. Universa Press, Wetteren, pp 25–46

Sorgeloos P, Kulasekarapandian S (1984) Production and use of *Artemia* in Aquaculture. In: Silas EG, Rengarajan K (eds) CMFRI Special Publication Number 15

Sorgeloos P, Tackaert W (1991) Roles and potentials of *Artemia* in coastal salt works. In Proceedings of the international symposium on biotechnology of salt ponds, International symposium on biotechnology of saltponds, Sept 18–21, 1990, Tanggu, Tianjin, pp 69–77

Sorgeloos P, Baeza-Mesa M, Benijts F et al (1975) Research on the culturing of the brine shrimp *Artemia salina L.* at the State University of Ghent (Belgium). In: 10th European symposium on marine biology, Ostend, Belgium, vol 1, pp 472–495

Sorgeloos P, Remiche-Van Der Wielen C, Persoone G (1978) The use of *Artemia* nauplii for toxicity tests—a critical analysis. Ecotoxicol Environ Saf 2(3–4):249–255

Sorgeloos P, Lavens P, Leger P et al (1991) State of the art in larviculture of fish and shellfish. Available via FAO-UN. http://agris.fao.org/agris-search/search.do?recordID=AV20120151346. Accessed 21 Jun 2019

Sorgeloos P, Dhert P, Condreva P (2001) Use of the brine shrimp, *Artemia* spp., in marine fish larviculture. Aquaculture 20:147–159
Stephens DW (1974) A summary of biological investigations concerning the Great Salt Lake, Utah (1861-1973). West North Am Nat 34(3):221–229
Stephens DW (1990) Changes in lake levels, salinity, and the biological community of Great Salt Lake (Utah, USA), 1847-1987. Hydrobiologia 197:139–146
Stephens D (1998) Salinity-induced changes in the aquatic ecosystem of Great Salt Lake, Utah. Available via AAPG/Datapages. http://archives.datapages.com/data/uga/data/069/069001/1_ugs690001.htm. Accessed 21 Jun 2019
Stephens DW, Birdsey PW (2002) Population dynamics of the brine shrimp, *Artemia franciscana*, Great Salt Lake, and regulation of commercial shrimp harvest. In: Gwynn JW (ed) Great Salt Lake: a scientific, historical and economic overview. Utah Geological and Mineral Survey, Salt Lake City, pp 327–336
Stephens DW, Gillespie DM (1972) Community structure and ecosystem analysis of the Great Salt Lake. In: The Great Salt Lake and Utah's Water Resources, pp 66–72
Stephens DW, Gillespie DM (1976) Phytoplankton production in the Great Salt Lake, Utah, and a laboratory study of algal response to enrichment. Limnol Oceanogr 21(1):74–87
Steven B, McCann S, Ward NL (2012) Pyrosequencing of plastid 23S rRNA genes reveals diverse and dynamic cyanobacterial and algal populations in two eutrophic lakes. FEMS Microb Ecol 82:607–615
Sturm PA, Sanders GC, Allen KA (1980) The brine shrimp industry on the Great Salt Lake. In: Gwynn JW (ed) Great Salt Lake: a scientific, historical and economic overview. Utah Geological and Mineral Survey, Salt Lake City, pp 243–248
Sun Y, Mansour M, Crack JA et al (2004) Oligomerization, chaperone activity, and nuclear localization of p26, a small heat shock protein from *Artemia franciscana*. J Biol Chem 279 (38):39999–40006
Tanguay JA, Reyes RC, Clegg JS (2004) Habitat diversity and adaptation to environmental stress in encysted embryos of the crustacean *Artemia*. J Biosci 29(4):489–501
Toi HT, Boeckx P, Sorgeloos P et al (2013) Bacteria contribute to *Artemia* nutrition in algae-limited conditions: a laboratory study. Aquaculture 388(391):1–7
Torrentera L, Dodson SI (2004) Ecology of the brine shrimp *Artemia* in the Yucatan, Mexico, Salterns. J Plankton Res 26(6):617–624
Triantaphyllidis GV, Poulopoulou K, Abatzopoulos TJ et al (1995) International study on *Artemia* XLIX. Salinity effects on survival, maturity, growth, biometrics, reproductive and lifespan characteristics of a bisexual and a parthenogenetic population of *Artemia*. Hydrobiologia 302 (3):215–227
Tyson R (1997) California under pressure to curb nation's largest PM source. Environ Sci Technol 31(11):504A–504A
Utah DEQ (2014) Core component 1: developing aquatic life criteria for priority pollutants. A Great Salt Lake Water Quality Strategy. Available via Utah DEQ. https://documents.deq.utah.gov/water-quality/standards-technical-services/gsl-website-docs/gsl-wq-strategy/DWQ-2019-000421.pdf. Accessed 12 May 2019
Vadeboncoeur Y, Vander Zanden MJ, Lodge DM (2002) Putting the lake back together: reintegrating benthic pathways into lake food web models. Bioscience 52(1):44–54
Valdes C, Black FJ, Stringham B et al (2017) Total mercury and methylmercury response in water, sediment, and biota to destratification of the Great Salt Lake, Utah, United States. Environ Sci Technol 51(9):4887–4896
Van de Meutter F, Trekels H, Green AJ (2010) The impact of the North American waterbug *Trichocorixa verticalis* (*Fieber*) on aquatic macroinvertebrate communities in southern Europe. Fundam Appl Limnol/Archiv für Hydrobiologie 177(4):283–292
Van Den Branden C, Bernaerts F, Decleir W (1980) The respiratory physiology of adult *Artemia salina* L. in relation to oxygen availability. In: Gilles R (ed) Animals and environmental fitness:

physiological and biochemical aspects of adaptation and ecology, vol 2. Pergamon, Oxford, pp 37–38

Van Stappen G (1996) Introduction, biology and ecology of *Artemia*. In: Lavens P, Sorgeloos P (eds) Manual on the production and use of live food for aquaculture, FAO Fisheries Technical Paper, No. 361. Rome, pp 79–129

Van Stappen G (2002) Zoogeography. In: Abatzopoulos TJ, Beadmore JA, Clegg JS, Sorgeloos P (eds) *Artemia*: basic and applied biology, vol 1. Kluwer Academic, London, pp 171–224

Van Stappen G (2008) *Artemia* biodiversity in Central and Eastern Asia. Ghent University Press, Ghent, pp 1–82

Van Stappen G, Fayazi G, Sorgeloos P (2001) International study on *Artemia* LXIII. Field study of the *Artemia urmiana* (Günther, 1890) population in Lake Urmiah, Iran. In: Melack JM, Jellison R, Herbst DB (eds) Saline lakes. Developments in hydrobiology, vol 162. Springer, Dordrecht, pp 133–143

Vanhaecke P, Persoone G, Claus C et al (1981) Proposal for a short-term toxicity test with *Artemia* nauplii. Ecotoxicol Environ Saf 5(3):382–387

Vanhaecke P, Siddall SE, Sorgeloos P (1984) International study on *Artemia*. XXXII. Combined effects of temperature and salinity on the survival of *Artemia* of various geographical origin. J Exp Mar Biol Ecol 80(3):259–275

Varo N, Green AJ, Sánchez MI et al (2011) Behavioural and population responses to changing availability of *Artemia* prey by moulting black-necked grebes, *Podiceps nigricollis*. Hydrobiologia 664(1):163–171

Vest JL, Conover MR (2011) Food habits of wintering waterfowl on the Great Salt Lake, Utah. Waterbirds 34(1):40–51

Vest JL, Conover MR, Perschon C et al (2009) Trace element concentrations in wintering waterfowl from the Great Salt Lake, Utah. Arch Environ Contam Toxicol 56(2):302–316

Viciano E, Monroig O, Salvador A et al (2013) Enriching *Artemia* nauplii with a high DHA-containing lipid emulsion: search for an optimal protocol. Aquac Res 46(5):1066–1077

Vitousek PM, Howarth RW (1991) Nitrogen limitation on land and in the sea: how can it occur? Biogeochemistry 13(2):87–115

von Rohden C, Ilmberger J, Boehrer B (2009) Assessing groundwater coupling and vertical exchange in a meromictic mining lake with an SF6-tracer experiment. J Hydrol 372 (1–4):102–108

Vörösmarty CJ, Green P, Salisbury J et al (2000) Global water resources: vulnerability from climate change and population growth. Science 289(5477):284–288

Walsby AE, Van Rijn J, Cohen Y (1983) The biology of a new gas-vacuolate cyanobacterium, *Dactylococcopsis salina* sp. nov., in Solar Lake. Proc R Soc Lond B 217:417–447

Wang Z, Asem A, Sun S (2017) Coupled effects of photoperiod, temperature, and salinity on diapause induction of the parthenogenetic *Artemia* (Crustacea: Anostraca) from Barkol Lake, China. North West J Zool 13(1):12–17

Warner AH, Finamore FJ (1967) Nucleotide metabolism during brine shrimp embryogenesis. J Biol Chem 242(8):1933–1937

Watanabe T (1993) Importance of docosahexaenoic acid in marine larval fish. J World Aquacult Soc 24(2):152–161

Watters C, Iwamura S, Ako H et al (2012) Nutrition considerations in aquaculture: the importance of omega-3 fatty acids in fish development and human health. College of Tropical Agriculture and Human Resources, University of Hawai'I at Manoa, FN-11

Wear RG, Haslett SJ (1986) Effects of temperature and salinity on the biology of *Artemia fransiscana Kellogg* from Lake Grassmere, New Zealand. 1. Growth and mortality. J Exp Mar Biol Ecol 98(1–2):153–166

Weiss KR (2018) Iran's Tarnished Gem. National Geographic. https://www.nationalgeographic.com/photography/proof/2018/04/lake-urmia-iran-drought/. Accessed 23 April 2019

WHSRN (2019) Great Salt Lake. https://whsrn.org/whsrn_sites/great-salt-lake/. Accessed 16 May 2019

Williams WD, Boulton AJ, Taaffe RG (1990) Salinity as a determinant of salt lake fauna: a question of scale. Hydrobiologia 197(1):257–266

Willsie JK, Clegg JS (2001) Nuclear p26, a small heat shock/alpha-crystallin protein, and its relationship to stress resistance in *Artemia franciscana* embryos. J Exp Biol 204:2239–2350

Wine ML, Rimmer A, Laronne JB (2019) Agriculture, diversions, and drought shrinking Galilee Sea. Sci Total Environ 651:70–83

Wirick CD (1972) *Dunaliella-Artemia* plankton community of the Great Salt Lake, Utah. Dissertation, University of Utah

Wunder MB, Jehl JR, Stricker CA (2012) The early bird gets the shrimp: confronting assumptions of isotopic equilibrium and homogeneity in a wild bird population. J Anim Ecol 81 (6):1223–1232

Wurtsbaugh WA (1988) Iron, molybdenum and phosphorus limitation of N_2 fixation maintains nitrogen deficiency of plankton in the Great Salt Lake drainage (Utah, USA). SIL Proceedings 1922-2010 23(1):121–130. https://doi.org/10.1080/03680770.1987.11897913

Wurtsbaugh WA (1992) Food-web modification by an invertebrate predator in the Great Salt Lake (USA). Oecologia 89(2):168–175

Wurtsbaugh WA (2007) Preliminary analyses of selenium bioaccumulation in benthic food webs of the Great Salt Lake, Utah. In: Final Report: Development of a Selenium Standard for the Open Waters of the Great Salt Lake. Utah Department of Environmental Quality. https://deq.utah.gov/legacy/destinations/g/great-salt-lake/steering-committee/docs/2008/03Mar/Preliminary_Analyses_of_Selenium_Bioaccumulation_in_Benthic_Food_Webs_of_the_GSL_FINAL_REPORT.pdf. Accessed 21 Jun 2019

Wurtsbaugh WA (2012) Paleolimnological analysis of the history of metals contamination in the Great Salt Lake, Utah. Watershed Sciences Faculty Publications. Paper 556. https://digitalcommons.usu.edu/wats_facpub/556. Accessed 21 Jun 2019

Wurtsbaugh WA (2014) Management of the Great Salt Lake ecosystem: water, economic values and competing interests. Watershed Sciences Faculty Publications. Paper 594. https://digitalcommons.usu.edu/wats_facpub/594. Accessed 21 Jun 2019

Wurtsbaugh WA, Berry TS (1990) Cascading effects of decreased salinity on the plankton, chemistry, and physics of the Great Salt Lake (Utah). Can J Fish Aquat Sci 47:100–109

Wurtsbaugh WA, Gliwicz ZM (2001) Limnological control of brine shrimp population dynamics and cyst production in the Great Salt Lake, Utah. In: Melack JM, Jellison R, Herbst DB (eds) Saline lakes. Springer, Dordrecht, pp 119–132

Wurtsbaugh WA, Horne AJ (1983) Iron in eutrophic Clear Lake, California: its importance for algal nitrogen fixation and growth. Can J Fish Aquat Sci 40(9):1419–1429

Wurtsbaugh WA, Marcarelli AM (2004) Analysis of phytoplankton nutrient limitation in Farmington Bay and the Great Salt Lake. Report to the Central Davis County Sewer Improvement District. https://digitalcommons.usu.edu/wats_facpub/542/. Accessed 21 Jun 2019

Wurtsbaugh WA, Marcarelli A, Gross D (2004) Eutrophication in Farmington Bay: an Urban Embayment of the Great Salt Lake. Paper presented at the Utah State University water initiative spring runoff conference, Utah State University, Logan, 25–26 March 2004

Wurtsbaugh WA, Gardberg J, Izdepski C (2011) Biostrome communities and mercury and selenium bioaccumulation in the Great Salt Lake (Utah, USA). Sci Total Environ 409 (20):4425–4434

Wurtsbaugh WA, Marcarelli AM, Boyer GL (2012) Eutrophication and metal concentrations in three bays of the Great Salt Lake (USA). Final report to the Utah Division of Water Quality. https://digitalcommons.usu.edu/cgi/viewcontent.cgi?article=1548&context=wats_facpub. Accessed 21 Jun 2019

Wurtsbaugh WA, Miller C, Null SE et al (2017) Decline of the world's saline lakes. Nat Geosci 10:816–821

Zadereev E (2018) Salt lakes, surrounding environments and environmental management. In: Introduction to salt lake sciences. Science Press, Beijing, pp 172–179

Zeinoddini M, Bakhtiari A, Ehteshami M (2015) Long-term impacts from damming and water level manipulation on flow and salinity regimes in Lake Urmia, Iran. Water Environ J 29(1):71–87

Chapter 8
Importance of Great Salt Lake to Pelagic Birds: Eared Grebes, Phalaropes, Gulls, Ducks, and White Pelicans

Michael R. Conover and Mark E. Bell

Abstract Utah's Great Salt Lake (GSL) is so saline that the only invertebrates that survive in the open water are brine fly larvae and brine shrimp. In the absence of competition from other invertebrates, they are incredibly abundant. Only a few avian species can take advantage of their abundance because a bird cannot eat them without also ingesting salt. Moreover, brine shrimp and brine flies are so tiny that only a few avian species can consume the massive number of brine shrimp and brine flies required to meet a bird's nutritional needs. For example, eared grebes need to consume 28,000 adult brine shrimp each day to survive. To achieve this, an eared grebe has to spend 7 h daily foraging and needs to harvest one shrimp per second during foraging.

Eared grebe (*Podiceps nigricollis*), red-necked phalaropes (*Phalaropus lobatus*), Wilson's phalaropes (*Phalaropus tricolor*), California gulls (*Larus californicus*), American avocets (*Recurvirostra americana*), black-necked stilts (*Himantopus mexicanus*), and snowy plovers (*Charadrius nivosus*) have all evolved the ability to survive by harvesting brine flies and brine shrimp. But this may prove to be a Faustian bargain because their fate is now tied to the fate of North America's hypersaline lakes, and people are diverting freshwater that used to flow into them to irrigate crops and for human development. This water diversion causes hypersaline lakes to decrease in size and become too salty for brine flies or brine shrimp. GSL is becoming the last bastion for the world's population of eared grebes and Wilson's phalaropes. But its future is not secure because freshwater that used to flow into GSL is diverted by humans, thus causing the water level to drop several meters below what it would be if water diversion did not occur. The fear is that in some future year, GSL will become so salty that its populations of brine shrimp and brine flies will collapse. This will produce a massive die-off of eared grebes and may threaten the species' survival, which in turn will force the US Fish and Wildlife Agency to step in and stop water diversions from GSL. Future infrastructure projects to divert and store freshwater that would have flowed into GSL should be

M. R. Conover (✉) · M. E. Bell
Department of Wildland Resources, Utah State University, Logan, UT, USA
e-mail: mike.conover@usu.edu

B. K. Baxter, J. K. Butler (eds.), *Great Salt Lake Biology*,
https://doi.org/10.1007/978-3-030-40352-2_8

reconsidered because it will be a drain on Utah's economy to invest money building infrastructure only to be prevented from using the structures after they are built.

Keywords Alkali lakes · Alkaline lakes · Arid West · California gull · Climate change · Eared grebe · Environmental threats · Great Basin · Great Salt Lake; hypersaline lakes · Mono Lake · Phalarope · Water diversion

8.1 Hypersaline Lakes

The western United States is scattered with hypersaline lakes that have high concentrations of dissolved salts and other minerals in the water. Hypersaline lakes are created because the water in them has no outflow to the sea (i.e., they are landlocked). Although water can escape through evaporation, the minerals cannot and become concentrated in the lake water. After thousands of years, the lakes become saline. The salinity of lakes is referenced against seawater, which has an average salinity of 3.5%. Salinity levels higher than that of seawater are considered hypersaline. In the formation of a saline lake from a freshwater body, the rising salinity causes first a decline in fish species abundance and richness, then an absence of fish (Molony and Parry 2006). The rise of salinity levels also produces a decline in the number of invertebrate species. The absence of fish and other invertebrate species that are not salt tolerant creates opportunities for those few invertebrates that are salt tolerant, including brine shrimp (*Artemia franciscana*) and brine flies (e.g., *Ephydra hians*, *Ephydra cinerea*). In the West, many lakes are both hypersaline and alkaline. Alkaline lakes are lakes that have a pH greater than 9.0: a pH so high that most fish cannot survive in them (Wilkie and Wood 1996).

8.2 Great Salt Lake

Great Salt Lake (GSL), located in northern Utah, is part of the Great Basin, so named because the Great Basin lacks an outlet to the sea. GSL is the largest hypersaline lake in the West with a surface area of approximately 4400 km^2 (Table 8.1). GSL consists of five bays: Carrington, Gunnison, Gilbert, Ogden, Farmington, and Bear River (Fig. 8.1). A railroad causeway separates Gunnison Bay from the rest of GSL; the causeway restricts the flow of water between Gunnison Bay and the rest of GSL (Cannon and Cannon 2002). Because there is little freshwater inflow directly into Gunnison Bay, this bay is now supersaturated with salt: a salinity so high that neither brine shrimp nor brine flies can exist there, except in small localized areas where there are springs or during unusually wet years. Likewise, birds are rarely found on Gunnison Bay due to the lack of brine shrimp and brine flies.

Table 8.1 Major hypersaline or alkaline lakes in North America[a]

Lake name	Location	Surface area (km^2)[b]	Salinity % (ppt)	Brine flies present?	Brine shrimp present?	Eared grebes Present?	References	Salinity % (ppt)
Great Salt Lake	Utah, USA	4402 6474[b]	22 (average for both arms)	Yes	Yes	Yes >1,000,000	White et al. (2014) Szalay (2016) Wilsey et al. (2017) Learn. Genetics, University of Utah (2019)	22.2% (average for both arms)
Lahontan Valley Wetlands[c]	Nevada, USA	1943[b]		?	?	?	Wilsey et al. (2017)	
Salton Sea	California, USA	958 930[b]	5	?	?	Yes	Salton Sea Authority (1997) Los Angeles Times (2017) Wilsey et al. (2017) Taylor (2018)	5.25%
Klamath Basin	Oregon/ California	728[b]					Wilsey et al. (2017)	
Pyramid Lake	Nevada, USA	490 445[b]	0.5	?	?	Yes 1–500	Wikipedia (2019) Wilsey et al. (2017)	
Goose Lake[d]	Oregon/ California	380		?	?	?	Wikipedia (2019)	
Owens Lake[d]	California, USA	285 336[b]	6	Yes	Yes	Yes 10–50,000	Herbst and Prather (2014) Wilsey et al. (2017)	6.4%
Honey Lake	California, USA	220[b]		?	?	Yes 1–500	Wilsey et al. (2017)	
Mono Lake[d]	California, USA	171 210[b]	8	Yes	Yes	Yes >1,000,000	Mono Lake Committee (2019)	8.1%

(continued)

Table 8.1 (continued)

Lake name	Location	Surface area (km^2)[b]	Salinity % (ppt)	Brine flies present?	Brine shrimp present?	Eared grebes Present?	References	Salinity % (ppt)
							Herbst and Prather (2014) Wilsey et al. (2017)	
Summer Lake[d]	Oregon, USA	154 316[b]		?	?	Yes 1–500	Wilsey et al. (2017) Atlas of Oregon Lakes (2019)	
Lake Abert[d]	Oregon, USA	148 259[b]	25	Yes	Yes	Yes 10–50,000	Larson et al. (2016) Larsen (2017) Wilsey et al. (2017)	25.0%
Little Manitou Lake	Saskatchewan, Canada	13	18	?	Yes		Canadian Encyclopedia (2014) Rawson and Moore (2011)	18.0%

Adapted from Roberts et al. (2013)

[a]Any list of major hypersaline or alkaline lakes is arbitrary. Hence, we included the same lakes as Wilsey et al. (2017)

[b]Surface area of these lakes vary greatly among years. The first value is the commonly reported surface area. The second value is from Wilsey et al. (2017), which the authors acknowledge is larger than many estimates for the same lake. But the authors argue that their numbers have value in comparing the relative size of the lakes to each other. We are with them on this point and thus provide both estimates

[c]The Lahontan Valley Wetlands contain a combination of lakes, playas, and freshwater marshes

[d]This is an alkaline lake (Larson et al. 2016)

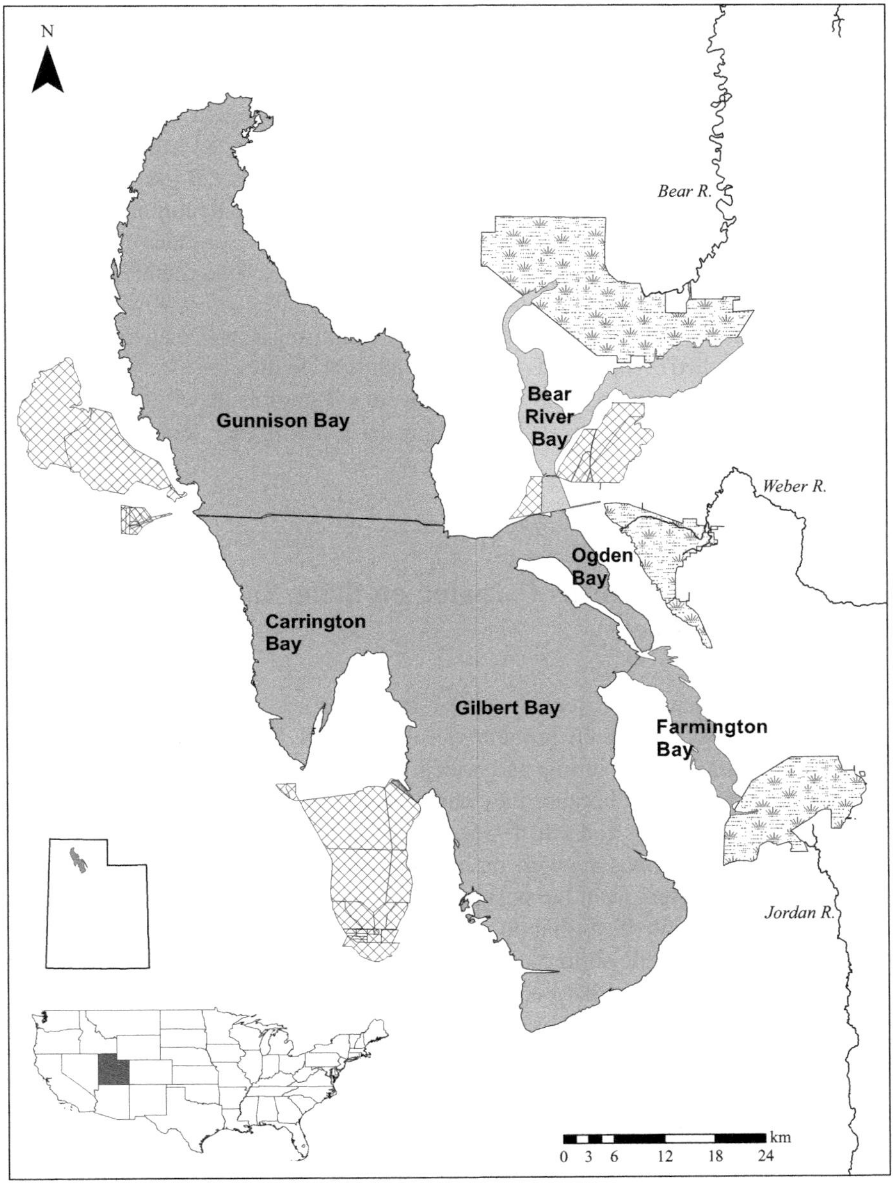

Fig. 8.1 Map of Great Salt Lake showing solar evaporation ponds in a checkboard pattern and freshwater impoundments used by nesting eared grebes are shown in a marsh symbol. The impoundments are on the eastern side of the lake (map by Leah Delahoussaye and used with permission)

In this chapter, we define the pelagic parts (also called the open waters) of the south arm as those areas where brine shrimp and brine flies are abundant. Gilbert, Ogden, and Carrington bays are all included in the pelagic part of GSL. The pelagic bays of GSL have a salinity of 12–15% (see Chap. 4). Farmington Bay, Bear River Bay, and parts of Ogden Bay and Bear River Bay have reduced salinities due to freshwater inflow, which allow other invertebrates to survive. Brine shrimp and brine flies are less abundant in these areas due to increased competition with other invertebrates and predation by water boatmen or corixids (Corixidae). But brine shrimp and brine fly abundance in GSL pelagic bays reach incredible densities. Adult brine shrimp densities were 0.62/L during the fall of 2017 when GSL had a volume of 75×10^{11} L of water, not counting Gunnison Bay (Delahoussaye and Conover unpublished). This equals 47×10^{11} adult brine flies for the southern arm of GSL. During the spring of 2019 when we were writing this paper, the density of adult brine flies had risen to 10/L or about 16 times higher than reported by Delahoussaye and Conover.

8.3 Challenges for Birds Foraging on Brine Shrimp and Brine Flies

Physiological challenges prevent most waterbird species from foraging on brine shrimp or brine flies. First is the problem of salt overload. Birds have to ingest some salt when foraging on brine shrimp and brine flies. Fortunately, pelagic birds are able to cope with salt ingestion because they have a salt gland capable of excreting large amounts of salt (Mahoney and Jehl 1985). Compounding the salinity problem is the fact that sources of freshwater are uncommon around hypersaline lakes and are often located many kilometers from the pelagic areas where brine shrimp and brine flies are abundant. Only waterbirds that do not need freshwater or only need it sparingly are able to forage on brine shrimp or brine flies.

The second challenge is that both brine shrimp and brine flies are very small so birds must be able to catch thousands of them each day. An adult brine shrimp weighs only 0.0019 g (wet weight), an adult brine fly weighs 0.0082 g, and brine fly larva weighs 0.0159 (Caudell and Conover 2006). Each brine shrimp and brine fly contain so little energy that an eared grebe has to catch and consume 28,000 adult brine shrimp daily to obtain enough energy to survive and migrate to wintering grounds (Conover and Caudell 2009). This is an impressive feat because eared grebes do not seine for food like flamingos (*Phoenicopterus* spp.) but must pluck each shrimp separately. Only a few bird species are able to capture so many individual prey items in a day. Eared grebes can do so, but just barely; to survive, an eared grebe on GSL must spend 7.5 h each day foraging and must catch one brine shrimp per second while foraging. To maintain such a high catch frequency, adult brine shrimp densities in GSL must remain above 380 adult brine shrimp per cubic meter of water.

8.4 Pelagic Birds

Eared grebes, red-necked phalaropes, Wilson's phalaropes, California gulls, and several duck species have evolved the ability to meet these challenges and exploit GSL's abundant brine shrimp and brine flies. These birds are able to forage while swimming or diving; this gives them access to the vast pelagic regions of GSL. This chapter covers these pelagic birds and examines the importance of GSL to them. Each of these species has evolved special adaptations that allow it to live in a hypersaline environment. These adaptations nonetheless come at a cost. These species are now dependent on hypersaline lakes for at least part of the year, leaving their fate tied to the fate of hypersaline lakes. American avocets, black-necked stilts, and snowy plovers are found along the shore and shallow parts of GSL where they consume adult brine flies or brine shrimp. These birds forage primarily while wading, so that they are concentrated on shorelines and shallow water. These birds will not be covered here because they are the subject of another chapter (Sorensen et al. 2020). We will also describe the importance of pelagic bays of GSL to American White Pelican (*Pelecanus erythrorhynchos*). This species uses GSL for nesting, not foraging.

8.5 Eared Grebes

Eared grebes (Fig. 8.2a) nest in lakes, marshes, and other wetlands within their breeding range. This range extends northward into southern Canada, southward into New Mexico and Utah, and westward into California and Oregon (Boe 1994; Cullen

Fig. 8.2 Eared grebes at Great Salt Lake. (**a**) Adult eared grebes (photo by Jimmie Grutzmacher and used with permission). (**b**) An eared grebe nest taken near Great Salt Lake (photo by Leah Delahoussaye and used with permission)

et al. 1999), including impoundments and marshes around GSL (Delahoussaye 2019). Grebes nest in colonies far from shore when vegetation is available in these areas (Breault 1990) (Fig. 8.2b). In Minnesota, grebe nesting colonies are located 0.1–1.5 km from shore where the average water depth is 50–120 cm (Boe 1993). In Minnesota, nests average 0.5–1.5 m in diameter with half of that diameter being above water (Bochenski 1961).

To determine why grebes nest on some water bodies around GSL and not others, Delahoussaye (2019) compared the characteristics of water bodies containing grebe colonies to those that the grebes avoided. Colonies were located in larger water bodies (mean = 1191 ha) than unoccupied wetlands (mean = 238 ha). Grebes also selected deeper water bodies, perhaps explained because grebes are diving waterbirds and dive to forage and avoid predators. By nesting far from shore and in deeper impoundments, common nest predators, such as raccoons (*Procyon lotor*), striped skunks (*Mephitis mephitis*), and red foxes (*Vulpes vulpes*), rarely threaten grebe colonies because they can only reach them by swimming.

Grebes around GSL build their nests by piling up large quantities of submergent vegetation, especially sago pondweed (*Stuckenia pectinata*) and spiral ditchgrass (*Ruppia cirrhosa*). In more northern latitudes, grebe nests are made of emergent vegetation, especially dried vegetation of cattails (*Typha* spp.), phragmites (*Phragmites* spp.), and other dried reeds from the prior year (Boe 1993, 1994; Hill et al. 1997; Cullen et al. 1999). Emergent vegetation in GSL impoundments was only located close to shore and, therefore, was not available to nesting grebes. Using submergent vegetation allows them to nest much farther from shore, but there are costs with its use. Growing sago pondweed and spiral ditchgrass are not very buoyant, requiring grebes to add to their nests daily to prevent the nest from sinking. The grebes also cannot start nesting until submergent vegetation has grown up to the water surface and formed mats. Hence, grebes around GSL begin to incubate their nests in late June and July while most waterbirds around GSL start nesting in April, May, or early June. This late start to nesting means that first-year grebes will have less time to grow before the onset of winter.

Food can be an important factor when it comes to selecting a colony nest site. Grebes eat primarily invertebrates and will not nest in a lake if invertebrates are scarce (Littlefield 1990; Cullen et al. 1999). Grebes usually nest in lakes with no or few fish, perhaps because they cannot compete with fish for their preferred invertebrate prey. Invertebrate density in the natal wetlands is critical to raising grebe chicks. These chicks cannot fly and are unable to regulate their own body temperature because their down feathers are not waterproof, meaning they can neither swim nor stay in a wet nest (McAllister 1963). Instead, one parent carries chicks on its back, while the other forages for food and feeds it to the young (Cullen et al. 1999). Parents and young are confined to the wetland where their nest is located. For this reason, it is not surprising that GSL water bodies used for nesting by grebes had higher densities of aquatic invertebrates than water bodies without colonies (Delahoussaye 2019).

Many grebes nest at other lakes and move to GSL after nesting for its rich food abundance. Starting in July, grebes begin leaving the nesting lakes and migrate

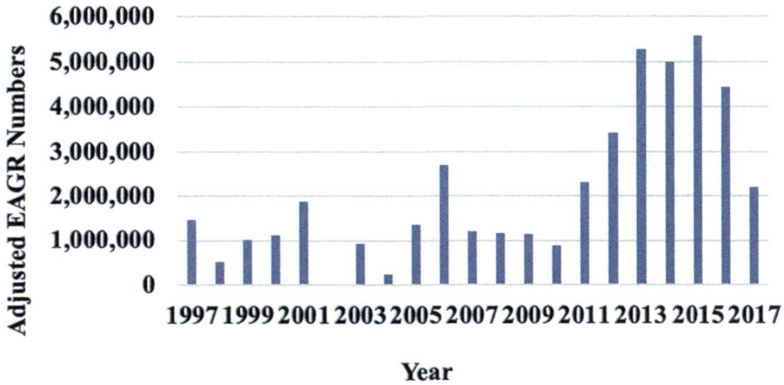

Fig. 8.3 Estimated population of eared grebes on Great Salt Lake during the fall based on aerial photo counts made by John Neill and Utah Division of Wildlife Resources' Great Salt Lake Ecosystem Program

to hypersaline waterbodies with most grebes going to Mono Lake in California and GSL where most will stay until November and December (Roberts et al. 2013). This is called the fall-staging period. During this period, grebes molt their feathers becoming flightless for most of the fall. Being flightless does not increase their risk of being killed by a predator because grebes dive when frightened and because few hawks, eagles, or owls venture over the pelagic bays of GSL. But there is one disadvantage of being flightless throughout the fall; grebes cannot leave GSL (or Mono Lake) if conditions become unfavorable. Even after grebes have completed their feather molt, they still cannot fly because their flight muscles atrophy while staging on GSL, and it takes weeks for the bird's muscles to grow strong enough for flight. Large numbers of grebes are attracted to GSL due to the high density of adult brine shrimp. Each October, Utah's Great Salt Lake Ecosystem Program (GSLEP) conducts aerial photographic surveys to count the number of grebes on GSL (Neill et al. 2016; GSLEP unpublished data). Population numbers have ranged from 0.5 to 5.5 million grebes, with the highest numbers counted from 2011 to 2015 (Fig. 8.3).

Grebes in pelagic bays of GSL during the fall-staging period feed almost exclusively on adult brine shrimp (Paul 1996; Conover and Vest 2009a). This is not true in less saline parts of GSL. Paul (1996) collected grebes in Farmington Bay during fall staging and found half of their diet to be brine fly larva, with brine shrimp making up the rest.

Grebes are not randomly located across the pelagic bays of GSL during fall staging but are concentrated over microbialites where brine shrimp adults and brine fly larva are most abundant (Roberts and Conover 2014b). Microbialites, which include subcategories of layered stromatolites or disorganized thrombolites, are organo-sedimentary carbonate structures that contain cyanobacteria and rich microbial mats that contribute to primary production. In Mono Lake, there are no stromatolites, but there are rocky shoals made of tufa, which may also be biologically precipitated (Scholl and Taft 1964). These are similar in geochemistry to the microbialites in GSL, and grebes are concentrated around these structures as are brine fly larva and adults.

Grebes leave GSL over a 30-day period in November and December to migrate to their wintering areas in the Gulf of California and Pacific Ocean. Grebes depart earlier during years when brine shrimp densities are high and water temperatures are above average. Frank and Conover (2017) concluded that grebes depart from GSL when they have gained sufficient mass to migrate successfully rather than lingering at GSL for as long as possible. Eared grebes departed from GSL over a period of 31 days each year, with departures occurring on 17 of those days.

Migration between GSL and the Gulf of California is dangerous for grebes because it must be made without stopping, and most of the trip occurs over the desert where lakes and reservoirs are uncommon. To increase the odds of a successful migration, grebes depart on nights when the barometric pressure is high and the local weather fair. High barometric pressure is related to low wind speeds and little precipitation (Frank and Conover 2017). Still, some grebes are unsuccessful in their migration attempt. Sometimes, large flocks of grebes crash-land (called a downing) in deserts of Utah and Arizona. Grebes cannot become airborne from land so most grebes involved in a downing die where they land. One downing in Cedar City occurred during a snowstorm and involved 7000 grebes. Grebes involved in a downing were similar in body condition and age as grebes that successfully migrate, leading Roberts et al. (2014) to conclude that most grebes involved in a downing were simply at the wrong place and the wrong time (i.e., caught in bad weather). But Roberts et al. (2014) also noted that the grebes that crash-landed in Cedar City had higher blood levels of mercury and selenium than most grebes.

As winter turns to spring, grebes start flying back to their nesting areas. Aerial surveys of GSL during the spring have found hundreds of thousands of grebes in the pelagic bays of GSL. We assume that these birds are migrating north to their breeding grounds and that they spend only a few days on GSL before continuing their northward migration. During April 2019, we collected 39 grebes on Gilbert Bay; 12 contained food in their upper digestive system. Of these, 11 contained adult brine shrimp and two contained adult brine flies. Gaffney (2009) collected grebes in Farmington Bay during the spring and found that 85% of grebe diet was brine fly adults and 8% brine shrimp (Roberts 2013).

8.6 Phalaropes

GSL is a critical fall-staging area for two species of phalaropes: Wilson's (*Phalaropus tricolor*) and red-necked (*Phalaropus lobatus*). Wilson's phalaropes nest primarily north of GSL in the western United States and Canada. This species is one of the few birds that engage in sex-role reversal where males incubate and care for the young. They are also polyandrous (i.e., a single female mate with several males). Some Wilson's phalaropes nest in the shallow wetlands surrounding GSL.

Fig. 8.4 A red-necked phalarope (photo by Jimmie Grutzmacher and used with permission)

Red-necked phalaropes (Fig. 8.4) nest in the Arctic in both the Old World and the New World and are more abundant than Wilson's phalaropes.

After nesting, both species of phalaropes stage on GSL to take advantage of abundant brine shrimp and brine fly populations (Fig. 8.5). In fact, GSL is the most important stopover site in the world for both phalarope species. In a single day, Aldrich and Paul (2002) counted more than 500,000 Wilson's phalaropes and 250,000 red-necked phalaropes on GSL. It is estimated that about one-third of all

Fig. 8.5 A flock of phalaropes on Great Salt Lake (photo by Wayne Wurtsbaugh and used with permission)

Wilson's phalaropes stage on GSL. Both phalarope species arrive at GSL during the summer with Wilson's arriving in July and leaving in August; red-necked phalaropes both arrive on GSL a little later and depart a little later than Wilson's.

There are over 11 million phalarope-use days on GSL annually (Paul and Manning 2002). A bird-use day is defined as one bird spending 24 h within a specific area; it provides information on the importance of an area during the entire year. One bird-use day means that the area is able to provide enough food to keep one bird alive there for 1 day.

From GSL, both phalarope species migrate to the Tropics for winter. Red-necked phalaropes winter in the seas around South America while Wilson's phalaropes can be found on South America's inland water bodies. While on GSL, both phalarope species flock together to form large flocks containing both species. While on GSL, they also forage in these mixed-species flocks, dining on the abundant invertebrate prey in pelagic bays as well as GSL's less saline bays. Both phalarope species can forage while swimming or wading but they cannot dive. Instead, they forage by picking food off the water surface, especially adult brine flies, and have the ability to upwell food from the upper 0.5 m of the water column by kicking their feet up while simultaneously spinning in place. Foraging flocks are very dense because foraging success increases when phalaropes are close together (Fig. 8.6). This allows one phalarope to catch an adult brine fly that has escaped from a neighboring phalarope. Dense flocks can also upwell food from greater depths than an individual phalarope can do by itself.

There are interspecific differences in how phalaropes obtain food. Foraging via surface picking was more common in red-necked phalaropes than Wilson's

Fig. 8.6 Phalaropes forage in dense flocks while feeding on Great Salt Lake (photo by Mia McPherson and used with permission)

phalaropes; 60% of red-necked phalaropes that were foraging used surface picking while on GSL, compared to 32% of Wilson's phalaropes. Spinning behavior to upwell prey was more common in Wilson's (19%) than red-necked (3%) phalaropes (Frank and Conover 2019). Phalaropes foraging in GSL's pelagic regions feed primarily on brine flies, while birds in Ogden and Farmington bays consume primarily adult brine flies and freshwater invertebrates (Roberts 2013).

8.7 Gulls

California Gulls congregate in large colonies on islands in inland lakes in western United States and Canada. The largest concentration of nesting California gulls occurs in colonies around GSL (Aldrich and Paul 2002). The gull colonies on GSL's Hat Island and Gunnison Island are the largest California Gull colonies in the world with more than 150,000 adults nesting in GSL colonies (Aldrich and Paul 2002; Paul and Manning 2002). During the nesting season, California Gulls forage on anthropogenic food where it is available, especially in landfills. In Utah, they also take advantage of the high densities of adult brine shrimp and dense swamps of adult brine flies during spring and summer months (Fig. 8.7). Conover and Caudell (2009) collected California gulls during May from Hat Island, Antelope Island, and Ogden Bay; 50–70% of the gulls' diet consisted of brine shrimp.

Ring-billed gulls (*Larus delawarensis*) also nest on inland lakes in the West but generally nest at more northern latitudes than GSL. However, there is a colony

Fig. 8.7 A California gull feeds on brine flies on the shores of Great Salt Lake (photo by Jimmie Grutzmacher and used with permission)

containing both ring-billed and California gulls on Neponset Reservoir in Rich County, Utah (M. R. Conover unpublished data).

While most California gulls winter along the Pacific Coast of the United States, many remain in Utah and forage in pelagic parts of GSL all year (Winkler 1996; Pollet et al. 2012). During winter, both California and ring-billed gulls forage for food items on the water surface of GSL or within a few centimeters beneath it (Roberts and Conover 2014b). California gulls also scavenge dead birds along the shoreline or floating in the water of GSL. During November and December, most dead grebes floating in GSL exhibit signs of being scavenged by California gulls; in most cases, the gulls only ate the grebes' liver (M. R. Conover unpublished data). Anthropogenic foods available at waste disposal sites and places where people eat outside also are a major diet component for both gull species (Greenhalgh 1952). What proportion of their diet during the fall and winter is obtained from GSL and what proportion is from anthropogenic food sources are unknown. There are over 23 million gull-use days on GSL annually (Paul and Manning 2002).

8.8 Ducks

Thousands of Gadwall (*Mareca strepera*), Mallards (*Anas platyrhynchos*), and Cinnamon Teal (*Spatula cyanoptera*) commonly nest in the freshwater impoundments on the eastern side of GSL (personal observation). Gadwall and Mallards nest throughout much of the northern United States and Canada; the Prairie Potholes constitute the heart of their range. Their populations number in the millions. The Great Basin and GSL are the heart of the breeding range of Cinnamon Teal. Cinnamon Teal nest densities averaged 7 pairs/km^2 on freshwater impoundment around Ogden Bay (Spencer 1953). The Cinnamon Teal population is small compared to other North American duck species, with a breeding population of about 200,000 and declining (Baldassarre 2014). Cinnamon Teal are one of the first duck species to migrate in the fall with most leaving GSL in August and September. For this reason, Utah duck hunters rarely harvest these birds.

During the fall, millions of other ducks spend at least part of the fall along the eastern edge of GSL, where there is freshwater or brackish water. Paul and Manning (2002) reported that there are over 22 million duck-use days on GSL and associated wetlands annually. Most of these ducks leave after these waters freeze. However, more than 200,000 Northern Shovelers (*Anas clypeata*), 150,000 Green-winged Teal (*Anas crecca*), and Common Goldeneyes (*Bucephala clangula*) spend the winter on GSL pelagic bays (Roberts and Conover 2014a). This raises the question of why they are out in the middle of GSL. One hypothesis is that they are avoiding hunters and are willing to fast rather than stay in freshwater marshes where both food and hunters are abundant. The second hypothesis is that they were foraging on something. Vest and Conover (2011) tested these hypotheses and found that the second hypothesis was correct. While on GSL, Northern Shovelers consume brine shrimp cysts (52% of diet) and adult brine shrimp (20%); Green-winged Teal consume mainly brine shrimp cysts (80%). Common Goldeneyes are diving ducks and ate brine fly larva (68%), which they pluck from the substrate, especially on microbialites.

While on GSL, Common Goldeneyes and Northern Shovelers are concentrated over areas with a mud substrate, likely because this substrate occurs where there is freshwater inflow. When all avian species were combined, densities were higher over microbialites (33 birds/km^2) than sand (29 birds/km^2), mud (12 birds/km^2), or deep brine layer (9 birds/km^2) (Roberts and Conover 2014b).

8.9 Pelicans

The world's second largest colony of American White Pelicans is on GSL's Gunnison Island. An average of 11,000 pelicans nest there (Paul and Manning 2002; Hoven 2017; Kijowski et al. 2020). During most years, this colony is surrounded by water and is safe from mammalian predators. In recent years, however, water levels in Gunnison Bay have dropped to the point that this island was connected to the mainland, and coyotes (*Canis latrans*) have gained access to the island (Kijowski et al. 2020).

While GSL is an important colony site for these pelicans, the birds do not use the salty parts of GSL for foraging. Pelicans feed on fish in shallow waters less than 2.5 m deep (Anderson 1991). From Gunnison Island, pelicans travel between 62 and 93 km to forage cooperatively in groups in freshwater lakes and impoundments scattered throughout Utah and southeastern Idaho and then return to nest sites to feed their young (Low et al. 1950). Pelicans target larger, slow-moving fish, which spend time in the shallows. This makes carp (*Cyprinus carpio*) a primary food source (Low et al. 1950).

8.10 Threats to Avian Use of Hypersaline Lakes in North America

There are several hypersaline lakes in North America that have been important to grebes, phalaropes, avocets, and stilts (Fig. 8.8). Almost all of these lakes are under threat, primarily from the diversion of freshwater that used to flow into them. Over half of them have shrunk from 50 to 95% since European settlement of the West (Wilsey et al. 2017).

Mono Lake, CA

The most important hypersaline lake for birds in America besides GSL is Mono Lake in California. It is smaller than GSL with an area of approximately 171 km^2 (Table 8.2) and differs in that it is lower in salinity (8–9%) and alkaline versus the neutral pH of GSL (Botkin et al. 1988). It is located along the western side of the

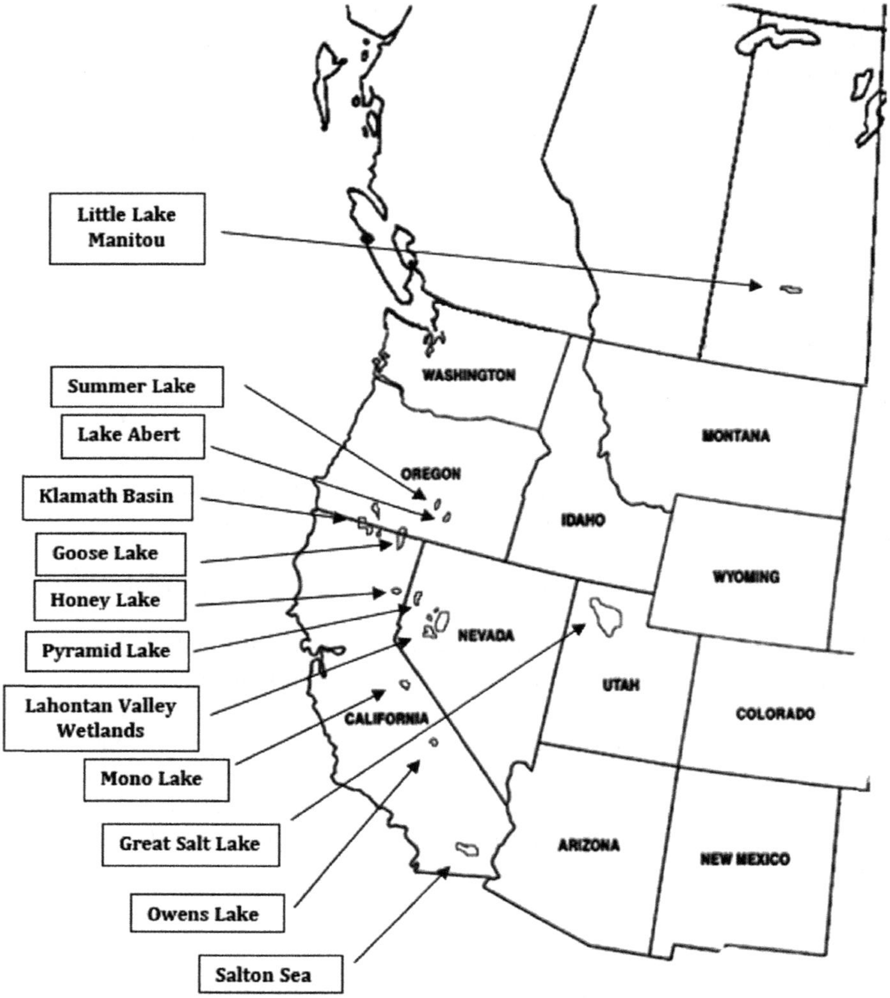

Fig. 8.8 Map of the western United States and Canada showing the location of hypersaline and alkaline lakes

Great Basin at the base of the Sierra Nevada mountain range. Mono Lake has an ecosystem similar to GSL, and Mono Lake has historically been home to many of the same species as GSL. But its ecosystem has suffered from water diversion by humans. Beginning in 1941, the city of Los Angeles diverted water from Mono Lake for its own use. The diversion caused the water level of Mono Lake to drop approximately 12.2 m, which doubled the salinity of the water (Botkin et al. 1988). Mono Lake is home to an endemic brine shrimp (*Artemia monica*) and brine fly (*Ephydra hians*), but the increasing salinity caused a loss in their productivity (Dana and Lenz 1986). On Mono Lake, brine flies are less tolerant than brine shrimp of

Table 8.2 Loss of surface area and volume of Great Salt Lake (GSL) south of the Union Pacific Railroad Causeway and north of the Antelope Island Causeway (the area where brine shrimp occur) as lake elevation decreases from 1278.6 to 1273.2 m above sea level (m.a.s.l.). At the lowest recorded lake level of 1274.1 m.a.s.l., which occurred during 1963, the surface area of GSL was 101,336 ha and the water volume was 2312 million m^3

Decrease in GSL elevation (m.a.s.l.)	Loss of surface area (km^2)	Loss of volume (million m^3)
1278.6–1278.3	49	513
1278.3–1278.0	53	500
1278.0–1277.7	45	481
1277.7–1277.4	40	470
1277.4–1277.1	40	459
1277.1–1276.8	53	444
1276.8–1276.5	45	429
1276.5–1276.2	57	412
1276.2–1275.9	40	398
1275.9–1275.6	49	386
1275.6–1275.3	45	371
1275.3–1275.0	45	358
1275.0–1274.7	45	342
1274.7–1274.4	49	330
1274.4–1274.1	45	316
1274.1–1237.8	54	300
1273.8–1273.5	49	285
1273.5–1273.2	49	270

Data for the table were adapted from Baskin (2005)

higher salinities. Consequently, Mono Lake's increasing salinity impacts them before it affects brine shrimp; birds that primarily feed on brine flies will be impacted more than birds that consume brine shrimp. For this reason, phalaropes, which primarily eat brine flies, decreased in population earlier on Mono Lake than grebes (Rubega and Inouye 1994). But eared grebes have also been impacted. Prior to 2000, more grebes staged on Mono Lake than GSL but this reversed after 2000 (Fig. 8.9). California gulls have also been impacted by Mono Lake's dropping water level. Negit Island, which they used for breeding colony sites, is no longer a safe place to nest as mammalian predators can now walk to the island along a land bridge that formed as water levels decreased (Ryan 2015).

Lake Abert, OR

Lake Abert is a hypersaline lake located in south central Oregon that covered an area of approximately 148 km^2, fed mainly from the Chewaucan River. However, the construction of a reservoir on the river and diversion of water for agriculture have led to a reduction in the size of Lake Abert and a corresponding increase in its salinity. Numbers of phalaropes, avocets, and gulls declined up to 83% as lake levels dropped (Senner et al. 2018).

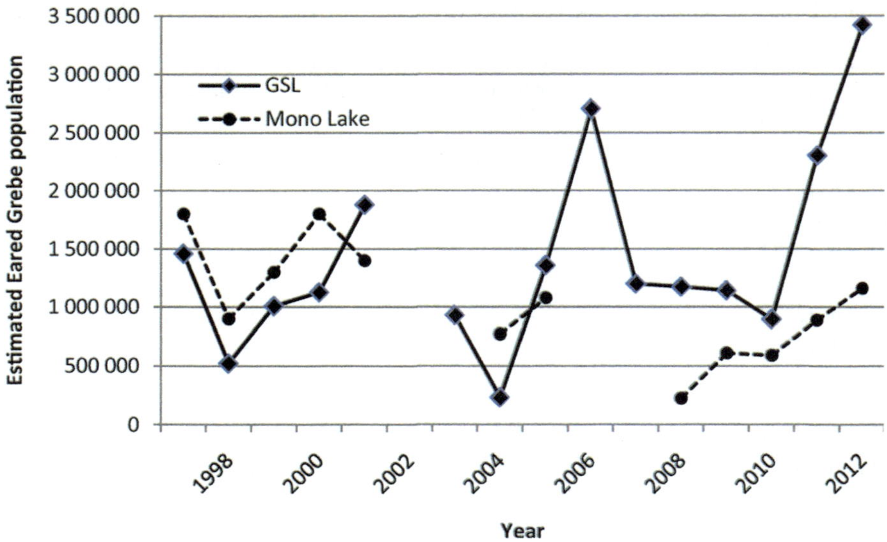

Fig. 8.9 Estimated population of eared grebes during the fall on Mono Lake and Great Salt Lake from 1997 through 2013

Salton Sea, CA
California's Salton Sea was an important wintering ground and migration stopover site where grebes were able to replenish their body condition by foraging on the abundant population of a polychaete, the pileworm (*Neanthes succinea*). Aerial surveys during 1988 documented 3.5 million grebes (Jehl and McKernan 2002), which was most of the grebes in North America (Roberts et al. 2013). But increasing salinity and the development of an anoxic water layer caused the pileworm population to crash; this, in turn, caused mass mortality of grebes on the Salton Sea (Jehl and McKernan 2002). By 2005, large numbers of grebes stopped using the Salton Sea (Anderson et al. 2007). Audubon Society biologists track grebe numbers on the Salton Sea by counting them from 14 observation posts scattered across the Salton Sea. During 2017–2018, biologists only counted three grebes (Orr 2018).

8.11 Threats to Avian Use of Great Salt Lake

GSL is fast becoming the last bastion in North America for large populations of pelagic birds that have evolved to forage in hypersaline lakes. Hence it is important to examine current and future threats to GSL. We can start by allaying concerns about one potential threat to birds. Brine shrimp cysts from GSL are commercially harvested and used by the aquaculture industry to feed baby fish and shrimp at aquaculture facilities across the globe. During 2010–2011, 10 million kg were harvested and sold for $34 million (Bioeconomics Inc. 2012). This harvest has not adversely affected avian populations because the harvest is closely monitored and regulated by the Utah Division of Wildlife's Great Salt Lake Ecosystem Program (GSLEP). This Utah state agency closes the harvest of cysts when cyst densities reach 21/L of water to make sure that there will be a sufficient number of cysts for GSL brine shrimp population to regenerate the next spring.

Another potential threat to GSL birds is high levels of selenium and mercury. Small quantities of selenium are essential for animal health, but high concentrations of selenium are toxic and cause embryonic defects, reduced egg hatchability, and lower survival rates of chicks and adults (Ohlendorf et al. 1989; Ohlendorf 2002). For example, birds foraging in California's Kesterson Reservoir, which was the disposal site for subsurface agricultural drainage from portions of the western San Joaquin Valley, accumulated high concentrations of selenium in their tissues (Ohlendorf 2002, 2003). The elevated concentrations of selenium impaired reproductive ability of several avian species nesting at Kesterson Reservoir and caused adult mortality.

Mercury is another element that harms avian health and reproduction (Eisler 1987, 2000; Yeardley et al. 1998). Grebes (Conover and Vest 2009a), California gulls (Conover and Vest 2009b), and ducks (Vest et al. 2009) that were collected on GSL had selenium and mercury concentrations above those found to impact the health and reproduction of birds collected elsewhere. Yet, no birds on GSL have been observed with the symptoms of selenium and mercury poisoning. It is unclear why GSL birds do not suffer from poisoning. But one hypothesis is that selenium and mercury interact with each other to form a chemical complex, which reduces the bioavailability or toxicity of both toxins (Belzile et al. 2005; Ralston et al. 2008).

A much greater threat to avian populations is posed from GSL's falling water level. GSL is located on a shallow basin; thus, a small change in the lake's surface elevation results in large changes in the surface area of the lake (Table 8.2). The area encompassed by the GSL varies annually due to changing precipitation amounts in its watershed. Yet, despite this annual variation in size, GSL has been steadily shrinking in size for decades due to the diversion of 1.7 km^3 of freshwater for anthropogenic uses. This diversion has decreased the level of GSL by 3.6 m from what it would have been its normal level if water had not been diverted (Wurtsbaugh et al. 2016). As GSL shrinks in size, its primary production decreases by about the same proportion; this, in turn, decreases brine shrimp and brine fly populations and GSL's ability to sustain avian populations.

Solar evaporation ponds cover 465 km^2 of GSL (Andrew Rupke, Utah Geological Survey, personal communication) and contribute to increased evaporation of GSL waters. These ponds are used by mineral companies to concentrate salts of economic value from the GSL. The ponds are so saline that neither brine flies nor brine shrimp can survive in them. Therefore, we hypothesized that few birds would use them for foraging. We confirmed this during several aerial flights over solar evaporation ponds; we failed to observe any foraging by birds, although there was a large California gull nesting colony on the dikes.

As GSL elevation decreases, its water volume decreases (Table 8.2). Water loss from GSL is through evaporation, and as its volume decreases, its salinity rises. This is the greatest threat to birds that are dependent on GSL because brine shrimp and brine flies have a limit to how much salinity they can tolerate. Thirty percent is the maximum salinity that adult brine shrimp can withstand, while the maximum salinity for newly hatched brine shrimp is 17% (Stephens and Birdsey 2002). Brine fly biomass is reduced by half when salinity increases to 21% (Barnes and Wurtsbaugh 2015).

The impact of increasing salinity can be observed today; large areas of GSL no longer support brine shrimp or brine flies due to the areas becoming too saline (M. R. Conover unpublished data). When GSL is at an elevation of 1280 m, Gunnison Bay constitutes ~39% of GSL. After a railroad causeway separated Gunnison Bay from the rest of GSL, this bay became so salty that its populations of both brine shrimp and brine flies collapsed (White et al. 2015). We have conducted aerial surveys over Gunnison Bay for the past 5 years and have not observed any grebes or ducks on Gunnison Bay. Before it became so salty, over half of all phalaropes on GSL (197,000) were counted there (White and Robinette 1993).

Most people in Utah consider freshwater that reaches GSL to be wasted and prefer that it be stored in reservoirs to support Utah's economy (Bedford 2009). Utah is one of the fastest growing states in the United States in terms of both human population and economic development. An increasing human population and economy will increase the need to divert freshwater for anthropogenic uses (Yidana et al. 2010). Most climate change models predict that Utah will become drier and warmer in the future (Cook et al. 2015), and these changes will likely increase the evaporation of water from GSL. The combination of increasing freshwater diversion and climate change will threaten the avian populations dependent on GSL by increasing water salinity. Unfortunately, GSL does not have a legal water right and only receives what water remains after humans have diverted water for agriculture and development. A high priority should be to secure an adequate amount of freshwater inflow into the GSL to sustain avian populations. This could be achieved if GSL was given a legal right to freshwater.

Giving water rights to GSL is both environmentally and economically prudent for Utah. If future conditions feature declining water levels and increased salinity, the brine shrimp population in GSL could collapse, leading to a dire scenario. This will result in a mass die-off of hundreds of thousands of grebes on GSL because they will be unable to leave due to their inability to fly during the fall. This die-off would be so extreme that it will garner national media coverage, and, given that most of the grebes in North America currently spend the fall on GSL, it would also threaten the survival of the species. In response, the US Fish and Wildlife Service would be forced to intervene and rule that a certain level of freshwater must flow into GSL each year. This would require that some water diversions will be stopped despite the infrastructure that has been built to use them, impacting communities upstream of GSL. For example, reservoirs built to hold freshwater will be emptied. This raises the question of whether it is prudent to build new dams and water infrastructure in GSL watershed when they cannot be used in the future. In addition, it argues that Utah should grant a water right for GSL.

References

Aldrich TW, Paul DS (2002) Avian ecology of Great Salt Lake. In: Gwynn JW (ed) Great Salt Lake: an overview of change. Special Publication of the Utah Department of Natural Resources, Salt Lake City, UT, pp 342–372

Anderson JGT (1991) Foraging behaviors of American White Pelican (*Pelecanus erythrorhynchos*) in Western Nevada. Colonial Waterbirds 14:166–172

Anderson TW, Tiffany MA, Hurlbert SH (2007) Stratification, sulfide, worms, and decline of the Eared Grebe (*Podiceps nigricollis*) at the Salton Sea, California. Lake Reservoir Manag 23:500–517

Atlas of Oregon Lakes (2019) Summer lake (lake). http://aol.research.pdx.edu/lakes/17120005004769. Accessed 7 June 2019

Baldassarre G (2014) Ducks, geese and swans of North America. John Hopkins University Press, Baltimore, MD

Barnes BD, Wurtsbaugh WA (2015) The effects of salinity on plankton and benthic communities in the Great Salt Lake, Utah, USA: a microcosm experiment. Can J Fish Aquat Sci 72:807–817

Baskin RL (2005) Calculation of area and volume for the south part of Great Salt Lake, Utah. Geological Survey, Washington, DC

Bedford D (2009) The Great Salt Lake America's Aral Sea? Environ Sci Policy Sustain Dev 51:8–21

Belzile N, Chen YW, Gunn JM, Tong J, Alarie Y, Delongchamp TL, Land CY (2005) The effect of selenium on mercury assimilation by freshwater organisms. Can J Fish Aquat Sci 63:1–10

Bioeconomics Inc (2012) Economic significance of the Great Salt Lake to the State of Utah. Final report to the Great Salt Lake Advisory Council, Missoula, MT

Bochenski Z (1961) Nesting biology of the black-necked grebe. Bird Study 8:6–15

Boe JS (1993) Colony site selection by eared grebes in Minnesota. Colonial Waterbirds 16:28–38

Boe JS (1994) Nest site selection by eared grebes in Minnesota. Condor 96:19–35

Botkin DB, Broecker WS, Everett LG, Shapiro J, Wiens JA (1988) The future of Mono Lake report of the cori "Blue Ribbon Panel" for the legislature of the state of California. Report-California Water Resources Center, University of California (USA)

Breault AM (1990) Breeding distribution, habitat selection and factors affecting coloniality in eared grebes in British Columbia. Thesis, University of Sherbrooke, Sherbrooke, BC, Canada

Canadian Encyclopedia (2014) Little Manitou Lake. https://www.thecanadianencyclopedia.ca/en/article/little-manitou-lake. Accessed 7 June 2019

Cannon JS, Cannon MA (2002) The Southern Pacific Railroad trestle—past and present. In: Gwynn JW (ed) Great Salt Lake: an overview of change. Special Publication of the Utah Department of Natural Resources, Salt Lake City, UT, pp 283–294

Caudell JN, Conover MR (2006) Energy content and digestibility of brine shrimp (*Artemia franciscana*) and other prey items of Eared Grebes (*Podiceps nigricollis*) on the Great Salt Lake, Utah. Biol Conserv 130:251–254

Conover MR, Caudell JN (2009) Energy budgets of eared grebes on the Great Salt Lake and implications for harvest of brine shrimp. J Wildl Manag 73:1134–1139

Conover MR, Vest JL (2009a) Concentrations of selenium and mercury in eared grebes (*Podiceps nigricollis*) from Utah's Great Salt Lake, USA. Environ Toxicol Chem 28:1319–1323

Conover MR, Vest JL (2009b) Selenium and mercury concentrations in California Gulls breeding on the Great Salt Lake, Utah, USA. Environ Toxicol Chem 28:324–329

Cook BI, Ault TR, Smerdon JE (2015) Unprecedented 21st century drought risk in the American Southwest and Central Plains. Sci Adv 1:e1400082

Cullen SA Jehl JR Jr., Nuechterlein GL (1999) Eared grebes (*Podiceps nigricollis*). Account 433. In Poole A, Gill F (eds) Birds of North America. Cornell Lab of Ornithology, Ithaca, NY

Dana GL, Lenz PH (1986) Effects of increasing salinity on an Artemia population from Mono Lake, California. Oecologia 68(3):428–436

Delahoussaye LM (2019) Eared grebe nesting ecology and chronology along the Great Salt Lake, Utah. Thesis, Utah State University, Logan, UT
Eisler R (1987) Mercury hazards to fish, wildlife, and invertebrates: a synoptic review. U.S. Fish and Wildlife Service Report No. 85, Washington, DC
Eisler R (2000) Handbook of chemical risk assessment: health and hazards to humans, plants and animals. Lewis, Boca Raton, FL
Frank MG, Conover MR (2017) Weather and prey availability affect the timing of fall migration of Eared Grebes (*Podiceps nigricollis*) from Great Salt Lake. Wilson Bull Ornithol 129:98–111
Frank MG, Conover MR (2019) Phalarope foraging behavior at sites with high prey abundance in Great Salt Lake, Utah. Condor 12(2). Doi 2005. https://doi.org/10.1093/condor/doi2005
Gaffney JM (2009) Tracking eared grebe (*Podiceps nigricollis*) migration using fatty acid signature analysis. Thesis, University of San Diego, San Diego, California, USA
Greenhalgh CM (1952) Food habits of the California Gull in Utah. Condor 54:302–308
Herbst DB, Prather M (2014) Owens Lake—from dustbowl to mosaic of salt water habitats. Lakeline 34:34–38
Hill WL, Jones KJ, Hardenbergh CL, Browne M (1997) Nest distance mediates the costs of coloniality in eared grebes. Colonial Waterbirds 20:470–477
Hoven H (2017) Gunnison Island: home of up to 20,000 nesting American White Pelicans. California Audubon, https://ca.audubon.org/news/gunnison-island-home-20000 nesting-american-white-pelicans. Accessed 2 July 2019
Jehl JR, McKernan RL (2002) Biology and migration of eared grebes at the Salton Sea. Hydrobiologia 473:245–253
Kijowski AK, Neill J, Wickline A, Swift J, Butler JK, Kimberly DA, VanLeeuwen J, Luft J, Stone K (2020) American White Pelicans of Gunnison Island. In: Baxter BK, Butler JK (eds) Great Salt Lake biology: a terminal lake in a time of change. Springer, Cham
Larsen L (2017) Oregon's drying Lake Abert could signal disaster for business on Great Salt Lake. https://www.standard.net/news/environment/oregon-s-drying-lake-abert-could-signal-disaster-for-business/article_b4f2966e-21b2-5e65-9d25-56f1e8178452.html. Accessed 4 June 2019
Larson R, Eilers JK, Kreuz K, Pecher WT, DasSarma S, Dougill S (2016) Recent desiccation-related ecosystem changes at Lake Abert, Oregon: a terminal alkaline salt lake. West N Am Nat 76:389–404
Learn. Genetics, University of Utah (2019) Brine flies. https://learn.genetics.utah.edu/content/gsl/foodweb/brine flies/. Accessed 12 June 2019
Littlefield CD (1990) Birds of Malheur National Wildlife Refuge, Oregon. Oregon State University Press, Corvallis, OR
Los Angeles Times (2017) Drawdowns and death of the Salton Sea. https://www.latimes.com/local/california/la-me-g-drought-drawdowns-and-death-of-the-salton-sea-20141021-htmlstory.html. Accessed 4 June 2019
Low JB, Kay L, Rasmussen DI (1950) Recent observations on the White Pelican on Gunnison Island, Great Salt Lake, Utah. Auk 67:345–356
Mahoney SA, Jehl JR (1985) Avoidance of salt-loading by a diving bird at a hypersaline and alkaline lake: eared grebe. Condor 87:389–397
McAllister NM (1963) Ontogeny of behavior in five species of grebes. Thesis, University of British Columbia, Vancouver, Canada
Molony BW, Parry GO (2006) Predicting and managing the effects of hypersalinity on the fish community in solar salt fields in north-western Australia. J Appl Ichthyol 22:109–118
Mono Lake Committee (2019) Quick facts. http://www.monolake.org/about/stats. Accessed 5 June 2019
Neill J, Leite B, Gonzales J, Sanchez J, Luft JT (2016) The 2015 Great Salt Lake Eared Grebe aerial photo survey. Utah Division of Wildlife Resources, Salt Lake City, Utah
Ohlendorf HM (2002) The birds of Kesterson Reservoir: a historical perspective. Aquat Toxicol 57:1–10

Ohlendorf HM (2003) Ecotoxicology of selenium. In: Hoffman DJ, Rattner BA, Burton GA Jr, Cairns J Jr (eds) Handbook of ecotoxicology. Lewis, Boca Raton, FL, pp 465–500

Ohlendorf HM, Hothem RL, Welsh D (1989) Nest success, cause specific nest failure, and hatchability of aquatic birds at selenium contaminated Kesterson Reservoir and a reference site. Condor 91:787–796

Orr D (2018) Where have all the Salton Sea's eared grebes gone? California Audubon. http://ca.audubon.org/news/where-have-all-salton-seas-eared-grebes-gone. Accessed 9 June 2019

Paul DS (1996) Eared grebe progress report. Utah Department of Natural Resources, Division of Wildlife Resources, Salt Lake City, UT

Paul DS, Manning AE (2002) Great Salt Lake waterbird survey five-year report (1997–2001). Publication Number 08-38, Utah Division of Wildlife Resources, Salt Lake City, UT

Pollet LL, Shutler D, Chardine J, Ryder JP (2012) Ring-billed Gull (*Larus delawarensis*). Birds of North America 33. Cornell Lab of Ornithology, Ithaca, NY

Ralston NVC, Ralston CR, Backwell JL III, Raymond LJ (2008) Dietary and tissue selenium in relation to methyl mercury toxicity. Neurotoxicology 29:802–811

Rawson DS, Moore JE (2011) The saline lakes of Saskatchewan. Can J Res 22:141–201

Roberts AJ (2013) Avian diets in a saline ecosystem: Great Salt Lake, Utah, USA. Hum Wildl Interact 7:158–168

Roberts AJ, Conover MR (2014a) Diet and body mass of ducks in the presence of commercial harvest of brine shrimp cysts in the Great Salt Lake, Utah. J Wildl Manag 78:1197–1205

Roberts AJ, Conover MR (2014b) Role of benthic substrate in waterfowl distribution on Great Salt Lake. Waterbirds 37:298–306

Roberts AJ, Conover MR, Luft J, Neill J (2013) Population fluctuations and distribution of staging Eared Grebes (*Podiceps nigricollis*) in North America. Can J Zool 91:906–913

Roberts AJ, Conover MR, Fusaro JL (2014) Factors influencing mortality of Eared Grebes (*Podiceps nigricollis*) during a mass downing. Wilson J Ornithol 126:584–591

Rubega M, Inouye C (1994) Prey switching in Red-necked Phalaropes *Phalaropus lobatus*: feeding limitations, the functional response and water management at Mono Lake, California, USA. Biol Conservat 70:205–210

Ryan E (2015) The public trust doctrine, private water allocation, and Mono Lake: the historic saga of National Audubon Society v. Superior Ct. Environmental Law (Northwestern School of Law), Chicago, IL

Salton Sea Authority (1997) The Salton Sea: a brief description of its current conditions and potential remediation projects. Basin-Delta Mothersite. http://www.sci.sdsu.edu/salton/Salton%20Sea%20Description.html. Accessed 4 June 2019

Scholl DW, Taft WH (1964) Algae, contributors to the formation of calcareous tufa, Mono Lake, California. J Sediment Res 34:309–319

Senner NR, Moore JN, Seager ST, Dougill S, Kreuz K, Senner SE (2018) A salt lake under stress: relationships among birds, water levels, and invertebrates at a Great Basin saline lake. Biol Conservat 220:320–329

Sorensen ED, Hoven HM, Neill J (2020) Great Salt Lake shorebirds, their habitats, and food base. In: Baxter BK, Butler JK (eds) Great Salt Lake biology: a terminal lake in a time of change. Springer, Cham

Spencer Jr HE (1953) The Cinnamon Teal, *Anas cyanoptera* (Vieillot): its life history, ecology, and management. Thesis, Utah State University, Logan, UT

Stephens D, Birdsey Jr. PW (2002) Population dynamics of the brine shrimp, *Artemia franciscana,* in Great Salt Lake and regulation of commercial shrimp harvest. In: Gwynn JW (ed) Great Salt Lake: an overview of change. Special Publication of the Utah Department of Natural Resources. Salt Lake City, UT, pp 342–372

Szalay J (2016) Facts about Great Salt Lake. Live Science. https://www.livescience.com/56941-great-salt-lake.html. Accessed 31 May 2019

Taylor M (2018) The Salton Sea: a status update. Legislative Analyst's Office. https://lao.ca.gov/reports/2018/3879/salton-sea-082918.pdf. Accessed 7 July 2019

Vest JL, Conover MR, Perschon C, Luft J, Hall JO (2009) Trace element concentrations in wintering waterfowl from the Great Salt Lake, Utah. Arch Environ Contam Toxicol 56:302–316

Vest JL, Conover MR (2011) Food habits of wintering waterfowl on the Great Salt Lake, Utah. Waterbirds 34:40–50

White PA Robinette K (1993) The 1993 Great Salt Lake Wilson's Phalarope Survey. Unpublished report to the Utah Division of Wildlife Resources, Salt Lake City, UT

White JS, Null SE, Tarboton D (2014) More than meets the eye: managing salinity in Great Salt Lake, Utah. Lakeline. http://hydrology.usu.edu/dtarb/White_etal_2014_Lakeline.pdf. Accessed 7 July 2019

White JS, Null SE, Tarboton DG (2015) How do changes to the railroad causeway in Utah's Great Salt Lake affect water and salt flow? PLoS One 10:e0144111

Wikipedia (2019) Goose Lake (Oregon-California). https://en.wikipedia.org/wiki/Goose_Lake_(Oregon%E2%80%93California). Accessed 12 June 2019

Wilkie MP, Wood CM (1996) The adaptations of fish to extremely alkaline environments. Comp Biochem Physiol 113B:665–673

Wilsey CB, Taylor C, Michel N, Stockdale K (2017) Water and birds in the Arid West: habitats in decline. National Audubon Society, New York

Winkler DW (1996) California Gull (*Larus californicus*). Birds of North America 259. Cornell Lab of Ornithology, Ithaca, NY

Wurtsbaugh W, Miller C, Null S, Wilcock P, Hahnenberger M, Howe F (2016) Impacts of water development on Great Salt Lake and the Wasatch Front. Watershed Sciences Faculty Publications Paper 875 Utah State University, Logan, UT

Yeardley RG Jr, Lazorchak JM, Paulsen SG (1998) Elemental fish tissue concentration in northeastern U.S. lakes: evaluation of an approach to regional assessment. Environ Toxicol Chem 17:1875–1884

Yidana SM, Lowe M, Emerson RL (2010) Wetlands in northern Salt Lake Valley, Salk Lake County, evaluation of the threats posed by groundwater development and drought. Utah Geological Survey Report of Investigation 268, Salt Lake City, UT

Chapter 9
Great Salt Lake Shorebirds, Their Habitats, and Food Base

Ella Dibble Sorensen, Heidi Morrill Hoven, and John Neill

Abstract Every year millions of shorebirds representing 42 species congregate on Great Salt Lake (GSL). It is one of the largest concentrations of shorebirds on Earth, and yet, compared to waterfowl and colonial nesting species at GSL, they have received relatively little attention. Some shorebirds nest and rear young, but many more use the lake as a fueling stopover during migration with some departing flights lasting thousands of nonstop kilometers. Three ecological parameters determine whether or not any given location is suitable shorebird habitat: water depth, type and extent of vegetation, and type of food items available. Although shorebirds are opportunistic, each species shows a preferential niche along the intersection of these parameters, which do not form distinct units but lie along overlapping continuums. We explain these continuums and describe how, in each specific shorebird habitat, salinity is a driver for both vegetation and macroinvertebrates, the primary food source. Playas and mudflats are important components of shorebird habitat, but the characteristics that define these features in the landscape have been mired in confusion. We clarify these major components of the GSL ecosystem. Finally, we provide species accounts for each of the 42 species of shorebird while at GSL, detailing status, abundance, range, and timing of arrival and departure and ecological preferences.

Keywords Shorebirds · Playa · Mudflat · Halophyte · Great Salt Lake · Macroinvertebrates · Great Salt Lake shorebird checklist

---

E. D. Sorensen (✉) · H. M. Hoven
Gillmor Sanctuary, National Audubon Society, Salt Lake County, UT, USA
e-mail: esorensen@audubon.org

J. Neill
Great Salt Lake Ecosystem Program, Utah Division of Wildlife Resources, Hooper, UT, USA

B. K. Baxter, J. K. Butler (eds.), *Great Salt Lake Biology*,
https://doi.org/10.1007/978-3-030-40352-2_9

9.1 Introduction

Every year millions of birds representing 338 species of shorebirds, waterfowl, and other birds (Neill and Sorensen unpublished list 2020) congregate on Great Salt Lake (GSL) and the vast complex of diverse wetlands and uplands associated with the lake. It is one of the largest concentrations of birds on earth.

Marbled Godwit. Photo Credit, Steve Earley

Shorebirds account for 42 of the 338 GSL species. They also represent a large percent of the total number of birds that depend on the lake for some part of their life cycle. Shorebirds are a diverse group of birds that include stilts, avocets, plovers, curlews, godwits, sandpipers, and phalaropes.

They include some of the world's greatest long-distance migrants, with some making biannual marathon flights from as far as their high Arctic nesting grounds to the tip of South America. Most shorebirds that fly long distances show a distinctive flight strategy, converging at a few specific sites like GSL that reliably provide an abundance of macroinvertebrates where they feed and store energy reserves to fuel their migratory journeys, which sometimes last thousands of nonstop kilometers.

Waterfowl and impounded freshwater wetland habitat have been the focus of much of the research and surveys of birds on GSL for over a century (Aldrich and Paul 2002; Vest 2013; Roberts 2013; Roberts and Conover 2014). Colonial nesting birds have also been studied, especially Eared Grebe and American White Pelican (Behle 1958; Ryder 1959; Knopf 1975; Jehl 1998; Jehl et al. 2003; Aldrich and Paul 2002; Roberts 2013; Cavitt et al. 2014; Ellis et al. 2016, Kijowski et al. 2020).

Shorebirds and their mudflat and shallow water habitats have received relatively little attention at GSL. Only six species of shorebirds have been the subject of research to varying amounts: Black-necked Stilt, American Avocet, Snowy Plover, Marbled

Godwit, and Wilson's and Red-necked phalaropes (Paton and Edwards 1992, 1996; Paton 1994; Paul and Manning 2002; Cavitt 2006a, b, 2008; Thomas et al. 2012; Frank 2016; Olson 2011; Ellis et al. 2015; Frank 2016; Frank and Conover 2017).

Sanderling. Photo Credit, Mike Schijf

The remaining 36 species are far less understood as described by Utah Division of Wildlife Resources biologists Tom Aldrich and Don Paul (2002):

> There is a plethora of shorebird species for which we have sketchy information, but for which coarse-scale observations lead one to believe that the GSL ecosystem plays an important role. . . .As the quilt work of migration is revealed through future studies, there will most likely be many new revelations concerning GSL's function and value to the peregrinations of shorebirds.

Besides the general longtime lack of emphasis on most shorebird species, other factors contribute to the poor understanding. Accurate shorebird identification comes with many pitfalls. Shorebirds such as Black-necked Stilt and American Avocet are large, visible, have distinctive plumage, and are easily identified, even by observers with little experience. Many of the other GSL shorebirds are prone to misidentification, simply overlooked, or located in the vast expanses of GSL where accessibility is prohibitive. Partially the result of dull plumages, many similar looking species, and the cryptic nature of some species, these misidentification and detection errors can lead to either overinflating population numbers or underreporting. Some species are so similar and separating them so difficult and time-consuming that it has resulted in the regular practice of lumping them together and reporting them as "yellowlegs species," "dowitcher species," "phalarope species," or "peeps." Each species has different ecological requirements, breeding and winter ranges, and timing of migration. The combining of these species obscures their individual identity and relative significance at GSL. Misinformation regarding numbers inhibits population monitoring and conservation management strategies.

Adding to identification errors, shorebirds have not one but three possible plumages: juvenile, basic (non-breeding), and alternate (breeding). In species such as Snowy Plover, Marbled Godwit, Lesser and Greater yellowlegs, and Willet, the differences between basic and alternate plumages are subtle. In other species such as Black-bellied Plover, Short- and Long-billed dowitchers, and Wilson's and Red-necked phalaropes, the bright colors of alternate plumage are replaced through molting into drab basic plumage making them more susceptible to misidentification in the fall. To complicate identification further, some shorebirds may be transitioning from one plumage to another during their time spent at GSL.

Some shorebirds are more adept at camouflage. The Snowy Plover can be especially difficult to detect because of its small size and the light gray and white plumage that blends so well into its salt flat habitat. Least Sandpipers often prefer to forage at the water's edge amidst scattered low vegetation. They typically hunch, almost crouch (Hayman et al. 1986), and can be overlooked by even experienced observers. Wilson's Snipe are typically only observed when flushed from the cover of thick vegetation.

Snowy Plover. Photo Credit, Lauri Taylor

9.2 Ecological Characterization of Shorebird Habitat of Great Salt Lake

9.2.1 Ecological Parameters of Shorebird Habitat

GSL and its associated wetlands form a vast complex of diverse habitats. The location and extent of specific habitat types change continually. In reality, habitat types are not distinct entities, but a continuum of transitional wetland types. Each shorebird species occurring on GSL has unique ecological requirements and locates the niche within the ecosystem that meets their specific needs. For some species, the niche may be rigid and narrow, while other species can exploit a wider range. In addition, nesting habitat may be different from foraging habitat. In this chapter, rather than define the habitat as commonly characterized, we define three components of productive foraging habitat that each shorebird species keys into when locating their preferred niche: water depth, vegetation, and macroinvertebrates.

Each of these components forms a continuum and each species occupies a range along that continuum, sometimes narrow and sometimes broad. Water and soil salinity play a vital role as drivers of location, type and extent of vegetation, and macroinvertebrates that occur within the system.

9.2.1.1 Foraging Water Depth

Shorebird species vary greatly in body size and leg length. A correlation exists between depth of water where a species of shorebird forages and their body size and leg length (Baker 1979; Norazlimi and Ramli 2015). Shorebirds are also uniquely adapted in both shape and length of their bill. They exhibit an array of foraging techniques such as picking, probing, gleaning, and sweeping, which are closely tied to their physical attributes. A species' bill structure and associated feeding behavior, as well as leg length, dictate preferential foraging areas.

Shorebirds are opportunistic and will forage wherever a food source is available, but each species shows an affinity for certain water depths. There is variability in the range of depths that each species utilizes. For example, the Snowy Plover with its short legs and short blunt bill seldom ventures far from moist mudflats or the interface with shallow water. Others like the American Avocet with its long legs and thin upturned scythe-like bill utilizes a variety of foraging methods that facilitate successful exploitation of different ecological niches including picking on moist mudflats, sweeping the long bill from side to side through water or loose sediment to locate hidden prey, or swimming in deep water and picking from the surface of the water.

Most shorebirds on GSL, with a few exceptions such as Wilson's and Red-necked phalaropes, forage with their feet planted firmly on a solid substrate such as the mud beneath the water surface, thus limiting the depth of the water they exploit. Some species such as Long-billed Dowitcher that probes the benthic layer for macroinvertebrates must have water shallow enough to access the bottom of the water body with their bills.

Defining a range of optimal water depths for the majority of shorebirds is somewhat arbitrary. A water depth range of roughly 0–18 cm is suggested by many authors (Thomas 2005; Helmers 1992; Olson et al. 2004; Sorensen et al. 2018).

Most shorebirds occurring on flooded mudflats preferentially segregate into guilds or groups of birds foraging at certain water depths. These depths are not rigid as shorebirds move around continually following the food source. This is especially true of the medium and large shorebirds as the smaller shorebirds are more restricted because of their short legs and bills. Table 9.1 subdivides mudflats

Table 9.1 Foraging water depths for mudflats

Mudflat water depth (cm)	Representative foraging species
0 (Moist to Dry)	Snowy Plover
<5	Baird's Sandpiper Least Sandpiper Western Sandpiper
5–10	Long-billed Dowitcher Lesser Yellowlegs Willet Greater Yellowlegs
10–18	Black-necked Stilt American Avocet Marbled Godwit
>18 (open water)	Wilson's Phalarope Red-necked Phalarope

Table 9.2 Foraging water depths for wet meadow, flooded fields, and uplands

Habitat type	Water depth (cm)	Representative foraging species
Wet meadow Flooded Agricultural fields	<18	Black-necked Stilt Killdeer Wilson's Snipe Lesser Yellowlegs Greater Yellowlegs
Uplands	0	Long-billed Curlew

into five ranges of water depth and lists several representative species that can regularly be found at these depths.

Some shorebird species also forage in flooded areas with vegetation such as wet meadows, agricultural fields, and pastures. There has been less attention to preferred water depth associated with each species in these areas. Since these species also forage on flooded mudflats, it is likely that their preferred depth of water in the vegetated areas is similar to the 0–18 cm depth on mudflats, but observation by the authors suggests that most foraging in these vegetated areas occurs in water depths in the more shallow range perhaps dictated by the availability and type of macroinvertebrates. Also, in these areas there is often much microtopography that can make the water depth extremely variable in a very small area. A few species of shorebirds forage in the adjacent upland areas. Table 9.2 gives water depths and representative species that forage in these areas.

9.2.1.2 Vegetation

A second parameter that influences the niche each species of shorebird prefers is the amount and type of vegetation present. Many species forage predominately on moist mudflats or shallow open water devoid of vascular vegetation. Shorebirds can often be found on mudflats vegetated with pickleweed *Salicornia* spp. Some use wet meadows with moderate to fairly dense saltgrass *Distichlis spicata*. Others still use flooded agricultural fields. A few species occasionally use areas sparsely covered with alkali bulrush *Bolboschoenus maritimus* if the water is shallow enough. Thick stands of emergent vegetation consisting of cattails *Typha* spp., hardstem bulrush *Schoenoplectus acutus*, and common reed *Phragmites australis* (*Phragmites*) are avoided.

GSL's wetland types often form a continuum as one wetland type transitions into another, based on allogenic processes such as length of flooding intervals, water depth, and salinity, which results in well-organized spatial heterogeneity (Mitsch and Gosselink 2007). GSL wetlands form a vast complex of diverse habitats in constant flux as they are influenced by changing hydrology and salinity. Changes in distribution and species composition of wetland vegetation help explain why shorebirds do not always exhibit site fidelity. They are not always found in the same location year after year or decade after decade but often shift dramatically to new locations as changing conditions make some areas unsuitable that formerly provided optimal habitat; in some cases, an area that was once unsuitable may become more appealing.

Because GSL is situated within a terminal basin with no natural outlet to help maintain a steady depth, its surface size and water depth have always been subject to the variability of evaporation and precipitation. The resulting advancing and retreating of the shoreline is amplified by the extreme shallowness of the lake and the flatness of much of the surrounding topography: a drop of a third of a meter in lake level can expose thousands of hectares of lake bottom, and a rise of a third of a meter can inundate an equal area of shore (Arnow 1984; Arnow and Stephens 1990).

Wetlands adjacent to GSL are dynamic, largely influenced by climatic fluctuations and the resulting rise and fall of GSL. Since 1963, water levels of GSL have varied about 6 m. At a low elevation of 1277.5 m in 1963, large portions of the lake bed were exposed and the adjacent wetlands largely dry except for some impounded wetlands and ephemeral ponds. At its most recent high point in 1986, when GSL water levels reached 1283.7 m, many of its wetlands were submerged with hypersaline GSL water, and only the highest landforms escaped inundation. The historic and geologic evidence strongly suggests that such fluctuations have been common in the GSL ecosystem. In fact, much of its character is probably a direct result of seasonal, annual, and long-term water level variations (Kadlec 1995). Recent studies suggest that the current trajectory of lake elevation is downward (Wurtsbaugh et al. 2017), raising concern about gradually losing a terminal lake with its natural ebb and tide that creates abundant shorebird habitat.

Salt is a fundamental driver in the diversity of wetland types on GSL, because it impacts vegetation, and relatively few species of plants can tolerate salt (Flowers 1934; Cronquist et al. 1972; Kienast-Brown 2003). When soil salt concentration is elevated enough, all vascular vegetative growth is inhibited. However, the shorelines of GSL feature plants that have specific mechanisms to grow in hypersaline soils (Bradbury and Parrott 2020). Salts concentrate in surface soils through several geological processes. Water from rivers flowing into GSL contains dissolved ions from weathering of rocks in the watershed. When the water evaporates, the dissolved ions remain; thus, the salinity of the lake increases through time. Some ions combine chemically with each other and precipitate out as minerals that accumulate as

Fig. 9.1 Great Salt Lake and its associated wetlands

sediments (such as calcium, Ca, and bicarbonate, HCO_3, combining as aragonite or calcite, $CaCO_3$). Other ions stay in solution (such as sodium, Na, and chloride, Cl) until the lake water becomes super concentrated in those ions (Gwynn 1998; Rupke and McDonald 2012). Highly saline waters of GSL periodically rise during flooding events, inundating the adjacent wetlands (Fig. 9.1). When the hypersaline floodwaters recede, such as after 1987, evaporative forces concentrate salt in the surface soils (Kienast-Brown 2003). As the soil dries when the water retreats, dissolved salt (from high water tables) wicks to the surface from subsurface sediment via capillary and evaporative processes, maintaining high salt levels in "discharge playas" such as those at GSL (Rosen 1994).

Movement of water and salt forms the basic dynamic that drives plant community shifts in distribution and abundance of wetlands adjacent to GSL. Once flood waters recede, the denuded landscape lies exposed with highly saline soil sometimes with scattered stubble from dead vegetation. Sheet flow of freshwater coming from precipitation, runoff, rivers, streams, drainage canals, springs, and groundwater gradually leaches salt from the surface soils, allowing vegetation to eventually regrow.

At any given time, vegetation or lack thereof occurs in distinct zones. The establishment of these zones in GSL wetlands is complex and dependent on many factors such as timing, duration, depth, and salinity of water. Nevertheless, a predicable dynamic follows inundation of GSL's associated wetlands that distribute vegetation types into zones as salts are leached and concentrated at lower elevations. The following describes four zones important to shorebirds as each species selects their preferred location for foraging. Each zone is described as a function of salt and water that follows the killing of vegetation by inundation of wetlands by GSL water. Each zone is characterized by the dominant vegetation. The species accounts describe preferred habitat niches for each species.

Unvegetated Mudflat When GSL water recedes, a vast, highly saline mudflat is exposed sometimes with stubble from dead plants. During dry periods, groundwater, which contains dissolved ions, rises up through the sediments and soils by capillary action. When the groundwater reaches the surface or is near the surface, it evaporates and leaves salts on the surface of the mudflat (Rosen 1994; Charles G. "Jack" Oviatt, expert on Pleistocene lake beds, retired, Kansas State University, in litt.). Because of the high soil salinity, these areas are devoid of vegetation.

Pickleweed As salts leach from the soil, the first pioneering vascular plant is pickleweed *Salicornia rubra,* the most salt-tolerant halophyte on GSL. In this zone, soils are typically moist to the surface but not saturated or ponded as often as in the unvegetated mudflat zone. The lower soil moisture and salinity allow pickleweed to grow. In some areas, especially the west side of GSL, iodine bush *Allenrolfea occidentalis* is the pioneering halophyte. Seepweed *Suaeda calceoliformis*, a halophyte slightly less salt tolerant than pickleweed, sometimes occurs in this zone.

Saltgrass Further leaching of soil salts or the infusion of less saline groundwater allows saltgrass to grow. The soil is often ponded and saturated.

Willet. Photo Credit, Matthew Pendleton

Sedges and Alkali Bulrush Sedges (e.g., Nebraska sedge *Carex nebrascensis*, clustered field sedge *C. praegracilis*, alkali bulrush, common spikerush *Eleocharis palustris*) and other plants such as Baltic rush *Juncus arcticus* grow in areas where soil is saturated most of the spring and frequently ponded.

Fluctuations of salt and water cause ecologically important shorebird habitats to continually change in distribution within GSL wetlands, thus having critical importance to where each species of shorebird occurs. Much anecdotal information of the dramatic shifting in vegetative distribution that occurs through time is well known to scientists, land managers, and others who have spent a prolonged time observing the interactions of salt, water, and vegetative communities as the levels of GSL waters fluctuate and how these processes affect locations of shorebird species and their abundance. However, there is relatively little documentation of these interactive processes of GSL in the literature. One of the best documented examples comes from the research on Snowy Plovers (Paton 1994) conducted at two focal sites about 10 km apart in Howard Slough Waterfowl Management Area and West Layton Marsh (part of GSL Shorelands Preserve), both on the eastern central shore of the lake, during the summers of 1990–1993. During this period, the lake was receding from recent flood years and vegetation was beginning to recover along the newly exposed shoreline.

For several decades prior to being flooded in 1980s, the focal sites were a vast expanse of hardstem and alkali bulrush with large colonies of nesting White-faced Ibis, Franklin's Gulls, and other marsh birds (Paul pers comm). Two years prior to the beginning of Paton's study, his focal sites were under water. Optimal Snowy Plover habitat consists of open mudflat areas with about 25% vegetation cover (Paton 1994). The moist salty soil and dead stubble of the emergent vegetation at the focal sites were abundant with brine flies, a primary food source for Snowy

Plovers (John Kadlec, former Chair, Fisheries and Wildlife, Utah State University, pers comm). During Paton's 4-year study, the sites became:

> vegetated with salt tolerant species within the Chenopodiaceae, including . . .iodine bush *Allenrolfea occidentalis*, seepweed *Suaeda* spp., summer cypress *Kochia scoparia*, bassia *Bassia hyssopifolia*, and pickleweed *Salicornia europaea (rubra)*. In addition, marsh vegetation increased dramatically over the course of the study, and consisted primarily of alkali bulrush *Scirpus (Bolboschoenus) maritmus*, cattail *Typha spp.*, and Phragmites *Phragmites australis*.

The distribution and availability of potential nesting habitat for Snowy Plovers at the sites changed substantially during the 4-year study period primarily due to the reestablishment of vegetation. At the Howard Slough study site, the amount of potential nesting habitat for Snowy Plovers declined 56% and at West Layton Marsh by 74.7% (Paton 1994). Aldrich and Paul (2002) describe this example in greater detail.

Today, the same area is dominated by dense stands of *Phragmites* (Justin Dolling, Rich Hansen, wetland managers, Utah Divsion of Wildlife Resourses, pers comm). Comparing the composition of shorebird species and abundance from 1990 to1993 (Peter Paton Waterbirds Using Great Salt Lake: 1990–1993 Raw Data, unpublished) to that of observations of the authors in 2019 powerfully illustrates the effect of vegetation on the location of shorebird species.

9.2.1.3 Macroinvertebrates

A third parameter that influences the niche each species of shorebird prefers is the availability of food sources. Shorebirds congregate in wetlands of GSL and its open waters to forage on macroinvertebrates and replenish their energy reserves. Macroinvertebrates are highly nutritious, providing easily digestible protein, fatty acids, carbohydrates, energy, and a source of freshwater in highly saline or dry environments for some species. The prevalence and abundance of brine flies and brine shrimp in the open waters of GSL and its edges as a predominant food source for avian species has received much focus in GSL discussions and the literature (Aldrich and Paul 2002; Roberts 2013; Conover and Bell 2020). However, the lake's wetlands feature a large diverse group of other macroinvertebrates that provides a reliable prey base for most shorebird species that forage at GSL.

Least sandpipers. Photo Credit, Cameron Cox

Macroinvertebrates act as housekeepers for all trophic level transfers in the wetland environment's food web. Many taxa are detritivores that break down decaying plant and other organic material into various nutrient components. Herbivore, gatherer, filterer, omnivore, and scavenger taxa provide additional critical ecosystem services via multiple pathways. Others scrape periphyton from surfaces of plants and other substrates and, of course, some are highly efficient predators. Macroinvertebrates transfer this accumulated energy and nutrition to shorebirds.

Recent studies have greatly expanded the current understanding of diversity of macroinvertebrate taxa and their tolerance to varying limiting factors found across wetland types of GSL (Vinson and Bushman 2005; Gray 2011; Richards 2014). But much remains to be learned.

Trophic structure within the macroinvertebrate community is an important component in dictating the abundance and diversity, particularly in saline shorebird habitats (Cantrell and McLachlan 1977; Herbst 2006; Andrei et al. 2009). Some taxa (e.g., those in the Chironomidae and Corixidae families) are known to be tolerant of a range of salinity and other conditions (Merritt and Cummins 1978; Murkin and Kadlec 1986; Herbst 1999; Andrei et al. 2009; Gray 2011; Richards 2014), which bodes well for many shorebirds of GSL as chironomids and corixids are often highly preferred in their diet (Wetmore 1925; Wilson 1973; Cavitt 2006a). Although species in the orders Ephemeroptera (e.g., mayflies), Odonata (e.g., dragonflies and damselflies), Hemiptera (e.g., corixid water boatmen and backswimmers), Coleoptera (e.g., carabid and predaceous diving beetles), Diptera (e.g., chironomid midges, deerflies, mosquitos), Amphipoda (e.g., amphipods or scuds), Gastropoda (e.g., snails), Oligochaeta (e.g., worms), and others are common, presence or absence of water, water depth, predation, and salinity tolerance are important factors controlling availability of aquatic macroinvertebrates to shorebirds at GSL.

9.2.2 A Discussion on Mudflats and Playas

Descriptions of shorebird habitat on GSL typically include the terms mudflat and playa. Mudflats whether dry or shallowly flooded are uncontestably the most important habitat for the majority of shorebirds occurring on GSL. Playas, distinct landforms often associated with the fringes of GSL, also play a critical role in providing shorebird habitat.

What characteristics define a mudflat or a playa? This question has long been mired in confusion. These terms are used in many different ways, often interchangeably, perhaps the result of overlapping definitions. Also, the terms are used across disciplines, for example biology and geology, and have evolved different usage patterns. The confusion of mudflats and playas has had an obscuring effect on the important and distinct role playas play as shorebird habitat on GSL, especially in spring migration and early summer nesting. In fact, GSL playas are disappearing at an escalating rate (authors' unpublished observations). Burgeoning human population growth, especially along the Wasatch Front east of GSL, has driven development pressures west toward the lake and its associated wetlands. Since these landforms provide a largely unrecognized, but distinct and critical function for shorebirds, and since they require clarification, we will define them herein and discuss their importance.

To better understand the current interpretations and perceptions of those with expertise and field experience on GSL wetlands, written input was solicited from soil scientists, hydrologists, land managers, academics, geologists, ecologists, researchers, and biologists. The responses confirmed the disparate interpretations and lack of agreement as to what constitutes a playa or a mudflat. The discussion below reflects these responses and also includes information from interviews, literature review, and personal experience of the authors.

Playa Playas are landforms. They are geomorphic features on the landscape.

Rosen (1994) reviewed the classification, sedimentology, and hydrology of playas. According to his definition, for a depression to be called a playa, three criteria must be met:

1. The regional shallow groundwater system does not directly discharge into the ocean.
2. The water balance (all sources of precipitation, surface water flow, and groundwater flow minus evaporation and evapotranspiration) must be negative for more than half the year.
3. The capillary fringe is close enough to the surface that evaporation will cause water to discharge to the surface.

Rosen defines a recharge playa the same as a discharge playa described above, except the recharge playa surface acts as a means for recharging water to the aquifer.

> In a discharge playa the water table is near the surface (on the order of a meter down), the playa mud is moist near the surface because of capillary rise from the water table, evaporite minerals (such as gypsum, halite, calcium carbonate, etc.) precipitate in the mud, and groundwater generally flows upward toward the playa surface. In contrast, in a recharge

> playa the water table is far below the surface (many meters), evaporite minerals are minor or non-existent in the dry mud, and groundwater generally flows downward away from the playa surface. Ephemeral water may occasionally be present on the surface of a recharge playa from atmospheric precipitation—some of that water evaporates and some soaks into the mud, but most of the time the surface of a recharge playa is dry. Most or all GSL playas are discharge playas. Deflation (that is, wind erosion) is common on both discharge and recharge playas (Oviatt in litt.).

One of the most important and unique landscape features providing foraging and nesting sites for many species of shorebirds mainly in spring and early summer are myriad low-lying depressions, or playas, scattered near the fringes of GSL. They are dry much of the year but when inundated in spring often receive considerable shorebird use. They draw down through evaporation to dryness through summer.

We have slightly modified Rosen's (1994) definition of a playa from a large-scale basin to a smaller scale. On GSL, the following criteria define a playa:

1. Shallow low-lying depression sometimes found some distance from the shore; can be surrounded by uplands.
2. Receives most water from precipitation, runoff, and groundwater.
3. Water is seasonal, temporary, ephemeral.

Inundation of playas generally follows the natural filling in spring and drawing down through evaporation to dryness by early summer. But there is great hydrological variation from year to year and on longer timescales with some playas remaining dry for years. Timing, depth, and duration of playa ponding are influenced by wet and dry cycles. This variability makes defining a playa by the duration of time flooded annually problematic. We consider Natural Resources Conservation Service's (NRCS) Web Soil Survey data describing the "Playa" component with frequent (>50% chance) springtime ponding and a seasonal high water table that fluxes within the soil profile to be a good generalized guide (from Randy Lewis, Soil Survey Major Land Resource Areas (MLRA) Office Project Leader at US Department of Agriculture (USDA)-NRCS, in litt.).

4. When dry, often have a crust of salt/alkali, indicative of a discharging system.

The importance of a groundwater connection is evident from the formation of evaporite minerals in some portion of the depression when dry.

> Around GSL and in the entire GSL Desert, the mud at and near the surface is moist at all times of the year. . ., even when it hasn't rained in a long time. When it's really hot in the summer, the mud at the surface (1 or 2 cm) is dry, but there's moist mud right below the surface crust. The only source of water to keep the mud moist is groundwater. The permeability of the mud is very low because it's so fine grained, but the permeability is not zero—water moves through it very slowly. The uppermost meter of mud is unsaturated, but water rises slowly through it (by capillary rise) to the surface where it evaporates, leaving salts—sodium chloride, calcium carbonate, calcium sulfate, etc. (Oviatt in litt.).

This point has been further validated by observations of Lewis (in litt.):

> Our soil sampling almost always has a water table present within the soil profile with the salts wicking through the capillary fringe to the surface leaving very high levels of salts and

alkali. I have seen water tables within the upper part of the soil profile even in drier cycles in the middle of the summer (Lewis in litt.).

From an aerial view, playas stand out in the landscape as mostly unvegetated, variably shaped white features. With a knowledgeable perspective of ancient hydrology of GSL, it becomes apparent that many of these playas were formed by hydrogeomorphic processes.

Mudflat A mudflat consists of a mixture of water and fine-grained soil, usually clay and silt occasionally mixed with sand. A mudflat is not a depression that pools water.

The mudflats around GSL are composed of moist GSL deposits/sediments (Oviatt in litt.). Mudflats are found in areas of fine-grained material that is alternately covered and uncovered as water inundates and recedes. Mudflats can be found widespread adjacent to the waterline of GSL. When GSL water levels are elevated, bare mudflat areas are restricted to a small fringe. When water levels are low, large expanses of lakebed mudflats are exposed. Often, mudflats of GSL appear dry but the water table is near the surface with salt and water wicking up through the capillary fringe, often depositing a crust/layer of mineral salts on the surface of the mudflat as the groundwater evaporates away.

Mudflats used by shorebirds are not restricted to the shoreline of GSL. They also occur along the shorelines of naturally occurring ponds, incoming waterways, or impoundments when hydrologic regimes subject to human strategies allow water shallow enough to expose bottom sediments. Mudflats are also found in playas during periods of inundation as water draws down by evaporation.

Mudflats and playas both are formed from the reworking of GSL sediments and deposits. The water of GSL is highly saline (sodium chloride, NaCl, being the dominant salt), and incoming water from rivers and streams has a high carbonate component. Deposition of carbonates in the sediments forming today's playas and mudflats has been ongoing for thousands of years and continues today (Oviatt 2014).

Approximately one-third of GSL sediments are small rod-shaped "ooids," which are brine shrimp fecal pellets that are thought to be recrystallization of ingested fine precipitated calcium and magnesium carbonate particles and occluded fine white clay interspersed with minute mineral fragments formed in the gut of brine shrimp; some are also coated with thin, concentric layers of calcium carbonate after they have been deposited (Eardley 1938). Eardley purports that since brine shrimp are filter feeders, they indiscriminately consume large volumes of inorganic materials from the water column as they search for microscopic algae. The carbonate coating, which is dominated by the mineral aragonite, consists of calcium and magnesium carbonate (77%), occluded fine white clay, and very small mineral fragments. The fecal pellet grains are one of the most important parts of the sediment of GSL (Eardley 1938; Piazza et al. 2016; Thompson et al. 2016).

In reality, playas and mudflats are part of one system, connected by hydrogeomorphological processes. When GSL most recently matched its historic high in 1986, most of the areas defined as playas in this chapter were inundated with GSL water and the overflowing of the tributary rivers and streams (Sorensen personal obs and unpublished survey data 1986).

After GSL waters receded following flooding that occurred during the 1980s, playas on the eastern and southern shore lost connectivity to water of GSL, once again functioning as depressions with ephemerally pooling water.

9.2.3 *Impounded Wetlands*

In addition to naturally occurring wetlands, artificially impounded wetlands extend over large areas of the eastern shores of GSL and provide significant shorebird habitat. Since the late 1800s, humans have altered the natural hydrology with dikes, canals, ditches, and water control structures to impound freshwater and create habitat for waterfowl, shorebirds, and other wetland species. Extensive areas of impounded wetlands exist on GSL including Utah Division of Wildlife Resources' Waterfowl Management Areas, a federal bird refuge, privately managed duck clubs, mitigation sites, and sites managed by nonprofit organizations. Management goals and objectives vary among landowners. Most impounded wetlands are managed for waterfowl, many having steep sides and water levels maintained too deep for shorebirds. Some impounded wetlands have more gradually sloping edges. Whenever water management strategies include areas of shallow flooding within the impoundment and water levels are about 18 cm or less, these areas are often extremely productive for benthic macroinvertebrates and receive considerable shorebird use, because the macroinvertebrates are accessible to the birds. Some impounded areas are managed specifically for shorebirds and follow a spring flood cycle with drawdown through evaporation.

When levels of GSL are elevated as they were in the 1980s, most impounded wetlands were inundated and thus influenced by the same hydrogeomorphological processes as described previously.

There is a diversity of shorebird habitat among the shallowly flooded, impounded wetlands around GSL that provide a range of salinity, water depth, vegetative cover, and an abundance of macroinvertebrates.

9.3 Great Salt Lake Shorebird Checklist

A total of 42 species of shorebirds have been recorded on GSL. For some species such as American Avocet, Snowy Plover, Marbled Godwit, and Wilson's Phalarope, a large percentage of the world's population occurs annually on the lake. Other species like Sharp-tailed Sandpiper, a vagrant from the Eastern Hemisphere, have but a single accepted record. Decisions by the Utah Ornithological Society (UOS) Records Committee were followed for those species with fewer than ten accepted records (UOS 2011).

The 42 species are listed in Table 9.3, which presents abundance for nesting, migration, and wintering. This is based on literature review, unpublished surveys,

Table 9.3 Status, abundance, nesting, and wintering range at Great Salt Lake

	Status and abundance			Western Hemisphere range	
Shorebird species	Nesting	Migration	Winter	Nesting	Winter
Black-necked Stilt	C	C		P/I	NA,SA[a]
American Avocet	C	C	R	P/I	NA
Black-bellied Plover		C		A/S[c]	NA,SA[c]
American Golden-Plover		R		A/S	SA
Pacific Golden-Plover		O		A/S[c]	[c]
Killdeer	C	C	U	P/I,A/S	NA,SA[a]
Semipalmated Plover		U		A/S	NA,SA
Mountain Plover		O		P/I	NA
Snowy Plover	C	C		P/I	NA,SA[a]
Upland Sandpiper		O		P/I,A/S	SA
Whimbrel		R		A/S[c]	NA,SA[c]
Long-billed Curlew	C	C		P/I	NA
Bar-tailed Godwit		O		A/S[c]	[c]
Hudsonian Godwit		O		A/S	SA
Marbled Godwit		C	O	P/I	NA,SA
Ruddy Turnstone		R		A/S[c]	NA,SA[c]
Red Knot		R		A/S[c]	NA,SA[c]
Ruff		O		[c]	[c]
Sharp-tailed Sandpiper		O		[c]	[c]
Stilt Sandpiper		R		A/S	NA,SA
Curlew Sandpiper		O		[c]	[c]
Sanderling		C	O	A/S[c]	NA,SA[c]
Dunlin		R	O	A/S[c]	NA[c]
Baird's Sandpiper		C		A/S	SA
Least Sandpiper		C	R	A/S	NA,SA
White-rumped Sandpiper		O		A/S	SA
Buff-breasted Sandpiper		O		A/S	SA
Pectoral Sandpiper		U		A/S[c]	SA[c]
Semipalmated Sandpiper		R		A/S	NA,SA
Western Sandpiper		C	O	A/S	NA,SA
Short-billed Dowitcher		R		A/S	NA,SA
Long-billed Dowitcher		C	O	A/S[c]	NA
Wilson's Snipe	C	C	U	P/I,A/S	NA,SA[b]
Spotted Sandpiper	U	C		P/I,A/S	NA,SA
Solitary Sandpiper		U		A/S	NA,SA
Wandering Tattler		O		A/S	NA,SA[c]
Lesser Yellowlegs		C	O	A/S	NA,SA
Willet	C	C		P/I	NA,SA
Greater Yellowlegs		C	R	A/S	NA,SA

(continued)

Table 9.3 (continued)

Shorebird species	Status and abundance			Western Hemisphere range	
	Nesting	Migration	Winter	Nesting	Winter
Wilson's Phalarope	U	C		P/I	SA
Red-necked Phalarope		C		A/S[c]	OO[c]
Red Phalarope		R		A/S[c]	OO[c]

Abbreviations and definitions. Abundance: Common (C), occurs consistently in large numbers in appropriate habitat and season; Uncommon (U), occurs consistently in small numbers in appropriate habitat and season; Rare (R), occurs infrequently but annually in very small numbers in appropriate habitat and season; Occasional (O), does not occur annually and in very small numbers. Status: Nesting, nests documented on GSL; Migration, passes through GSL in spring and/or fall; Winter, present at GSL in January and/or February. Range: Arctic/Subarctic (A/S); Prairie/Intermountain (P/I); North America, including Central America (NA); South America (SA; [a]resident population; [b]only in extreme north); Open Ocean (OO)

Note: Winter ranges reflect the location of the vast majority of the population in winter

[c]Also occurs in Eastern Hemisphere

and the authors' personal observations. See Sect. 9.3.2. It also distinguishes long-distance migrants that nest in the Arctic and Subarctic from those nesting in the Prairie region and Intermountain West that typically travel shorter distances. Some species of shorebirds nest outside of these regions as detailed in the species accounts. Winter range is given as North America (includes Central America), South America, and open ocean in the case of Red-necked and Red phalaropes that do not winter on land. Many GSL shorebird species occur only in the Western Hemisphere, but those that also breed and/or winter in the Eastern Hemisphere are indicated although these are generally distinct populations from those occurring on GSL. Three vagrant species of shorebirds recorded on GSL nest only in the Eastern Hemisphere. A species summary from those listed in Table 9.3, sorted by their use of GSL, will be presented below in Sect. 9.3.1.

9.3.1 Species Summary

Nesting Nine species of shorebirds breed annually in GSL wetlands (Table 9.4). All nest widespread and are common in their appropriate habitat except for the Wilson's Phalarope and Spotted Sandpiper, which are uncommon nesters. Nesting and foraging habitat can be the same such as Snowy Plover, which nests and forages on mudflats, and Wilson's Snipe, which nests and typically forages in dense vegetation. Others, like American Avocet, nest on mudflats and islands and forage usually in various depths of shallow water. Specific details of nesting habitat for each species are given in the species accounts.

Migrant (or Passage) Twenty-two species of shorebirds occur annually at GSL in spring and fall only as migrants as they fly between breeding and wintering areas (Table 9.5). They utilize diverse foraging habitats from moist mudflats to open saline water of GSL. Shorebirds such as the two species of yellowlegs are distributed over

Table 9.4 Abundance of nesting species at Great Salt Lake

Nesting species (9)	Abundance
Black-necked Stilt	Common
American Avocet	Common
Killdeer	Common
Snowy Plover	Common
Long-billed Curlew	Common
Wilson's Snipe	Common
Spotted Sandpiper	Uncommon
Willet	Common
Wilson's Phalarope	Uncommon

Table 9.5 Abundance of annual migrants at Great Salt Lake

Annual migrant species (22)	Abundance
Black-bellied Plover	Common
American Golden-Plover	Rare
Semipalmated Plover	Uncommon
Whimbrel	Rare
Marbled Godwit	Common
Ruddy Turnstone	Rare
Red Knot	Rare
Stilt Sandpiper	Rare
Sanderling	Common
Dunlin	Rare
Baird's Sandpiper	Common
Least Sandpiper	Common
Pectoral Sandpiper	Uncommon
Semipalmated Sandpiper	Rare
Western Sandpiper	Common
Short-billed Dowitcher	Rare
Long-billed Dowitcher	Common
Solitary Sandpiper	Uncommon
Lesser Yellowlegs	Common
Greater Yellowlegs	Common
Red-necked Phalarope	Common
Red Phalarope	Rare

widespread locations throughout the entire GSL wetland complex, while others such as Black-bellied Plover and Sanderling tend to be more localized. Population numbers vary from Red-necked Phalarope, which occur in the hundreds of thousands, to species such as Dunlin, Ruddy Turnstone, and Stilt Sandpiper, which occur regularly but in very small numbers. In addition to migrant species that do not nest in GSL wetlands, the populations of some local breeding species swell as birds breeding further to the north stage at GSL to fuel up and add fat to sustain a long migration. A notable example is Wilson's Phalarope. The annual numbers from a small breeding population increase to the hundreds of thousands as they are joined by fall migrants staging on GSL.

Wilson's phalaropes. Photo Credit, Cameron Cox

Occasional Eleven species of shorebirds do not occur annually and have fewer than ten records accepted by the UOS Records Committee (Table 9.6). Three of the species do not nest in North America and are often classified as accidental or vagrant.

Table 9.6 Occasional species at Great Salt Lake

Occasional species (11)
Pacific Golden-Plover
Mountain Plover
Upland Sandpiper
Bar-tailed Godwit
Hudsonian Godwit
Ruff[a]
Sharp-tailed Sandpiper[a]
Curlew Sandpiper[a]
White-rumped Sandpiper
Buff-breasted Sandpiper
Wandering Tattler

[a]Nests only in Eastern Hemisphere

9.3.2 Species Accounts

The following species accounts, listed in taxonomic order, describe breeding and wintering areas based on ranges given in *The Birds of North America* (2019), Hayman (1986), and eBird (2019). Status and abundance of the species when present on GSL is given. A general temporal timeline that spans the annual occurrence of each species on GSL was derived from a review of thousands of GSL shorebird observations. While not inclusive of every shorebird ever reported, the sources provided a large collection of data from which to deduce typical timing and numbers for most species. These observations come from multiple sources and have received varying levels of review, or in some cases, no review. Data were provided by individuals with varying degrees of expertise in bird identification and accurate determination of number of birds present. Based on the experience of the authors who have decades of critical shorebird evaluation, outliers and all atypical occurrence or numbers without adequate documentation have been disregarded. The data reviewed include a combination of published studies, unpublished survey reports, and personal observations:

- US Fish and Wildlife Service Bear River Migratory Bird Refuge Survey Data: 1946–1950, 1956–1960, 1991–2009 (unpublished)
- Peter Paton Waterbirds Using Great Salt Lake Raw Data: 1990–1993 (unpublished)
- Utah Division of Wildlife Resources Great Salt Lake Waterbird Survey Five-Year Report: 1997–2001 (Paul and Manning 2002)
- Salt Lake City Airport Authority Wetland Mitigation Bird Survey: 1994–2000 (unpublished)
- Kennecott Inland Sea Shorebird Reserve Avian Data: 1995–2002 (unpublished)
- Audubon Gillmor Sanctuary Bird Surveys: 2009–2014, 2017 to present (unpublished)
- Personal observations John Neill: 2001 to present
- Personal observations Ella Sorensen: 1979 to present
- eBird
- UOS Records Committee (UOS 2011)

Research on shorebird diets at GSL are for the most part lacking (Wetmore 1925; Wilson 1973; Knudsen 1970; Cavitt 2006a, 2006b; Cavitt and Stone 2007; Barber and Cavitt 2013). Dietary information for only those shorebirds that have been studied at GSL has been included in the accounts. For these species, additional data have been included in their species accounts (i.e., Black-necked Stilt, American Avocet, Snowy Plover, Marbled Godwit, Wilson's Phalarope, and Red-necked Phalarope).

9.3.2.1 Black-Necked Stilt *Himantopus mexicanus*, Common Nesting and Passage

Around GSL, Black-necked Stilts prefer to nest near shallow, freshwater wetlands in salt-tolerant vegetation near or within managed wetlands, on dikes, on islands, and in playas near and adjacent to GSL. Their breeding range extends to these preferred habitats often near saline lakes primarily within the western United States and southern Canada, Hawaii, and the Caribbean to southern South America east of the Andes Mountains and outside of the Amazon River basin. Wintering stilts reside in much of the same range as they breed, except breeding populations in the United States and Canada migrate south to areas of California, Arizona, the Texas Gulf Coast, and into Central America (Robinson et al. 1999).

Habitat at GSL is abundant for breeding and migrating stilts. They begin returning to the lake at the end of March or early April. Breeding begins shortly thereafter from late April to mid-July often within or next to managed wetlands and playas on the east side of the lake within areas of pickleweed or saltgrass but also on bare ground (Sordahl 1996).

Stilts are visual feeders and generally pick their food off the surface of the water, mud, or vegetation, but they will also probe for food. In general, macroinvertebrates from the families Corixidae, Chironomidae, and Hydrophilidae are the most abundant food items consumed along with unidentified Coleoptera parts, but stilts are opportunistic and feed on a wide variety of food items from other families such as Ephydridae, Odonata, Syrphidae, Muscidae, and some seeds. Although Chironomidae may be more abundant in the environment, stilts consume more Corixidae, which tend to be more active in the water column rather than living in the mud (Cavitt 2006a).

Nest success and daily survival is high for stilts in areas where there is active predator control, but predation is still the leading cause of nest failure (67–90%). Abandonment and flooding also contributed to nest failure (Cavitt 2006a).

During migration, Black-necked Stilts congregate in wetlands and freshwater outflows into GSL with numbers peaking during the second week of August. As much as 40% of the continental population will stop at the lake to feed on the abundant food resources (Shuford et al. 1994). Winter migrants depart through the rest of August and September with most leaving by October (Paul and Manning 2002).

9.3.2.2 American Avocet *Recurvirostra americana*, Common Nesting and Passage, Rare Winter

The breeding range of American Avocets extends from central Alberta and Saskatchewan, Canada, south along the western edge of the Great Plains to central Texas and west to central Washington and Oregon and to the Pacific Coast in California. There are also permanent populations in central Mexico and the southern

Texas Gulf Coast. Winter range is in the northern interior of Mexico and along both coasts extending north along the Pacific Coast and Central Valley of California and along the Gulf and Atlantic Coasts up to North Carolina (Ackerman et al. 2013).

GSL provides abundant habitat for breeding, migrating, and even for a few, wintering flocks of up to several hundred of avocets. Wintering birds start returning to the lake in late March and over 52,000 stay and breed from early April to early July often within or next to managed wetlands on the east side of the lake or playas. Avocets nest next to small clumps of pickleweed or debris in mostly bare mudflats and playas next to water, though some nest in saltgrass and on bare islands or dikes (Aldrich and Paul 2002; Paul and Manning 2002; Ackerman et al. 2013).

Avocets are indiscriminate feeders, and at GSL, their diet during the breeding season is proportionate to the food resources available near their specific nesting site. In general, macroinvertebrates from the families Chironomidae and Corixidae are the most abundant food items consumed. Seeds are also abundant followed by Ephydridae, Hydrophilidae, and Carabidae. Nest success and daily survival is higher than other similar studies conducted outside of Utah. Active predator control at managed wetlands contributes to higher survival, but predation is still the leading cause of nest failure (67–90%) followed by flooding, and abandonment (Cavitt 2006a).

During migration, American Avocets shift from wetlands out to the shallow, more saline bays of GSL with numbers peaking in mid-August. A 2-day, lake-wide shorebird survey in August 1994 observed over 250,000 avocets, 56% of the continental population (Shuford et al. 1994). Undoubtedly, additional avocets passed through before or after the survey, but new monitoring is needed to understand how important GSL is to migrating avocets. Winter migrants gradually depart through the rest of August and September with most leaving by late October (Ackerman et al. 2013). Satellite telemetry of breeding avocets captured at GSL showed wintering locations along the coast of Sinaloa, Mexico (Avian Ecology Laboratory 2014).

9.3.2.3 Black-Bellied Plover *Pluvialis squatarola*, Common Passage

Black-bellied Plovers nest in the high Arctic of both Eastern and Western hemispheres. It winters farther north than most shorebirds mainly along Atlantic, Pacific, and Gulf coasts south of Canada to South America (Poole et al. 2016). On GSL, it is common but local in spring migration, reported as single individuals up to flocks of thousands. They prefer shallowly flooded mudflats often near GSL shoreline where up to 5000 regularly occur at single locations. They also occur in short vegetation including flooded agricultural fields and pickleweed flats. Most migrate through from mid-April to the last week in May with rare reports in June and July. Fall migration is protracted from August through October, but numbers are much smaller than in spring with flocks seldom exceeding 100 individuals. A few linger into November or December.

Black-bellied plovers. Photo Credit, Cameron Cox

9.3.2.4 American Golden-Plover *Pluvialis dominica*, Rare Passage

This long-distance migrant nests on the Arctic tundra from Baffin Island, Canada, to western Alaska and winters primarily in Argentina and Uruguay. Most migration follows an elliptical migratory pattern with an offshore, nonstop, trans-Atlantic route in fall and a mid-continental flyway in spring but some travel through the Intermountain West (Johnson et al. 2018a). Most GSL spring records occur in late April though the first three weeks of May and consist of one or two birds mixed with Black-bellied Plovers on mudflats or short grassy flooded fields. Fall records are distributed over a longer period. The majority, occurring in late September through October, are likely juveniles, which migrate later than adults. Scattered records in August and early September are likely adults. Most reports are of 1–3 birds, but flocks of up to 14 have been observed in the fall on the Antelope Island Causeway. They forage in shallow water, adjacent mudflats, and sparsely vegetated moist areas.

9.3.2.5 Pacific Golden-Plover *Pluvialis fulva*, Occasional

The breeding range of the Pacific Golden-Plover from western Alaska to the Yamal Peninsula, Russia, lies farther north than the American Golden-Plover with only a small area of overlap. The majority winter across Oceania and Southeast Asia with only small numbers following the Pacific Coast and overwintering primarily in

California. Prior to 1993, American and Pacific Golden-Plovers were considered by the American Ornithological Union to be subspecies of Lesser Golden-Plover (Johnson et al. 2018b). Previous GSL observations did not distinguish between the subspecies. The only accepted record for GSL is of a single bird observed and photographed August 25–September 16, 2006, on the Antelope Island Causeway.

9.3.2.6 Killdeer *Charadrius vociferus*, Common Nesting and Passage, Uncommon Winter

Killdeer are probably the most familiar shorebird. They nest through much of southern Canada south into Mexico including at GSL. A resident population extends along the Pacific Coast of Ecuador and Peru. In winter, northern birds drift south to at least the southern half of the United States with some wintering in Central and South America (Jackson and Jackson 2000). Killdeer are present at GSL year round. Numbers begin swelling in March. Nesting occurs ubiquitously wherever the ground is bare or covered with short vegetation. They show an affinity for roads and dikes, especially if graveled. Killdeer are not usually associated with flocking behavior. They occur in small dispersed numbers during the nesting season; however, in late July through November, groups of up to many hundreds congregate to forage in areas of prey abundance. In migration, Killdeer breeding north of Utah augment the numbers of those that breed on GSL. Occurrence from December through February is low but variable depending on the availability of open water. Foraging habitat includes all bare to shallowly flooded and sparsely vegetated habitats.

9.3.2.7 Semipalmated Plover *Charadrius semipalmatus*, Uncommon Passage

Semipalmated Plover nests in the Arctic and Subarctic from Alaska across northern Canada and winters in coastal areas from southern United States to South America (Nol and Blanken 2014). On GSL, occurrence is regular in both spring and fall migration. Often seen in small numbers of 1–3 individuals, they sometimes occur in loose dispersed flocks rarely exceeding 25 birds. Most pass through the last week of April through mid-May. Few have been reported from June and early July. Fall migration predominantly spans mid-July through mid-September with a few lingering into mid-October. Like other plovers, they are visual hunters requiring open moist to shallowly flooded mudflats to facilitate running to capture prey.

9.3.2.8 Mountain Plover *Charadrius montanus*, Occasional

Mountain Plovers breed on the western Great Plains and Colorado Plateau and winter in California and northern Mexico (Knopf and Wunder 2006). With GSL's location between wintering and breeding ranges, the paucity of records suggests the birds are overflying or taking routes that circumvent the lake. There are five accepted records, two historical and only three in the last 60 years. Records include three on August 26, 1915, near the mouth of the Weber River; one on March 25, 1946, west of Brigham City; one on March 31–April 3, 1992, at Layton Marsh; two on April 10 and 11, 2013, at Antelope Island; and two on December 25–27, 2012, at Harold Crane Waterfowl Management Area. On GSL, they forage along the edges of flooded wetlands.

9.3.2.9 Snowy Plover *Charadrius nivosus*, Common Nesting and Passage

Snowy Plovers prefer saline lakes, playas, beaches, salt flats, evaporation ponds, river bars, and dredge spoils with little to no vegetation across their range during all seasons. Nesting in these localized habitats is mainly concentrated in the western half of the interior United States and extends from southern Canada south to central Mexico. Coastal nesting occurs from Washington State to Baja California Sur, Mexico, along the Gulf Coast and the Caribbean, and from southern Ecuador to central Chile (Page et al. 2009). The largest concentration of breeding Snowy Plovers within its range nests at GSL, Utah, with over 5500 plovers or 21% of the continental population. Together with Salt Plains National Wildlife Refuge in Oklahoma, these two areas account for nearly 42% of the Snowy Plover breeding population (Thomas et al. 2012). Winter range is similar to breeding range along the coasts but more widespread. The interior wintering locations concentrate in southern California, especially the Salton Sea, and central Mexico (Page et al. 2009).

Banded breeding Snowy Plovers at GSL are known to winter on the shores of the Baja Peninsula and the Gulf of California in Mexico, but other returns have come from the northern Pacific Coast, an area designated as a Threatened population under the US Endangered Species Act (Paton 1995; Page et al. 1995; Cavitt pers comm). The majority of the plovers leave GSL by mid-October and the return trip begins mid-March with the breeding season starting the second week of April and incubation finishing around mid-August (Paton 1995; Ellis et al. 2015; Cavitt pers comm). Concentrations of plovers are generally disbursed around the lake during the breeding period, but before and after they do congregate in diffuse flocks numbering in the low hundreds.

Nest success (0.05–0.49) at GSL did not change much between studies by Paton (1995) and Ellis et al. (2015). Both also concluded predation was the most common cause of nest failure over weather-related causes or abandonment. A remote camera study documenting nest failure at GSL and other sites in northern Utah by Ellis et al. (2018) showed gulls and Common Raven depredated the most nests followed by fox, coyote, and small mammals.

9.3.2.10 Upland Sandpiper *Bartramia longicauda*, Occasional

Upland Sandpiper breeds widely in the central and northern plains with small populations farther north and winters in east-central South America. As the name implies, the species is typically found in uplands, only rarely in wetlands (Houston et al. 2011). Two accepted records from GSL include one in typical open grassland habitat near Inland Sea Shorebird Reserve June 4, 2003, and one in the pickleweed flats off of the Antelope Island Causeway on August 28, 2009.

9.3.2.11 Whimbrel *Numenius phaeopus*, Rare Passage

Whimbrels breed in both Eastern and Western hemispheres. Two distinct populations in North America nest primarily in Subarctic tundra and alpine areas, with those in the western population wintering mainly coastally from California to the southern tip of South America (Skeel and Mallory 1996). On GSL, Whimbrels occur in both spring and fall migrations. Reports in spring are more numerous occurring mid-April through early June, with most passing through the first three weeks in May and include single birds or sometimes reaching flocks of up to twenty. Most fall records are of single birds, and they are largely absent from October through mid-April. Whimbrels are usually found foraging along the shallow margins of wetlands or in short grassy upland areas.

9.3.2.12 Long-Billed Curlew *Numenius americanus*, Common Nesting and Passage

Long-billed Curlews nest in the grasslands of the northern Intermountain West and western Great Plains from southern Canada to northern Texas. Wintering birds concentrate along the coasts and nearby inland areas of California, Texas, Louisiana, and northern Mexico. They are less common extending along the coasts to British Columbia, Canada, South Carolina, and Panama. Some curlews winter in the interior of the southwestern United States and northern Mexico (Dugger and Dugger 2002). Wintering curlews begin returning to GSL in small flocks usually fewer than 20 birds during the final week of March, and they initiate nests from the middle of April through the middle of May. Nests are typically found in upland areas within patches of salt-tolerant vegetation no taller than 7 cm in height and are often near barren areas like the playas and mudflats surrounding GSL. Local birds depart the area rather quickly as numbers decrease by the second week of June. Most residents and transients leave by mid-August but some linger into October. Total counts for GSL can peak around 400 individuals (Paton et al. 1992; Paton and Dalton 1994; Paul and Manning 2002).

Long-billed curlew. Photo Credit, Cameron Cox

9.3.2.13 Bar-Tailed Godwit *Limosa lapponica*, Occasional

Bar-tailed Godwit nests from northern Alaska west to Scandinavia. It is suspected that the population nesting in Alaska usually flies nonstop over the ocean to their wintering grounds in Australia and New Zealand (McCaffery and Gill 2001). There is only one accepted GSL record August 5 and 11, 2013, at Willard Spur Waterfowl Management Area. It forages in shallow water. This state first record is a sight record not confirmed by a photograph.

9.3.2.14 Hudsonian Godwit *Limosa haemastica*, Occasional

Hudsonian Godwit nests in scattered breeding locations from western Alaska to Hudson Bay. This long-distance migrant winters in southern South America. It is one of the least common shorebirds in the Western Hemisphere (Walker et al. 2011). There are eight accepted records for GSL, five records in spring occurring between April 20 and May 11, and three records in fall from September 3 to November 6. They forage in shallow water.

9.3.2.15 Marbled Godwit *Limosa fedoa*, Common Passage, Occasional Winter

The vast majority of Marbled Godwits breed mid-continent on the prairies of north-central United States, and south-central Canada. Two small disparate nesting populations also occur in James Bay, Canada, and in Alaska. This medium-distance migrant winters mainly at coastal locations in Mexico and to a lesser degree at coastal locations in the southern United States (Gratto-Trevor 2000).

As part of a Pacific Flyway project initiated by Point Reyes Bird Observatory to survey Intermountain West wetlands for shorebirds, an intensive 2-day ground, airboat, and aerial survey was conducted on GSL on August 10 and 11, 1994. This led to the realization that GSL is the largest interior site for staging Marbled Godwits in western North America. In fall, GSL has 86% of all Intermountain West Marbled Godwits (Shuford et al. 1994, 2002). Bear River Migratory Bird Refuge has long been recognized as an important concentration or stopover area for the species. Surveys with counts exceeding thousands of birds are common during migration with a single day peak count of 33,000 on October 9, 1996 (US Fish and Wildlife Service unpublished census data). Refuge records indicate spring populations peak the third week in April with an average for that week of 8100 godwits. As part of a Master's Thesis, Bridget Olson (2011) fitted 13 marbled godwits at Bear River Migratory Bird Refuge with satellite transmitters and followed their individual movements from 49 to 522 days. They remained an average of 4–5 days in spring before dispersing to breeding sites in Alberta, Saskatchewan, Montana, and North Dakota. After nesting, they returned to GSL with the mean arrival date of July 14. They staged at GSL an average of 38 days with a variable range of 8–72 days. Wintering was along the Baja Peninsula and west coast of mainland Mexico. The closest wintering site for the tagged birds was Salton Sea, 925 km distance from GSL, and the farthest was Sinaloa, Mexico, at 2190 km.

Marbled Godwit occurrence is widely distributed on GSL wetlands with flocks from a few birds to thousands. Birds start arriving from wintering sites in small numbers in late March with numbers swelling before reaching a peak the last two weeks in April. Numbers drop sharply in May with few records until fall migrants begin returning the last week in June. Godwits are present in large numbers, generally July through September with numbers continuing to drop through October and November. Waterbird surveys on GSL detected higher numbers of godwits during fall migration as compared to spring. December through February have few reports. Marbled Godwits forage primarily in open unvegetated flooded mudflats but are occasionally observed foraging in adjacent upland areas of low moist grass.

9.3.2.16 Ruddy Turnstone *Arenaria interpres*, Rare Passage

Ruddy Turnstone nests widespread in tundra regions of the Arctic in both Eastern and Western hemispheres. Wintering range in the Western Hemisphere includes coastal areas from the United States to Tierra del Fuego in South America (Nettleship 2000). On GSL, Ruddy Turnstone is a rare migrant in both spring and fall migrations. Most spring records are for mid-May with birds sometimes lingering in the same area for a week or so. Most spring reports are fewer than five birds but occasionally more. Most fall records span from late July through mid-October and usually include only one or two birds. Foraging habitat includes shallowly flooded mudflats and drier areas adjacent to the shoreline where they probe, jab, and flip small rocks or dried bits of algal mats. They often perch on rocks if present.

9.3.2.17 Red Knot *Calidris canutus*, Rare Passage

Red Knots nest in the high Arctic on both hemispheres. In the Western Hemisphere it winters coastally from the United States to southern South America. The Red Knot is probably best known for its extremely long migrations and for the spring convergence on Delaware Bay to fatten on emerging horseshoe crab eggs (Baker et al. 2013). On GSL, it is present in migration in small numbers with spring reports far outnumbering those in fall. Most spring records occur the last week of April through May in variable flocks usually numbering less than 50 birds. Fall reports span late July through mid-October. Red Knots forage, often with Black-bellied Plovers, in shallowly flooded mudflats and adjacent moist uplands often with low vegetation.

9.3.2.18 Ruff *Calidris pugnax*, Occasional

Ruff, a bird of the Eastern Hemisphere, nests across northern Eurasia and winters widespread in southern and western Europe, Africa, southern Asia, and Australia. It is one of the most common shorebird vagrants to the Western Hemisphere. There are seven accepted records for GSL all in fall and winter including one each for August, October, December, and February and three in September.

9.3.2.19 Sharp-Tailed Sandpiper *Calidris acuminata*, Occasional

Sharp-tailed Sandpiper, a bird of the Eastern Hemisphere, nests across Siberia and winters in Australia and New Zealand. A few of these long-distance migrants reach North America each year, but inland records are few. The only record on GSL is a single juvenile October 13–14, 2007, on the Antelope Island Causeway.

9.3.2.20 Stilt Sandpiper *Calidris himantopus*, Rare Passage

Stilt Sandpiper breeds in low-Arctic and Subarctic areas of North America. Most of these long-distance migrants move through the Great Plains and winter primarily in the interior of central South America (Klima and Jehl 2012). On GSL, it occurs in both spring and fall migration. Most spring records are for the first three weeks in May, and most fall records span late July through October. Stilt Sandpipers occur in flocks, usually fewer than 15 birds. They are typically observed foraging in near belly-deep water or loafing on adjacent mudflats.

9.3.2.21 Curlew Sandpiper *Calidris ferruginea*, Occasional

Curlew Sandpiper, a bird of the Eastern Hemisphere, nests in the Arctic of Siberia and winters mainly in Africa. A few vagrants reach North America every year. There are four accepted records for GSL, three in spring and one in fall. All are of single birds seen on May 18, 1991, at Harold Crane Waterfowl Management Area, May 2–18, 1994, on the Antelope Island Causeway, May 6 and 8, 2001, on the Antelope Island Causeway, and from August 14 to September 12, 2005, also on the Antelope Island Causeway. They forage on shallowly flooded mudflats or adjacent edges.

9.3.2.22 Sanderling *Calidris alba*, Common Passage, Occasional Winter

Sanderling breeding range is circumpolar. In North America, most nest in the Canadian Artic archipelago and winter widespread along coastal areas from the United States throughout South America (Macwhirter et al. 2002). On GSL, Sanderlings occur commonly in spring migration with variable flocks up to 5000 or more regularly reported. They start arriving in early April with numbers increasing to a peak in mid-May. All have departed by June. Fall migration is prolonged from July to November with most passing through in August, September, and October. Fall flocks seldom number more than 100. The period of December through March has few records. Sanderlings on GSL tend to be rather local showing a preference for shorelines of GSL. Large numbers have been reported from the Antelope Island Causeway, Great Salt Lake Shorelands Preserve, and along the south shore. They are seldom reported in freshwater at managed wetlands. Sanderlings forage in shallow water and adjacent mudflats and often loaf on sandy bars, spits, and rocks, if present.

9.3.2.23 Dunlin *Calidris alpina*, Rare Passage, Occasional Winter

Dunlin is a circumpolar nesting species. In North America, they breed from southwestern Alaska north and east to James Bay and winter mostly along the Pacific and Atlantic coasts of the United States and northern Mexico (Warnock and Gill 1996). At GSL, Dunlins pass through in both spring and fall migrations usually singly or in flocks numbering less than 25 birds. Most spring reports occur from mid-April to mid-May. In fall, Dunlins are among the later migrants with most reports in October and November with some lingering into December. There are a few reports from January through March. Dunlins forage near the shoreline in shallow water or on adjacent mudflats.

9.3.2.24 Baird's Sandpiper *Calidris bairdii*, Common Passage

Baird's Sandpiper breeds over a broad expanse of the high Arctic in North America and northeastern Russia, wintering in South America from Ecuador south to Tierra del Fuego. It undertakes one of the longest of bird migrations (Moskoff and Montgomerie 2002). On GSL, it is regular but less often reported in both spring and fall migration than either Western or Least sandpipers. Baird's is especially prone to misidentification. Other peeps are often identified as Baird's, and Baird's is sometimes misidentified as Western or Least sandpipers. This confusion adds to the difficulty of reliably evaluating the species. There are a few March and early April records, but most begin arriving the last week of April and are gone by the last week in May. The majority of fall migrants pass through from mid-July through mid-October with some lingering through November. Most migrate in small groups of less than five birds but variable flocks up to a hundred are regularly reported, rarely several hundred.

Winter records on GSL are problematic (Jehl 1979, pers comm); an authority on Baird's Sandpiper movements was skeptical of North American winter records. In evaluating Baird's winter range, Moskoff and Montgomerie's (2002) species account referenced Jehl's article; however, from January 29 to February 2, 1988 a Baird's Sandpiper was present on the south shore of GSL. A photograph of the bird was confirmed by avian ecologist Richard Veit, who called it a "most unusual record." Another record from December 28, 2012, on the Antelope Island Causeway was also documented by a photograph. Baird's Sandpipers forage on edges of shallowly flooded mudflats often on moist areas some distance from the waterline.

9.3.2.25 Least Sandpiper *Calidris minutilla*, Common Passage, Rare Winter

Least Sandpipers breed in a broad band across the Subarctic tundra and far northern boreal forests of North America. They migrate widespread across the continent to winter in southern United States, Central America, the Caribbean, and northern South America (Nebel and Cooper 2008). On GSL, Least Sandpipers occur in both spring and fall migrations, but numbers reported are considerably less than Western Sandpiper. Spring migrants begin arriving in early April peaking the last week of April through mid-May. They occur in small flocks usually numbering less than 100 birds but have been occasionally reported in the 1000s. They are largely absent from late May through June. Fall migration lasts from July through November with flocks, as in spring migration, only occasionally exceeding more than a hundred birds. Least Sandpiper occurs regularly in small numbers in winter months, December through March. It is the most commonly reported winter peep. They show a preference for shallow flooded mudflats but show a propensity to forage along the edges of more vegetated areas making them easily overlooked and undoubtedly underreported.

9.3.2.26 White-Rumped Sandpiper, *Calidris fuscicollis*, Occasional

White-rumped Sandpipers nest primarily in the Canadian Arctic. This long-distance migrant winters in South America primarily south of the equator and east of the Andes (Parmelee 1992). There is only one accepted record at GSL of two birds on June 1, 1993, approximately two kilometers south of the eastern end of the Antelope Island Causeway. They were flushed from shallow water habitat. The June 1st date of this observation is consistent with the late spring migratory behavior of the species.

9.3.2.27 Buff-Breasted Sandpiper *Calidris subruficollis*, Occasional

Buff-breasted Sandpiper breeds along Arctic coastlines from central Alaska to Devon Island, Canada. This long-distance migrant winters mainly inland and along the coasts in Argentina, Uruguay, and Brazil (McCarty et al. 2017). On GSL, there are three accepted records all for September: one, September 6–12, 1990, at Bear River Migratory Bird Refuge; one, September 27, 1992, at Salt Wells Flats; and two, September 3–10, 1995, on Antelope Island Causeway. GSL wetlands are atypical habitat for Buff-breasted Sandpipers, which are most commonly observed in short grass upland sites during migration.

9.3.2.28 Pectoral Sandpiper *Calidris melanotos*, Uncommon Passage

Pectoral Sandpiper breeds in the Arctic of both North America and Siberia with most wintering in south-central and southern South America (Farmer et al. 2013). On GSL, Pectoral Sandpiper is extremely rare in spring, a time when most pass through central North America. In fall, there is a more widespread dispersal with far more records on GSL. A few migrants start arriving in late July increasing through August, peaking in September and October before dwindling in November. They occur singly or in small flocks rarely exceeding more than 50 birds. Its habitat preference is on or near shoreline edges with shallow water or adjacent moist mudflats where it forages amidst vegetation or on algal mats.

9.3.2.29 Semipalmated Sandpiper *Calidris pusilla*, Rare Passage

Semipalmated Sandpipers nest in the Arctic. It is a long-distance migrant that winters along the northern and central coasts of South America. Most migrate east of the Rocky Mountains (Hicklin and Gratto-Trevor 2010). On GSL, it is reported in both spring and fall in small numbers usually fewer than 10 birds. They often occur mixed with Western and Least Sandpipers. Most spring reports are from late April through mid-May. Most fall records span late July through September with a few records in

October and November. Semipalmated and Western sandpipers are similar and require experience to accurately identify. Western Sandpipers occur in flocks often in the thousands and Semipalmated Sandpipers are probably being overlooked and underreported. Semipalmated Sandpipers forage in shallow water and adjacent moist mudflats.

9.3.2.30 Western Sandpiper *Calidris mauri*, Common Passage, Occasional Winter

Western Sandpiper is one of the most common shorebirds in the Western Hemisphere. It breeds in western Alaska and eastern Siberia. Most of these long-distance migrants follow the Pacific Coast, but significant numbers move through the interior including GSL. They winter mainly along the coasts from the United States into northern South America (Franks et al. 2014). Migrants start arriving mid-April, peaking the last week in April through the first week in May. Records are few from late May to late June when southern bound migrants begin arriving. Most pass through in July through September with dwindling numbers through October to November. Fall migration is more prolonged and numbers greater than during spring. Records for December through mid-April are few. Western Sandpipers forage on moist to shallowly flooded mudflats. Of the common GSL peeps, they prefer deeper water, less vegetated habitats.

9.3.2.31 Short-Billed Dowitcher *Limnodromus griseus*, Rare Passage

Short-billed Dowitcher breeds further south than Long-billed Dowitcher from coast to coast across boreal and Subarctic regions of Canada and Alaska with three disparate nesting populations. It winters coastally from the United States south to northern South America (Jehl et al. 2001). On GSL, it is regularly reported in small numbers usually 1–3 birds in both spring and fall migrations with most reports from fall. Accurately evaluating the status of Short-billed Dowitcher is problematic because of the great similarity to Long-billed, which occurs in far greater numbers. Behle (1985) accepted only a single valid specimen of Short-billed for Utah taken at the mouth of Bear River May 20, 1915. Behle lists several specimens in collections at the University of Utah and Brigham Young University, which were identified originally as Short-billed Dowitchers but later determined to be misidentified Long-billed. The calls of the two dowitcher species are distinctive and help confirm their identification, especially for birds in basic plumage. Because of the difficulty in separating the two species, it is likely that Short-billed occur more often on GSL than the sparse reports would indicate. Short-billed show a preference for saltwater habitats in contrast to the Long-billed that prefer freshwater wetlands. They probe with a sewing-machine motion in the mud in open water sometimes belly deep.

9.3.2.32 Long-Billed Dowitcher *Limnodromus scolopaceus*, Common Passage, Occasional Winter

Long-billed Dowitcher's breeding range extends from northwestern Canada into northeastern Russia. It winters primarily along the Pacific and Gulf coasts into Mexico (Takekawa and Warnock 2000). On GSL, it is one of the more common shorebird species occurring in both spring and fall migration. Spring migrant numbers start increasing mid-March peaking the last week in April though the first three weeks of May. Numbers for late May through early July are sparse. Fall migration is prolonged from mid-July through early November when numbers decline rapidly, with only a few records from December through February. Long-billed Dowitchers occur in flocks of variable size from a few individuals to many thousands. The large numbers that pass though GSL wetlands in dispersed areas can be illustrated by a few examples of single day, single site counts: 33,000 on October 9, 1996, at Bear River Migratory Bird Refuge; 10,000 on April 30, 1992, at Harold Crane Waterfowl Management Area; 11,150 on August 26, 1992, at Howard Slough Waterfowl Management Area; and 4500 on October 26, 1994, at the wetland mitigation site on the south shore for the Salt Lake City International Airport. Long-billed Dowitchers prefer to forage in fresher habitats than Short-billed. They forage in flooded mudflats without vegetation using a sewing-machine motion often submerging their head in water as they probe the substrate for prey.

Long-billed Dowitcher. Photo Credit, Mike Schijf

9.3.2.33 Wilson's Snipe *Gallinago delicata*, Common Nesting and Passage, Uncommon Winter

Wilson's Snipe breeds widespread in Canada and northern United States with a southern extension into interior parts of the West. It winters primarily from southern United States through Central America to the northern-most part of South America (Mueller 1999). GSL lies on the southern boundary of the breeding range and near the northern edge of the winter range. Birds occurring on GSL during the appropriate season can be local breeders, wintering birds from farther north, or birds passing through in spring and fall migration.

Snipes are elusive, usually observed when flushed from heavy vegetation. They tend to be solitary, and most reports are of one or two birds although congregations up to several dozen have been reported. Snipes occur at GSL all year round, but numbers are low when wetlands freeze, especially in January and February. Increasing numbers starting in late March are likely passage birds returning from southern wintering grounds. Snipes nest commonly in dense vegetation. Numbers decrease through fall into December as most local breeders and passage birds depart to more southern wintering grounds. Preferred foraging habitat is near cover, concealed in vegetation but occasionally in the open or in shallow water at the edges of vegetation.

9.3.2.34 Spotted Sandpiper *Actitis macularius*, Uncommon Nesting and Common Passage

Spotted Sandpiper is the most widespread-breeding sandpiper in North America. It nests across most of Canada and northern United States. The main wintering range extends from southern United States to central South America (Reed et al. 2013). In Utah, it is also the most widespread breeding shorebird, nesting near water of lakes and streams from timberline to the low valleys. On GSL, Spotted Sandpipers begin arriving the last week in April. They usually occur singly or in groups of fewer than three during spring migration. A small number of Spotted Sandpipers will nest on the wetlands of GSL and most June records are of one or two birds. Numbers start increasing in July as migrating birds arrive with most passing though in August and early September. These sandpipers will often congregate during fall migration at areas with good foraging opportunities in variable flocks sometimes reported up to 50 birds. Numbers decrease from late September to November. They are absent in winter months of December through March. Spotted Sandpipers show a strong preference for freshwater wetlands where they teeter and forage along the edges, often where the water is flowing.

9.3.2.35 Solitary Sandpiper *Tringa solitaria*, Uncommon Passage

Solitary Sandpiper nests in the boreal forests of Canada and Alaska and winters from southern Texas south through much of South America (Moskoff 2011). On GSL, it is an uncommon but regular migrant in both spring and fall. Spring migration is late April through mid-May. Fall migrants start appearing mid-July peaking in August and early September with a few lingering into early October. They are absent from November through March. As the name implies most reports are of one or two birds occasionally three and rarely up to a dozen. They show a preference for freshwater and forage in shallow water near the edges of ponds or channels.

9.3.2.36 Wandering Tattler *Tringa incana*, Occasional

Wandering Tattler nests mainly in parts of Alaska extending into a small part of northwest Canada. It winters along coastlines of the Pacific Ocean from southern Canada to Chile and west to Australia, including islands in the southwest Pacific. It is one of the least common shorebirds in North America (Gill et al. 2002). On GSL, there are two accepted records both of single birds on the Antelope Island Causeway during May 25–June 3, 2001, and on September 11–18, 2005. As typical for this species, they were observed along rocky areas.

9.3.2.37 Lesser Yellowlegs *Tringa flavipes*, Common Passage, Occasional Winter

Lesser Yellowlegs nest further north than Greater Yellowlegs in the boreal zone from northwestern Alaska to central Québec and winter throughout Central and South America, the West Indies, and the southern United States (Tibbitts and Moskoff 2014). On GSL, Lesser Yellowlegs are common in both spring and fall migration and like Greater Yellowlegs move through in small flocks usually fewer than 100 birds but sometimes numbering in the hundreds. They peak during spring from mid-April through mid-May. Fall migration is much more protracted beginning in late June and extending into November. They are observed in small numbers every month November through March in areas of unfrozen water, but reports are fewer than Greater Yellowlegs during winter months. Lesser Yellowlegs prefer similar shallow, freshwater habitats often with vegetation as does Greater Yellowlegs, and they often occur together.

9.3.2.38 Willet *Tringa semipalmata*, Common Nesting and Passage

Willets have two distinct breeding ranges. On GSL, willets are part of the western population that nests locally from northern Utah, Nevada, and California north into south-central Canada (Lowther et al. 2001). Many GSL Willets winter near San Francisco Bay. Willets use freshwater habitats throughout their nesting range, but those on GSL also utilize shallow saline wetlands. Largely absent from November through March, migrants appear in early April, with most passing through the last half of April into the first week of May. In spring, flocks of foraging Willets sometimes number into the hundreds especially in areas along the shoreline of GSL. Some will remain at GSL to nest near shallowly flooded wetlands. Willet nesting habitat is the most upland of any GSL shorebird except Long-billed Curlew and Killdeer. The nest is usually concealed in low vegetation often some distance from water. Birds will perch on greasewood, sagebrush, or fence posts, but foraging occurs along shorelines and in shallow water sometimes with scattered vegetation. Starting in late June, Willets depart their nesting sites and begin gathering in post-breeding feeding flocks, but the numbers dwindle rapidly through early August and continue to early November. As in spring, fall populations may include both GSL nesting birds and migrants.

9.3.2.39 Greater Yellowlegs *Tringa melanoleuca*, Common Passage, Rare Winter

The similar appearance of Greater and Lesser yellowlegs complicates the evaluation of these two species. Many reliable records exist for both, as they are easily separated by skilled observers, but they are often lumped together or misidentified by those with less experience. Greater Yellowlegs nest farther south than Lesser Yellowlegs in a band across central Canada and southern Alaska and winter in the southern United States and throughout Central and South America. It is one of the most widespread and ubiquitous shorebirds (Elphick and Tibbitts 1998). On GSL, it occurs in greatest numbers in spring and fall migration, but it has been recorded in increasing numbers throughout the late fall and in all winter months. Greater Yellowlegs' occurrence in migration is widely distributed throughout GSL wetlands locations. They are often recorded in groups of 1–5 birds, but variable flocks, some up to several hundred, are regularly reported. Most spring migrants pass through from mid-March through April, dwindling rapidly in early May. Records are few between mid-May and mid-June. Birds begin returning the last week in June and occur regularly through November. Greater Yellowlegs forage in shallow open water, often with low vegetation, wet meadows, flooded pastures, and agricultural fields. They show a preference for wetland habitats with fresher water.

9.3.2.40 Wilson's Phalarope *Phalaropus tricolor*, Uncommon Nesting, Common Passage

Few Wilson's Phalaropes nest at GSL, which is near the southern extent of their breeding range. Most breed in the prairies north to the Northwest Territories, Canada, east to Minnesota, and west to the Cascade Mountains. Some also nest around the Great Lakes region. Nests are concealed in rushes and grasses of wetlands and nearby uplands. Of the three species of phalarope, Wilson's Phalarope winters inland rather than on the open ocean. They prefer saline lakes of South America outside of and south of Brazil to Tierra del Fuego with some wintering in the north part of the continent within and west of the Andes Mountains from Colombia to Chile (Colwell and Jehl 1994).

Wintering phalaropes begin appearing at GSL during late April as they travel to the breeding grounds (Paul and Manning 2002). The sex roles are reversed in phalaropes with females being more brightly colored than the males, and once the eggs are laid, one or possibly two clutches, the female leaves while the male incubates the eggs and cares for the brood. The return trip south begins early compared to other shorebirds since the females are free to leave once egg laying is complete. Females arrive first to GSL in mid-June followed by the males and then the juveniles.

The staging population at the lake peaks in late July with around one-third of the entire population of Wilson's Phalaropes. They take advantage of the abundant food supplies at GSL to undergo a quick body, tail, and partial primary molt and prepare for their nonstop flight to South America by doubling their weight (Jehl 1988). Wilson's phalarope is one of only two species of shorebird (Wood Sandpiper) that are known to undergo a molt migration (Jehl 1987). Most phalaropes depart GSL by late August (Paul and Manning 2002).

Diet studies for phalaropes at GSL show variable results based on sex or age class. Since adult females, adult males, and juveniles tend to flock together, these findings may be an artifact of where the birds are collected on the lake in conjunction with local salinity and macroinvertebrate abundance rather than actual preferences of prey. Wilson's Phalaropes near shore in the main body of the south part of GSL consumed brine fly adults. They consumed brine fly larvae in GSL north of the Antelope Island causeway and chironomid larvae south of the causeway with smaller proportions of *Daphnia* and brine shrimp adults/juveniles (Frank unpublished data).

9.3.2.41 Red-Necked Phalarope *Phalaropus lobatus*, Common Passage

The breeding range of Red-necked Phalaropes is circumpolar. In North America, they nest in northern Canada, Alaska, and southern Greenland but are generally absent from the high Arctic. Nesting habitat is in areas with few shrubs, near water, and among sedges, moss, and grasses. Wintering phalaropes in the Western Hemisphere concentrate off the Pacific Coast of South America south to northern Chile (Rubega et al. 2000).

Following winter, Red-necked Phalaropes arrive to GSL at the end of April with most spring migrants observed mid-May departing by June. Migrants return mid-July, spending most of their time on the open water of the lake well into the fall, with most departing by early October though some will stay into November (Paul and Manning 2002; Frank unpublished data).

While at GSL, the majority of Red-necked Phalaropes prefer expansive areas of open water. Even though there are only two abundant macroinvertebrates present in the southern half of the lake, brine shrimp and brine flies, phalaropes almost solely eat brine fly adults, larvae, and pupae. Phalaropes consume more brine fly adults as compared to brine fly larvae and pupae, which is likely attributed to the higher abundance and greater accessibility of adult flies (Aldrich and Paul 2002; Frank 2016; Frank and Conover 2019).

Habitats vary even in the south part of GSL with changes in salinity and water depth. Phalaropes north of the Antelope Island causeway consumed a higher proportion of brine fly larvae compared to adult flies. South of the causeway, which is much fresher at low lake elevations, phalaropes consumed mostly brine fly larvae with some brine fly adults and corixids even though chironomid larvae and corixids were the most abundant food items. Preferred food items also vary by season. As compared to adult flies, more brine fly larvae are consumed during midsummer when larvae are more available (Frank 2016; Frank unpublished data).

9.3.2.42 Red Phalarope *Phalaropus fulicarius*, Rare Passage

Red Phalarope breeds in the high Artic. It is the most pelagic of the three phalarope species migrating almost completely over the ocean and winters at sea in the southern oceans (Tracy et al. 2002). On GSL, it is far less common than the Red-necked or Wilson's phalarope. Spring records are few, with most occurring in fall, September through early December, which is consistent with the species as one of the later fall migrants. A few records exist for August. Most GSL records are in deep, open water.

9.4 Insights

In this chapter, the authors focus on shorebirds, a large diverse group of birds that occurs at GSL, which have received relatively little attention compared to other groups such as waterfowl and colonial nesting birds. We have departed from a typical classification of shorebird habitat and present instead three necessary ecological parameters (water depth, vegetation, and macroinvertebrates) that define which niche a species prefers among the myriad habitat choices present on the lake. Defining characteristics of mudflats and playas were discussed to clarify these often confusing terms. Finally, we presented the first annotated checklist that describes seasonal abundance, status, range, and habitat preference for all species with accepted records specific to GSL.

Shorebirds, their habitats, and food base face many challenges. The expanding human population that is driving development further toward the lake disproportionately destroys or fragments the shallow flooded playas, mudflats, and wet meadows upon which many shorebirds depend. These areas are especially critical when lake levels are high, and the lake constricts shorebirds into narrow bands of habitat and upland areas closer to development. GSL is inherently dynamic leading to large fluctuations in lake level. When lake levels recede, water diversions and droughts limit the amount of freshwater flowing into the lake, thereby reducing productive, water-dependent habitat.

One of the most concerning threats to shorebirds at GSL is the exploding *Phragmites* infestation. Open, unvegetated mudflat habitat of shorebirds at GSL is vanishing at an alarming rate beneath the rapid expansion of dense forests of *Phragmites*, the antithesis of productive shorebird habitat. Adding to these hazards facing shorebirds at the lake are the uncertainties of climate change that will affect timing and availability of water and potentially the life cycles of the macroinvertebrates upon which shorebirds depend. Greater understanding of shorebirds and their needs will lead to a better understanding of the impacts of a changing GSL.

Acknowledgements Many people contributed to the concepts presented in this chapter. We owe a special thanks to Jack Oviatt and Randy Lewis for many discussions and comments that helped formulate the sections on GSL dynamic processes that contribute to the diversity of shorebird habitat and a scientific understanding of the terms playa and mudflat. Ann Neville reviewed most sections of the chapter and greatly improved both content and organization. Richard Mingo's inspiration and ideas helped shape the creation of the map of GSL and associated wetlands. Max Malmquist provided input and support. Bryant Olsen reviewed most species accounts. Colton Norman, Bridget Olson, and Peter Paton helped with accessing databases. We would like to thank the following photographers for capturing the beauty of shorebirds and sharing their photos: Cameron Cox, Matthew Pendleton, Steve Earley, Mike Schijf, and Lauri Taylor. We owe special thanks to John Cavitt for his many years of shorebird research on the lake and his valuable input as we developed ideas for this chapter.

We greatly appreciate the suggestions and comments from the following people who reviewed sections relevant to their expertise, attended meetings or participated in discussion of the concepts covered in this chapter, or helped in other ways: Don Paul, Rob Clay, David Richards, Jeff DenBleyker, Marcelle Shoop, Brian Tavernia, Justin Dolling, Becka Downard, Diane Menuz, Bridget Olson, Maureen Frank, Pam Kramer, Cianna Wyshnytzky, Eric McCulley, Chris Brown, Jack Ray, Rich Hansen, and Heather Dove. We especially want to thank the editors, Bonnie and Jaimi, for their vision for this book and Alison Holloran, John Kloster-Prew, and John Luft for understanding the importance of shorebirds and supporting our efforts in writing this chapter.

References

Ackerman JT, Hartman CA, Herzog MP, Takekawa JY, Robinson JA, Oring LW, Skorupa JP, Boettcher R (2013) American Avocet *Recurvirostra americana*, version 2.0. In: Poole AF (ed) The birds of North America. Cornell Lab of Ornithology, Ithaca, NY. https://doi.org/10.2173/bna.275

Aldrich TW, Paul DS (2002) Avian ecology of the Great Salt Lake. In: Gwynn JW (ed) Great Salt Lake: an overview of change. Special publication, Utah Department of Natural Resources, Salt Lake City, UT, pp 343–374

Andrei AE, Smith LM, Haukos DA, Surles JG, Johnson WP (2009) Foraging ecology of migrant shorebirds in saline lakes of the Southern Great Plains. Waterbirds 32(1):138–148

Arnow T (1984) Water-level and water-quality changes in Great Salt Lake, Utah, 1843–1985. US Geological Survey Circular 913

Arnow T, Stephens DW (1990) Hydrologic characteristics of the Great Salt Lake, Utah, 1847–1986. US Geological Survey Water-Supply Paper 2332

Avian Ecology Laboratory (2014) American Avocet migration project. http://departments.weber.edu/avianecologylab/AMAV/home.html. Accessed 2 Aug 2019

Baker MC (1979) Morphological correlates of habitat selection on a community of shorebirds (Charadriiformes). Oikos 33(1):121–126

Baker AP, Gonzalez P, Morrison RIG, Harrington BA (2013) Red Knot *Calidris canutus*, version 2.0. In: Poole AF (ed) The birds of North America. Cornell Lab of Ornithology, Ithaca, NY. https://doi.org/10.2173/bna.563

Barber B, Cavitt J (2013) Dietary review for aquatic birds utilizing Willard Spur, Great Salt Lake. Ogden, UT, Avian Ecology Laboratory, Weber State University, p 36

Behle WH (1958) The bird life of Great Salt Lake. University of Utah Press, Salt Lake City, UT, p 203

Behle WH (1985) Utah birds: geographic distribution and systematics. Occasional publication no 5, Utah Museum of Natural History. University of Utah, Salt Lake City, UT, p 147

Bradbury ES, Parrott DL (2020) Shoreline plants of Great Salt Lake. In: Baxter BK, Butler JK (eds) Great Salt Lake biology: a terminal lake in a time of change. Springer, Cham

Cantrell AM, McLachlan AJ (1977) Competition and chironomid distribution patterns in a newly flooded lake. Oikos 29:429–433

Cavitt JF (2006a) Productivity and foraging ecology of two co-existing shorebird species breeding at Great Salt Lake, UT: 2005–2006 Report. Avian Ecology Laboratory Technical Report AEL 06-03 Weber State University, Ogden, UT, p 38

Cavitt JF (2006b) Concentration and effects of selenium on breeding shorebirds at Great Salt Lake. In: Development of a selenium standard for the open waters of the Great Salt Lake. State of Utah Department of Environmental Quality, Salt Lake City, UT, p 15

Cavitt JF (2008) Snowy Plover survey—Great Salt Lake, UT summary report. Avian Ecology Laboratory, Weber State University, Ogden, UT, p 6

Cavitt JF, Stone K (2007) Selenium and mercury concentrations in breeding female American Avocets at Ogden Bay, Great Salt Lake, Utah. In: Development of a selenium standard for the open waters of the Great Salt Lake, State of Utah Department of Environmental Quality, Salt Lake City, UT, p 15

Cavitt JF, Jones SL, Wilson NM, Dieni JS, Zimmerman TS, Doster RH, Howe WH (2014) Atlas of breeding colonial waterbirds in the interior western United States. Research report, United States Department of the Interior, Fish and Wildlife Service, Denver p 26

Colwell MA, Jehl JR Jr (1994) Wilson's Phalarope *Phalaropus tricolor*, version 2.0. In: Poole AF, Gill FB (eds) The birds of North America. Cornell Lab of Ornithology, Ithaca, NY. https://doi.org/10.2173/bna.83

Conover MR, Bell ME (2020) Importance of Great Salt Lake to pelagic birds: eared grebes, phalaropes, gulls, ducks, and white pelicans. In: Baxter BK, Butler JK (eds) Great Salt Lake biology: a terminal lake in a time of change. Springer, Cham

Cronquist A, Holmgren AH, Holmgren NH, Reveal JL, Holmgren P (1972) Intermountain flora, vascular plants of the Intermountain West, U.S.A. Volume one, geological and botanical history of the region, its plant geography and glossary. The vascular cryptograms and gymnosperms. The New York Botanical Garden, Hafner, New York

Dugger BD, Dugger KM (2002) Long-billed Curlew *Numenius americanus*, version 2.0. In: Poole AF, Gill FB (eds) The birds of North America. Cornell Lab of Ornithology, Ithaca, NY. https://doi.org/10.2173/bna.628

Eardley AJ (1938) Sediments of Great Salt Lake, Utah. Am Assoc Petrol Geol Bull 22 (10):1305–1411

eBird (2019) eBird: an online database of bird distribution and abundance [web application]. eBird, Ithaca, NY. Available http://www.ebird.org. Accessed 29 July 2019

Ellis KS, Cavitt JF, Larson RT (2015) Factors influencing Snowy Plover *Charadrius nivosus* nest survival at Great Salt Lake, Utah. Waterbirds 38(1):58–67

Ellis KS, Joseph RJ, Knight RN, Marvel KS, Larsen RT (2016) Spring downings clarify the migration biology of Eared Grebes *Podiceps nigricollis*. Waterbirds 39(4):338–345

Ellis KS, Cavitt JF, Larson RT, Koons DN (2018) Using remote cameras to validate estimates of nest fate in shorebirds. Ibis 160:681–687. https://doi.org/10.1111/ibi.12607

Elphick CS, Tibbitts TL (1998) Greater Yellowlegs *Tringa melanoleuca*, version 2.0. In: Poole AF, Gill FB (eds) The birds of North America. Cornell Lab of Ornithology, Ithaca, NY. https://doi.org/10.2173/bna.355

Farmer A, Holmes RT, Pitelka FA (2013) Pectoral Sandpiper *Calidris melanotos*, version 2.0. In: Poole AF (ed) The birds of North America. Cornell Lab of Ornithology, Ithaca, NY. https://doi.org/10.2173/bna.348

Flowers S (1934) Vegetation of the Great Salt Lake region. Bot Gaz 95:353–418

Frank MG (2016) Migratory waterbird ecology at a critical staging area, Great Salt Lake, Utah. Dissertation, Utah State University https://digitalcommons.usu.edu/etd/4940

Frank MG, Conover MR (2017) Weather and prey availability affect the timing of fall migration of Eared Grebes *Podiceps nigricollis* from Great Salt Lake. Wilson J Ornithol 129(1):98–111

Frank MG, Conover MR (2019) Threatened habitat at Great Salt Lake: importance of shallow-water and brackish habitats to Wilson's and Red-necked phalaropes. Condor 121(2):duz005. https://doi.org/10.1093/condor/duz005

Franks S, Lank DB, Wilson WH Jr (2014) Western Sandpiper *Calidris mauri*, version 2.0. In: Poole AF (ed) The birds of North America. Cornell Lab of Ornithology, Ithaca, NY. https://doi.org/10.2173/bna.90

Gill RE, McCaffery BJ, Tomkovich PS (2002) Wandering Tattler *Tringa incana*, version 2.0. In: Poole AF, Gill FB (eds) The birds of North America. Cornell Lab of Ornithology, Ithaca, NY. https://doi.org/10.2173/bna.642

Gratto-Trevor CL (2000) Marbled Godwit *Limosa fedoa*, version 2.0. In: Poole AF, Gill FB (eds) The birds of North America. Cornell Lab of Ornithology, Ithaca, NY. https://doi.org/10.2173/bna.492

Gray LJ (2011) Macroinvertebrate and zooplankton communities in the impounded wetlands of the Great Salt Lake, May-November 2010. Prepared for the UT Department of Environmental Quality, Division of Water Quality, Salt Lake City, UT p 35

Gwynn JW (1998) Great Salt Lake, Utah: chemical and physical variations of the brine and effects of the SPRR Causeway, 1966-1996. In: Pitman JK, Carroll AR (eds) Modern and ancient lake systems: new problems and perspectives. AAPG, Tulsa, OK, pp 71–90

Hayman P, Marchant J, Prater T (1986) Shorebirds: an identification guide to the waders of the world. Houghton Mifflin, London

Helmers DA (1992) Shorebird management manual. Western Hemisphere Shorebird Reserve Network, Manomet, p 58

Herbst DB (1999) Biogeography and physiological adaptations of the brine fly genus *Ephydra* (Diptera: Ephydridae) in saline waters of the Great Basin. Great Basin Nat 59(2):Article 3. Available at https://scholarsarchive.byu.edu/gbn/vol59/iss2/3

Herbst DB (2006) Salinity controls on trophic interactions among invertebrates and algae of solar evaporation ponds in the Mojave Desert and relation to shorebird foraging and selenium risk. Wetlands 26:475–485. https://escholarship.org/uc/item/3hg27142

Hicklin P, Gratto-Trevor CL (2010) Semipalmated Sandpiper *Calidris pusilla*, version 2.0. In: Poole AF (ed) The birds of North America. Cornell Lab of Ornithology, Ithaca, NY. https://doi.org/10.2173/bna.6

Houston CS, Jackson C, Bowen DE Jr (2011) Upland Sandpiper *Bartramia longicauda*, version 2.0. In: Poole AF (ed) The birds of North America. Cornell Lab of Ornithology, Ithaca, NY. https://doi.org/10.2173/bna.580

Jackson BJ, Jackson JA (2000) Killdeer *Charadrius vociferus*, version 2.0. In: Poole AF, Gill FB (eds) The birds of North America. Cornell Lab of Ornithology, Ithaca, NY. https://doi.org/10.2173/bna.517

Jehl JR Jr (1979) The autumn migration of Baird's Sandpiper. Stud Avian Biol 2:55–68

Jehl JR Jr (1987) Moult and moult migration in a transequatorially migrating shorebird: Wilson's Phalarope. Ornis Scandinavica (Scand J Ornithol) 18(3):173–178. https://doi.org/10.2307/3676763

Jehl JR Jr (1988) Biology of the eared grebe and Wilson's phalarope in the nonbreeding season: a study of adaptations to saline lakes. Stud Avian Biol 12:1–74

Jehl JR Jr (1998) Conspecific collisions can precipitate migrating eared grebes. Wilson Bull 110 (3):409–411

Jehl JR Jr, Klima J, Harris RE (2001) Short-billed Dowitcher *Limnodromus griseus*, version 2.0. In: Poole AF, Gill FB (eds) The birds of North America. Cornell Lab of Ornithology, Ithaca, NY. https://doi.org/10.2173/bna.564

Jehl JR Jr, Henry AE, Ellis HI (2003) Optimizing migration in a reluctant and inefficient flier: the Eared Grebe In: Berthold P, Gwinner E, Sonnenschein E (eds) Avian migration. Springer, Berlin

Johnson OW, Connors PG, Pyle P (2018a) American Golden-Plover *Pluvialis dominica*, version 3.0. In: Rodewald PG (ed) The birds of North America. Cornell Lab of Ornithology, Ithaca, NY. https://doi.org/10.2173/bna.amgplo.03

Johnson OW, Connors PG, Pyle P (2018b) Pacific Golden-Plover *Pluvialis fulva*, version 3.0. In: Rodewald PG (ed) The birds of North America. Cornell Lab of Ornithology, Ithaca, NY. https://doi.org/10.2173/bna.pagplo.03

Kadlec J (1995) Water requirements and conceptual plan for the South Shore Ecological Reserve and suggestions for restoring seasonal and permanent wetlands In: Feasibility study for the South Shore Wetlands Ecological Reserve of the Great Salt Lake. National Audubon Society

Kienast-Brown S (2003) Remote sensing in soil survey updates—east shore of the Great Salt Lake. RS6750 MLRA-NRCS

Kijowski AK, Neill J, Wickline A, Swift J, Butler JK, Kimberly DA, VanLeeuwen J, Luft J, Stone K (2020) American White Pelicans of Gunnison Island. In: Baxter BK, Butler JK (eds) Great Salt Lake biology: a terminal lake in a time of change. Springer, Cham

Klima J, Jehl JR Jr (2012) Stilt sandpiper *Calidris himantopus*, version 2.0. In: Poole AF (ed) The birds of North America. Cornell Lab of Ornithology, Ithaca, NY. https://doi.org/10.2173/bna.341

Knopf FL (1975) Spatial and temporal aspects of colonial nesting of the white pelican *Pelecanus erythrorhynchos*. Dissertation, Utah State University

Knopf FL, Wunder MB (2006) Mountain plover *Charadrius montanus*, version 2.0. In: Poole AF (ed) The birds of North America. Cornell Lab of Ornithology, Ithaca, NY. https://doi.org/10.2173/bna.211

Knudsen AD (1970) The biology and control of the deer fly, *Chrysops discalis* Williston, and other tabanids in marshlands near the Great Salt Lake, Utah. Dissertation, University of Utah

Lowther PE, Douglas HD III, Gratto-Trevor CL (2001) Willet *Tringa semipalmata*, version 2.0. In: Poole AF, Gill FB (eds) The birds of North America. Cornell Lab of Ornithology, Ithaca, NY. https://doi.org/10.2173/bna.579

Macwhirter RB, Austin-Smith P Jr, Kroodsma DE (2002) Sanderling *Calidris alba*, version 2.0. In: Poole AF, Gill FB (eds) The birds of North America. Cornell Lab of Ornithology, Ithaca, NY. https://doi.org/10.2173/bna.653

McCaffery BJ, Gill RE (2001) Bar-tailed godwit *Limosa lapponica*, version 2.0. In: Poole AF, Gill FB (eds) The birds of North America. Cornell Lab of Ornithology, Ithaca, NY. https://doi.org/10.2173/bna.581

McCarty JP, Wolfenbarger LL, Laredo CD, Pyle P, Lanctot RB (2017) Buff-breasted sandpiper *Calidris subruficollis*, version 2.0. In: Rodewald PG (ed) The birds of North America. Cornell Lab of Ornithology, Ithaca, NY. https://doi.org/10.2173/bna.bubsan.02

Merritt RW, Cummins KW (1978) An introduction to the aquatic insects of North America. Kendahk-Hunt, Dubuque

Mitsch W, Gosselink JG (2007) Wetlands, 4th edn. Wiley, New York

Moskoff W (2011) Solitary sandpiper *Tringa solitaria*, version 2.0. In: Poole AF (ed) The birds of North America. Cornell Lab of Ornithology, Ithaca, NY. https://doi.org/10.2173/bna.156

Moskoff W, Montgomerie R (2002) Baird's sandpiper *Calidris bairdii*, version 2.0. In: Poole AF, Gill FB (eds) The birds of North America. Cornell Lab of Ornithology, Ithaca, NY. https://doi.org/10.2173/bna.661

Mueller H (1999) Wilson's snipe *Gallinago delicata*, version 2.0. In: Poole AF, Gill FB (eds) The birds of North America. Cornell Lab of Ornithology, Ithaca, NY. https://doi.org/10.2173/bna.417

Murkin HR, Kadlec JA (1986) Responses by benthic macroinvertebrate densities to prolonged flooding of marsh habitat. Can J Zool 64:65–72

Nebel S, Cooper JM (2008) Least sandpiper *Calidris minutilla*, version 2.0. In: Poole AF (ed) The birds of North America. Cornell Lab of Ornithology, Ithaca, NY. https://doi.org/10.2173/bna.115

Nettleship DN (2000) Ruddy turnstone *Arenaria interpres*, version 2.0. In: Poole AF, Gill FB (eds) The birds of North America. Cornell Lab of Ornithology, Ithaca, NY. https://doi.org/10.2173/bna.537

Nol E, Blanken MS (2014) Semipalmated plover *Charadrius semipalmatus*, version 2.0. In: Poole AF (ed) The birds of North America. Cornell Lab of Ornithology, Ithaca, NY. https://doi.org/10.2173/bna.444

Norazlimi NA, Ramli R (2015) The relationships between morphological characteristics and foraging behavior in four selected species of shorebirds and water birds utilizing tropical mudflats. Sci World J 2015:Article ID 105296. https://doi.org/10.1155/2015/105296. Accessed 27 July 2019

Olson BE (2011) The biogeography of Marbled Godwit *Limosa fedoa* populations in North America. All graduate theses and dissertations, 1119, Utah State University. https://digitalcommons.usu.edu/etd/1119

Olson BE, Lindsey K, Hirschboeck V (2004) Bear River migratory bird refuge habitat management plan. U.S. Department of the Interior Fish and Wildlife Service

Oviatt CG (2014) The Gilbert episode in the Great Salt Lake Basin, Utah. Miscellaneous publication 14-3 Utah Geological Survey, Utah Department of Natural Resources. isbn: 978-1-55791-893-2

Page GW, Stern MA, Paton PWC (1995) Differences in wintering areas of Snowy Plovers from inland breeding sites in western North America. Condor 97:258–262

Page GW, Stenzel LE, Warriner JS, Warriner JC, Paton PW (2009) Snowy plover *Charadrius nivosus*, version 2.0. In: Poole AF (ed) The birds of North America. Cornell Lab of Ornithology, Ithaca, NY. https://doi.org/10.2173/bna.154

Parmelee DF (1992) White-rumped sandpiper *Calidris fuscicollis*, version 2.0. In: Poole AF, Stettenheim PR, Gill FB (eds) The birds of North America. Cornell Lab of Ornithology, Ithaca, NY. https://doi.org/10.2173/bna.29

Paton PWC (1994) Breeding ecology of snowy plovers at Great Salt Lake, Utah. Dissertation, Utah State University

Paton PWC (1995) Breeding biology of snowy plovers at Great Salt Lake, Utah. Wilson Bull 107:275–288

Paton PWC, Dalton J (1994) Breeding ecology of long-billed curlews at Great Salt Lake, Utah. Great Basin Nat 54(1):79–85

Paton PWC, Edwards TC Jr (1992) Nesting ecology of the snowy plover at Great Salt Lake, Utah 1992 breeding season. Dept. Fish and Wildlife, Utah State University

Paton PWC, Edwards TC Jr (1996) Factors affecting interannual movements of Snowy Plovers. Auk 113(3):534–543

Paton PWC, Kneedy C, Sorensen E (1992) Chronology of shorebird and ibis use of selected marshes at Great Salt Lake. Utah Birds 8:1–19

Paul DS, AE Manning (2002) Great Salt Lake waterbird survey five-year report (1997-2001). Publication number 08-38, Great Salt Lake Ecosystem Program, Utah Division of Wildlife Resources, Salt Lake City, UT

Piazza O, Corsetti FA, Stamps BW, Stevenson BS, Bardsley A, Hammond DE, Nunn HS, Berelson W, Spear JR (2016) Microbial composition and preliminary age of ooids from the Great Salt Lake, Utah. In: AGU Fall meeting abstracts, San Francisco, CA, 12–16 Dec 2016

Poole AF, Pyle P, Patten MA, Poole AF, Pyle P, Patten MA, Paulson DR (2016) Black-bellied plover *Pluvialis squatarola*, version 3.0. In: Rosewald PG (ed) The birds of North America. Cornell Lab of Ornithology, Ithaca, NY. https://doi.org/10.2173/bna.bkbplo.03

Reed JM, Oring LW, Gray EM (2013) Spotted sandpiper *Actitis macularius*, version 2.0. In: Poole AF (ed) The birds of North America. Cornell Lab of Ornithology, Ithaca, NY. https://doi.org/10.2173/bna.289

Richards DC (2014) Development of a macroinvertebrate index of biological integrity (MIBI) for impounded freshwater wetland ponds of Great Salt Lake, Utah. Prepared for Jordan River/Farmington Bay Water Quality Council, Oreohelix, Moab, p 71

Roberts AJ (2013) Winter waterbird ecology on the Great Salt Lake, Utah, and interactions with commercial harvest of brine shrimp cysts. All graduate theses and dissertations, 2042 Utah State University. https://digitalcommons.usu.edu/etd/2042

Roberts AJ, Conover MR (2014) Role of benthic substrate in waterbird distribution on Great Salt Lake, Utah. Waterbirds 37(3):298–306

Robinson JA, Reed JM, Skorupa JP, Oring LW (1999) Black-necked stilt *Himantopus mexicanus*, version 2.0. In: Poole AF, Gill FB (eds) The birds of North America. Cornell Lab of Ornithology, Ithaca, NY. https://doi.org/10.2173/bna.449

Rosen MR (1994) The importance of groundwater in playas: a review of playa classifications and the sedimentology and hydrology of playas. In: Rosen MR (ed) Paleoclimate and basin evolution of playa systems. Geological Society of America Special Paper, Boulder, CO, p 289

Rubega MA, Schamel D, Tracy DM (2000) Red-necked phalarope *Phalaropus lobatus*, version 2.0. In: Poole AF, Gill FB (eds) The birds of North America. Cornell Lab of Ornithology, Ithaca, NY. https://doi.org/10.2173/bna.538

Rupke AL, McDonald A (2012) Great Salt Lake brine chemistry database, 1966-2011. Utah Geological Survey, State of Utah Department of Natural Resources, Salt Lake City, UT

Ryder RA (1959) Interspecific intolerance of the American Coot in Utah. Auk 76:424–442

Shuford WD, Roy VL, Page GW, Paul DS (1994) A comprehensive survey of shorebirds in wetlands at Great Salt Lake, Utah, 10–11 August. Point Reyes Bird Observatory Contribution No 655, p 9

Shuford WD, Page GW, Stenzel LE (2002) Patterns of distribution and abundance of migratory shorebirds in the Intermountain West of the United States. West Birds 33:134–174

Skeel MA, Mallory EP (1996) Whimbrel *Numenius phaeopus*, version 2.0. In: Poole AF, Gill FB (eds) The birds of North America. Cornell Lab of Ornithology, Ithaca, NY. https://doi.org/10.2173/bna.219

Sordahl TA (1996) Breeding biology of the American avocet and black-necked Stilt in northern Utah. Southwest Nat 41:348–354

Sorensen E, Hoven H, Trusty B (2018) National audubon society's gillmor sanctuary management and conservation plan. In: National Audubon society, pp 39–42

Takekawa JY, Warnock ND (2000) Long-billed dowitcher *Limnodromus scolopaceus*, version 2.0. In: Poole AF, Gill FB (eds) The birds of North America. Cornell Lab of Ornithology, Ithaca, NY. https://doi.org/10.2173/bna.493

The Birds of North America (2019) Cornell Laboratory of Ornithology, Ithaca, NY. https://birdsna.org

Thomas S (2005) Recent Intermountain West conservation projects that benefit shorebirds. Intermountain West Shorebird Working Group, Intermountain West Joint Venture

Thomas SM, Lyons JE, Andres BA, T-Smith EE, Palacios E, Cavitt JF, Royle JA, Fellows SD, Maty K, Howe WH, Mellink E (2012) Population size of Snowy Plovers breeding in North America. Waterbirds 35(1):1–14

Thompson RS, Oviatt CG, Honke JS, McGeehin JP (2016) Late quaternary changes in lakes, vegetation, and climate in the Bonneville Basin reconstructed from sediment cores from Great Salt Lake. In: Oviatt CG, Shroder JF Jr (eds) Lake Bonneville: a scientific update, Developments in earth surface processes, vol 20. Elsevier, Amsterdam, pp 221–291

Tibbitts TL, Moskoff W (2014) Lesser yellowlegs *Tringa flavipes*, version 2.0. In: Poole AF (ed) The birds of North America. Cornell Lab of Ornithology, Ithaca, NY. https://doi.org/10.2173/bna.427

Tracy DM, Schamel D, Dale J (2002) Red phalarope *Phalaropus fulicarius*, version 2.0. In: Poole AF, Gill FB (eds) The birds of North America. Cornell Lab of Ornithology, Ithaca, NY. https://doi.org/10.2173/bna.698

UOS (2011) Utah Ornithological Society. http://www.utahbirds.org/uos/index.htm. Accessed 30 Aug 2019

Vest JL (2013) Winter ecology of waterfowl on the Great Salt Lake, Utah. All graduate theses and dissertations, 2051, Utah State University. https://digitalcommons.usu.edu/cgi/viewcontent.cgi?article=3054&context=etd

Vinson M, Bushman B (2005) An inventory of aquatic invertebrate assemblages in wetlands in Utah. National Aquatic Monitoring Center, Utah State University, Logan, UT, p 224

Walker BM, Senner NR, Elphick CS, Klima J (2011) Hudsonian Godwit *Limosa haemastica*, version 2.0. In: Poole AF (ed) The birds of North America. Cornell Lab of Ornithology, Ithaca, NY. https://doi.org/10.2173/bna.629

Warnock ND, Gill RE (1996) Dunlin *Calidris alpina*, version 2.0. In: Poole AF, Gill FB (eds) The birds of North America. Cornell Lab of Ornithology, Ithaca, NY. https://doi.org/10.2173/bna.203

Wetmore A (1925) Food of American phalaropes, avocets, and stilts. United States Department of Agriculture Bulletin 1359

Wilson GW (1973) The foods and feeding habits of botulism intoxicated and healthy waterbirds on the Bear River Refuge, Utah, with emphasis on the American Avocet and Black-necked Stilt. All Graduate Theses and Dissertations, 3140, Utah State University. https://digitalcommons.usu.edu/etd/3140

Wurtsbaugh WA, Miller C, Null SE, DeRose RJ, Wilcock P, Hahnenberger M, Howe F, Moore J (2017) Decline of the world's saline lakes. Nat Geosci 10:816–821

Chapter 10
American White Pelicans of Gunnison Island, Great Salt Lake, Utah

Ashley M. Kijowski, John Neill, Adam Wickline, Jessica Swift, Jaimi K. Butler, David A. Kimberly, Jim Van Leeuwen, John Luft, and Kyle Stone

Abstract Great Salt Lake (GSL) is recognized as a site of "Hemispheric Importance" for shorebirds by the Western Hemisphere Shorebird Reserve Network. An estimated ten million birds visit GSL every year for breeding, staging, and for some species, as a wintering destination. American white pelicans (*Pelecanus erythrorhynchos*) rely on GSL for both breeding and foraging habitat. Surveys conducted by the Utah Division of Wildlife Resources (UDWR) during mid-September 1997 estimated over 85,000 pelicans using GSL wetlands for foraging and loafing. Gunnison Island, situated in the northwestern section of GSL, is home to one of the largest breeding colonies of American white pelicans in North America. Aerial counts completed by the UDWR have shown up to 20,000 breeding pelicans on the island. Naturally protected by water and the island's remoteness, pelicans have been able to breed and raise their young free from predation and disturbance from red fox (*Vulpes vulpes*), coyote (*Canis latrans*), and humans. Lower water availability and threats of increasing pressure on water resources in recent years has caused increased attention to, preparation for, and response to losses of aquatic habitat. The population of American white pelicans in Utah has remained stable over time, but the potential effects of local and regional stressors on pelicans and their habitat are poorly understood. Recent research provides an eye into the lives of American white pelicans in Utah and to the broader watershed and flyway dynamics.

Keywords Climate change · Migration · Remote island · Breeding · Pelicans · Flyway · Airstrikes · Avian surveys · American white pelican · Waterbird · PELI

A. M. Kijowski (✉) · J. Neill · A. Wickline · J. Swift · J. Van Leeuwen · J. Luft · K. Stone
Great Salt Lake Ecosystem Program, Utah Division of Wildlife Resources, Utah Department of Natural Resources, Hooper, UT, USA
e-mail: akijowski@utah.gov

J. K. Butler
Great Salt Lake Institute, Westminster College, Salt Lake City, UT, USA

D. A. Kimberly
Department of Biology, Westminster College, Salt Lake City, UT, USA

B. K. Baxter, J. K. Butler (eds.), *Great Salt Lake Biology*,
https://doi.org/10.1007/978-3-030-40352-2_10

project · Surface elevation · Saline lakes · Mark-resight · Great Salt Lake · Satellite tracking · Utah

10.1 Introduction

The American white pelican (*Pelecanus erythrorhynchos,* hereafter referred to as "pelican") soars high in the skies throughout North America using its large wingspan (2–3 m) to carry its heavy (5–9 kg) body along rivers, lakes, and reservoirs (Knopf and Evans 2004). Gregarious birds by nature, white pelicans often fly in flocks, spiraling in unison high in the sky with the help of the rising column of warm air in thermals. Pelicans are easily recognized by their large, orange-colored pouched bills. Contrary to popular belief, they do not use their bill to carry food, but rather they use it like a net to scoop up fish, crustaceans, and occasionally small mammals. Unlike brown pelicans that plunge into the water to capture fish on their own, American white pelicans are cooperative hunters. They can often be seen swimming and dipping their bills into the water together as they drive fish into shallow areas for capture (Fig. 10.1).

Fossil evidence suggests that pelicans have held a place in the Great Salt Lake (GSL) region for at least 125,000 years (Mengel 1952; Howard 1955; Wetmore 1933). They have witnessed glacial accretion and recession, the rise and fall of Lake Bonneville (Oviatt et al. 1992), the arrival of the first humans in North America

Fig. 10.1 American white pelicans herding fish (Photo credit: Mia McPherson)

approximately 16,000 years ago (Goebel et al. 2008), as well as current human development along the Wasatch Front on the eastern shores of GSL (Trentelman 2020).

A paraphrased quote by William L. Dawson (1923), a notable ornithologist, provides us with a colorful description of pelicans, their habitat, and a forecast for their fate:

> Well did the psalmist of old choose the pelican as the symbol of wilderness. Salty seas and the mighty evaporation pans of the desert gave rise to a bird as majestic and graceful as any that have adorned the pages of time. For eons the great white birds have circled and soared over the desert wastes and have fished the waters of Lake Lahontan, Lake Bonneville, and their successors. The pelican and the wilderness stand together in their mute appeal. When one is fully reclaimed, the other certainly must perish.

It might be surprising to think of piscivorous (fish eating) birds living in a desert landscape devoid of fish, but historical documentation has proven pelicans have lived and flourished at GSL in large numbers. Thompson (1933) was the first to conduct extensive surveys of pelican nesting areas, and he described the Utah breeding colony as one of the largest in North America. We will explore the history of this significant pelican population, discuss data from current research, and provide insights into future conservation and management.

10.2 Great Salt Lake

GSL is the largest body of water west of the Mississippi River in the United States and the largest terminal saline lake in the Western Hemisphere (Cohenour and Thompson 1966). It lies in northern Utah, an area known for its arid climate and mountainous landscapes. Located along the Pacific and Central flyways, GSL is one of the most significant stopover sites for migratory waterbirds in North America and is often referred to as an "oasis in the desert" (Science Friday 2018).

GSL is broken into different bays by impermeable causeways and dikes, which create vastly different microenvironments within the lake (Fig. 10.2). The majority of water entering GSL comes from three river systems. This freshwater flows into the lake through wetlands and impoundments lining the eastern shores of the south arm of the lake, but is prevented from flowing freely into the north arm by a largely impermeable railroad causeway. The Union Pacific railroad causeway splits the lake in half (Cannon and Cannon 2002), and the two separate bodies of water are commonly referred to as the north arm and south arm. The causeway to Antelope Island slows water inputs and helps define Farmington Bay.

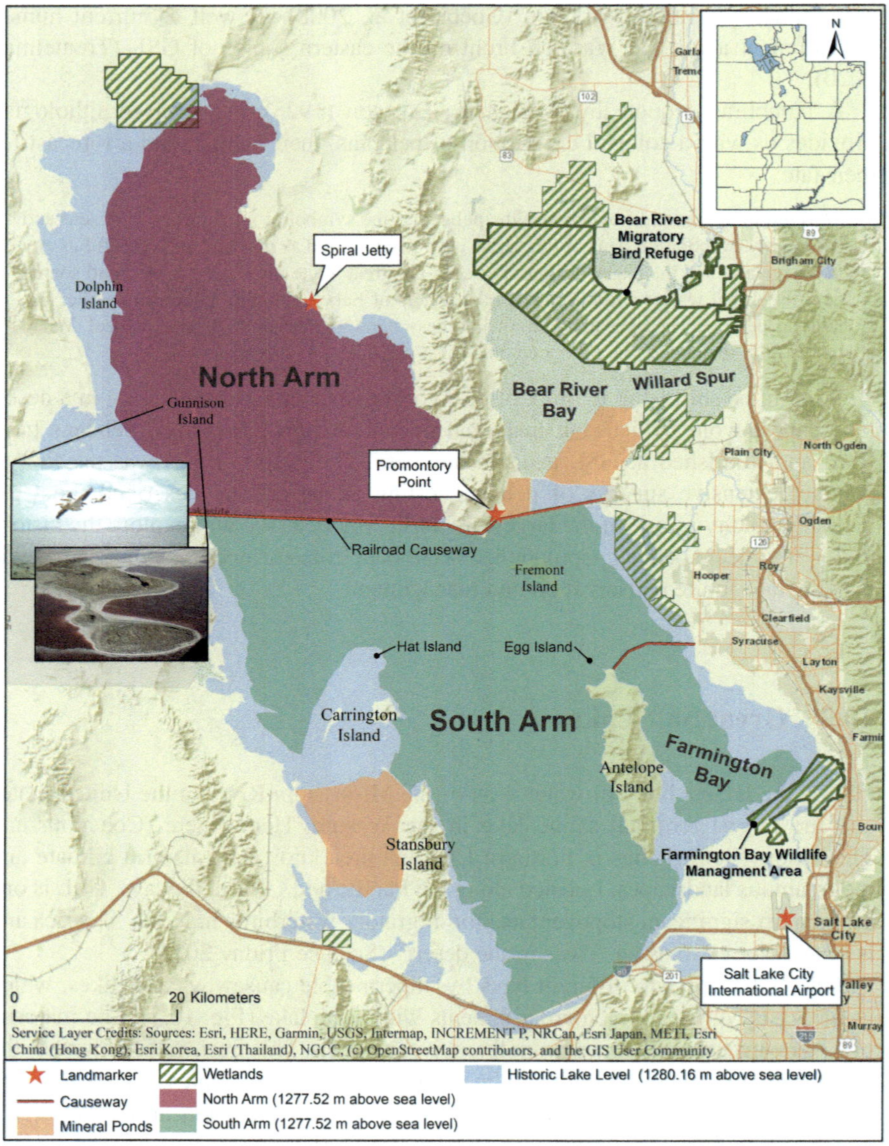

Fig. 10.2 Map of GSL showing the segmented bays (north and south arm), the location of Gunnison Island, and various points of interest (Map credit: Jessica Swift, Utah Division of Wildlife Resources, Great Salt Lake Ecosystem Program)

10.2.1 Wetlands

The wetlands surrounding GSL are vast and diverse, containing approximately 142,000 hectares of submergent, emergent, meadow, and playa wetlands (Downard et al. 2017). The salinity and depth of these wetlands follow a salinity gradient from fresh to brackish to hypersaline and alkaline, providing a diverse continuum of habitat for migratory shorebirds (Sorenson et al. 2020) and other waterbirds. The wetlands support vegetation, macroinvertebrates, and fish for wildlife and birds to feed on.

10.2.2 South Arm of Great Salt Lake

The south arm of GSL receives water through precipitation, snow runoff, sewage treatment effluent, streams, and springs. The salinity of the south arm typically varies from 9 to 17% depending on the lake elevation (Belovsky et al. 2011). The water is too saline for aquatic vertebrates to survive, but highly productive phytoplankton (Belovsky et al. 2011) and microbialite communities (Lindsay et al. 2019) support incredible amounts of invertebrate biomass including brine shrimp (*Artemia franciscana*), two species of brine flies (*Ephydra cinerea* and *E. hians*) and many shore fly species. The invertebrates in GSL provide food for hundreds of thousands of shorebirds like Wilson's phalarope (*Phalaropus tricolor*) and American avocet (*Recurvirostra americana*) (Sorenson et al. 2020).

Where other macroorganisms struggle with the extreme osmotic challenges in hypersaline waters, brine shrimp thrive with their unparalleled osmoregulation capabilities (Croghan 1958a, b; Marden et al. 2020). With little competition for resources and predation from other aquatic organisms, brine shrimp are ubiquitous throughout the south arm of GSL. For example, the biomass of brine shrimp in the south arm during peak production can surpass the equivalent biomass of 1.8 million humans of average weight (UDWR unpublished). As such, they are a valuable food source for avifauna, particularly waterbirds that are adapted to hypersaline waters. Eared grebes (*Podiceps nigricollis*), for example, eat an average of two brine shrimps per second while underwater, which equates to about 26,500–29,600 brine shrimp per day per grebe (Conover and Caudell 2009). GSL typically hosts anywhere from 2 to 5 million eared grebes per year, which at times is almost the entire North American population (Neill et al. 2016).

Brine shrimp also support a multimillion-dollar commercial fishery in GSL, where their cysts (overwintering eggs) are harvested, dried, and shipped around the world for use in aquaculture (Marden et al. 2020). Through adaptive management, researchers working with the Utah Division of Wildlife Resources (UDWR) in cooperation with the brine shrimp industry and other GSL stakeholders have worked for more than 20 years to create a stable and predictable brine shrimp population that benefits both harvesters and birds alike (Belovsky et al. 2011; Belovsky and Perschon 2019; Marden et al. 2020). Additionally, the brine shrimp harvesting industry, through fees and royalites, funds much of the ecological research about many aspects of the lake (Marden et al. 2020), including pelican monitoring and research.

Brine flies are also present in tremendous numbers at GSL. They begin their lives after hatching from eggs in the lake as benthic larvae that are mostly congregated on microbialites, organosedimentary carbonate structures with photosynthetic mats (Barrett and Belovsky 2020). Studies on the density of benthic brine fly larvae have shown an average density of 9140 per m^2 (Wurtsbaugh 2009) and peak benthic larval numbers can be as high as 25,000 per m^2 (UDWR unpublished). Following the pupation of larvae in the spring and summer, adult brine flies emerge and jettison to the water's surface where they become terrestrial organisms (Oldroyd 1964). Brine flies provide food for the largest staging concentration of Wilson's phalaropes and red-necked phalaropes (*Phalaropus lobatus*) in the world (Frank 2016).

10.2.3 North Arm of Great Salt Lake

The north arm of GSL is cut off from most direct freshwater inputs other than precipitation, and its salinity is between 25 and 29% or above (Baxter 2018). Brine shrimp and brine flies cannot survive at this salinity in great numbers, but it is a perfect environment for halophilic microorganisms to thrive (Baxter et al. 2005; Almeida-Dalmet and Baxter 2020). The pigments in these microorganisms give the water its pink color (Baxter 2018).

10.3 American White Pelicans of Gunnison Island

Within the pink hypersaline waters of the north arm lies a remote island that is home to one of the largest breeding colonies of American white pelicans in the world (Fig. 10.3). Gunnison Island is about 1.6 km long and 0.8 km wide, has very little vegetation, no freshwater, and is about 48 km away from the closest foraging habitat for pelicans (Fig. 10.1). This island might seem like an impractical place for a fish-eating waterbird to live and breed; however, the island is naturally protected from potential land-based disturbance from humans or predators, such as red fox (*Vulpes vulpes*) and coyote (*Canis latrans*).

10.3.1 Pelicans Make History

Written accounts of pelicans in the GSL region date back to the early- to mid-1800s. Peter Skene Ogden, a fur trapper and trader, captured a pelican in a beaver trap in the Bear River Valley circa 1825, and famed western United States explorer John C. Fremont caught a pelican near GSL in 1843 (Behle 1958). One of the first detailed observations of pelicans, particularly those nesting at GSL, is from the Stansbury expedition in 1850 (Stansbury 1852, Fig. 10.4). Tasked by the US Army with surveying and describing GSL valley flora, and fauna, Captain Howard Stansbury and his team left detailed descriptions of the location and behavior of pelicans

Fig. 10.3 Gunnison Island is surrounded by pink hypersaline water

Fig. 10.4 The eastern shore of Gunnison Island from the Stansbury expedition (1852) showing flocks of pelicans and gulls. Image is public domain

throughout the lake. In May of that year, the crew camped on what they called "Pelican Island," now known as Gunnison Island. John Hudson, a member of the crew, wrote that they saw "innumerable gulls and pelicans whose nests were found strewing the ground in every direction" (Madsen 1981), and the men began supplementing their diets with pelican and gull eggs. Pelican eggs were described as "very fine" with sweet butter by Albert Carrington, the crew's lead surveyor. They were able to capture a single adult pelican, attempting to keep it alive in the boat for the rest of the expedition. This unlucky subject "afforded [them] immense amusement, his indignation was excessive and expressed in the most ludicrous manner, by snapping in a very spiteful but impotent way..." (Madsen 1981). While this bird may have amused its captors, it did not fare well in captivity. After failing to keep it alive, they boiled it to extract its skeleton for their collection.

The crew observed pelican rookeries primarily on Gunnison and Egg islands during their expedition (Madsen 1981). There they witnessed the highly colonial nature of pelicans and how pre-fledglings cluster in groups called pods for protection when the adults are absent. In these colonies, they found the remains of fish, and determined that pelicans fly to the marshes and rivers of GSL and Utah Lake to catch fish and bring them back to the islands for their young. Stansbury made notes throughout the expedition that pelicans and gulls seemed to be the one sign of life in an otherwise very quiet and forlorn setting. Official numbers were scant in his final report, but observations were in the "thousands," or noted as "exceedingly abundant" (Stansbury 1852).

After the Stansbury Expedition, conflicts between pelicans and settlers in the region became more common. Many people thought that the pelicans were competing for valued game fish, such as trout, in the rivers and lakes around the Salt Lake Valley. There was suspicion that the Utah Game Commission, the precursor to the UDWR, was providing ammunition for anglers to shoot pelicans (Sugden 1936). Compared to the massive breeding populations observed by Stansbury and population estimates from the 1870s, populations in the early twentieth century were much lower and no longer present on Gunnison and Egg islands. Few pelicans were seen at their traditional feeding grounds on wetlands around GSL and Utah Lake (Thompson 1933). Later, the extermination of colonies by clubbing young pelicans on Gunnison Island was reported in local news (Deseret News and Telegram 1963, Fig. 10.5).

Aside from the overt results of extermination, pelicans were also under pressure from indirect human disturbances. In the 1890s, guano extraction moved onto the nesting islands (Carter 1996). Collecting, sifting, and packaging pelican and gull guano to sell as fertilizer kept adult pelicans away from their nests and exposed defenseless eggs and nestlings to avian predators and the elements (Lambourne 1909). However, the industry did not last long, and by the early 1900s, most companies abandoned their operations at GSL, leaving the islands available once again for nesting (Carter 1996). Small tourism industries also grew up around the lake at this time, with boats taking people out for the day on the lake, including landing at sensitive nesting locations like Hat Island (previously known as Bird Island). When people stepped on shore, adult pelicans would take flight, leaving young pelicans unprotected and vulnerable.

Pelicans also faced threats to reproductive success from poisoning and egg shell thinning from heavy pesticide use in the mid-twentieth century (Keith 2005). Knopf and Street (1974) collected 100 pelican eggs on five separate days throughout the breeding season (April and May) in 1972 from Gunnison Island in order to survey the incidence of pesticide residues in the breeding population. Results from

Vandals Destroy Pelicans

If backing a whole colony of young pelicans against a rock and shooting them with 22 rifles is any sport someone must have had a lot of fun recently.

Two men, using Tote Gotes to get to the island, killed 135 pelicans, wiping out an entire colony.

C. C. Sanders of Ogden, who has several brine shrimp stations around the lake, discovered the dead birds.

"Last year one man on a Tote Gote went to the island and shot up 85 birds.

"State and federal law enforcement agencies are assisting the sports men in keeping a close check on the birds," Mr. Sanders said. "We intend that this will never happen again or we will soon lose these rare birds."

"The pelican colonies have been photographed by some of the world's most famous naturalists. Our Utah colonies are rated attractions and we should do all we can to preserve them.

"Young pelicans are defenseless and make an easy target," Mr. Sanders said. "I cannot believe anyone would purposely destroy these birds in such numbers.

"People have a way about bragging on their ventures. We would like any information we can get as to who might have killed the birds. Last killing was about a week ago!"

This picture shows the destruction of a colony of pelicans on Gunnison Island in Great Salt Lake— Vandals used 22 rifles.

Fig. 10.5 Deseret News and Telegram (1963). From the William H. Behle Collection at the University of Utah Marriott Library Special Collections

55 randomly analyzed yolks (11 from each of 5 collection dates) showed measurable quantities of pesticide residues.

Pelican numbers began to increase once the persecution and disturbance decreased. Hat Island had an estimated 8000 breeding pelicans by the early 1930s (Thompson 1933), and William H. Behle (1935) counted 3300 nests, or 6600 breeding adults, on Gunnison Island in 1932. A drought in the mid-1930s dropped lake levels and dried up freshwater marshes and ponds throughout the GSL region (Gwynn 2002), reducing forage habitat and giving predators access to nesting islands via land bridges. These two factors led to a precipitous drop in the GSL pelican population, with only 2000 total adults counted on Gunnison Island and no documented nesting on Hat Island. Numbers rebounded to 3700 adults counted on Gunnison Island in the early 1940s, but the Hat Island colony never truly recovered; its former inhabitants were presumed to have moved to Gunnison Island (Behle 1958).

The consolidation of pelicans into a small number of large breeding colonies such as found on Gunnison Island raised concerns, particularly because Gunnison Island had no federal or state protections at the time to ensure the colony's success (Lies and Behle 1966). Knowing this, the Utah Audubon Society worked diligently with the cooperation of The Nature Conservancy to have Gunnison Island made into a refuge for pelicans. In 1944, due to the urging for protection, the UDWR closed the island and surrounding waters to human visitors from April to October; however, this was an emergency closure subject to year-to-year renewal. In 1977, Utah State Senator Thorpe Waddingham introduced the Pelican Management Act (Utah Code Annotated, Section 23-21a), to protect the pelicans on Gunnison and Hat islands for good. During his appeal for an amendment to his bill, officials from Salt Lake City's Hogle Zoo delivered a 5-year old, full-grown pelican to surprise the senator. Sen. Darrel Renstrom quipped, "This is the grossest lobbying I've ever seen." The purpose of Sen. Waddingham's bill was to provide a sanctuary for pelicans on Gunnison Island, and it passed (Salt Lake Tribune 1977). Because of the successful passage of this bill, Gunnison Island and Hat Island became state Wildlife Management Areas, and access to the islands was restricted to prevent people from disturbing the pelicans.

10.3.2 Life History of Great Salt Lake Pelicans

Pelicans typically reach sexual maturity around three years of age. Both male and female pelicans grow a large breeding horn on top of their bills and a crest of feathers on the back of their head (Fig. 10.6). Once a nesting pair mates and lays eggs, the breeding horn and crest are lost, and the feathers on the nape and crown become grayish (Dunn et al. 2017).

Breeding pelicans start to arrive on Gunnison Island in mid- to late-March, and they quickly form monogamous pairs and begin constructing their nests (Knopf 1975; Behle 1958). They typically construct their nests on the ground in an area that is relatively flat and consists of gravel, sand, or soil. Once a mate and nest site have been selected, both the male and the female rake the ground with their bills forming a shallow depression in the ground or a slightly raised mound. The depression is typically 0.6 m in diameter and normally no taller than 20 cm. The bottom or edges of the nest may include nearby vegetation (The Cornell Lab of Ornithology 2019). The pelican pairs will fiercely guard their nests from other pelicans or from co-nesting California gulls (*Larus Californicus*), which will prey upon unprotected eggs and young. To avoid tension with nest neighbors, the distance between each nest is about the length of two adult pelican bills (Behle 1958).

In order to forage and bring back food to their young on Gunnison Island, the pelicans fly at least 90 km round-trip to the freshwater wetlands along the eastern shores of GSL (Fig. 10.7). These foraging trips can often exceed 320 km round-trip, with pelicans known to fly to Utah Lake (270 km round-trip), Strawberry Reservoir (373 km), and American Falls Reservoir in Idaho (325 km) (UDWR 2019c). This

Fig. 10.6 Pelican with breeding horn taken with a motion-activated wildlife camera (PELIcam image credit: Utah Division of Wildlife Resources and Great Salt Lake Institute at Westminster College)

behavior highlights the importance of disturbance-free nesting grounds to pelicans, as flying so far for food is energetically costly.

10.3.3 Conflicts with Humans

10.3.3.1 Fisheries Conflicts

Being such a large bird, pelicans require a lot of calories to maintain a healthy weight. Pelicans eat mostly fish, but they are also known to eat salamanders, small mammals, and crayfish (Knopf and Evans 2004). Studies of daily fish consumption by adult pelicans near two breeding colonies at Yellowstone Lake in Wyoming and Pyramid Lake in Nevada estimated 0.85–1.8 kg of fish were eaten per day to sustain themselves (Hall 1925; Stapp and Hayward 2002). With an average of 12,541 pelicans around GSL from April through September, it is easy to assume that they are eating a lot of fish (Paul and Manning 2002).

Additional food is necessary to raise a young pelican from chick to fledgling. Young pelicans are completely dependent on their parents and require an estimated 68 kg of fish from the time they leave the nest until they can forage on their own (Hall 1925).

A pelican's diet of mainly fish is one of the reasons they can find themselves at odds with humans. The Bonneville cutthroat trout (*Oncorhynchus clarkii utah*) is a subspecies of cutthroat trout native to the Bonneville Basin of Utah, Wyoming,

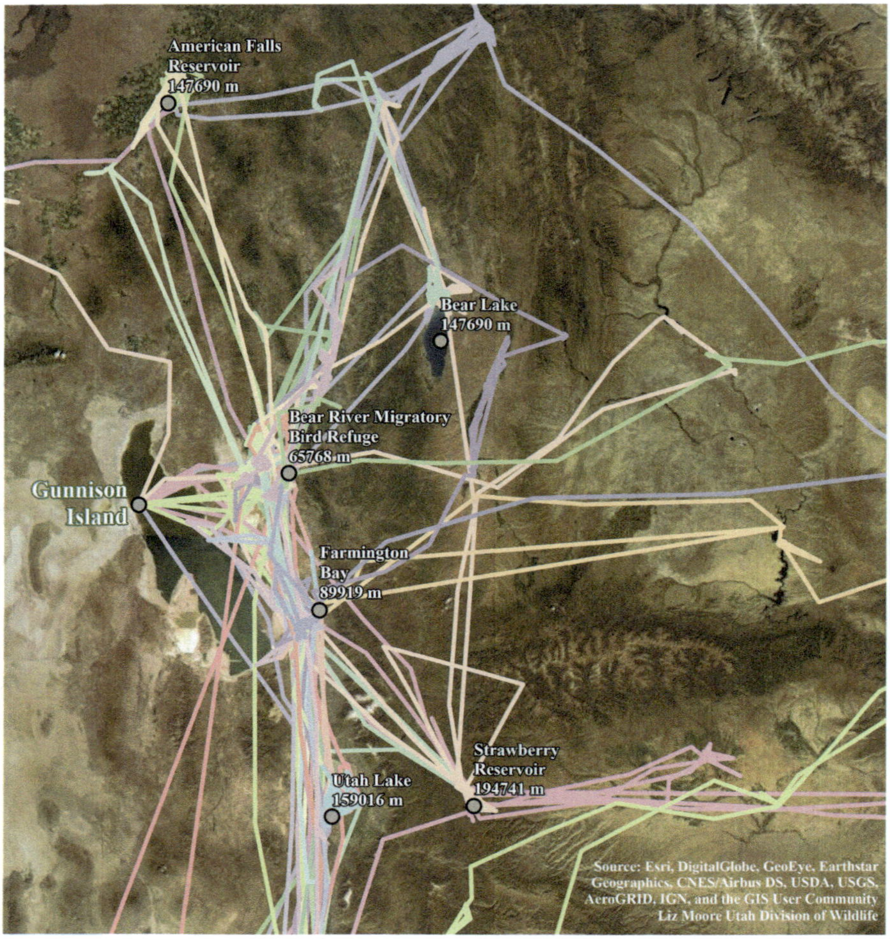

Fig. 10.7 Individual American white pelican flight patterns around Great Salt Lake tracked by telemetry by Utah Division of Wildlife Resources in conjunction with the Salt Lake City International Airport and Tracy Aviary. Each color denotes the track of a different pelican (UDWR 2019c)

Idaho, and Nevada. Genetically pure Bonneville cutthroat trout are rare throughout their historic range, but several Utah populations exist, including populations in Bear Lake and Strawberry Reservoir. Because of the many threats to this fish, such as competition and hybridization, the local subspecies is included on the *Utah Sensitive Species List* (UDWR 2017). Concerns over trout consumption by pelicans make a conundrum for wildlife managers and anglers. In fact, this has been a dilemma for over 100 years, even though the belief that Utah pelicans consume a large amount of game fish has been discredited.

Fortunately, for biologists it is relatively easy to assess the diet of pelicans, because they regurgitate their stomach contents when disturbed. Even as early as the 1930s, Behle (1958) found that the stomach contents of pelicans contained slow-moving, nongame fish such as carp, chubs, and suckers. His analysis of the regurgitated materials only occasionally showed the remains of trout and bass. Morgan (1947) wrote of Behle's research:

> A better public attitude toward the fish-eating birds of Great Salt Lake is likely to result from the studies of scholars like Dr. Behle, who has learned from stomach contents that the Salt Lake cormorants and pelicans today feed almost entirely on trash fish like carp, chubs, and suckers, not upon trout and other game fish. As the sportsmen themselves inveigh against the trash fish, they may come to forget their long hostility and accept the fisher-birds as allies.

A more recent study was conducted by Utah State University to investigate the role that pelicans play in Bonneville cutthroat trout predation at Strawberry Reservoir, a location popular with anglers (UDWR 2019a). This study showed that over 90% of pelican diet samples consisted of nongame fish: Utah sucker (85%, *Catostomus ardens*) and Utah chub (6%, *Gila atraria*). The pelicans consumed an average of 19 nongame fish for every spawning cutthroat over a 2-year period. It is estimated that pelicans remove 50,000 Utah suckers per year from Strawberry Reservoir. Consumption of suckers is a potential ecosystem service that pelicans provide by removing competition to Bonneville cutthroat trout (Budy et al. 2016). This research provides more evidence that pelicans are generalist, opportunistic predators, and typically prey on nongame fish (Hall 1925; Marshall and Giles 1953; Evans et al. 2016; Budy et al. 2016).

Additionally, habitat where pelicans feed does not overlap with most game fish habitat. Pelicans feed in shallow, warm water, and game fish are often found in deep, cool water (Budy et al. 2016); however, there is evidence that pelicans can impact Bonneville cutthroat trout movement and reproduction during the spawning season. Pelicans are known to forage in the shallow water where streams empty into the reservoir, and this is where the trout pass to spawn. Pelicans can directly affect spawning by eating the fish or deterring trout from entering the streams thereby impeding their reproduction. This effect has been demonstrated to be density dependent. If the density of pelicans is below a certain threshold, it does not affect trout movement (Budy et al. 2016). Questions remain on whether Bonneville cutthroat trout return to enter the stream again after being deterred by the pelicans. This research has important management implications as hazing pelicans during a short period of time may be an effective management strategy when pelican densities exceed the critical threshold impacting spawning game fish (Budy et al. 2016).

10.3.3.2 Airplane Collisions

American white pelicans are a serious concern for airport hazard control. They are large, heavy, and abundant birds that move frequently in social flocks and commonly fly at the same altitudes as descending and ascending aircraft. Recent data gathered by the UDWR shows that pelicans can ride thermals to an altitude of 9100 m near the Salt Lake City International Airport, which is located adjacent to wetlands that serve

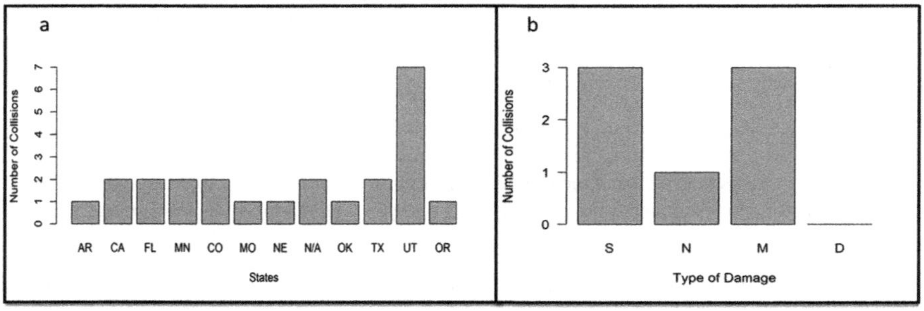

Fig. 10.8 (**a**) Number of American white pelican collisions with airplanes nationwide (FAA Wildlife Strike Database 2019). Considering states with reported collisions, Utah is ranked highest (AR Arkansas, CA California, FL Florida, MN Minnesota, CO Colorado, MO Missouri, NE Nebraska, N/A Unknown, OK Oklahoma, TX Texas, UT Utah, OR Oregon). (**b**) Reported damage following incidents regarding airstrikes with American white pelicans in Utah (S Substantial, N None, M Minor, D Destroyed) (FAA Wildlife Strike Database 2019)

as a central hub of pelicans and other migrating birds for both annual and seasonal movements.

According to the Federal Aviation Administration (FAA) airstrike data, the number of bird strikes in the United States has increased 62% from the 1990s to 2007 (Caudell 2009). Utah has reported seven airplane collisions with pelicans at the Salt Lake City International Airport, three of which have caused "Substantial Damage" to aircraft (Fig. 10.8) (FAA Wildlife Strike Database 2019). This risk obviously poses a direct threat to human lives.

10.3.3.3 Water Availability

Human development within the GSL watershed has had dramatic effects on both the lake and the pelicans that utilize its various aquatic habitats. The diversion and storage of natural streamflow to supply water for human use are attributed to ~3.4 m of elevation loss (~48% of total volume) in GSL since European settlement began in the mid-1800s (Wurtsbaugh et al. 2016, 2017). This loss of water, compounded by extended regional droughts (Gywnn 2002), uncovers a land bridge to Gunnison Island when the surface elevation of the north arm reaches 1278.2 m and below (Fig. 10.9). Access to the island via the land bridge leaves pelican nests and young vulnerable to disturbance by human visitors and even predation by coyotes. Improper water management has resulted in reduced habitat for pelicans in the past (Thompson 1933) and is still challenging wildlife managers throughout their range (Moreno-Matiella and Anderson 2005).

Additionally, water diversions upstream of the lake (Null and Wurtsbaugh 2020) reduce the availability of freshwater in natural marshes and impounded wetlands that pelicans utilize as foraging grounds. Given the uncertainty of the amount of

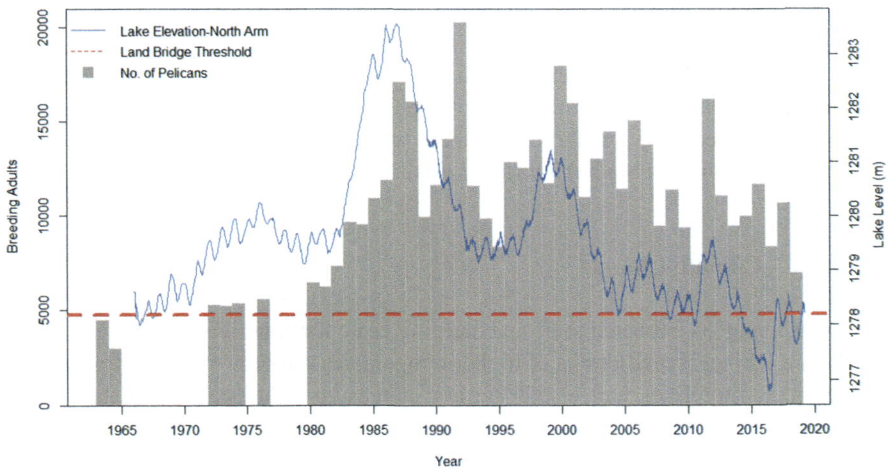

Fig. 10.9 The total number of breeding American white pelicans on Gunnison Island recorded by the Utah Division of Wildlife Resources. The blue line represents the elevation of GSL's north arm (United States Geological Survey 2019) over time, and the dotted line represents the lake level below which the island becomes connected to the land

precipitation under current climate change scenarios (Cook et al. 2015; Gillies et al. 2015; Baxter and Butler 2020), both GSL and surrounding wetlands could be at risk of further degradation or loss.

There are other saline lake systems in the world that have decreased significantly in size or dried up completely, and the ramifications of this on wildlife have been documented. For example, the salinity of the Salton Sea has risen from 0.4% in 1905 to over 5% in recent years (Lyons et al. 2018). As a consequence, the fish population has decreased by over 95%, and pelican use is now dwindling. In fact, according to the California Chapter of the Audubon Society, the number of pelicans decreased from a range of 500–2000 pelicans to zero by 2017. GSL pelicans are known to stop at the Salton Sea during their migration (UDWR 2019c). If these pelicans have to fly elsewhere for food, it could potentially affect their survival. These data reflect the importance of pelican feeding grounds to their survival indicating that sufficient water levels in the wetlands of GSL are just as important as ensuring their breeding grounds are inaccessible to predation and disturbance.

An example showing the effects of reduced inflows into GSL took place during the GSL Waterbird Survey from 1997 to 2001. Following three years of normal (~0.6 m) or above normal increases in spring lake level, the lake rose only 0.3 m in 2000 and merely 0.15 m in 2001. During the low water inflow years, waterbird surveyors often observed both fish mortality and loss of shallow water habitat around GSL. In 2001, the available wetland habitat at the Bear River Migratory Bird Refuge was less than 27% of their capacity, and the Willard Spur, an important foraging ground for pelicans in the northwest portion of GSL, turned completely dry during the summer (Paul and Manning 2002).

Although the North American population of pelicans is currently stable and has increased from past lows in the 1960s, biologists and researchers alike are unsure of how the local and regional stressors will affect the population of pelicans in Utah. To maintain the current nesting population of pelicans at GSL, as much as 16% of the continental breeding population, managers will need input in the state's water use planning to ensure viable habitat (Neill et al. 2017).

10.3.3.4 Climate Change

Climate change is expected to be an emerging issue for American white pelicans and GSL as a whole. If predictions of higher temperatures and lower snowpack amounts are correct, shorelines will recede further increasing the duration land bridges are exposed, decreasing the size and number of feeding areas, and reducing food abundance (Baxter and Butler 2020).

Weather patterns are also expected to change and pelicans in other areas of the country have begun to arrive earlier and earlier to their breeding grounds. An increase in volatile spring storms could lead to higher chick mortality (Sovada et al. 2014). Adults may abandon their nests for self-protection more frequently during these storms, exposing eggs and chicks to environmental stress and predation. California gulls, which co-nest with pelicans on Gunnison Island, are persistent and aggressive nuisances to pelican nests when they are unattended (Behle 1958; Luft 2019). A single gull has been seen to hop from nest to nest breaking unattended eggs without stopping to ingest their contents (Behle 1958). The threats from these avian predators compound those of coyotes, as nests and young are exposed more often due to erratic weather patterns. If an adult is kept away from the nest for an extended period of time, eggs or chicks are exposed to the elements and can become too cold or can overheat (Behle 1958).

Even those young pelicans lucky enough to make it past the helpless phase still have to be wary of threats to their well-being. According to observations made by Behle (1958), when juvenile pelicans are disturbed while they are huddled together in a pod, they will be startled into a stampede. Stampedes can cause birds to become trampled, fall from rocks or cliffs, and can even cause birds to get hung up in the bushes. Even if a bird can survive the stampede, they normally get disoriented and wander too far from their nest, and if not found by their parents, they starve to death. Those individuals who become isolated from the colony are often times attacked by gulls and, if not killed by their attacker, succumb to their injuries later (Behle 1958).

10.4 Current Pelican Research

The UDWR studies many aspects of pelican life history in Utah. The studies described below are often completed with help from a variety of community partners.

10.4.1 Aerial Surveys of Gunnison Island

Aerial photographic surveys of Gunnison Island provide an efficient and reliable census of the pelican breeding population with minimal disturbance to the colony. This survey technique was first described by Knopf (1975) as an efficient and consistent approach to monitoring of the pelican colony. His ideas were refined and adopted as a regular annual survey by the UDWR in 1980 after an initial trial run in 1976. During a survey flight conducted around May 20th each year, a biologist takes photographs of each individual pelican subcolony along with wide-angle reference photographs to capture the relationship of subcolonies within the whole colony on the island to avoid counting pelicans twice. Pictures are compared and analyzed with each nonflying adult assumed to be on one nest. This census technique gives a reliable estimate of nests and total breeding adults on Gunnison Island even though some adults are not on nests and some are yet to arrive for the season. Another aerial photographic survey is held around June 20th each year with the goal of capturing late arriving pelicans. In 2004, a mid-July survey was added to assess annual productivity of the colony by counting the number of juveniles produced during the nesting season (Neill et al. 2017).

The aerial pelican census with the highest recorded count on Gunnison Island occurred in 1992 with 20,270 breeding adults, while the lowest was in 1976, the first year of the aerial census, with 5590 breeding adults. The most recent census from 2019 had a total of 6928 breeding adults (Fig. 10.9). Over the last 10 years from 2010 to 2019, the breeding population averaged 10,090 adult pelicans.

Nest success can be measured by comparing the number of young pelicans produced with the number of nests counted during the May census. By the time of the July survey, most of the young pelicans are old enough to be easily identifiable by their size and pale gray color of their bills and feet, yet few are capable of flying off the island. Rawley (1976), who conducted the first aerial survey of Gunnison Island, used Knopf's (1975) dissertation data on the island's pelicans to calculate a nest success rate of 0.69 juvenile pelicans per nest, or 69% of all nests produced one pelican. July count data since 2004 show that nest success fluctuates from year-to-year and tends to be lower than initially calculated by Rawley (1976; Fig. 10.10).

10.4.2 Mark-Resight Research

The UDWR has been conducting mark-resight studies since 2011 in order to understand the survivorship and movement patterns of pelicans from Gunnison Island. The goal of this method is to capture and mark 500 juvenile pelicans every year. The flightless pelicans are herded into a pen, where biologists and volunteers fit each juvenile pelican with a metal leg band and a pair of patagial wing tags with unique numbers to identify individual pelicans. This operation is done early in the morning and as quickly as possible with the goal of minimizing the stress placed on

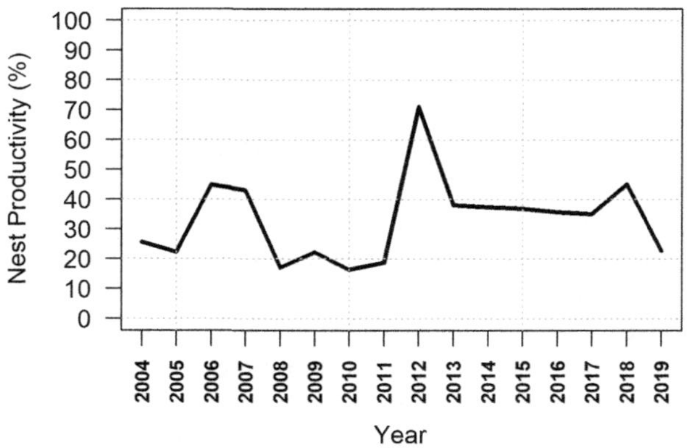

Fig. 10.10 Nest success at Gunnison Island, Utah

the juvenile pelicans. After the juvenile pelicans leave Gunnison Island, the hope is that the general public and perhaps other scientists working near or around pelican habitat will report sightings of marked pelicans to the UDWR or the National Banding Lab run by the US Geological Survey (reportband.gov). Reports of marked pelicans allow the UDWR and other researchers to increase their understanding of pelican movement patterns and survivorship (Fig. 10.11).

From 2011 to 2019, 3714 juvenile pelicans have been banded and tagged by the UDWR on Gunnison Island. This project has resulted in the resight of 432 unique individual pelicans after leaving Gunnison Island or 12% of the total pelicans banded. Some of the pelicans resighted alive were seen on multiple occasions, with one pelican being reported 33 times, so there are a total of 750 resights of pelicans away from Gunnison Island (Fig. 10.12). Of the 432 unique pelican resights, 187 were seen alive, but eight of these were later found dead. Another 238 unique pelicans were also found dead, and for seven pelicans, the resight report was unclear if the pelican was dead or alive (UDWR Unpublished data). Additional resights will improve modeling of the pelican population, survivorship, and other variables. Outreach efforts described below strive to increase community awareness and increase the resight rate of pelicans by researchers, birders, photographers, and the general public.

10.4.3 Satellite Tracking

In 2014, the UDWR began efforts to place Platform Transmitter Terminals (PTT) on American white pelicans. The PTTs record periodic location data (latitude, longitude, and altitude) of each pelican and transmit the data to a satellite network, which are then relayed to a ground-based database. This research is a partnership of ongoing cooperative efforts among numerous organizations, including Salt Lake City International Airport, Endangered Species Mitigation Fund of the Utah

Fig. 10.11 Tagging and banding of American white pelicans. (**a**) Biologists from Utah Division of Wildlife Resources working quickly to attach patagial wing tags and a leg band to a juvenile pelican on Gunnison Island, Utah. (**b**) A juvenile pelican walking away after it has been tagged with unique patagial wing tags. (**c**) A tag found on Gunnison Island during the routine tag sweeps performed by Westminster College's Great Salt Lake Institute, Tracy Aviary, Utah Division of Wildlife Resources and volunteers each October. (**d**) A deceased pelican found on Gunnison Island during routine tag sweeps in October

Department of Natural Resources, Utah Division of Wildlife Resources, US Army Dugway Proving Ground, and Tracy Aviary.

In order to attach the PTTs, crews capture adult pelicans with foothold traps hidden on land or just underwater near the shore of a roosting island or mudflat. The traps are modified with padded jaws, a light spring, and anchored by a bungee cord to lessen the chance of injury to the pelican. Once a pelican is captured, it is fitted with a 70-gram PTT, similar to a small backpack, in a way that allows it to swim, fly, and forage. These satellite transmitters allow the tracking of pelican movements hourly throughout the year by satellites from the Argos location and data collection

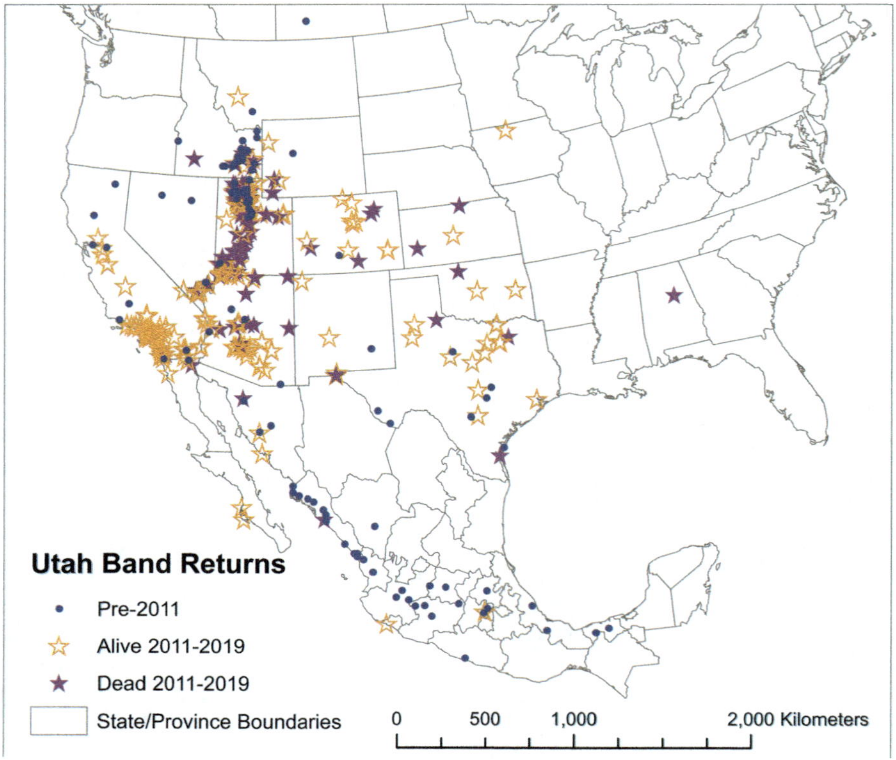

Fig. 10.12 Sightings of tagged and banded pelicans, from the Gunnison Island colony, reported to the Utah Division of Wildlife Resources

system. In 2012 and 2015–2018, 68 pelicans were fitted with GPS backpacks. Each pelican captured was given a name and can be individually tracked on the Pelitrack website created and hosted by the UDWR (2019c).

The telemetry data have provided researchers with a better understanding of the complicated patterns of pelican arrival to and departure from GSL wetlands. These data show the center of activity lies over the eastern wetlands of GSL and along the most frequently used locations between preferred water bodies (Fig. 10.7). The results also show that there is a high level of near-daily interconnectedness among colonies and foraging grounds, movements across multiple states during the breeding season, and common daily movement distances of over 320 linear kilometers (UDWR 2019c).

Not only do the telemetry results provide much of the detailed insight into the regularity, seasonality, location, and altitude of pelican movements needed to assess airframe strike hazard, but the data reflect the importance of freshwater to pelican survival. As pelicans migrate each spring and fall, they are tethered to freshwater sources along their journey. They stop to forage at freshwater bodies as people stop at gas stations during a long car trip.

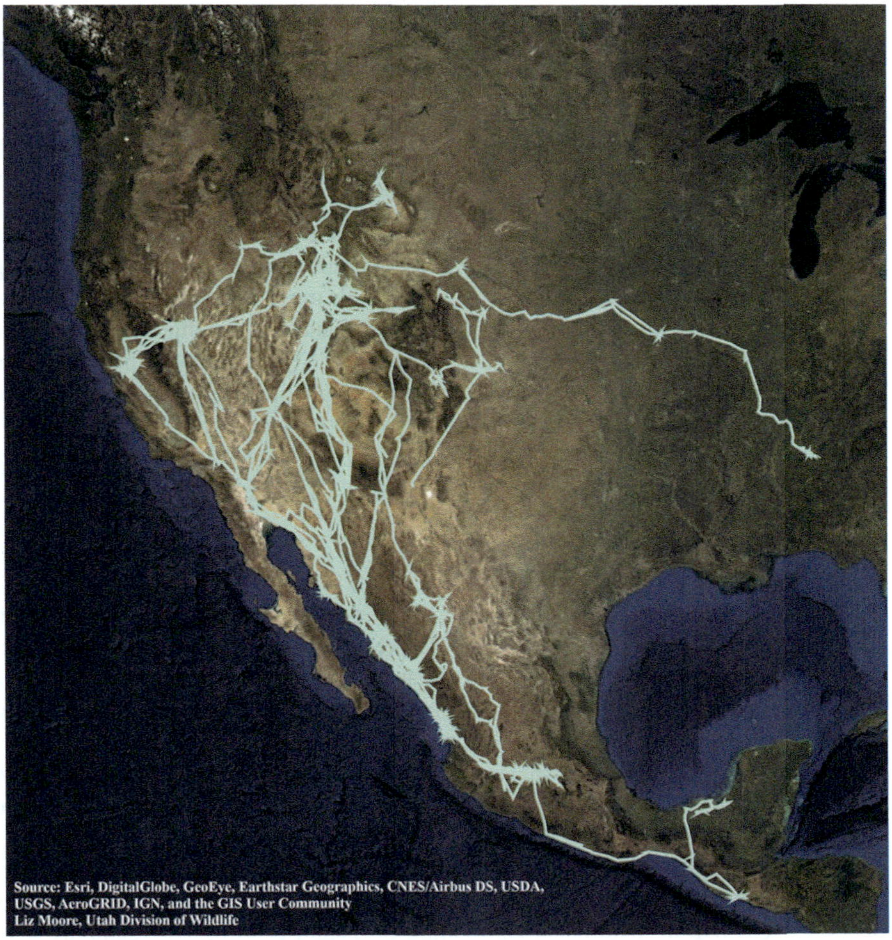

Fig. 10.13 Movement patterns of pelicans over a 1-year period captured at Great Salt Lake, Utah (April 2018 to April 2019)

The PTT data have also uncovered important information about migration routes. American white pelicans have been known to range as far north as Canada during the breeding season and migrate as far south as southern Guatemala during the winter, yet little was known about the path and pattern of these migrations. It was thought that pelicans could easily be separated into two separate geographical populations; generally, those that breed east of the Rocky Mountains and those that breed west of the Rocky Mountains. The eastern population typically migrates to the Gulf of Mexico for winter, while the western population migrates to the Pacific Coast of southern California and Mexico (Knopf and Evans 2004). However, the results of the UDWR movement study paints a different picture (Fig. 10.13), with regular movements across the Rocky Mountains each fall.

Most importantly, GSL is a hub for migratory movements of pelicans across the region. Migratory waterbird survey data by the UDWR (e.g., Paul and Manning 2002) and mark-resight data of wing-tagged pelicans (UDWR 2019c) concur that the value of GSL wetlands is not just for breeding resident pelicans; vastly more birds visit GSL's wetlands than nest on Gunnison Island. Groups of migratory and nonbreeding pelicans contribute the majority of wetland activity near the Salt Lake City International Airport.

While telemetry data clearly show breeding adults keep regular daily habits focused on productive wetlands and nesting, juveniles, subadults, and nonbreeding adults tend to be more mobile and irregular in their movements. Population control actions sometimes proposed for the Gunnison Island colony that seek to reduce potential pelican–plane collisions have the potential to backfire in the short term (by scattering thousands of breeding individuals into the wetlands closer to the airport) or have a disproportionate long-term effect through colony collapse.

10.4.4 The PELI Project

The "**P**roject in **E**ducation and **L**ongitudinal **I**nvestigation of pelicans on Gunnison Island" was created in 2016 by the UDWR, Great Salt Lake Institute at Westminster College, Tracy Aviary, MesoWest, and Great Salt Lake Audubon. Dubbed the PELI Project, it had the goals of (1) increasing community involvement with GSL and pelicans; (2) increasing the understanding of pelican life history strategies such as migration, feeding, and survival; and (3) understanding the consequences of low water levels for GSL and resulting land bridges that allow humans and mammalian predators onto Gunnison Island.

The project goals are accomplished through a variety of methods. Most importantly, observations of pelicans were needed on Gunnison Island. Because pelicans are very sensitive to disturbance and Gunnison Island is so remote, cameras were chosen as a way to have eyes on the island and the pelicans. The PELIcams, as they are affectionately called, now record all aspects of life on Gunnison Island.

The PELIcams consist of one real-time camera, which was installed on Gunnison Island on March 8, 2017. This camera is trained on the pelicans breeding in Lambourne Bay, eastern shore of the island, and takes a long-range image of multiple pelican subcolonies every three minutes (Fig. 10.14). Each image is transmitted with cell phone technology from the island directly to a database where it is automatically uploaded to a website for viewing. These images allow researchers to study the timing of arrival, nest initiation, and nesting strategies, such as formation of subcolonies. The real-time camera also photographically records weather events and is operated and maintained by MesoWest at the University of Utah (Horel et al. 2002; MesoWest 2019; Great Salt Lake Institute 2019).

Along with the real-time camera, 14 motion-activated cameras (6 cameras removed in 2019 to refine data collection) were strategically placed throughout Gunnison Island to get close-up views of pelican behavior (mating, nesting, chick

Fig. 10.14 Long-range camera view of Gunnison Island's Lambourne Bay taken by the live PELIcam (Photo credit: MesoWest at the University of Utah)

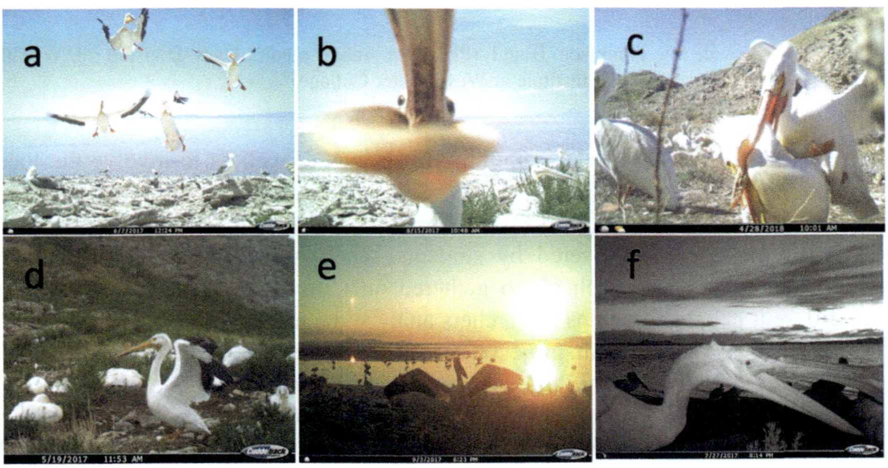

Fig. 10.15 Close-up images of pelicans taken with motion-activated cameras. (**a**) Landing on the island, (**b**) curious about the camera, (**c**) mating, (**d**) standing over eggs, (**e**) wing stretching, (**f**) other interactions (PELIcam image credit: Utah Divison of Wildlife Resources and Great Salt Lake Institute at Westminster College)

rearing) and capture possible predation by gulls or mammals. Some cameras face directly at nesting sites while others are aimed toward potential predator access points and provide data collection throughout the year. Since 2017, scientists have captured three breeding seasons of data with over 500,000 images and counting (Fig. 10.15).

Fig. 10.16 Coyote on Gunnison Island (PELIcam image credit: Utah Division of Wildlife Resources and Great Salt Lake Institute at Westmisnter College)

In order to analyze the large number of images as well as to involve the local community in the study of pelicans, Great Salt Lake Institute at Westminster College initiated a citizen science project on Zooniverse, which is a popular worldwide platform for research powered by students, teachers, and community members from their computers. Each photo gathered on Gunnison Island is looked at by 15 different people and helps researchers analyze the multitude of pictures taken by the PELIcams. The PELIcam project has 3391 volunteers (Zooniverse 2019) and has been featured on a National Public Radio program (Science Friday 2018).

While it will take years to analyze all of the data collected through the PELIcam project, researchers were able to confirm that coyotes were on Gunnison Island at all times of the year in 2018 (Fig. 10.16). Direct nest or young predation by coyotes has not been detected on the images. However, if coyotes were preying on young pelicans, biologists would expect to see the adult pelicans abandon the entire colony and expose eggs and young pelicans to nesting gulls or harmful weather conditions that could lead to their death. It is hypothesized that the coyotes focus their hunting on their preferred food source, the island's abundant rodent population, and they limit impacts to nesting pelicans by staying on the south end of the island where the land bridge connects (Luft 2019). Recent censuses show pelicans prefer to nest on the northern half of Gunnison Island, likely in response to the presence of coyotes on the southern half; however, several cases of pelican subcolonies being lost to

Fig. 10.17 (**a**) Pouch stretching, (**b**) swimming in the hypersaline water, (**c, d**) flying practice during wind events (PELIcam image credit: Utah Division of Wildlife Resources and Great Salt Lake Institute at Westminster College)

predation and disturbance have been documented in the past after the formation of a land bridge to breeding islands (Lies and Behle 1966; Low et al. 1950).

PELIcam images are responsible for debunking the "salt death theory" of pelican mortality dating to the 1930s. When dead pelicans were found encrusted in salt along routes from Gunnison Island to foraging areas, people assumed that it was caused by swimming in the hypersaline north arm water. It was thought that when they landed on the water salt would immediately build upon their feathers and lead to death by weighing them down and preventing them from flying. The salt encrusted pelicans found dead on the east side of the north arm are juvenile pelicans that did not have enough energy or fat reserves to make it on their maiden voyage to the feeding grounds in wetlands east of GSL and not birds that were killed by the buildup of salt on their feathers. While it is possible that spending too much time in the hypersaline water could lead to juveniles becoming encrusted with salt, the PELIcams on Gunnison Island regularly capture pelicans swimming in the water throughout the breeding season without the significant buildup of salt (Fig. 10.17b). Coincidentally, when the pelicans migrate, some of the encrusted salt may be transferring halophilic microorganisms to other saline environments (Kemp et al. 2018).

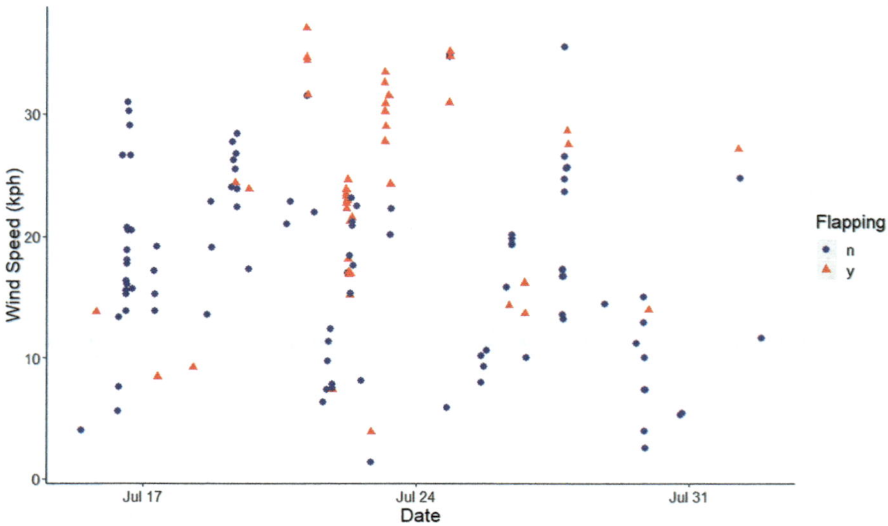

Fig. 10.18 Juvenile pelican flapping behavior. Each point represents a juvenile pelican seen in PELIcam images on the beach during summer, (red triangle if flapping was noted, blue circle when no flapping was detected). The frequency of flapping increases with increasing wind speeds, confirming the theory that young pelicans use wind events as practice for fledging

Other PELIcams captured juvenile pelicans stretching and flapping their wings to assist in the strengthening of flight muscles. A recent UDWR analysis from July 2017 shows a significant correlation between increased wind speed and flapping behavior, presumably to practice flying by achieving lift with less effort (Figs. 10.17c, d and 10.18). These data agree with Schaller (1964) who found increased wing beats and flight attempts in pelican pre-fledglings with higher winds near Yellowstone Lake, Wyoming; however, fine-scale parsing of these flight behaviors (coordinated/uncoordinated wing beats, hopping, gliding) is difficult using still images from motion-activated cameras.

The PELIcams are also responsible for documenting six species of birds that have never been recorded on Gunnison Island before. These species include red-tailed hawk (*Buteo jamaicensis*), ferruginous hawk (*Buteo regalis*), barn owl (*Tyto alba*), great horned owl (*Bubo virginianus*), mountain bluebird (*Sialia currucoides*), and European starling (*Sturnus vulgaris*) (UDWR 2019b). The ability to capture these birds visiting the island has given researchers an understanding into the importance of Gunnison Island for raptors and songbirds.

Another aspect of the PELI Project was collecting morphometric data from juvenile pelicans on Gunnison Island during the banding events (see Sect. 10.4.2) in 2017 and 2018. Very little is known regarding the development of young pelicans, especially those about to fledge and leave the island; therefore, the PELI Project measured juvenile, tarsal-metatarsal diameter, tarsal length, culmen length, weight, and parasite load (Fig. 10.19; Schreiber and Schreiber 1978). Parasite load was a

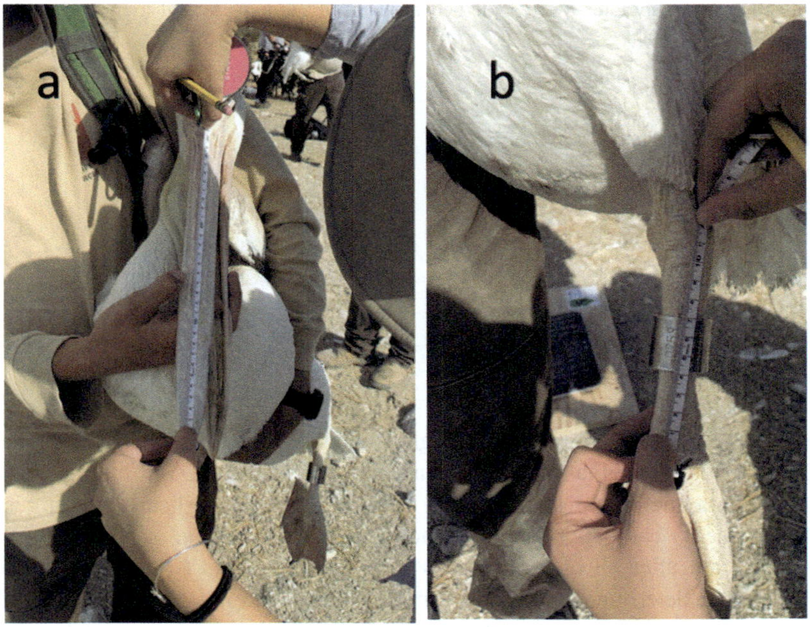

Fig. 10.19 Measuring young pelicans on Gunnison Island. (**a**) Culmen length measurement. (**b**) Tarsal length measurement

Table 10.1 Morphometric data collected from 509 fledgling pelicans during the 2017 and 2018 banding events

	Tarsal-metatarsal diameter (cm)	Tarsal length (cm)	Culmen length (cm)	Weight (kg)	Parasite load (0–2)
Banded ($n = 488$)	5.4 (0.9)	8.56 (1.51)	22.2 (5.9)	6.3 (1.5)	1.47 (0.48)
Non-banded ($n = 21$)	4.65 (1.9)	7.74 (1.99)	12.7 (2.8)	4.3 (1.1)	1.72 (0.46)
Combined ($n = 509$)	5.3 (1.7)	8.01 (2.4)	19.3 (4.7)	5.7 (1.5)	1.5 (0.55)

Numbers are averages (±) one standard deviation. Qualitative rating of parasite load inside the pouch ranged from none (0) to complete coverage (2)

qualitative rating from 0 to 2, with a rating of 2 signifying complete coverage of the interior portion of a pelican's pouch by lice. Over the 2-year project, data were collected on 509 banded and non-banded juvenile pelicans. Non-banded juveniles were suspected of being too small or weak to be banded and removed from the group of captured juveniles prior to banding (Table 10.1).

In general, banded pelicans are heavier, have longer culmen and tarsal lengths, and have a lower average parasite load. This could be the inherent variability in fledgling health, or it could be simply age differences, with the younger birds having

Table 10.2 Mortality rates for juvenile pelicans banded and wing tagged from 2011 to 2018 on Gunnison Island, Utah

Year	Number banded	Number dead	Mortality rate (%)
2011	200	4	2.0
2012	400	40	10.0
2013	500	32	6.4
2014	541	86	15.9
2015	500	25	5.0
2016	500	34	6.8
2017	500	12	2.4
2018	499	57	11.4
Totals	3640	290	8.0

smaller overall measurements and higher parasite loads. The decision to band a pelican or not is a qualitative one but made by experienced UDWR biologists. Based on these data, the biologists are clearly choosing fledgling pelicans of a larger size, which in theory can better handle the added stress of banding.

In the fall, after all of the pelicans have left the island (October), a survey is completed to determine how many banded pelicans did not make it off the island. Comparing morphometric data to banded pelicans that died prior to leaving the island may help us better understand juvenile mortality on the island.

The mortality of young pelicans was assessed in 2017 and 2018 by scanning the island during the postbreeding season to look for marked pelicans. Of the 500 pelicans banded in 2017, the remains of 12 deceased pelicans have been found on Gunnison Island. This result constitutes a pre-fledgling mortality of at least 2.4%, and 97.6% of the banded pelicans either made it off the island or have not been found on the island yet.

Many factors likely contributed to the increased mortality from 2017 to 2018. The biggest potential contributor to data variability may be that juveniles were banded four weeks later in 2017, as compared to 2018, and older, larger birds made up a greater proportion of the pre-fledgling population. In 2018, smaller pelicans made up a greater proportion of the overall banded population when compared to 2017. Smaller pelicans may be more susceptible to natural stressors like sibling competition, weather exposure, and predation. Additionally, drought conditions intensified in 2018 as compared to previous years. Continued exposure of the land bridge provided predators reliable access to the island while food availability for pelicans may be impacted by drought conditions as the amount of available water for pelican foraging habitat decreased. The additive stressors experienced in 2018 may have contributed to the higher fledgling mortality on Gunnison Island. Deciphering the causes of high and low mortality rates are complex as the causes may have fluctuated since banding and wing tagging began in 2011 (Table 10.2). More research is needed in order to fully describe and interpret the observed patterns.

10.4.5 Other Research

Pelicans are known to be susceptible to West Nile Virus and contamination from the bioaccumulation of pesticide residues and heavy metals due to eating high on the food chain (Knopf and Street 1974; Benson et al. 1976; US Fish and Wildlife Service 1984; Sovada et al. 2008). On June 23, 2009, 20 juvenile pelicans on Gunnison Island were tested for West Nile Virus and heavy metal concentrations in blood and feathers; however, testing was likely too early in the season to observe an outbreak of the virus in the birds. In Utah, July is typically when the mosquito (*Culex* spp.) is first observed and tests positive for the disease. Mosquitoes were present and biting during the ground research in 2009, but specimens were not collected. The West Nile Virus mosquito vector is more active at night and two overnight carbon dioxide traps were placed on the north end of Gunnison Island on July 14–15, 2014. A total of six *Culex tarsalis* mosquitoes were identified from both traps combined, but at least ten mosquitoes are needed for a reliable diagnostic test. No other species of mosquitos were collected.

A full metal scan on the pelican blood and feathers tested for 30 constituents. The blood samples ($n = 15$) showed lower metal concentrations than breast feathers ($n = 20$) for all metals except potassium in 14 pelicans and iron in one pelican. Total mercury concentrations had a geometric mean of 0.330 parts per million (ppm) wet weight ($\sigma = 0.252$) for blood (range 0.131–0.929) and 4.413 ppm dry weight ($\sigma = 2.130$) for breast feathers (range 2.586–10.144) (Neill et al. 2009).

A geometric mean of 0.330 ppm wet weight total mercury for juvenile pelican blood from Gunnison Island could potentially fall within the high-risk category (0.3–0.4 ppm wet weight) for methylmercury, as defined in Evers et al. (2004) with their work on known impacts to common loons (*Gavia immer*). High-risk denotes exposure could lead to impacts at the molecular, individual, or population level. These risk levels pertain to common loon juveniles, aged three to five weeks old. Blood risk categories are different for adults with <1.0 ppm wet weight considered low risk (NOAEL, no observed adverse effect level). Juveniles and molting birds have the ability to depurate mercury into their feathers. The juvenile feathers would fall in the low-risk category (0–9 ppm dry weight; Evers et al. 2004). This apparent discrepancy between the level of risk for blood and feathers may be explained by analyzing the percent methylmercury in each medium rather than using total mercury values (Neill et al. 2009).

10.5 The Future

Pelicans are one of GSL's most charismatic waterbirds. While their populations have rebounded in recent years, the future of one of their largest breeding grounds at GSL is not certain. The elevation and surface area of the lake naturally ebbs and flows, but an overall trend of decreasing size has become apparent (Wurtsbaugh et al. 2017). When the lake level is low, land bridges form to Gunnison Island allowing predators access to pelican breeding grounds. Less water availability for wetlands also impacts food resources for pelicans.

Long-term monitoring of pelican populations by the UDWR and others demonstrates that pelicans are very sensitive to disturbance and food availability. Pelicans rely on adequate water resources not just for food resources and safe areas to breed and raise young successfully in Utah, but they also need freshwater habitats across the western United States that link their migration routes to Mexico and beyond.

Pelican populations at GSL are at a historic crossroads. Their life-cycle necessitates peace and seclusion, along with nearby freshwater resources, both of which are threatened by long-term drought conditions and a booming urban environment along the Wasatch Front. The future of GSL's pelicans was forecast by Dawson (1923) and remains true today.

> The pelican and the wilderness stand together in their mute appeal. When one is fully reclaimed, the other certainly must perish.

If the past is an indication, we have hope for GSL's future. The wilderness of the GSL ecosystem has a place for both humans and wildlife, including pelicans. We look forward to continuing partnerships that benefit both the lake and all of its components.

Acknowledgments We would like to acknowledge the hard work of many people who have completed research on pelicans. William H. Behle, biologist and professor, documented in great detail both GSL and its pelicans through his writing and his extensive collection of newspaper clippings and interview notes. Allison Elbrader at the Marriott Library Special Collections helped with navigating the collection. Florence (Krall) Shepherd was a strong and formidable ally for getting Gunnison Island protected by the State of Utah and has spent many hours talking and writing about pelicans. Don Paul taught, inspired, and encouraged many of the authors, and we thank him. Alex Jacques and Eric Crosman from MesoWest at the University of Utah maintain the real-time PELIcam and provide priceless weather monitoring for biologists in boats on the lake. Liz Moore from the UDWR created applications for fieldwork that will help future data collection on pelicans. Thanks to the volunteers, technicians, and students who helped band pelicans, count pelicans, and look at pelican images. Pilots with the UDWR make the pelican aerial census possible. Utah Division of Parks and Recreation and the US Geological Survey provided boats critical for Gunnison Island research. Compass Minerals and the Utah Division of Forestry Fire and State Lands provide land access to Gunnison Island.

Funding was provided by the Salt Lake City International Airport, Dugway Proving Ground, Eccles Family Foundation, W.M. Keck Foundation, Stephen G. and Susan E. Denkers Family Foundation, Lawrence T. and Janet T. Family Foundation, Tracy Aviary, and Great Salt Lake Audubon. Zooniverse.org has made data analysis of PELIcam images possible, development of which is funded by generous support, including a Global Impact Award from Google and funding from the Alfred P. Sloan Foundation.

References

Almeida-Dalmet S, Baxter BK (2020) Unexpected complexity at salinity saturation: microbial diversity of the north arm of Great Salt Lake, Utah. In: Baxter BK, Butler JK (eds) Great Salt Lake biology: a terminal lake in a time of change. Springer, Cham

Barrett KL, Belovsky GE (2020) Invertebrates and phytoplankton of Great Salt Lake: is salinity the driving factor? In: Baxter BK, Butler JK (eds) Great Salt Lake biology: a terminal lake in a time of change. Springer, Cham

Baxter BK (2018) Great Salt Lake microbiology: a historical perspective. Int Microbiol 21 (3):79–95

Baxter BK, Butler JK (2020) Climate change and Great Salt Lake. In: Baxter BK, Butler JK (eds) Great Salt lake biology: a terminal lake in a time of change. Springer, Cham

Baxter BK, Litchfield CD, Sowers K, Griffith JD, Dassarma PA, Dassarma S (2005) Microbial diversity of Great Salt Lake. In: Gunde-Cimerman N, Oren A, Plemenitaš A (eds) Adaptation to life at high salt concentrations in archaea, bacteria, and eukarya. Cellular origin, life in extreme habitats and astrobiology, vol 9. Springer, Dordrecht

Behle WH (1935) A history of the bird colonies of Great Salt Lake. Condor 37:24–35

Behle WH (1958) The bird life of Great Salt Lake. University of Utah Press, Salt Lake City, UT, 203 pp

Behle WH Collection. Ms576, Bx 64, Fd8, University of Utah, J Willard Marriott Library Special Collections

Belovsky GE, Perschon WC (2019) A management case study for a new commercial fishery: brine shrimp harvesting in Great Salt Lake, Utah, USA. Ecol Appl 29:e01864

Belovsky GE, Stephens D, Perschon C, Birdsey P, Paul D, Naftz D, Baskin R, Larson C, Mellison C, Luft J, Mosley R (2011) The Great Salt Lake ecosystem: long term data and a structural equation approach. Ecosphere 2:1–40. https://doi.org/10.1890/ES10-00091.1

Benson WW, Brock DW, Gabica J, Loomis M (1976) Pesticide and mercury levels in pelicans in Idaho. Bull Environ Contam Toxicol 15:543–546

Budy P, Thiede GP, Chapman K (2016) Pelican predation effects on the fish community in Strawberry Reservoir, Utah. Final progress report to Utah Division of Wildlife Resources. Sport Fish Restoration, Grant number: F-134-R, Segment 2. UTCFWRU 2017(2):1–59

Cannon JS, Cannon MA (2002) The Southern Pacific Railroad trestle - past and present. In: Gwynn JW (ed) Great Salt Lake: an overview of change. Special Publication of the Department of Natural Resources, Salt Lake City, pp 283–294

Carter L (1996) Guano Sifters on Gunnison Island. The History Blazer, pp 4–6

Caudell JN (2009) In the news. Hum Wildl Interact 3(2):article 2. https://doi.org/10.26077/f2qa-y192

Cohenour RE, Thompson KC (1966) Geologic setting of Great Salt Lake. In: The Great Salt Lake: guidebook to geology of Utah, vol 20. Utah Geological Society, Salt Lake City, UT, pp 35–46

Conover MR, Caudell JN (2009) Energy budgets for eared grebes on the Great Salt Lake and implications for harvest of brine shrimp. J Wildl Manag 73(7):1134–1139

Cook BI, Ault TR, Smerdon JE (2015) Unprecedented 21st century drought risk in the American Southwest and Central Plains. Sci Adv 1:e140082

Croghan PC (1958a) The mechanism of osmotic regulation in *Artemia salina* (L.): the physiology of the branchiae. J Exp Biol 35(1):234–242

Croghan PC (1958b) The mechanism of osmotic regulation in *Artemia salina* (L.): the physiology of the gut. J Exp Biol 35(1):243–249

Dawson WL (1923) The birds of California. South Moulton Company, San Francisco

Deseret News and Telegram (1963) MS576 BX64 FD8. From the William H. Behle Collection at University of Utah Marriott Library Special Collections

Downard R, Frank M, Perkins J, Kettenring K, Larese-Casanova M (2017) Wetland plants of Great Salt Lake, a guide to identification, communities, and bird habitat. Utah State University Extension, Logan

Dunn JL, Alderfer J, Lehman PE (2017) National Geographic field guide to the birds of North America. National Geographic, Washington

Evans AF, Payton Q, Turecek A, Cramer B, Collis K, Roby DD, Loschl PJ, Sullivan L, Skalski J, Weiland M, Dotson C (2016) Avian predation on juvenile salmonids: spatial and temporal analysis based on acoustic and passive integrated transponder tags. Trans Am Fish Soc 145 (4):860–877. https://doi.org/10.1080/00028487.2016.1150881

Evers DC, Lane OP, Savoy L, Goodale W (2004) Assessing the impacts of methylmercury on piscivorous wildlife using a wildlife criterion value based on the common loon, 1998-2003. Report BRI 2004–05 submitted to the Maine Department of Environmental Protection. BioDiversity Research Institute, Gorham, Maine

FAA Wildlife Strike Database (2019) Federal Aviation Administration, Washington, DC. https://wildlife.faa.gov. Accessed 20 Sep 2019

Frank MG (2016) Migratory waterbird ecology at a critical staging area, Great Salt Lake, Utah. Thesis, Utah State University

Gillies RR, Chung O-Y, Simon Wang S-Y, DeRose RJ, Sun Y (2015) Added value from 576 years of tree-ring records in the prediction of the Great Salt Lake level. J Hydrol 529(3):962–968

Goebel T, Waters MR, O'Rourke DH (2008) The late Pleistocene dispersal of modern humans in the Americas. Science 319(5869):1497–1502

Great Salt Lake Institute, Westminster College (2019) PELIProject website. www.GSLPELIProject.org. Accessed 18 Sep 2019

Gwynn JW (ed) (2002) Great Salt Lake: an overview of change. Utah Geological Survey, Salt Lake City, UT

Hall ER (1925) Pelicans versus fishes in Pyramid Lake. Condor 27(4):147–160

Horel J, Splitt M, Dunn L, Pechmann J, White B, Ciliberti C, Lazarus S, Slemmer J, Zaff D, Burks J (2002) Mesowest: cooperative mesonets in the western United States. Bull Am Meteorol Soc 83 (2):211–226

Howard H (1955) Fossil birds from Manix Lake. US Government Printing Office, California

Keith JO (2005) An overview of the American White Pelican. Waterbirds 28(sp 1):9–17

Kemp B, Tabish E, Wolford A, Jones D, Butler J, Baxter B (2018) The biogeography of Great Salt Lake Halophilic Archaea: testing the hypothesis of Avian mechanical carriers. Diversity 10 (4):124

Knopf FL (1975) Spatial and temporal aspects of colonial nesting of the white pelican (*Pelecanus erythrorhynchos*). Ph.D. Dissertation. Utah State University, Logan, pp 1–86

Knopf FL, Evans RM (2004) American White Pelican (*Pelecanus erythrorhynchos*), version 2.0. In: Poole AF (ed) The birds of North America. Cornell Lab of Ornithology, Ithaca, NY. https://doi.org/10.2173/bna.57

Knopf F, Street J (1974) Insecticide residues in White Pelican eggs from Utah. Wilson Bull 86 (4):428–434. Retrieved from http://www.jstor.org.dist.lib.usu.edu/stable/4160542

Lambourne A (1909) Our Inland Sea: the story of a homestead. Deseret News

Lies MF, Behle WH (1966) Status of the White Pelican in the United States and Canada through 1964. Condor 68:279–292

Lindsay MR, Johnston RE, Baxter BK, Boyd ES (2019) Effects of salinity on microbialite-associated production in Great Salt Lake, Utah. Ecology 100(3):e02611

Low JB, Kay L, Rasmussen DI (1950) Recent observations on the White Pelican on Gunnison Island, Great Salt Lake, Utah. Auk 67:345–356

Luft J (2019) Pelicans of Gunnison Island; where are the birds? Friends of Great Salt Lake Newsletter. Autumn 2019 Volume 27 Number 2.

Lyons DE, Patterson AG, Tennyson J, Lawes TJ, Roby DD (2018) The Salton Sea: critical migratory stopover habitat for Caspian Terns (*Hydroprogne caspia*) in the north American Pacific flyway. Waterbirds 41(2):154–166

Madsen BD (ed) (1981) A forty-niner in Utah with the Stansbury exploration of Great Salt Lake: letters and journal of John Hudson, 1848–50. University of Utah, Salt Lake City, UT

Marden B, Brown P, Bosteels T (2020) Great Salt Lake Artemia: ecosystem functions and services with a global reach. In: Baxter BK, Butler JK (eds) Great Salt Lake biology: a terminal lake in a time of change. Springer, Cham

Marshall DB, Giles LW (1953) Recent observations on birds of Anaho Island, Pyramid Lake, Nevada. Condor 55(3):105–116

Mengel RM (1952) White pelican from the Pleistocene of Oklahoma. Auk 69(1):81–82
MesoWest (2019) MesoWest at the University of Utah. http://meso1.chpc.utah.edu/station_cameras/gunnison_cam/gunnison_cam_day.mp4. Accessed 18 Sep 2019
Moreno-Matiella LA, Anderson DW (2005) Water level variation and its effects on nesting habitat configuration and availability for the American white pelican at Clear Lake Reservoir, California. Waterbirds 28(sp1):73–83
Morgan DL (1947) The Great Salt Lake. University of Utah, Salt Lake City
Neill J, Hall JO, Luft J, Bankhead E, Bartosz B, Dalton M, Juran C, Rackham B (2009) American white pelican census, Gunnison Island, Utah. Utah Division of Wildlife Resources, Salt Lake City. Unpublished report, 20 pp
Neill J, Leite B, Gonzales J, Sanchez K (2016) Great Salt Lake eared grebe aerial photo survey. Utah Division of Wildlife Resources, Salt Lake City. Unpublished report, 10 pp
Neill J, Brewerton A, Davidson M., Leite B, Gonzales J, Sanchez K (2017) American white pelican census, Gunnison Island, Utah. Utah Division of Wildlife Resources, Salt Lake City. Unpublished report, 23 pp
Null SE, Wurtsbaugh WA (2020) Water development, consumptive uses, and Great Salt Lake. In: Baxter BK, Butler JK (eds) Great Salt Lake biology: a terminal lake in a time of change. Springer, Cham
Oldroyd H (1964) The natural history of flies. Weidenfeld and Nicolson, London, xiv, 324 pp
Oviatt CG, Currey DR, Sack D (1992) Radiocarbon chronology of Lake Bonneville, eastern Great Basin, USA. Palaeogeogr Palaeoclimatol Palaeoecol 99(3–4):225–241
Paul DS, Manning AE (2002) Great Salt Lake waterbird survey five-year report (1997-2001). Publication Number 08-38, Great Salt Lake Ecosystem Program, Utah Division of Wildlife Resources, Salt Lake City
Rawley EV (1976) Great Salt Lake wildlife report (no. 2). State of Utah, Department of Natural Resources, Division of Wildlife Resources
Salt Lake Tribune (1977) MS576 Bx64 Fd8. From the William H. Behle Collection at University of Utah Marriott Library Special Collections
Schaller G (1964) Breeding behavior of the White Pelican at Yellowstone Lake, Wyoming. Condor 66(1):3–23. https://doi.org/10.2307/1365232
Schreiber RW, Schreiber EA (1978) Colonial bird use and plant succession on dredged material islands in Florida. Vol. I, Sea and wading bird colonies. US Army Engineers Waterways Experiment Station Tech Rep D-78–14
Science Friday (2018) The Great Salt Lake is an oasis for migratory birds. National Public Radio. https://www.sciencefriday.com/segments/the-great-salt-lake-is-an-oasis-for-migratory-birds/. Accessed 24 Sep 2019
Sorenson ED, Hoven HM, Neill J (2020) Great Salt Lake shorebirds, their habitats and food base. In: Baxter BK, Butler JK (eds) Great Salt Lake biology: a terminal lake in a time of change. Springer, Cham
Sovada MA, Pietz PJ, Converse KA, King DT, Hofmeister EK, Scherr P, Ip HS (2008) Impact of West Nile virus and other mortality factors on American white pelicans at breeding colonies in the northern plains of North America. Biol Conserv 141:1021–1031
Sovada MA, Igl LD, Pietz PJ, Bartos AJ (2014) Influence of climate change in productivity of American White Pelicans, *Pelecanus erythrorhynchos*. PLoS One 9(1):e83430. https://doi.org/10.1371/journal.pone.0083430
Stansbury H (1852) Exploration and survey of the valley of the Great Salt Lake of Utah, including a reconnaissance of a new route through the Rocky Mountains. Report prepared for the U.S. Senate. Lippincott, Grambo and Co., Philadelphia
Stapp P, Hayward GD (2002) Estimates of predator consumption of Yellowstone cutthroat trout (*Oncorhynchus clarki bouvieri*) in Yellowstone Lake. J Freshw Ecol 17(2):319–329
Sugden JW (1936) Sanctuaries of the bird islands of Great Salt Lake. Auk 53(3):288–294

The Cornell Lab of Ornithology (n.d.) American White Pelican life history, All About Birds, Cornell Lab of Ornithology. https://www.allaboutbirds.org/guide/American_White_Pelican/lifehistory. Retrieved 20 Feb 2019

Thompson BH (1933) History and present status of the breeding colonies of the White Pelican in the United States. U.S. Department of Interior, National Park Service, Control Wildlife Division, Occasional Paper No. 1

Trentelman CK (2020) The relationship between humans and Great salt Lake: dynamics of change. In: Baxter BK, Butler JK (eds) Great salt Lake biology: a terminal lake in a time of change. Springer, Cham

U.S. Fish and Wildlife Service (1984) Guidelines for the management of the American White Pelican, western population. U.S. Fish Wildlife Serv, Portland, OR

U.S. Geological Survey (2019) National Water Information System: U.S. Geological Survey database available online at https://waterdata.usgs.gov/ut/nwis/uv?site_no=10010000. Accessed 17 July 2019

Utah Division of Wildlife Resources (2017) Utah sensitive species list. Retrieved from http://dwrcdc.nr.utah.gov/ucdc/ViewReports/sslist.htm

Utah Division of Wildlife Resources (2019a) Fish Utah application. https://dwrapps.utah.gov/fishing/fStart?id=12. Accessed 24 Sep 2019

Utah Division of Wildlife Resources (2019b) Gunnison Island wildlife management area habitat management plan. Utah Division of Wildlife Resources, Salt Lake City

Utah Division of Wildlife Resources (2019c) Pelitrack website. https://wildlife.utah.gov/pelican_webmap/. Accessed 18 Sep 2019

Wetmore A (1933) Pliocene bird remains from Idaho. Smithsonian Miscellaneous Collections. 87 (20):1–12) at Smithsonian Website (https://repository.si.edu/handle/10088/23955)

Wurtsbaugh WA (2009) Biostromes, brine flies, birds and the bioaccumulation of selenium in Great Salt Lake, Utah. Nat Resour Environ Issues 15(1):2

Wurtsbaugh W, Miller C, Null S, Wilcock P, Hahnenberger M, Howe F (2016) Impacts of water development on Great Salt Lake and the Wasatch Front. White Paper. Utah State University. https://works.bepress.com/wayne_wurtsbaugh/171/

Wurtsbaugh W, Miller C, Null S, Wilcock P, Hahnenberger M, Howe F (2017) Decline of the world's saline lakes. Nat Geosci 10(11):816. https://doi.org/10.1038/ngeo3052

Zooniverse (2019) Zooniverse.org. www.zooniverse.org/projects/jaimibutler/pelicams/about/research. Accessed 19 Sep 2019

Chapter 11
Amphibians and Reptiles of Antelope Island, Great Salt Lake, Utah

David A. Kimberly and Chloe L. Fender

Abstract Antelope Island, situated in Great Salt Lake, Utah, is predominantly grassland (~90%), which supports a variety of organisms from grazers to predators. The island is also home to multiple freshwater springs and rocky sagebrush highlands, as well as rolling sandy dunes. Reptiles are an oft overlooked, but incredibly important, contributor of local food webs. Lizards dominate dune habitats while snakes are often observed in rocky upland habitats searching for small mammals. This chapter seeks to describe the paleogeographic history of amphibians and reptiles in the western United States generally and in Utah specifically. It is also our goal to provide context to the rich herpetological research conducted in Utah over the last 75 years that has contributed to our current understanding. This chapter will also include the species accounts for the known amphibians and reptiles inhabiting Antelope Island as well as a dichotomous key for accurate field identification. Lastly, in addition to describing how these organisms have adapted to their current habitats, we will discuss how those populations may transform as a result of a changing climate.

Keywords Amphibians · Reptiles · Herpetology · Island ecology · Climate change · Desert ecology · Dispersal · Paleoecology

11.1 Introduction

11.1.1 What Are Amphibians and Reptiles?

The study of amphibians and reptiles is called herpetology, from the Greek word *herpein*, meaning "to creep," which was applied to *herpeton*, meaning "reptile." This root conjures a picture of a slimy, slithering, and *creepy* organism. Obviously, this

D. A. Kimberly (✉) · C. L. Fender
Department of Biology, Westminster College, Salt Lake City, UT, USA
e-mail: dkimberly@westminstercollege.edu

B. K. Baxter, J. K. Butler (eds.), *Great Salt Lake Biology*,
https://doi.org/10.1007/978-3-030-40352-2_11

name has done little to help the public image of amphibians and reptiles. Regardless, this group lovingly referred to as "herps" by scientists and enthusiasts alike, are incredibly diverse, occupy a wide range of ecosystems, and provide a multitude of ecosystem services.

The grouping of amphibians and reptiles is one of pure convenience since modern day snakes, for example, are more closely related to a field mouse than to a frog. However, both groups are members of Subphylum Vertebrata, in which all representatives have a central vertebral column. Additionally, amphibians and reptiles are considered ectotherms, which means they use environmental sources of heat (e.g., the sun or a warm rock) to maintain their internal temperature (Pough 2016). Amphibians and reptiles can also generally be found together in similar habitats, which include waterways, moist areas below fallen logs and stones, and under the canopy of low growing shrubs and trees. The differences, though, are plenty. They include the fact that reptile skin is made of keratinized scales and/or scutes, which originate in the epidermis and provide protection against predation and desiccation in arid habitats. Amphibians on the other hand, tend to have moist, smooth, or rough skin that is semipermeable to allow osmosis and gas exchange. Therefore, amphibians can "breathe" through their skin (some salamanders have no lungs at all!) while all reptiles, with the exception of some aquatic turtles must use lungs for respiration. Reptiles are further diverged from amphibians in being part of the amniotes. Amniotes are organisms that produce an amniotic egg during reproduction. The amniotic egg is a special vascularized structure and is accompanied by external (shelled eggs) or internal development. Amphibians lack this characteristic and are tied to aquatic or very moist habitats for reproduction (Pough 2016).

11.1.2 Ecoregions of Utah, USA

Utah is home to a modest diversity of reptiles and amphibians, ranking 27th in comparison to the rest of the United States. While Utah boasts 57 reptile species (US average is 47), only 17 amphibians can be found within the state (US average is 31) (Duellman 1999). This imbalance is of little surprise when considering the dominant ecoregions of the intermountain West, which characterize much of Utah's landscapes.

Utah is divided into five separate ecoregions, Northern Basin and Range, Southern Basin and Range, Colorado Plateau, Wasatch-Uinta Mountains, and Wyoming Plateau (Fig. 11.1) (Omernik 1987). The Northern Basin and Range and the Colorado Plateau region cover the majority of Utah. While these ecoregions are similar in soil and vegetation composition, the Colorado Plateau is dominated by grasslands and open woodlands. Interestingly, the Northern Basin and Range vegetation include more salt-tolerant shrub species. The Colorado Plateau is more recognizable by its tablelands, canyons, and high mountains (Ramsey et al. 1995; Harper 1994). Antelope Island is positioned within the Northern Basin and Range ecoregion, but also near the foothills of the Wasatch-Uinta Mountains ecoregion. The access to variable

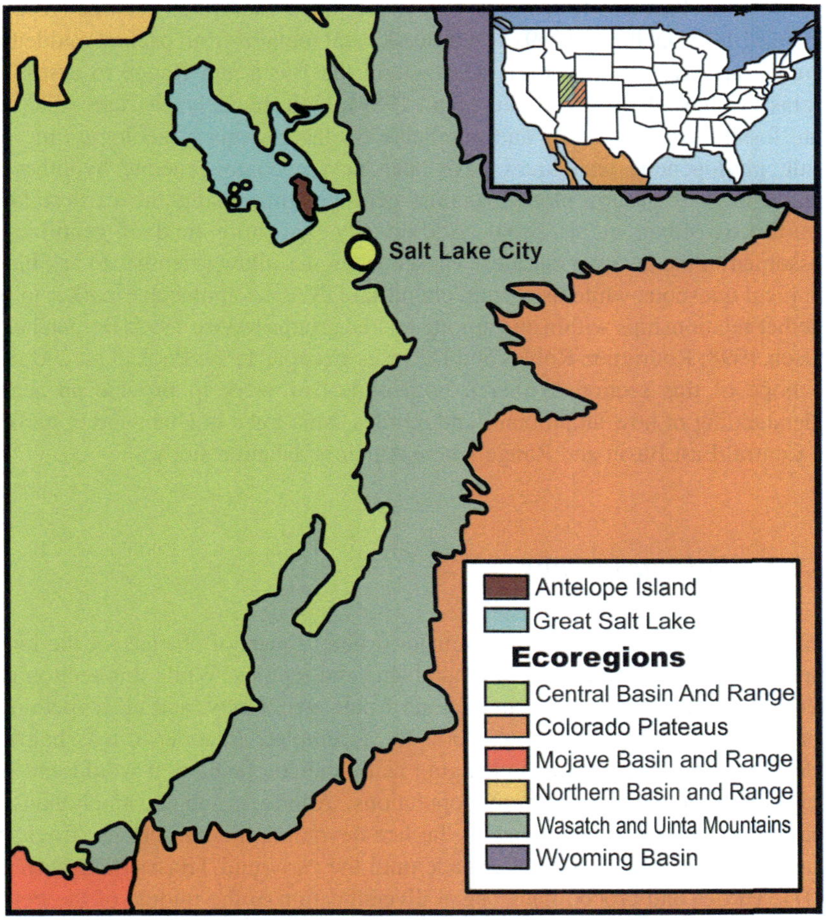

Fig. 11.1 Ecoregions of Utah. Image credit: Johanna Bossart

ecoregions has potential implications on organism interactions as well as historic movement within and among habitats. Although the island is surrounded by saline waters and brackish habitat, there are freshwater springs available to the fauna here.

11.2 Biogeography of Amphibians and Reptiles in the West

It has been previously suggested that Utah has a very low degree of herpetological endemism (Tanner 1978; Hovingh 1986). This means that the amphibians and reptiles found within the state are also found elsewhere, and that organisms within Utah are a result of range extensions from another origin. This may seem trivial but when considering how and from where an organism originates, this information can be important. Previous thinkers on this matter have used fossil records (Estes and Sanchíz 1982; Duellman 1999) as well as skeletal remains from packrat middens and natural trap caves from various sites within the Basin and Range to assess their origins (Tanner 1978; Mead and Mead 1989). Unfortunately, the terrestrial vertebrate fossil record is patchy and unreliable for large spans of geologic time. As a result, phylogenetic techniques have been developed to generate hypotheses of relationships of groups plotted against geologic time using the earliest known occurrence (Milner et al. 2000). Additionally, an entire field of genomics has blossomed in the last two decades. These techniques allow scientists to ask biogeographical questions while using mitochondrial DNA as a molecular marker to piece together relationships within and among related groups (Avise 1995; Bernatchez and Wilson 1998; Rodriguez-Robles and De Jesus-Escobar 1999; Pook et al. 2000). It is the hope of this section to merge both bodies of work to provide an accurate understanding of how amphibians and reptiles came to be in Utah, and in particular the Central/East Basin and Range where Antelope Island resides.

11.2.1 Paleogeography

Global plate tectonics is one of the major driving forces of change for the biogeographical distribution patterns of amphibians and reptiles. While this section is not concerned with a complete examination of paleogeography, and all herpetological taxa involved (See Milner et al. 2000 for a complete treatment), it is helpful to understand how the earth was changing as the herpetofauna that would end up in Utah was evolving into modern populations. Although ancient amphibians first diverged from lobe-finned fishes in the late devonian, modern clades (frogs, salamanders, caecilians) did not take shape until the early/mid-Triassic (Roelants et al. 2007). Turtles and crocodilians began diversification in the middle of the Permian and the beginning of the Triassic, respectively (Wang et al. 2013). Lastly, squamata reptiles, which include all modern lizards and snakes, diverged around 200 million years ago, toward the end of the Triassic (Vitt et al. 2003).

By the Cretaceous, all groups were diversifying and split between the Laurasian and Gondwanan landmasses. The breakup of Gondwana in the mid-Cretaceous created extreme isolation, which contributes much to the modern endemism and diversity seen in New Zealand, Africa, Madagascar, Seychelles, India, South America, and New Zealand (Milner et al. 2000). While Gondwana was splitting throughout the Cretaceous, corridors were forming between North America and other continents. The Bering Corridor between East Asia and North America created a unique area bound by the Mid-continental Seaway, which effectively divided western and eastern North America. As such, the continents share some salamander and many lizard faunas (Russell 1993; Milner et al. 2000). Importantly for salamander movement, the North Atlantic Land Bridge (NALB) formed allowing for European fauna to migrate into North America (Milner 1983; Zhang et al. 2008). Recent molecular data provide more support for the NALB hypothesis as the main source for North American salamander diversity (Zhang et al. 2008). According to Milner (1983), the formation of the Mid-Continental Sea drove diversification of North American salamanders. By the end of the Cretaceous, the sea had withdrawn and eastern and western populations once again merged. However, it has been suggested that modern eastern and western fauna are likely relicts of their respective endemic populations. Interestingly, many eastern taxa would go on to successfully extend these ranges westward.

At the end of the Cretaceous and the beginning of the Paleocene, many organisms faced extinction, including the majority of dinosaurs. And while it might be reasonable to expect even extinctions across taxa, Archibald and Bryant (1990) found that extinctions were highly selective. In North America, most herpetofauna were unaffected, with few exceptions. Curiously, lizards in the Teiidae family (Whiptails and Racerunners) became extinct from North America and Asia (Gao and Fox 1991). Teiidae representatives likely repopulated North America during the Miocene when North and South America were close enough that overwater herpetofaunal dispersals could occur (Estes and Baez 1985). However, the Panama Isthmus formed well into the Pliocene (~5 ma) and the "Great American Interchange" began.

During the late Miocene and Pliocene, global diversity of herpetofauna dropped as climates cooled. Turtles and crocodilians felt the brunt of the impact of increased glaciation during the Pleistocene, while only two amphibian species went extinct during the same time (Holman 1995). Many have argued that in addition to southern regions of North America acting as source populations, due to a high degree of stability during this time, mountain regions with North to South ranges (Appalachian, Rocky Mountains, etc.) also likely provided protected hibernacula (Tanner 1978; Holman 1995; Duellman 1999).

11.2.2 Biogeography of Modern Day Amphibians and Reptiles in Utah

Seventeen species of amphibians occur in Utah (Table 11.1). The majority of these amphibians can be placed into one of three groups. The first group is suggested to have arrived in the Intermountain plateau from Central or South America, having had Pleistocene refugia in Mexico and the Southwest of North America (Hovingh 1986; Pauly et al. 2004; Lemmon et al. 2007; Hua et al. 2009). These species likely included, *Hyla arenicolor*, *Pseudacris triseriata*, *P. regilla*, *Spea intermontana*, and *S. multiplicata*. The second group includes all members from the *Rana* (*Lithobates*) genus. This group is thought to have dispersed from Asia in single or multiple dispersal events. Therefore, those Ranids found in Utah came from the North West, and include *Rana* (*Lithobates*) *onca*, *R.* (*Lithobates*) *fisheri* (Extinct), *R. pretiosa*, *R. luteiventris,* and *R.* (*Lithobates*) *pipeans*. The third group includes those amphibians that arrived from the East. *Ambystoma tigrinum* is the only salamander found in Utah and dispersed from east to western North America (Zhang et al. 2008). The true toads of Utah include, *Anaxyrus cognatus*, *A. punctatus*, *A. woodhouseii*,

Table 11.1 Amphibians of Utah

Common name	Scientific name
Western toad	*Anaxyrus boreas*
Tiger salamander	*Ambystoma tigrinum*
Great Plains toad	*Bufo cognatus*
Arizona toad	*Bufo microscaphus*
Red-Spotted toad	*Bufo punctatus*
Woodhouse's toad	*Bufo woodhousii*
Canyon treefrog	*Hyla arenicolor*
Pacific treefrog	*Pseudacris regilla*
Western chorus frog	*Pseudacris triseriata*
American bullfrog	*Rana catesbeiana*
Green frog	*Rana clamitans*
Columbia spotted frog	*Rana luteiventris*
Relict leopard frog	*Rana onca*
Northern leopard frog	*Rana pipiens*
Plains spadefoot	*Spea bombifrons*
Great Basin spadefoot	*Spea intermontana*
Mexican spadefoot	*Spea multiplicata*

Produced from Stebbins (2003)

A. microscaphus, A. boreas, and their biogeography is complicated. Old World Bufonids arose in South America in the Upper Cretaceous, which then led to global distribution. In the Eocene, these organisms returned to North America via Eurasia, spreading eastward and southward, giving rise to our contemporary *Anaxyrus* genus (Pauly et al. 2004; Pramuk et al. 2008). Of this group, only *A. tigrinum, A. woodhousei,* and *S. intermontana* extend beyond the Colorado River Basin and into the Northern Basin and Range and the majority of the Colorado Plateau (Hovingh 1986; AmphibiaWeb 2019).

There are 57 reptile species found in Utah. Well over half the species are snakes (31), followed by 22 lizard species, and finally, 4turtle species (Table 11.2). The once popular theory for the origin of great basin reptiles being northern expansions from the South, specifically the Mexican plateau (Tanner 1978; Holman 1995; Rodriguez-Robles and De Jesus-Escobar 1999), has found only some support. Threadsnakes from the *Rena* genus likely originated in South America and dispersed northward in North America prior to the permanent landmass that now connects South to Central America (Adalsteinsson et al. 2009). *Thamnophis* is also thought to have origins in Mexico with potentially two dispersals into North America (Alfaro and Arnold 2001). Most great basin lizards, including those from the genus *Xantusia*, *Uta*, *Sceloporus*, and *Aspidoscelis* also have origins in Central or South America, from which they dispersed into North America (Crother et al. 1986; Upton and Murphy 1997; Macey et al. 2006). In addition to a Southern origin and dispersal, researchers have found origins of Great Basin reptiles in the North West. Burbrink and Lawson (2007) suggest that all New World Lampropeltini snakes, which include great basin genera *Arizona,* New World *Elaphe, Lampropeltis,* and *Pituophis,*

Table 11.2 Reptiles of Utah

Common name	Scientific name
Spiny softshell	*Apalone spinifera*
Glossy snake	*Arizona elegans*
New Mexico whiptail	*Aspidoscelis neomexicana*
Tiger whiptail	*Aspidoscelis tigris*
Plateau striped whiptail	*Aspidoscelis velox*
Zebra-tailed lizard	*Callisaurus draconoides*
Rubber boa	*Charina bottae*
Snapping turtle	*Chelydra serpentina*
Painted turtle	*Chrysemys picta*
Western banded gecko	*Coleonyx variegatus*
Eastern racer	*Coluber constrictor*
Sidewinder	*Crotalus cerastes*
Speckled rattlesnake	*Crotalus mitchellii*
Midget faded (Western) rattlesnake	*Crotalus oreganus concolor*
Great Basin (Western) rattlesnake	*Crotalus oreganus lutosus*
Mojave rattlesnake	*Crotalus scutulatus*
Hopi (Prairie) rattlesnake	*Crotalus viridis nuntius*
Green Prairie (Prairie) rattlesnake	*Crotalus viridis viridis*
Great Basin collared lizard	*Crotaphytus bicinctores*
Eastern collared lizard	*Crotaphytus collaris*
Ring-necked snake	*Diadophis punctatus*
Desert iguana	*Dipsosaurus dorsalis*
Cornsnake	*Elaphe guttata*
Many-lined skink	*Eumeces multivirgatus*
Western skink	*Eumeces skiltonianus*
Long-nosed leopard lizard	*Gambelia wislizenii*
Mojave desert tortoise	*Gopherus agassizii*
Gila monster	*Heloderma suspectum*
Common lesser earless lizard	*Holbrookia maculata*
Nightsnake	*Hypsiglena torquata*
Common kingsnake	*Lampropeltis getula*
Sonoran Mountain kingsnake	*Lampropeltis pyromelana*
Milksnake	*Lampropeltis triangulum*
Western threadsnake	*Leptotyphlops humilis*
Coachwhip	*Masticophis flagellum*
Striped whipsnake	*Masticophis taeniatus*
Smooth green snake	*Opheodrys vernalis*
Greater short-horned lizard	*Phrynosoma hernandesi*
Desert horned lizard	*Phrynosoma platyrhinos*
Spotted leaf-nosed snake	*Phyllorhynchus decurtatus*
Gopher snake	*Pituophis catenifer*

(continued)

Table 11.2 (continued)

Common name	Scientific name
Long-nosed snake	*Rhinocheilus lecontei*
Western patch-nosed snake	*Salvadora hexalepis*
Common chuckwalla	*Sauromalus ater*
Common sagebrush lizard	*Sceloporus graciosus*
Desert spiny lizard	*Sceloporus magister*
Western fence lizard	*Sceloporus occidentalis*
Eastern fence lizard	*Sceloporus undulatus*
Ground snake	*Sonora semiannulata*
Smith's black-headed snake	*Tantilla hobartsmithi*
Black-necked garter snake	*Thamnophis cyrtopsis*
Terrestrial garter snake	*Thamnophis elegans*
Common garter snake	*Thamnophis sirtalis*
Western lyre snake	*Trimorphodon biscutatus*
Ornate tree lizard	*Urosaurus ornatus*
Common side-blotched lizard	*Uta stansburiana*
Desert night lizard	*Xantusia vigilis*

Produced from Stebbins (2003)

diverged from Old World members in Asia. Having crossed the Beringia landmass in the North West, this group then diversified throughout North America. Similarly, *Crotalus* has old-world divergence in Asia and subsequent diversification into North and then Central/South America (Parkinson 1999; Pook et al. 2000). Horned lizards from the genus *Phrynosoma*, as well as *Eumeces*, also originated from the North West (Zamudio et al. 1997; Macey et al. 2006). The Gopher tortoise (Genus *Gopherus*) originated in the Central Plains of North America followed by a Southerly dispersal (Reynoso and Montellano-Ballesteros 2004). Lastly, the snakes *Masticophis* and *Coluber* both originated from the East, likely from the Floridian peninsula (Burbrink et al. 2008).

Certainly, understanding the biogeography of herpetofauna of the Great Basin is not an easy task. And since origination or arrival to the area, climatic factors have been great barriers for dispersal as well as strong selective pressures. Species that were highly water-dependent were impacted heavily by the complex pluvial cycles during the middle and late Pleistocene. This likely resulted in range contractions and expansions for herpetofauna in the Great Basin (Hovingh 1997; Hewitt 1996, 2000; Masta et al. 2003). Currently, there is a general trend toward drought-tolerant species in light of the desert habitats that characterize much of the Basin and Range ecoregion. Moreover, those organisms that occupy Antelope Island must not only contend with arid environments but also high saline conditions.

11.3 History of Herpetofaunal Collection in the Great Salt Lake Region

The very first official collection and description of herpetofauna in the Salt Lake valley was done by Captain Howard Stansbury, who led an expedition of the West between 1849 and 1852, which included Great Salt Lake. This expedition resulted in an enormous amount of data on the geology, flora, and fauna of the region (Stansbury 1852). In the report of the survey, Girard and Baird describe the specimens collected by the Stansbury company to be *Uta stansburiana*, *Sceloporus graciosus*, *Cnemidophorus tigris*, *Phrynosoma douglassii*, *Phrynosoma platyrhinos*, and *Coluber mormon* [*Coluber constrictor mormon*] (Figs. 11.2 and 11.3).

Almost 100 years later, Woodbury (1952) conducted a more thorough accounting of the Herpetofauna in the Salt Lake valley, suggesting nine amphibians, 11 lizards,

Fig. 11.2 Plate II of *Cnemidophorus tigris* [*Aspidoscelis tigris*] from Stansbury (1852), public domain

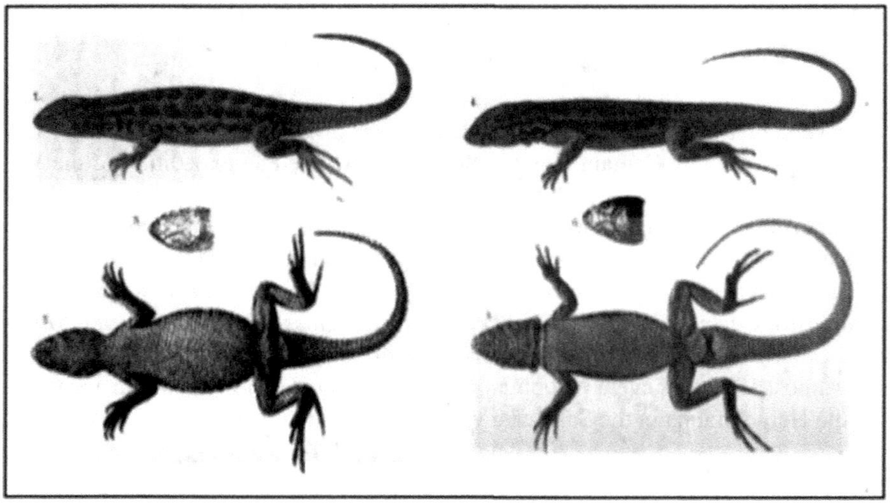

Fig. 11.3 Plate V of *Sceloporus graciosus* (Left) and *Uta stansburiana* (Right) from Stansbury (1852), public domain

and 11 snakes exist therein. Certainly, the time between these two publications was rich with herpetological collecting. Wilmer Tanner from Brigham Young University (BYU), for example, published extensively on the herpetofauna of the Great Basin from the mid-1920s to the late 1960s. One such effort by Tanner and Banta (1966) provided an in-depth review of reptile collections at BYU and the University of Utah. Based on these museum data, Tanner and Banta suggest evidence for eight species of lizard and 11 species of snake being found in the Salt Lake valley. Interestingly, some species within the genera *Crotaphytus*, *Sceloporus*, and *Masticophis* are described as absent from the Salt Lake valley even though records exist in counties to the North, West, and South. Moreover, two snake genera, *Arizona* and *Salvadora* are only described from Nevada portions of the Great Basin at that time.

In the most recent decades, herpetological research has remained robust in Utah, however, it would likely be characterized as more theoretical. For example, researchers at Utah State University, including Edmund Brodie, Susannah French, and Alan Savitsky study broad questions ranging from antipredator behaviors in neotropical amphibians to the effects of urbanization on reptiles, to defense toxins in snakes. Interestingly, early century herpetological research in Utah was generally done on Utah fauna. Moreover, since Dave Ross in the early 1990s, when the Division of Wildlife was shifting its scope to game species, there has not been a dedicated state herpetologist. Lastly, the herpetologist Todd Esque worked for the Bureau of Land Management from 1991 to 1998 out of St. George, without replacement. Regardless of the lack of state and federal representation, there exist many state conservation societies, including the Wasatch Herpetological Society and city-specific groups like the St. George Herp Society.

11.4 Herpetofauna of Antelope Island

Antelope Island is approximately 24 km long and 8 km wide, depending on lake level. As discussed in other parts of this book, the island is primarily (90%) a grassland community, oolitic sand beaches and dunes along the northern (Bridger Bay) and western (White Rock Bay) shores, and shrubland typically found in higher elevation sections of the park. Lastly, well over 40 freshwater springs are found on the island, which flows from the Bonneville Terrace level, before disappearing into alluvium and then reappearing as a seep at lake level.

In addition to the published surveys and theoretical experiments described above, internal reports have been generated by state and federal biologists over the years. One such report titled by wildlife biologist Greg Mortenson (2004) provides insight into the diversity and abundance of herpetofauna on the island. Mortenson surveyed for herpetofauna using active means, from walking transects across the island, to night road observations. In his report, Mortenson confirms six species of reptiles and zero observations of amphibians. The list of observed reptiles, along with abundance can be seen in Table 11.3.

During the summer months of 2015, 2016, and 2017 a group from Westminster College, led by Dr. David Kimberly surveyed the dune and sagebrush habitats for herpetofauna, with the ultimate goal of understanding the fate and distribution of mercury in reptiles. Unlike Mortenson, these efforts employed passive methods such as drift fence and pitfall traps (Figs. 11.4 and 11.5). It should be noted that while drift fences are usually built of thatched plastic, which material proved easy for lizards to

Table 11.3 Type and abundance of reptiles species observed on Antelope Island by Mortensen (2004)

Species	Observed	% of total
Coluber constrictor (Racer)	18	5
Masticophis taeniatus (Striped whipsnake)	5	1
Pituophis catenifer (Gopher snake)	33	9
Cnemidophorus tigris (Western whiptail)	24	7
Sceloporus graciosus (Sagebrush lizard)	9	3
Uta stansburiana (Side-blotched lizard)	260	74
Total	349	99[a]

[a]Net 100 % due to rounding effects

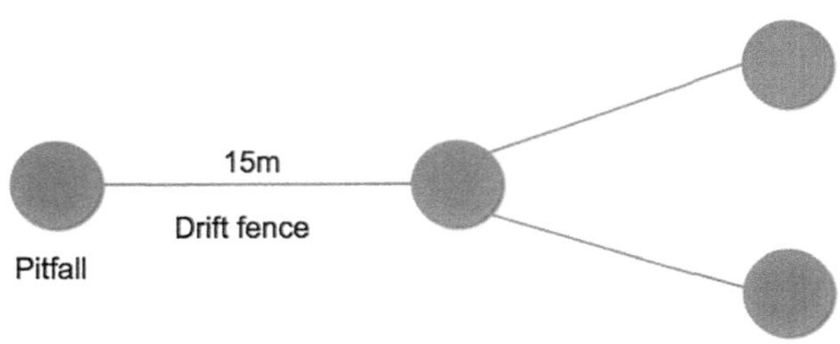

Fig. 11.4 Diagram of drift fence and pitfall trap configuration

Fig. 11.5 Picture of drift fence and pitfall trap in dune habitat on Antelope Island

climb up and over. Therefore, the fence pictured in Fig. 11.5 is made of aluminum flashing. In addition to drift fence and pitfall traps, funnel traps were placed strategically throughout sagebrush habitats because the ground was generally too rocky to dig trenches for the drift fence.

The surveys using passive forms of sampling confirmed the presence of the six reptiles found by Mortenson (Kimberly, unpublished data). The only addition was *Thamnophis elegans*, which was retrieved from a funnel trap in sagebrush habitat on the east side of the island. Drift fence and pitfall traps were very successful at collecting lizards, especially in the dune habitats on the west side of the island. Only snakes were found in funnel traps, leading one to conclude that this passive form of sampling is insufficient to sample lizards that occupy sagebrush habitats. The species accounts below describe seven reptiles confirmed on Antelope Island.

Fig. 11.6 *Sceloporus graciosus*, sagebrush lizard illustration. Image credit: Johanna Bossart

11.4.1 Species Accounts

Species accounts were derived from the report generated by Greg Mortensen in 2004, and Stebbins (2003). Measurements are given as snout to tip of the tail.

11.4.1.1 *Sceloporus graciosus*, Sagebrush Lizard; Fig. 11.6

An 11.5–14 cm primarily brown or gray colored lizard. The underside of the lizard is usually a light color and males will have a blue patch on each side of the belly (hence the nickname of "blue belly") and on the throat/chin when breeding. Females may have a light patch or no coloration on their belly even when breeding. Dorsal scales are keeled and pointed (mucronate), giving the lizard a rough or spiny appearance. Sagebrush lizards have an incomplete gular fold (skin flap on the throat which does not go all the way across the throat). They often have a black stripe on the shoulder and rust in the axilla (armpit). The keeled dorsal scales, incomplete gular fold differentiate this lizard from the Side-blotched lizards. At times they will have two or three fairly distinct stripes on their back, but this is not always the case. Of interest, these lizards have a "third eye" (a photoreceptor) located on a large scale in between and slightly behind its eyes. This lizard is found all over the island, from higher elevations, associated with rock outcrops and junipers to sagebrush and dune habitat. Mating usually occurs in late May through early June. Thereafter, the female lays eggs (2–8) in a 15–20 cm deep burrow in May through July. Eggs usually hatch in 48 days and young appear in "late July through August" (Bosworth 2001).

11.4.1.2 *Uta stansburiana elegans*, Western Side-Blotched Lizard; Fig. 11.7

A small (8–13 cm), mostly light brown or dirty-gray lizard. Side-blotched lizards have a complete gular fold and smooth (granular) dorsal scales (compare to sagebrush lizard account). Typically they will have a dark blotch just behind their forelimbs, however, the blotch can be very light or not present at all and is frequently covered by the forelimbs. Side-blotched lizards also have a photoreceptor "eye" on

Fig. 11.7 *Uta stansburiana elegans*, western side-blotched lizard illustration. Image credit: Johanna Bossart

Fig. 11.8 *Ascpidocelis tigris tigris,* Great Basin whiptail illustration. Image credit: Johanna Bossart

the top of their head. Side-blotched lizards occur throughout Antelope Island. They have the widest distribution of any lizard and can be found from the lakeshore (1280 m) to Frary Peak (2010.5 m). Andre and MacMahon (1980) found that lizards began breeding shortly after hibernation ended (around mid-April) in Utah's Tule Valley. Eggs were laid from April to mid-July and clutch sizes range from two to five eggs and two clutches seem to be typical. Eggs are buried in the sand and young emerge in early to mid-July.

11.4.1.3 *Ascpidocelis tigris tigris,* Great Basin Whiptail; Fig. 11.8

This is the island's largest lizard, generally ranging from 11 to 28 cm in length. Western whiptails have an obvious neck with a narrow head that ends at a point.

Their bodies usually have four, yellowish-cream stripes (which fade with age) on the back bordered by black blotches, at times giving the lizard a leopard-like appearance. Adult tails are exceptionally long and black to gray, whereas juveniles have blue tails. Whiptails move rapidly into cover and are very difficult to approach. This species is mostly found in the island's lowlands and shoreline (1280 m), among the dunes. Andre and MacMahon (1980) reported that mating began shortly after emergence from hibernation. Thereafter, about 2–8 eggs (per clutch with 1 or 2 clutches likely) are laid in April through August. Andre and MacMahon (1980) observed some hatchlings in early August.

11.4.1.4 *Coluber constrictor mormon*, Western Yellow-Bellied Racer; Figs. 11.9 and 11.10

A brownish-olive colored snake with a yellow belly, most ranging between 30 and 100 cm in length. Racers have smooth scales and 15–17 scalerows at midbody and 15 rows just in front of the vent (Stebbins 2003). This species has large eyes due to

Fig. 11.9 *Coluber constrictor mormon*, western yellow-bellied racer illustration. Image credit: Johanna Bossart

Fig. 11.10 Juvenile *Coluber constrictor mormon*, western yellow-bellied racer illustration. Image credit: Johanna Bossart

its diurnal hunting habits. Juveniles have a brown blotched pattern (including a row of brown blotches on the back) on a light background causing them to resemble a small rattle snake or Gopher snake at first glance. Racers are found from the island's lowlands (1280 m) up to the ridgelines (1615.5 m) and in habitats dominated by bunch grasses or cheatgrass. Courtship begins in mid-May and 4–8 eggs are deposited in burrows in late June to early July. Young will emerge in mid- to late-August. Males mature after 1 year, while females mature after 2 or 3 years (Brown 1973).

11.4.1.5 *Masticophis taeniatus*, Striped Whipsnake; Fig. 11.11

A bluish snake with cream sides bisected by a black line as well as a few thin, black lines. The belly is usually a cream color turning pink toward the tail in adults. A long, slender, extremely fast-moving snake with large eye characteristics of its diurnal nature. Those found on island ranged from 30.5 to 82 cm but the typical range elsewhere is 91.5–182 cm. Striped whipsnakes have 15 scalerows at midbody and scales are smooth. Striped whipsnakes were found from the island's shoreline area, to the island's highest point (Frary Peak—2010.5 m). Striped whipsnakes tend to mate in late April or early May. Thereafter, 3–12 eggs are typically laid in a small mammal burrow which maybe shared by eggs from Gopher snakes and Racers (Parker and Brown 1973).

11.4.1.6 *Pituophis catenifer deserticola*, Great Basin Gopher Snake

This is a heavy-bodied snake that tends to be cream colored with a patchwork of dark blotches (including a single row along the back), giving the snake a checkerboard look. A dark line near the eyes gives the appearance of a mask. Dorsal scales are keeled and typically numbered 27–37 rows at midbody. When threatened, Gopher snakes will hiss, vibrate their tail, and/or flatten their heads to mimic a viper's head shape, but these snakes are non-venomous! Gopher snakes found on Antelope Island ranged from 25 to 122 cm in length, but are known to attain up to 280 cm elsewhere (Stebbins 2003). Gopher snakes are found in habitats ranging from cheatgrass dominated fields to brush dominated areas to sandy dunes near the beach. About

Fig. 11.11 *Masticophis taeniatus*, striped whipsnake illustration. Image credit: Johanna Bossart

Fig. 11.12 *Pituophis catenifer deserticola*, Great Basin Gopher snake illustration. Image credit: Johanna Bossart

4–15 eggs are left in small mammal burrows (which may be shared with the eggs of Racers and Striped whipsnakes) in late June or mid-July and hatch in late August to September (Parker 1976) (Fig. 11.12).

11.4.1.7 *Thamnophis elegans vagrans*, Wandering Garter snake

This snake gets its name from the idea that it wanders farther from water than most other garter snakes. Moreover, this subspecies tends to be found in some of the more arid habitats, in comparison to other garter snake species. This garter snake measures 45–110 cm in length. Ground color is gray, brown, or greenish and there are typically light dorsal and lateral stripes. The dorsal stripe is yellow, brown, or orangish, but black markings on the edges may make it appear irregular or a series of dark and light dots. The sides are checkered with black markings. The underside is light with scattered black markings, often concentrated in the center (Stebbins 2003). On the island, this species can be found in sagebrush and grassy fields, usually within close proximity to one of the many freshwater springs (Kimberly, unpublished data). Of note, like many North American garter snakes, the wandering garter snake has mildly venomous saliva. Bites to humans can cause very mild local edema, but no instances of systemic symptoms have been observed (Gomez et al. 1994). *T. elegans vagrans* breeds primarily in spring, giving birth to live young from July to September (Fig. 11.13).

11.4.2 Barriers to Herpetofaunal Dispersal and Colonization on the Island

Earlier publications regarding the distribution of the herpetofauna of Antelope Island noted with surprise the absence of a few species (Mortensen 2004). *Crotalus viridis* (Western rattlesnake), *Hypsiglena torquata* (Night snake), *Phrynosoma platyrhinos*

Fig. 11.13 *Thamnophis elegans vagrans*, Wandering garter snake illustration. Image credit: Johanna Bossart

(Desert horned lizard), *Spea intermontana* (Great Basin spadefoot toad) are found in all habitats surrounding Great Salt Lake but have never been confirmed on Antelope Island. The reader of this book is well aware of the inhospitality of this salty lake to vertebrates, with the exception of migratory and resident birds, some of whom have salt glands that allow extrusion of salt (Fänge et al. 1958). Smaller reptiles without wings depend on overland locomotion for dispersal and do not have anatomical features which allow them to successfully swim in saline waters. Because of the salinity of Great Salt Lake, a natural barrier exists to amphibians in particular, but also to reptiles (Hopkins and Brodie 2015). However, many times since the Pleistocene water levels dropped low enough to form land bridges (Atwood et al. 2016). These low levels were created in the past by climatic cycling and today by increased water use upstream. There is also an access causeway running from eastern side of the valley to the Northern tip of the island (Gwynn 2000). While these "bridges" provide some relief from the saltwater barrier they are insufficient to support the dispersal for all herpetofauna that live in the Salt Lake valley.

The salinity may not be the only barrier for rattlesnakes. Wetlands are the dominant ecosystem over much of the shoreline, which includes dense emergent and submergent aquatic vegetation. Because this is considered to be a poor rattlesnake habitat, these snakes may avoid it, thereby removing all access to the island. Even if organisms disperse through the wetland habitats they often have to contend with large sections of salt flats, which are exposed, hot, and as the name implies, covered in salt crystals. While amphibians may find the wetlands hospitable, the salt flats are almost certain death (Hopkins and Brodie 2015). Swift moving reptiles, like *Masticophis* and *Coluber*, can disperse to the island by limiting time spent in the exposed spaces. Rattlesnakes and other slow-moving reptiles may have a more difficult time dispersing from the shortline to the island.

Once on the island, there exist barriers to colonization, the biggest being a changing vegetative landscape. More specifically, cheatgrass (*Bromus tectorum*) populations have expanded into all areas on the island. Cheatgrass is an invasive annual from Eurasia (Novak and Mack 2001) and is implicated in the decline of small mammals (Gitzen et al. 2001) and birds (Wiens and Rotenberry 1985). Hall et al. (2009) found that as cheatgrass coverage increased on Antelope Island, snake abundance decreased. Bosworth (2001) described that in dunes with high cheatgrass coverage, Whiptails and other lizard species were almost completely absent. Although there certainly exists a suitable habitat on the island for herpetofauna, it may be decreasing due to a number of factors, both natural and manmade.

Lastly, while several reptile species have successfully colonized the island, conditions are only predicted to get more challenging for high desert habitats. Arid and semiarid ecosystems across the western United States face widespread ecological shifts as a consequence of global climate change (Archer and Predick 2008). The rate of change is predicted to be much faster than the three centuries prior to industrialization (North et al. 2006). The patterns of change vary widely by geography, but in general fluctuations in temperature and precipitation project earlier spring snow runoff (Stewart et al. 2004), declines in snowpack in the Rocky Mountain (Plummer et al. 2006) and Great Basin regions (Mote et al. 2005), and increased frequency, duration, and range of drought events (Sheffield and Wood 2008). Globally, studies have described adverse effects and population declines among organisms exposed to climate change conditions (Sinervo et al. 2010; Reading et al. 2010). In arid environments, the story is more complicated for many amphibian and reptile populations. Flesch et al. (2017) discussed that while some populations may decline, such whiptail lizards, other populations may increase, like side-blotched lizards. Even though both lizards are terrestrial, whiptail lizards breed during warmer months of the season while side-blotched lizards breed earlier, during the cooler months. Interestingly, the adverse effects of climate change may be more pronounced for whiptail lizards. On Antelope Island, we might not expect to see an abrupt decline in reptile populations, instead a shift in community structure could be predicted. Therefore, the continued study of island populations will be crucial in understanding the long-term impacts of climate change on this sensitive and unique ecosystem.

11.5 Antelope Island Reptile Identification Key [Modified from Mortensen (2004)]

1a. Lizard.. Go to 2
1b. Snake.. Go to 4
2a. Dorsal scales smooth.. Go to 3
2b. Dorsal scales keeled, partial gular fold, about 4½–5½" length, stripes on back—
Sceloporus graciosus

3a. 3½–5″ length, complete gular fold, blotch behind forelimbs (not always), dusty or grayish color—*Uta stansburiana*

3b. Large (4½–7″—if smaller, blue tail may be present), narrow head, yellow/cream stripes on the back bordered by black blotches—*Ascpidocelis tigris tigris*

4a. Dorsal scales smooth... Go to 5

4b. Dorsal scales keeled... Go to 6

5a. Brownish-olive colored snake with yellow belly. If juvenile, brown blotched pattern on a light background—*Coluber constrictor mormon*

5b. Bluish snake with cream sides bisected by a black line as well as a few thin, black lines, belly cream color turning pink toward the tail in adults; long, slender, extremely fast moving—*Masticophis taeniatus*

6a. Heavy bodied, with a patchwork of dark blotches, dark line near the eyes gives a masked appearance—*Pituophis catenifer*

6b. Generally smaller, more slender than *Pituophis*, and with a patchwork of blotches resembling a checkerboard. Has a yellow stripe running dorsally from neck to tail—*Thamnophis elegans vagrans*

References

Adalsteinsson SA, Branch WR, Trape S, Vitt LJ, Hedges SB (2009) Molecular phylogeny, classification, and biogeography of snakes of the Family Leptotyphlopidae (Reptilia, Squamata). Zootaxa 2244(2244):1–50

Alfaro ME, Arnold SJ (2001) Molecular systematics and evolution of Regina and the thamnophiine snakes. Mol Phylogenet Evol 21(3):408–423

Amphibiaweb (2019) AmphibiaWeb: information on amphibian biology and conservation. University of California, Berkeley, California. Available from http://amphibiaweb.org/

Andre JB, MacMahon JA (1980) Reproduction in three sympatric lizard species from west-Central Utah. Great Basin Nat 40:68–72

Archer SR, Predick KI (2008) Climate change and ecosystems of the southwestern United States. Rangelands 30(3):23–29

Archibald JD, Bryant LJ (1990) Differential cretaceous/tertiary extinctions of nonmarine vertebrates; evidence from northeastern Montana. Geol Soc Am Spec Pap 247:549–562

Atwood G, Wambeam TJ, Anderson NJ (2016) The present as a key to the past: Paleoshoreline correlation insights from Great Salt Lake. In: Developments in earth surface processes, vol 20. Elsevier, Amsterdam, pp 1–27

Avise JC (1995) Mitochondrial DNA polymorphism and a connection between genetics and demography of relevance to conservation. Conserv Biol 9(3):686–690

Bernatchez L, Wilson CC (1998) Comparative phylogeography of nearctic and palearctic fishes. Mol Ecol 7(4):431–452

Bosworth W (2001) Inventory of amphibians and reptiles in the Utah test and training range and the associated military operations area. Final report prepared for Environmental Management, Natural Resources, Ogden Air Logistics Center

Brown WS (1973) Ecology of the racer, *Coluber constrictor mormon* (Serpentes, Colubridae), in a cold temperate desert in northern Utah

Burbrink FT, Lawson R (2007) How and when did Old World ratsnakes disperse into the New World? Mol Phylogenet Evol 43(1):173–189

Burbrink FT, Fontanella F, Pyron RA, Guiher TJ, Jimenez C (2008) Phylogeography across a continent: the evolutionary and demographic history of the North American racer (Serpentes: Colubridae: *Coluber constrictor*). Mol Phylogenet Evol 47(1):274–288
Crother BI, Miyamoto MM, Presch WF (1986) Phylogeny and biogeography of the lizard family Xantusiidae. Syst Biol 35(1):37–45
Duellman WE (1999) Global distribution of amphibians: patterns, conservation, and future challenges. Johns Hopkins University Press, Baltimore, MD
Estes R, Baez A (1985) Herpetofaunas of North and South America during the Late Cretaceous and Cenozoic: evidence for interchange? In: The great American biotic interchange. Springer, Boston, MA, pp 139–197
Estes R, Sanchíz B (1982) Early cretaceous lower vertebrates from Galve (Teruel), Spain. J Vertebr Paleontol 2(1):21–39
Fänge R, Schmidt-Nielsen K, Robinson M (1958) Control of secretion from the avian salt gland. Am J Physiol Legacy Content 195(2):321–326
Flesch AD, Rosen PC, Holm P (2017) Long-term changes in abundances of Sonoran Desert lizards reveal complex responses to climatic variation. Glob Chang Biol 23(12):5492–5508
Gao K, Fox RC (1991) New teiid lizards from the upper cretaceous Oldman formation (Judithian) of southeastern Alberta, Canada, with a review of the cretaceous record of teiids. Ann Carnegie Mus 60(2):145–162
Gitzen RA, West SD, Trim BE (2001) Additional information on the distributions of small mammals at the Hanford site, Washington. Northwest Sci 75:350–362
Gomez HF, Davis M, Phillips S, McKinney P, Brent J (1994) Human envenomation from a wandering garter snake. Ann Emerg Med 23(5):1119–1122
Gwynn JW (2000) The waters surrounding Antelope Island, Great Salt Lake, Utah. In: The geology of Antelope Island, Davis County, Utah, vol 1. Utah Geological Survey, Salt Lake City, UT, p 117
Hall LK, Mull JF, Cavitt JF (2009) Relationship between cheatgrass coverage and the relative abundance of snakes on Antelope Island, Utah. Western North Am Nat 69(1):88–96
Harper KT (1994) Natural history of the Colorado plateau and Great Basin. University Press of Colorado, Boulder, CO
Hewitt GM (1996) Some genetic consequences of ice ages, and their role in divergence and speciation. Biol J Linn Soc 58(3):247–276
Hewitt GM (2000) The genetic legacy of the quaternary ice ages. Nature 405(6789):907
Holman JA (1995) Pleistocene amphibians and reptiles in North America. Oxford University Press, New York
Hopkins GR, Brodie ED Jr (2015) Occurrence of amphibians in saline habitats: a review and evolutionary perspective. Herpetol Monogr 29(1):1–27
Hovingh P (1986) Biogeographic aspects of leeches, mollusks, and amphibians in the intermountain region. Great Basin Nat 46:736–744
Hovingh P (1997) Amphibians in the eastern Great Basin (Nevada and Utah, USA): a geographical study with paleozoological models and conservation implications. Herpetol Nat Hist 5 (2):97–134
Hua X, Fu C, Li J, De Oca ANM, Wiens JJ (2009) A revised phylogeny of Holarctic treefrogs (genus Hyla) based on nuclear and mitochondrial DNA sequences. Herpetologica 65 (3):246–259
Lemmon EM, Lemmon AR, Cannatella DC (2007) Geological and climatic forces driving speciation in the continentally distributed trilling chorus frogs (Pseudacris). Evolution 61 (9):2086–2103
Macey RJ, Schulte JA, Strasburg JL, Brisson JA, Larson A, Ananjeva NB, Wang Y, Parham JF, Papenfuss TJ (2006) Assembly of the eastern North American herpetofauna: new evidence from lizards and frogs. Biol Lett 2(3):388–392

Masta SE, Laurent NM, Routman EJ (2003) Population genetic structure of the toad *Bufo woodhousii*: an empirical assessment of the effects of haplotype extinction on nested cladistic analysis. Mol Ecol 12(6):1541–1554

Mead EM, Mead JI (1989) Snake Creek burial cave and a review of the quaternary mustelids of the Great Basin. Great Basin Nat 143–154

Milner AR (1983) The biogeography of salamanders in the Mesozoic and early Caenozoic: a cladistic-vicariance model. In: Evolution, time, and space. The emergence of the biosphere. Academic Press, New York, pp 431–468

Milner AC, Milner AR, Evans SE (2000) Amphibians, reptiles and birds: a biogeographical review. In: Biotic response to global change: the last, vol 145. Cambridge University Press, Cambridge, pp 316–332

Mortensen G (2004) Antelope Island herptile study. Technical report submitted to Antelope Island wildlife personnel. Utah Division of Parks and Recreation, Salt Lake City, UT

Mote PW, Hamlet AF, Clark MP, Lettenmaier DP (2005) Declining mountain snowpack in western North America. Bull Am Meteorol Soc 86(1):39–50

North GR, Biondi F, Bloomfield P, Christy JR (2006) Surface temperature reconstructions for the last 2,000 years. The National Academies Press, Washington, DC, p 160

Novak SJ, Mack RN (2001) Tracing plant introduction and spread: genetic evidence from *Bromus tectorum* (Cheatgrass) introductions of the invasive grass *Bromus tectorum* worldwide were broadly similar and closely tied to patterns of European human immigration. Bioscience 51 (2):114–122

Omernik JM (1987) Ecoregions of the conterminous United States. Ann Assoc Am Geogr 77 (1):118–125

Parker WS (1976) Population estimates, age structure, and denning habits of whipsnakes, Masticophis *t. taeniatus*, in a northern Utah Atriplex-Sarcobatus community. Herpetologica 32:53–57

Parker WS, Brown WS (1973) Species composition and population changes in two complexes of snake hibernacula in northern Utah. Herpetologica 28:319–326

Parkinson CL (1999) Molecular systematics and biogeographical history of pitvipers as determined by mitochondrial ribosomal DNA sequences. Copeia 3:576–586

Pauly GB, Hillis DM, Cannatella DC (2004) The history of Nearctic colonization: molecular phylogenetics and biogeography of the Nearctic toads (Bufo). Evolution 58:2517–2535

Plummer DA, Caya D, Frigon A, Côté H, Giguère M, Paquin D, Biner S, Harvey R, De Elia R (2006) Climate and climate change over North America as simulated by the Canadian regional climate model. J Clim 19:3112–3132

Pook CE, Wüster W, Thorpe RS (2000) Historical biogeography of the western rattlesnake (Serpentes: Viperidae: *Crotalus viridis*), inferred from mitochondrial DNA sequence information. Mol Phylogenet Evol 15(2):269–282

Pough FH (2016) Herpetology. Sinauer Associates, Inc., Sunderland, MA

Pramuk JB, Robertson T, Sites JW Jr, Noonan BP (2008) Around the world in 10 million years: biogeography of the nearly cosmopolitan true toads (Anura: Bufonidae). Glob Ecol Biogeogr 17 (1):72–83

Ramsey RD, Falconer A, Jensen JR (1995) The relationship between NOAA-AVHRR NDVI and ecoregions in Utah. Remote Sens Environ 53(3):188–198

Reading CJ, Luiselli LM, Akani GC, Bonnet X, Amori G, Ballouard JM, Filippi E, Naulleau G, Pearson D, Rugiero L (2010) Are snake populations in widespread decline? Biol Lett 6 (6):777–780

Reynoso VH, Montellano-Ballesteros M (2004) A new giant turtle of the genus Gopherus (Chelonia: Testudinidae) from the Pleistocene of Tamaulipas, México, and a review of the phylogeny and biogeography of gopher tortoises. J Vertebr Paleontol 24(4):822–837

Rodriguez-Robles JA, De Jesus-Escobar JM (1999) Molecular systematics of New World lampropeltinine snakes (Colubridae): implications for biogeography and evolution of food habits. Biol J Linn Soc 68(3):355–385

Roelants K, Gower DJ, Wilkinson M, Loader SP, Biju SD, Guillaume K, Moriau L, Bossuyt F (2007) Global patterns of diversification in the history of modern amphibians. Proc Natl Acad Sci 104(3):887–892
Russell DA (1993) The role of Central Asia in dinosaurian biogeography. Can J Earth Sci 30 (10):2002–2012
Sheffield J, Wood EF (2008) Projected changes in drought occurrence under future global warming from multi-model, multi-scenario, IPCC AR4 simulations. Clim Dyn 31:79–105
Sinervo B, Mendez-de-la-Cruz F, Miles DB (2010) Erosion of lizard diversity by climate change and altered thermal niches. Science 328:894–899
Stansbury H (1852) Exploration and survey of the valley of the Great Salt Lake of Utah, including reconnaissance of a new route through the rocky mountains. Lippincott, Grambo & Co, Philadelphia
Stebbins RS (2003) Field guide to Western reptiles and amphibians. Houghton Mifflin, Boston
Stewart IT, Cayan DR, Dettinger MD (2004) Changes in snowmelt runoff timing in western North America under a 'business as usual' climate change scenario. Clim Chang (62):217–232
Tanner WW (1978) Zoogeography of reptiles and amphibians in the intermountain region. Great Basin Nat Memoirs 43–53
Tanner WW, Banta BH (1966) A systematic review of the Great Basin reptiles in the collections of Brigham Young University and the University of Utah. Great Basin Nat 26(3/4):87–135
Upton DE, Murphy RW (1997) Phylogeny of the side-blotched lizards (Phrynosomatidae: Uta) based on mtDNA sequences: support for a midpeninsular seaway in Baja California. Mol Phylogenet Evol 8(1):104–113
Vitt LJ, Pianka ER, Cooper WE Jr, Schwenk K (2003) History and the global ecology of squamate reptiles. Am Nat 162(1):44–60
Wang Z, Pascual-Anaya J, Zadissa A, Li W, Niimura Y, Huang Z, Li C, White S, Xiong Z, Fang D, Wang B (2013) The draft genomes of soft-shell turtle and green sea turtle yield insights into the development and evolution of the turtle-specific body plan. Nat Genet 45(6):701
Wiens JA, Rotenberry JT (1985) Response of breeding passerine birds to rangeland alteration in a North American shrubsteppe locality. J Appl Ecol 22:655–668
Woodbury AM (1952) Amphibians and reptiles of the Great Salt Lake Valley. Herpetologica 8 (2):42–50
Zamudio KR, Jones KB, Ward RH (1997) Molecular systematics of short-horned lizards: biogeography and taxonomy of a widespread species complex. Syst Biol 46(2):284–305
Zhang P, Papenfuss TJ, Wake MH, Qu L, Wake DB (2008) Phylogeny and biogeography of the family Salamandridae (Amphibia: Caudata) inferred from complete mitochondrial genomes. Mol Phylogenet Evol 49(2):586–597

Chapter 12
Shoreline Plants of Great Salt Lake

Ember S. Bradbury and David L. Parrott Jr.

Abstract A catalogue and description of the major plant species inhabiting Great Salt Lake (GSL) shoreline is presented in this chapter. The mechanisms by which these plant species interact with the harshly saline shoreline environment are also described. The history of research surrounding these mechanisms and survey work will give readers a picture of the scientific body of knowledge that pertains to the GSL shoreline flora and its significance. Until now, there has not been a comprehensive review of the shoreline plants, which have typically been lumped in with plants of the Great Basin or other broad taxonomic studies, making it difficult for lake visitors to identify them or for scientists to understand the role they play. While there are a limited number of plant species inhabiting the shoreline, a guide to their identity and their adaptations to the harsh saline soils is presented here, which will provide a highly informative resource to inform future work.

Keywords Great Salt Lake · Utah · Shoreline · Plants · Halophyte · Soil · Salinity

12.1 Introduction

Great Salt Lake (GSL) is the largest salt lake in the Western Hemisphere (Fig. 12.1a). The lake spans across ~4400 square kilometers of Northern Utah and is approximately 120 km wide (Stephens and Gardener 2007). Over the past 780,000 years, this terminal lake location has been home to four deep lakes including Lake Bonneville (30–13,000 years ago) which covered nearly 52,000 square kilometers of western Utah, eastern Nevada and southern Idaho (Oviatt et al. 1999; Shroder et al. 2016). However, over the last several million years, the Bonneville basin mainly held shallow lakes such as GSL, or mudflats and playa (Atwood et al. 2016). GSL sits in a large depression created by the Lake Bonneville, and for this reason, it

E. S. Bradbury · D. L. Parrott Jr. (✉)
Department of Biology, Westminster College, Salt Lake City, UT, USA
e-mail: dparrott@westminstercollege.edu

B. K. Baxter, J. K. Butler (eds.), *Great Salt Lake Biology*,
https://doi.org/10.1007/978-3-030-40352-2_12

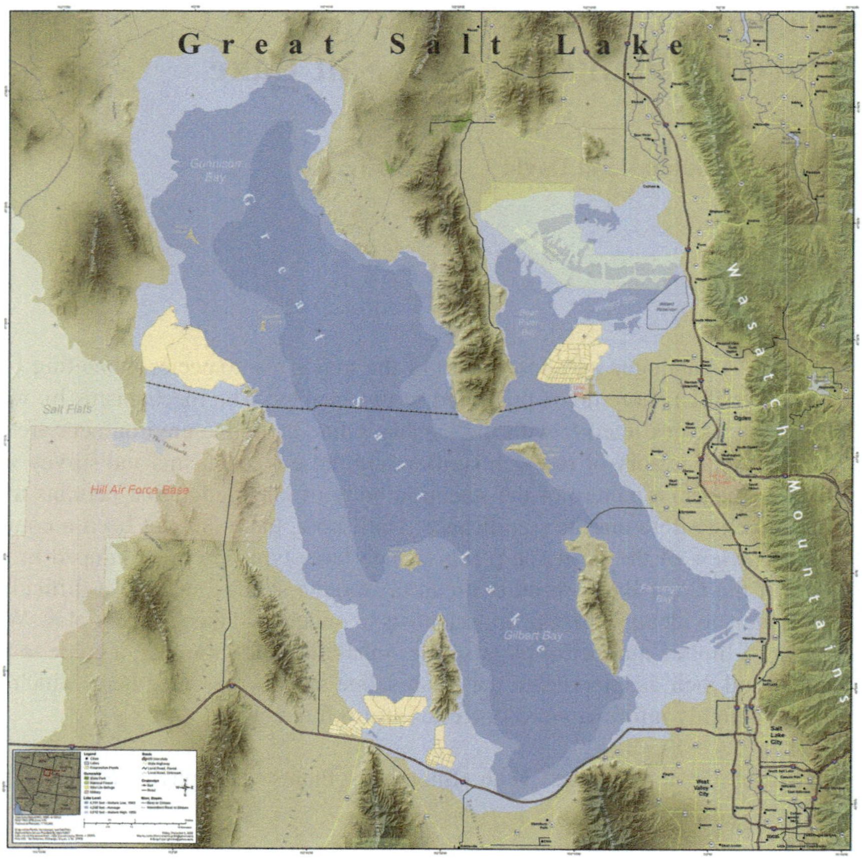

a

Fig. 12.1 Great Salt Lake, Utah. (**a**) Map of Great Salt Lake. Image credit: Justin Morris—morris.justin@gmail.com. (**b**) Satellite image of Great Salt Lake, August 2018. Image credit: Copernicus Sentinel-2 data 2018

does not have any outlet. Salt and minerals that enter the closed basin have no escape and its waters become more saline over time (Gwynn 1996). Currently, there is anywhere from 4.5 to 4.9 billion tons of dissolved salt contained within the lake. This amount changes depending on the water level, as well as other environmental factors that contribute to salinity.

The salinity of GSL and its shorelines vary throughout the lake. A rock filled rail line prevents movement of water separates the major bays of the lake (Cannon and Cannon 2002; Madison 1970). This division has segregated the nutrient and mineral cycling of the lake, and in the present day, the north side (known as the north arm) has a much higher concentration of salt than the south (Fig. 12.1b). The south arm receives a majority of the freshwater input, as it is fed directly by the inflowing rivers. This physical divide in the ecosystem of GSL compounds show unique and interesting the lake truly is and what life can thrive in each spot.

b

Fig. 12.1 (continued)

The high salinity of the main water body makes the surrounding shoreline an incredibly dramatic soil climate and forces specialization for the biotic communities that thrive in the area. For this reason, the halophytic (salt-tolerant) plant species that live around the lake are not only important to document, but to understand mechanistically. The objective of this chapter is to succinctly describe the floral biota of the GSL shoreline and outline their evolutionary salt tolerance mechanisms. This will add to the much broader picture of ecological interaction and give readers a sense of the sensitive ecosystem that surrounds this incredible feature. To best create this understanding, the history of botanical research surrounding the GSL shoreline will first be discussed. This will not only include work that has been done in recent history (i.e., Internet databases) but will go through the botanical history as well. The features of the GSL shoreline soil will then be outlined, along with a summary of various mechanisms that plant species use to cope with such extreme soil compositions. Finally, a comprehensive guide to the individual species that are found on the shoreline will be presented in an organized format to serve as a resource for future work.

12.2 History of Botanical Research of Great Salt Lake

12.2.1 Early Survey Work and Findings

Botanists have been working since 1871 to appropriately characterize the plant and algae species found at GSL. Although many of these researchers were incredibly important to adding to the scientific body of knowledge, the name that is perhaps most prevalent is that of Seville Flowers (Fig. 12.2). Flowers began his professional career in 1929, and throughout his lifetime became one of the most well-respected authorities on Utah's flora. His early work surrounded the survey of the GSL region, with one of his first publications being in the March of 1934 issue of the Botanical Gazette entitled: *Vegetation of the Great Salt Lake Region with Contributions from the Hull Botanical Library* (Flowers 1934). This survey was ground-breaking as it provided a true synthesis of the biotic and abiotic factors that comprise the make-up of the flora of GSL. In the article, Flowers breaks down the communities of the lake into five categories: (1) the main body, (2) the strand (including the inlets, swamps, and springs), (3) river deltas, (4) playas and alkaline plains, and (5) salt marshes and sloughs. Each of these is then broken down further into communities, and the soil composition, water composition, soil salinity, and floristics were all listed and compared for each community (Fig. 12.3).

This review focuses on the strand, which is defined by Flowers as: "Any ground left exposed between the margin of water and highest level attained by the lake since records have been kept. This level is well defined by the old shore of 1868," (Flowers 1934). To this day, the strand is where the most robust halophytic plant species are found, and because of the harsh conditions in which they survive, the community as not changed dramatically since 1934. This assertion is demonstrated in the limited literature that was published after Flowers' initial survey focused on this region, including his own updated work (Flowers and Evans 1966) and additional surveys (Vest 1962; Wyckoff 1973).

Each of these prior studies demonstrated that the major flora found on the GSL strand is highly alkali-tolerant, halophytic species. All found similar halophyte

Fig. 12.2 Seville Flowers. Image credit: Johanna Bossart

Fig. 12.3 The Great Salt Lake shoreline cross section showing soil salinity levels (%) and representative plant communities at each salinity level. Image credit: Johanna Bossart, adapted from Vest (1962)

species that were present. Our current observations suggest that these species continue to compose the majority of the floral biotic community and will be listed in Sect. 12.3.2. Our modern work, when compared to Flowers and others, serves to create a timeline reflecting stability and trends in the flora communities.

12.2.2 Current Survey Work and Findings

Though the data from Flowers' surveys are still some of the most referenced and prominent regarding the shoreline plants of GSL, there are several other influential studies conducted in more modern times that have greatly added to the scientific body of knowledge. In 1980, the Utah Geological Survey published a large volume detailing the intricacies of GSL's geology and ecology (Gwynn 1980). Though the chapter that discusses plant life solely cites Flowers' studies, this document was influential because it combined with other fields of study to visualize the complex interactions that halophytic flora has with the environment (Rawley 1980).

Another study that aids in this visualization was drafted by a professor at the University of Utah, Angus M. Woodbury (1956). Throughout his career, he catalogued the biota of several ecological communities in Utah as simple "check-lists" which were succinct versions of his field notes. Most notable to this review was a checklist he published in 1956, which surveyed all the biota that Woodbury saw in an overlook through the Great Salt Lake Desert, the area west of the lake. Though this checklist is not centered around the shoreline (and in fact has a very little survey on the shoreline), it is still incredibly useful for contrasting to other work.

Perhaps the most comprehensive and current source of data that pertains to the plant life of the shoreline of GSL, however, is that of the SEINet Data Portal (SEINet Portal Network 2019). The data portal contains geotagged instances of each species found and submitted by individuals, museums, collectors, and agencies with the I.D. number, identification (order, family, and scientific name), and specimen status attached to each data point. Users can specify the region in which they want to query, and the system creates a downloadable spreadsheet file containing all of the data associated with that region. Through this system, the list of species in Sect. 12.4.2 was obtained. For the purpose of this review, a search was done that outlined the dimensions of the lake including the shoreline (41.700607 N, 40.662936 S, 113.037442 W, 111.963529 E) and a file was created from the huge amount of data that is available on the database. The data was then sifted through and categorized, and the species with the highest number of instances on the shoreline were noted. The species were then contrasted with historical data to create the comprehensive species list found in Sect. 12.4. Although the primary species found on the shoreline of the lake appear to stay constant over time, species that were less prominent historically but noted by multiple sources have also been included. All community factions were pulled from both historical and SEINet data as the most prominent speciation and grouping.

12.3 General Ecology of Great Salt Lake

12.3.1 Soil System and Abiotic Factors

Soil type, texture, and salinity vary depending on location around GSL, with soil types named based on the biotic community they are associated with. For the purpose of this chapter, we will focus on those soils associated with the shoreline or playa areas. Surface soil deposits at lake level consist primarily of mud and salt flats (Fig. 12.4), with areas of alluvial and colluvial deposits, Lake Bonneville deposits and salt marshes. The general composition of the mud and salt flats shoreline soils is a mix of sand, clay, and silt, with varying proportions of each depending on location. Vest (1962) describes the soil texture in the pickleweed and saltgrass communities as clay (10% sand, 10% silt and 80% clay and 15% sand, 25% silt and 60% clay, respectively), while the greasewood and shadscale community are clay loam (40% sand, 20% silt, 40% clay and 48% sand, 15% silt, 37% clay, respectively). It was observed, however, that the soil texture differs around plant roots, with the percentage of sand generally increasing and the percentage of clay generally decreasing. This is caused by plants modifying soils by extracting minerals, dropping organic matter, and trapping windblown particles. By enhancing soil texture, plants are able to modify both the water retention from precipitation and the availability of water to their roots.

Shoreline soil salinity varies based on distance from the water's edge, soil composition and texture, precipitation penetration (if any), and depth sample was taken (Flowers 1955; Vest 1962). Soil salinity ranges from well over 20% at 2 inches (salt crust) to nearly 5% at 1 foot (root depth) in the pickleweed zone nearest the water, to just under 1% at 2 feet (root depth) in the Greasewood zone (Table 12.1).

Depending on the community and related soil type, there is a "salt zone" where a dramatic increase in salt concentration occurs (Vest 1962). This can be seen in Table 12.1, and directly correlates to the plants, which grow in each zone. Shallow rooted plants can thrive in soils with deep salt zones, but deeper-rooted plants or plants in communities with shallow salt zones will only survive if they can tolerate the saline environment. Taken together, soil type, soil texture, and soil salinity have a profound effect on the diversity of plant communities along the shore of GSL.

12.3.2 Biotic Communities

Though the biodiversity of the GSL shoreline is limited, there are several types of plant communities that form consistently. The communities can be separated out as follows: Pickleweed, Shadscale, Saltgrass, and Greasewood. The primary species in each community will serve to lend the community name, though there is a large amount of overlap between communities in species diversity (Fig. 12.5). The

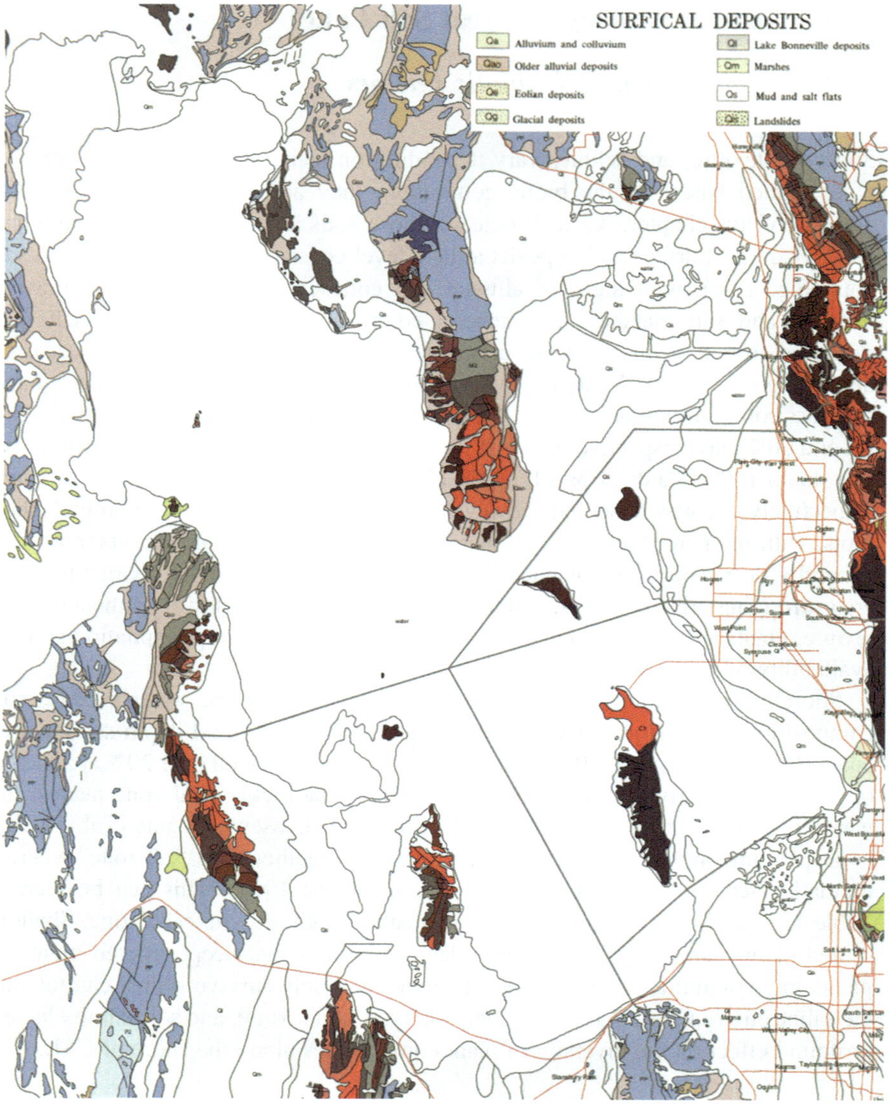

Fig. 12.4 Map of the Great Salt Lake surface soil types (Modified from Digital Geological Map of Utah, 2000, Utah Geological Survey https://geology.utah.gov/map-pub/maps/geologic-maps/state-of-utah-geologic-maps/)

purpose of this section is to give an outline of the interaction between species of the GSL shoreline, and how that interaction is seen through speciation.

Pickleweed Community Generally found in sections of the lake with a higher amount of clay content than silt or sand between plants and generally creates a

Table 12.1 Soil salinity in the Great Salt Lake shoreline soils

Sample area	Depth	NaCl	Na_2SO_4	Na_2CO_3	$NaHCO_3$	Total
Greasewood	2″	0.040	0.026	0	0.106	0.136
	1′	0.330	0.168	0.042	0.080	0.496
	2′	0.965	0.532	0.043	0.140	1.634
	3′	1.210	0.294	0.020	0.081	1.834
Shadescale	2″	0.110	0.026	0	0.053	0.350
	1′	1.03	0.328	0	0.080	1.384
	2′	1.693	0.365	0	0.056	1.200
	3′	1.049	0.523	0	0.031	1.756
Pickleweed	2″	20.80	0.842	0	0.018	21.600
	1′	4.90	0.204	0.021	0.087	5.300
	2′	4.05	0.275	0.037	0.081	4.240
	3′	5.40	0.204	0.023	0.070	5.47

Values shown are percent oven dry weight. Based on Vest (1962) and Flowers (1955)

sandy–clay–loam under the plants (Vest 1962, Sect. 12.3.1). The most prominent species in the community is *Salicornia utahensis*. This community is generally found closest to the lakeshore, as it is not only highly salt and alkali tolerant but corresponds well to the soil texture of the near-lake soil. The second most prominent species in the Pickleweed community is Greasewood (*Sarcobatus vermiculatus*), both of which are succulent in nature. Species that are also included in this community, though to a much lesser extent include Budsage (*Artemisia spinescens*), Shadscale (*Atriplex confertifolia*), Big Sagebrush (*Artemisia tridentata*), Red Swampfire (*Salicornia rubra*), and Alkali sacaton (*Sporobolus airoides*).

Shadscale Community Generally found slightly farther away from the water's edge, or in soil that is composed of more sand than clay. This community is made up of fewer succulents and contains more plants that are woody and pubescent in nature. These key differences make the communities easier to differentiate, though there is some overlap in the boundary between them. As the community's namesake, Shadscale (*Atriplex confertifolia*) is the most abundant species within the community. However, Gray molly (*Kochia americana*) is also incredibly plentiful, as the two have similar survival characteristics and prefer similar soil texture. Other species found in this community are Seepweed (*Suaeda erecta*) and Greasewood (*Sarcobatus vermiculatus*).

Saltgrass Community Primarily dominated by representatives of the family Poaceae (the grass family). This community also prefers soil with high amounts of clay, and most members employ the exuding salt tolerance mechanism (to be discussed in Sect. 12.3.3). This community is found close to the water's edge, but often spans across semi-marsh regions as well. Besides Desert Saltgrass (*Distichlis stricta*), this community is also abundant in Alkali sacaton (*Sporobolus airoides*) and Dropseed (*Sporobolus cryptandrus*).

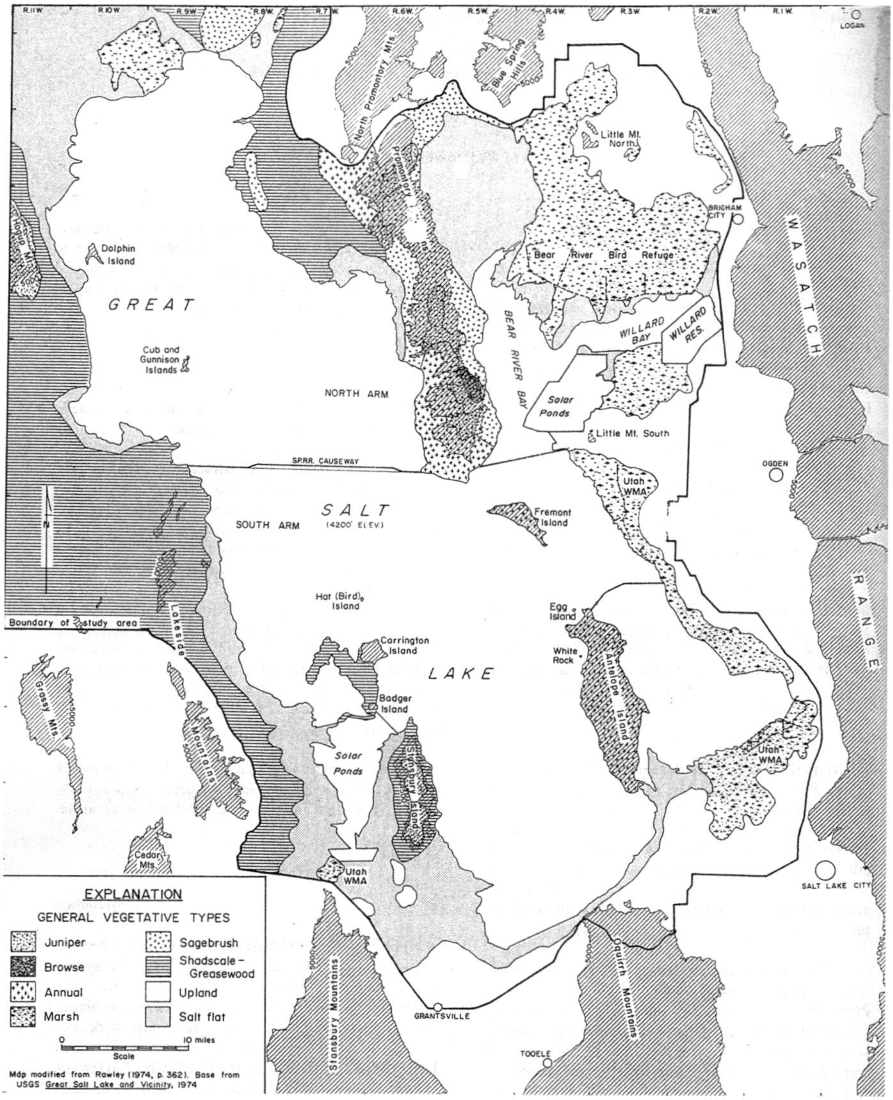

Fig. 12.5 General vegetation types found around Great Salt Lake, adapted from Rawley (1980)

Greasewood Community Similar to the Pickleweed community, though it generally is a bit farther from the water's edge and prefers soils with an almost even distribution of sand, silt, and clay. This community is the least distinct, though it does comprise areas where Greasewood (*Sarcobatus vermiculatus*) is the more prevalent species. Other species that are dominant are Seepweed (*Suaeda erecta*) and Iodine Bush (*Allenrolfea occidentalis*).

12.3.3 Overview of Salt Tolerance Mechanisms Employed by Flora

GSL has several different biotic communities surrounding it. This review specifically focuses on the shoreline surrounding the lake water, which can have soil salt concentrations ranging from 20% in the salt crusts at the water's edge to less than 1% in the Greasewood community. Even the lowest shoreline salt concentration is considerably higher than most plants can tolerate. For this reason, for a species to thrive in the shoreline community, it has to have an incredibly evolved salt tolerance mechanism. To understand which species are able to thrive in the harsh environment that is the GSL shoreline, one must first comprehend the mechanisms by which the species respond to salt tolerance. These mechanisms will be discussed and expanded upon.

Halophytes have the ability to adjust to high salinity due in part to unique anatomical and morphological adaptations and/or mechanisms to avoid salt uptake (Flowers et al. 1986). The presence of high concentrations of salt lowers soil water potential, thus making it harder for roots to take up water, while the ionic stress is associated with the gradual accumulation of salts in plant tissues over time (Munns and Tester 2008). One of the key responses to salt stress is to maintain cellular ion homeostasis by restricting the accumulation of toxic sodium (Na^+) (Clarkson and Hanson 1980; Tester and Davenport 2003). Typically, ions in saline soils which are toxic to plants, commonly Na^+ and Cl^-, are sequestered into the vacuole and used to regulate osmosis (Blumwald et al. 2000; Niu et al. 1995). Not surprisingly, many of the cellular mechanisms involved in ion homeostasis and salt stress signaling are similar in all plants (Hasegawa et al. 2000). Plants cope with salt stress either by avoiding salt by remaining dormant during high salt exposure or tolerating salt exposure by adjusting cellular processes to accommodate the hypersaline environment. These tolerance mechanisms fall into two main categories; those that function to minimize ion disequilibrium and osmotic stress or those that mitigate salt exposure mechanically (Yokoi et al. 2002).

Plants are continuously regulating ion concentration in response to turgor pressure variations using osmotic adjustment. Because cells of both halophytes and glycophytes (salt-sensitive) are equally sensitive to Na^+ and Cl^- ions, these ions are sequestered in the central vacuole, which leads to ion disequilibrium within the cell. Osmotic adjustment is thus achieved by accumulation of compatible osmolytes and osmoprotectants (Bohnert et al. 1995; Bohnert and Jensen 1996). Salt tolerance requires that these compatible solutes accumulate in the cytosol and organelles rather than Na^+ and Cl^- ions, maintaining osmotic balance and turgor pressure, but preventing damage to the plant cell (Rhodes and Hanson 1993). Generally, these organic osmotic solutes consist of simple sugars (fructose and glucose), sugar alcohols (glycerol and methylated inositols) and complex sugars (trehalose, raffinose, and fructans) (Bohnert and Jensen 1996), and can accumulate at high levels in the cell without disturbing intracellular biochemistry (Bohnert and Jensen 1996). Compatible solutes persevere enzyme activity and have a negligible impact on pH or charge balance in the cytosol or organelles. These compounds are synthesized by alteration of metabolic pathways, triggered by the perception of salt or salt stress (Yokoi et al. 2002).

The majority of halophytes growing along the shores of GSL filter salt out at the roots using sodium–potassium pumps within the cell membrane. Although these pumps filter out the majority of Na^+ ions, there is still a great deal that "leaks" into the plant and must be sequestered into vacuoles. Pumping and sequestration of sodium come at a cost to the plant, consuming ATP, and creating ion disequilibrium, leading to turgor decrease and ultimately limiting growth (Bohnert et al. 1995). However, in succulent halophytes like *Allenrolfea occidentalis* (iodine bush) and *Salicornia rubra* (pickleweed), it has been postulated that the overall process of transporting excess sodium into the vacuole results in ATP production. This ATP can be used for additional work, thus explaining the growth stimulation caused by sodium in halophytes (Jennings 1968). When the vacuole cannot hold any more salt, the cell breaks down and dies. Younger cells then take over the job of pumping and storing salt. This process results in what is called "salt-succulence," where the photosynthetic tissue of the plant has many large cells holding massive amounts of salt in the vacuoles.

Some halophytes, such as *Distichlis spicata* (saltgrass), have specialized glands that secrete excess salt out of the cells and deposit it on the leaf surface (Fig. 12.6a, b). Saltgrass requires abundant water and grows rapidly in areas where the soil salinity is between 0.1 and 1.5%, although it can tolerate salinity of up to 3%. Most plants, halophyte or glycophyte, sequester salt in the vacuole. But in plant with salt glands, either active or passive, salt is transported symplastically via plasmodesmata to the salt gland, where it is stored and eventually released to the surface (Thomson 1975). In the case of *Distichlis spicata,* epidermal salt glands are found in direct association with water-storing parenchyma cells that act as salt collectors. The vacuoles in these collector cells appear to actively pump their contents into the salt glands via a network of microtubules in the cell (Semenova et al. 2010), where the salt is released to the leaf surface. Salt crystals on the surface of the leaves are easily observed through the use of a hand lens or microscope.

Detecting changes in ion concentration and providing the appropriate response is required to sustain plant growth in hypersaline environments. Because plant growth is challenged by both osmotic and ionic stress, salt perception, and signaling between and within plant cells is crucial in coordinating osmotic adjustment, salt sequestration and movement of salt into salt glands. Salt tolerance has been shown to be controlled by many genes through complex genetic and biochemical regulatory networks. Over the past several decades, many of the molecular pathways that mediate Na^+ and K^+ homeostasis has been identified (Yokoi et al. 2002). One of these pathways, the Salt Overly Sensitive (SOS) signaling pathway, is composed of a gene family that has been proposed to mediate cellular signaling under salt stress (Ji et al. 2013). The SOS stress signaling pathway has been identified to be an important regulator of ion homeostasis and salt tolerance (Hasegawa et al. 2000; Sanders 2000), controlling Na^+ transport systems that are involved in net flux across the plasma membrane (Amtmann and Sanders 1998; Blumwald et al. 2000; Hasegawa et al. 2000).

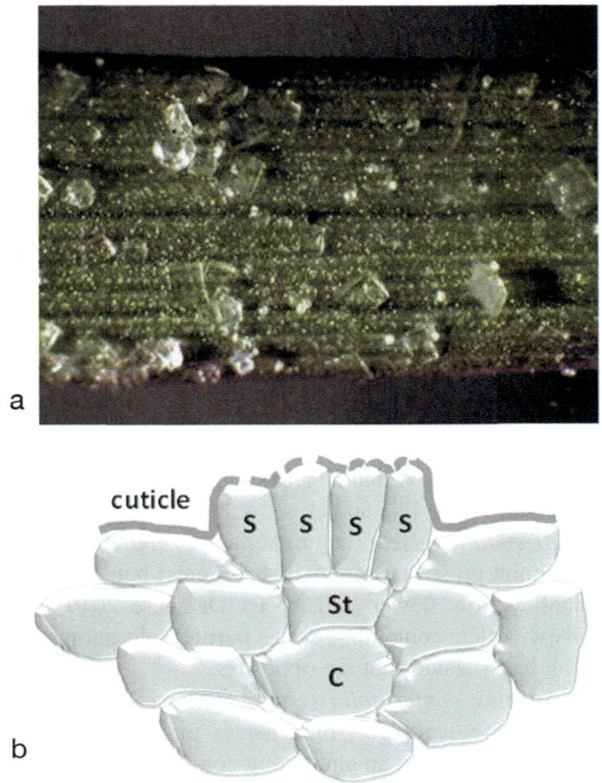

Fig. 12.6 *Distichlis spicata*. (**a**) Salt extrusion via multiple glands on its leaves. Note the formation of salt crystals (courtesy of Justin Klitzes). (**b**) Schematic diagram of a salt gland showing a collective cell (C), a stalk cell (St), and secretory cells (S). A cuticle is present with perforations to allow the excreted salt to exit the symplast (From Maathuis et al. 2014)

Increasingly, research suggests that the rhizosphere microbiome plays is a crucial role in plant productivity and stress tolerance, including salt tolerance (Hingole and Pathak 2016). For example, it has been hypothesized that the halo-tolerance of *Suaeda salsa* (seepweed) is tightly linked to a specialized soil microbiome. 16S rRNA gene sequencing data identified α- and γ-proteobacterial communities in bulk soil and the rhizosphere whose genomes were enriched in genes contributing to salt stress acclimatization, nutrient solubilization, and competitive root colonization (Yuan et al. 2016). A wide diversity of rhizobacteria with similarity to known halotolerant taxa further supported this interpretation. These findings suggest that an ecological patterned root–microbial interaction strategy has been adopted in *S. salsa* system to confront soil salinity (Yuan et al. 2016). As it has been demonstrated that halotolerant bacteria isolated from saline environments have the potential to enhance plant growth under saline stress through direct or indirect mechanisms (Hingole and Pathak 2016), it is likely that plant-based salt tolerance mechanisms and the soil microbiome work in concert to allow GSL shoreline plants to thrive in this high salt environment. Current studies aim to elucidate the rhizosphere microbiome of several GSL strand-inhabiting plants (Calhoun et al. 2019).

12.4 Major Speciation and Botanical Communities of Great Salt Lake

12.4.1 *Prominent Species of the Great Salt Lake Shoreline Description*

Through the compilation of both current and historical data, it was found that the following species are the most common flora found along the shoreline of GSL. By understanding the interaction of these species with their environments, as well as giving a larger characterization of them, an understanding of how the GSL shoreline ecosystem interacts can be seen. In the following section, a description of each of these plant types will be given.

Allenrolfea occidentalis (common name, Iodine Bush) (Fig. 12.7a) is a low-lying and glaucous with a woody stem which is much darker in coloration than the green succulent nodes which appear similar to the *Salicornia* species in their jointing. The leaves are scale like and close to the stem (Flora of North America Database 2019).

Aristida longiseta (common name, Fendler Threeawn) (Fig. 12.7b) is a monocot perennial grass that grows in small erect bunches. It blooms in mid-spring and has small yellow flowers (PLANTS Database 2019). The seed bracts are large and wispy, often containing some purple coloration. The stems are ridged and thin with basal leaves that are opposite.

Atriplex confertifolia (common name, Shadscale, Saltbush) (Fig. 12.7c) has lightly pubescent gray-blue leaves that have an entire margin and are ovate in shape. They are in an alternating conformation along the spiny branches. Shadscale is a dioecious shrub that has a rounded crown and yellow inflorescence (Range Plants of Utah Database 2019). Leaves are salty to the taste.

Distichlis stricta/spicata (common name, desert saltgrass) is a monocot species belonging to the family Poaceae (Fig. 12.7d). It is a warm-season grass with a short-seed head and dark green leafy stems. Spikelets are smooth and short, and bracts are tightly packed together. Leaves are alternate and acicular. It is highly salt-tolerant and grows in large mats (Banner et al. 2011).

Kochia americana (S. Watson); also *Bassia Americana* S. Watson, *Neokochia americana* (S. Watson) (common name, green molly/gray molly) is a dwarf dicot shrub belonging to the family Chenopodiaceae (Fig. 12.7e). Found in alkaline soils, flats, dry lake margins, it consists of mostly unbranched stems with knobby glabrous, fleshy leaves with white wooly flowers either single or in clusters (Hrusa and Wilken 2012).

Phlox longifolia (common name, Longleaf Phlox) (Fig. 12.7f). It is a woody perennial dwarf species. It has long dark green leaves that have an ovular and complete margin. The leaves are arranged oppositely. Perhaps the most distinct feature of the Longleaf Phlox is its small light-pink five-lobed flower (Parker 2019). These flowers generally grow in small clumps, and are an incredibly hardy wildflower, demonstrating the capability to grow in high alkaline and saline environments such as GSL (PLANTS Database 2019).

Fig. 12.7 Representative plants surrounding Great Salt Lake. (**a**) *Allenrolfea occidentalis,* Image credit: Stan Shebs. (**b**) *Aristida longiseta*, Image credit: Max Licher. (**c**) *Atriplex confertfolia,* Image credit: Matt Lavin. (**d**) *Distichlis stricta/spicata*, Image credit: Sheri Hagwood. (**e**) *Kochia americana*, Image credit: David Parrott. (**f**) *Phlox longifolia*, Image credit: Sheri Hagwood. (**g**) *Salicornia rubra*, Image credit: Brian Gratwicke. (**h**) *Salicornia utahensis*, Image credit: Robert Sivinski. (**i**) *Sarcobatus vermiculatus*, Image credit: Cory Maylett. (**j**) *Suaeda erecta*, Image credit: Glen Lee

Fig. 12.7 (continued)

Salicornia rubra (common name, Red Swampfire) is a dicot belonging to the family Amaranthaceae (Fig. 12.7g). It is characterized by scale-like bracts that are red in color when mature. Its flowers are opposite and grow in 3-flowered clusters at each node. It is a hermaphrodite species and is highly tolerant to alkaline soil. Due to its succulent and halophytic nature, it can withstand incredibly high salinity content in the soil (Minnesota Wildflowers 2019). The salt-tolerant mechanism involves the plant redistributing the salt ions to balance the vacuole and cytoplasm osmotic pressure, allowing for it to uptake large quantities of salt ions and still maintain a balanced ratio of water-to-salt concentration (McNulty 1985).

Salicornia utahensis (common name, Utah Pickleweed or Utah Swampfire) (Fig. 12.7h) is a salt-tolerant succulent species much like *S. rubra,* belonging to the family Amaranthaceae. Unlike *S. rubra*, it is a perennial, and keeps its dark green coloration throughout its lifetime. It has a creeping rhizome root system that anchors it to the often sandy habitat of the GSL shoreline. It is phenotypically quite similar to *S. rubra* as its flowers are opposite in 3-flowered clusters at each node. Its leaves are scale like and the flowers are covered by pointy bracts. It grows in densely clumped mats near the shoreline and is hermaphroditic (PLANTS Database 2019).

Sarcobatus vermiculatus (common name, Greasewood) (Fig. 12.7i) has bright yellowish-green long leaves on tall and erect branches. Inflorescence is pistillate and surrounding leaves are usually slightly pubescent. Greasewood is classified as a noxious weed and is poisonous to livestock. It is often codominant in areas with iodine bush as it is also highly alkaline and saline tolerant (Downard et al. 2017).

Suaeda erecta (common name, Seepweed) (Fig. 12.7j). It is a small forb that has thick alternate lanceolate leaves, margin entire. The leaves are gray-green and succulent in nature. Stems are bunched and woody throughout (PLANTS Database 2019).

12.4.2 Full Species List and Keys

Though the species found in Sect. 12.4.1 are the most prominent (and therefore the most imperative to be able to understand and identify), there are a multitude of plants that can be found throughout the vast shorelines of GSL. All plants listed were found on several occasions through many sources of literature and will be separated by family. Descriptions of families and individual species were found in *Flora of Utah* by Dr. Welsh, Atwood, Goodrich, and Higgins in 1987 (Welsh et al. 1987). This is a comprehensive botany handbook for the identification of the flora of Utah, listed by phylum.

<u>Amaranthaceae</u>
Allenrolfea occidentalis—Iodine Bush
Atriplex canescens—Fourwing Saltbush
Atriplex confertifolia—Shadscale
Grayia spinosa—Spiny Hopsage
Kochia americana—Gray molly
Salicornia utahensis—Utah Swampfire
Salicornia rubra—Red Swampfire
Suaeda erecta—Seepweed

<u>Asteraceae</u>
Artemisia tridentate—Big Sagebrush
Artemisia spinescens—Budsage
Chrysothamnus viscidiflorus—Rabbitbrush
Gutierrezia sarothrae—Broom Snakeweed
Tetradymia glabrata—Littleleaf Horsebrush

<u>Boraginaceae</u>
Cryptantha flavoculata—Roughseed Cryptantha

<u>Brassicaceae</u>
Descurainia pinnata—Western Tansymustard

<u>Cupressaceae</u>
Juniperus osteosperma—Utah Juniper

<u>Ephedraceae</u>
Ephedra nevadensis—Mormon Tea

<u>Geraniaceae</u>
Erodium cicutarum—Stork's Bill

<u>Malvaceae</u>
Sphaeralcea coccinea—Scarlet Globemallow

<u>Onagraceae</u>
Oenothera pallida—Pale Evening Primrose

<u>Poaceae</u>
Aristida longiseta—Fendler's Threeawn
Bromus tectorum—Cheatgrass
Distichlis stricta—Desert Saltgrass
Hilaria jamesii—James' Galleta
Hordeum jubatum—Foxtail Barley
Oryzopsis hymenoides—Indian Ricegrass
Poa secunda—Sandberg Bluegrass
Sporobolus airoides—Alkali sacaton
Sporobolus cryptandrus—Sand Dropseed

<u>Polemoniaceae</u>
Phlox longifolia—Longleaf Phlox
Leptodactylon caespitosum—Mat Prickly Phlox

<u>Polygoniaceae</u>
Eriogonum cernuum—Nodding Buckwheat
Eriogonum ovalifolium—Cushion Buckwheat
<u>Sarcobataceae</u>
Sarcobatus vermiculatus—Greasewood

<u>Tropaeolaceae</u>
Cymopterus longipes—Longstalk Spring Parsley

Key 1—Families

1. Seeds naked on the surface of individual scales. Scales individual and not organized in a cone
   a. See 3
2. Seeds in carpels, not scales. Plants have flowers with leaves that are not needle like or scale like
   a. See 5
3. Stems jointed, leaves scale-like, branches green, and photosynthetic
   a. Ephedraceae
4. Stems not jointed, come bearing, leaves scale-like or awn shaped
   a. Cupressaceae
5. Monocot
   a. Poaceae
6. Dicot
   a. See 7
7. Perianth consisting of a single whorl
   a. See 9
8. Perianth dichlamydeous, two whorls clearly separate
   a. See 14
9. Ovary inferior
   a. Asteraceae
10. Ovary superior
    a. See 11
11. Plants trailing vines, flowers perfect
    a. Polygonaceae
12. Plants not trailing vines, flowers imperfect, no thorns
    a. Amaranthaceae
13. Thorns present
    a. Sarcobataceae
14. Corolla of united petals
    a. See 18

15. Corolla of separate petals
    a. See 16
16. Ovary has four separate pollen sacs and four lobes
    a. Boraginaceae
17. Ovary has one to three pollen sacs and four lobes
    a. Polemoniaceae
18. Number of stamens not more than twice as many as the number of petals
    a. See 20
19. Number of stamens more than twice as many as the number of petals
    a. Malvaceae
20. Ovary inferior
    a. Onagraceae
21. Ovary superior
    a. See 22
22. Fruits stipitate
    a. Brassicaceae
23. Fruits separating into 3, indehiscent
    a. Tropaeolaceae

Key 2—Species

I. Amaranthaceae: Annual herbs or perennial subshrubs; leaves simple with an entire margin, usually alternate and more rarely opposite; flowers inconspicuous and usually perfect with 3 dry, pungent, bracts; sepals scarious; perianth of one to five joined tepals; petals none; stamens in same number as tepals; one pistil; ovary superior with one pollen-sac and two or three stigmas; fruit a utricle.

1. Woody clustered branches with long thin thorns
    a. *Grayia Spinosa*
2. Woody clustered branches without long thorns
    a. Leaves succulent
        i. See 5
    b. Leaves not succulent
        i. See 3

3. Leaves glabrous
   a. *Kochia americana*
4. Leaves pubescent
   a. 4 prominent wings extending bract length, united throughout
      i. *Atriplex canescens*
   b. Bracts not winged
      i. *Atriplex confertifolia*
5. Leaves alternate
   a. Leaves reddish in color
      i. *Salicornia rubra*
   b. Leaves green in color
      i. *Salicornia utahensis*
6. Leaves opposite
   a. Leaves appear segmented
      i. *Allenrolfea occidentalis*
   b. Leaves do not appear segmented
      i. *Suaeda erecta*

II. <u>Asteraceae</u>: Herbs or shrubs; leaves in all conformations; flowers composite and composed of central disk florets surrounded by few to many rays; head surrounded by cup-shaped green bracts.

i. Leaves gray green
   1. See 3.
ii. Leaves green to olive green
   1. See 7
iii. Leaves glabrous
   1. See 5
iv. Leaves pubescent
   1. *Artemisia spinescens*
v. Leaves ovate and small
   1. *Tetradymia glabrata*

vi. Leaves long and thin, lanceolate

1. *Gutierrezia sarothrae*

vii. Leaves end in three lobes

1. *Artemisia tridentate*

viii. Leaves lanceolate, margin entire

1. *Chrysothamnus viscidiflorus*

III. Boraginaceae: Herbaceous or shrubby, leaves in all conformations but always with coarse pubescence; flowers perfect, in a scorpiod, usually with bracts between; calyx usually 5-lobed and persistent, the lobes valvate; corolla 5-lobed, 5 stamens all on corolla tube, ovary superior and bicarpellate, 4-ovules, entire or lobbed, that at maturity are tough and dry, fruit commonly in 4-lobed nutlets.

i. *Cryptantha flavoculata*

IV. Brassicaceae: Herbaceous or shrubby; leaves simple to compound and either in a basal rosette or alternate up the stem; flowers in racemes, each having 4 clawed sepals, blade of each sepal in the shape of a cross.

i. *Descurainia pinnata*

V. Cupressaceae: Monoecious trees or shrubs; leaves scale or needle like, evergreen, in an opposite or whorl conformation; seeds borne in cones that are small and either terminal or axillary; microsporophylls decussate; each scale bearing at least one ovule which are either dry or fleshy at maturity.

i. *Juniperus osteosperma*

VI. Ephedraceae: Dioecious shrub; leaves scale-like in an opposite or whorled conformation; branches dark green and jointed throughout; male cones compound and found at nodes (including apex) with 2–8 microsporophylls, involucre surrounding each stalk of microsporophylls; female cones whorled and surrounded by firm or scarious bracts; seeds angled within cones.

i. *Ephedra nevadensis*

VII. Geraniaceae: Annual or perennial herbs; leaves in all conformations and are simple or compound usually lobed and stipulate; inflorescence in umbels with perfect and flowers; each flower having 5 sepals and petals with up to twice as many stamen; each stamen filament united at the base; one pistil, ovary superior with five pollen sacs, fruit dry, one seed per locule, valves of each coil at maturity, and separate from base.

i. *Erodium cicutarum*

VIII. Malvaceae: Herbaceous or occasionally shrubby; leaves pubescent with coarse hairs, arranged alternately and palmate and stipule in character; perfect flowers arranged in a raceme or panicle, sometimes in thyrsoid cymes, sometimes with an involucel of sepaloid bractlettes, five persistent sepals, five petals that are separate and adnate to staminal sheath; many stamen and united by filaments; ovary superior.

i. *Sphaeralcea coccinea*

IX. Onagraceae: Herbaceous or occasionally woody plants; leaves in alternate, opposite, or basal rosette conformations; perfect flowers with hypanthium adnate to inferior ovary; two to four sepals and as many petals; up to twice as many stamens; ovary composed of four pollen sacs; fruit a capsule, nut, or berry

i. *Oenothera pallida*

X. Poaceae: Perennial or annual herbaceous plants, stems have swollen or depressed nodes that are different from stem in color, internodal regions usually hollow; leaves alternate, lower portion usually membranous with overlapping acute, or rounded margins at base (auricles) with a narrow blade further up the leaf, parallel venation, often scabrous due to the prescience spicules which are sharp.

i. Each node with spikelets arranged in trios in 3s, tufted culms, glums awn-like

1. *Hordeum jubatum*

i. Spikelets diversely arranged, glumes not awn-like

1. Spikelets disarticulating above glumes, 2-many flowered

a. See 3

2. Each spikelet with a single perfect or pistilate floret

a. See 6

iii. Scale-like leaf blades at culm base

1. *Distichlis stricta*

iv. Leaf blades not scale-like, ovary bearing an apical tuft of short hairs

1. *Bromus tectorum*

v. Ovary lacking apical tuft of hair

1. *Poa secunda*

vi. Spikelets sessile, lemmas laterally compressed

1. *Hilaria jamesii*

vii. Spikelets pedicelled, awns deciduous, lemma glaborous

1. *Oryzopsis hymenoides*

viii. Awns persistent, awn of lemma 3-branched

1. *Aristida longiseta*

ix. Awn of lemma simple, lemma and glumes rounded on the back, and smooth at the midnerve

1. *Sporobolus airoides*

x. Lemma and glumes keeled and typically scaberulous on midnerve

1. *Sporobolus cryptandrus*

XI. Polemoniaceae: Herbaceous; leaves in alternate or opposite conformations, simple, margin entire, pinnate, or compound; flowers perfect, ovary superior and composed of three fused carpels; corolla five fused petals, same number of stamen, fruit a capsule.

a. Flowers white

i. *Leptodactylon caespitosum*

b. Flowers pink

i. *Phlox longifolia*

XII. Polygoniaceae: Herbaceous, shrubby, subshrubs, or vines; nodes swollen in stem, leaves alternate, opposite, or whorled, simple with stipules forming a sheath; flowers perfect and regular with a perianth that is two to six segments, ovary superior, two to nine stamens, one pollen sac, fruit an achene.

a. Tufted bracts with pulvinate ends

i. *Eriogonum ovalifolium*

b. Bracts not tufted, glaborose, and outer perianth segments violin shaped

i. *Eriogonum cernuum*

XIII. Sarcobataceae: Shrub with prominent thorns; leaves alternate and sessile; flowers imperfect and found in the outer spikes, staminate flowers arranged in a whorl; staminate spiked and like a catkin.

i. *Sarcobatus vermiculatus*

XIV. Tropaeolaceae: Herbaceous; leaves alternate and angled or peltate, occasionally lobed; flowers perfect with five sepals and petals, irregular, often petals appear fringed and upper flowers are usually smaller on the inside of a spur; eight stamens; one pistil; ovary superior and three pollen sacs with one apical style; fruit dehisced at maturity into three segments.

i. *Cymopterus longipes*

12.5 Concluding Remarks

From the earliest plant surveys in 1871, to Flowers' landmark work in 1934, to modern-day surveys, there has been a surprising stability in the make-up of the unique halophyte communities at GSL. However, one thing that has changed dramatically in recent years is the number of people in the field making observations. The advent of SEInet has allowed many more plant enthusiasts to post their own findings, massively increasing the data available. This does have a downside; while the data generally appears reliable, the massive quantity of data makes verifying entries difficult. Regardless, it appears that, even with fluctuations in climate and lake size, the hypersaline-tolerant plant communities (including the plant and soil zones shown in Fig. 12.3) appear to be as consistent today as they were when first surveyed. This is likely due in large part to the saline soils, harsh climate, and a lack of major disruption to the shoreline of GSL.

With the pressure from diversion of inlet water for human uses, the lake, like many other terminal lakes, has been shrinking (Wurtsbaugh et al. 2017). Over the course of these plant surveys, the lake level has fluctuated dramatically. In the 1930s and again in the in the 1960s, the lake level was at its lowest (2460 km^2) while in the 1980s it was at its highest (8500 km^2) with an average of 4400 km^2 (Freeman 2018). Construction of the railroad causeway dramatically decreased water flow to the north arm of the lake thus increasing salinity. And what impact might climate change have on perception patterns and annual rainfall and seasonal snowpack? What does this mean for the GSL plants? Changing lake levels correspond to changes in the location of the shoreline, the plants that are present at any given time, and the salinity those plants experience. Areas may be submerged for long periods of time, only to be dry at others. The ebb and flow of lake water level can change the overall soil salinity, thus reshaping plant community locations. But while these changes can at times be dramatic, it appears that the overall species composition has not varied over time, just the location of their habitable space. A shrinking lake will, over time, expose more shoreline. The newly exposed shoreline will be populated with halophytes and plant communities will encroach on the receding lake, while rainwater may leech soil salts, changing the composition of shoreline plant communities further from the water's edge. Only when the waters recede completely might there be a massive disruption of the current plant communities.

Through the identification and understanding of the plants that reside on the shoreline of GSL, a broader picture of the ecosystem can be envisioned. Though Seville Flowers' work is perhaps the most well-cited on this topic, a huge amount of data is available categorizing the biotic communities on the shores of this remarkable lake. These data drive home the point that several halophytic species can be found in this niche area and are consistent over time. These halophytic species have representatives from a multitude of orders and families, however, are primarily either grass like, shrubby, or succulent in nature. In addition, the diversity of mechanisms by which shoreline plants interact with their environment remind us about the molecular evolution of life in extreme places. Through this understanding of how each individual plant species maintains its cellular growth in the face of the high stress environment that is GSL, visitors and stewards of the lake alike can understand the intricacies by which the whole system is connected.

References

Amtmann A, Sanders D (1998) Mechanisms of Na+ uptake by plant cells. Adv Bot Res 29:75–112

Atwood G, Wambeam TJ, Anderson NJ (2016) The present as a key to the past: paleoshoreline correlation insights from Great Salt Lake. In: Oviatt CG, Shroder JF (eds) Lake Bonneville: a scientific update. Elsevier, Dordrecht

Banner RE, Pratt M, Bowns JE (2011) Grasses and grasslike plants of Utah: a field guide. Utah State University Extension, Logan, UT

Blumwald E, Aharon GS, Apse MP (2000) Sodium transport in plant cells. Biochem Biophys Acta 1465:140–151

Bohnert HJ, Jensen RG (1996) Strategies for engineering waterstress tolerance in plants. Trends Biotechnol 14:89–97

Bohnert HJ, Nelson DE, Jensen RG (1995) Adaptations to environmental stresses. Plant Cell 7:1099–1111

Calhoun EG, Staab AM, Parrott DL (2019) Characterization of halophyte rhizosphere microbiomes at Great Salt Lake, Utah. Studia UBB Biologia 64:1

Cannon JS, Cannon MA (2002) The southern Pacific Railroad trestle - past and present. In: Gwynn JW (ed) Great Salt Lake: an overview of change. Utah Geological Survey, Salt Lake City, UT, pp 283–294

Clarkson DT, Hanson JB (1980) The mineral nutrition of higher-plants. Annu Rev Plant Phys 31:239–298

Downard R, Frank M, Perkins J, Kettenring K, Larese-Casanova M (2017) Wetland plants of the Great Salt Lake: a guide to identification, communities, and bird habitat. USU Extension, Logan, UT

Flora of North America Database (2019) http://www.efloras.org/browse.aspx?flora_id=1. Accessed 21 Apr 2019

Flowers S (1934) Vegetation of the Great Salt Lake region. Bot Gaz 95(3):353–418. https://doi.org/10.1086/334398

Flowers S (1955) Ecological sample areas established. In: Symposium on ecology of disease transmission in native animals. Ecol Res University of Utah, pp 11–12

Flowers S, Evans FR (1966) The flora and fauna of the Great Salt Lake region, Utah. In: Salinity and aridity. Springer, Dordrecht, pp 367–393

Flowers TJ, Hajibagheri MA, Clipson NJW (1986) Halophytes. Quart Rev Biol 61:313–337

Freeman M (2018) Great Salt Lake elevations. United States Geological Survey, Utah Water Science Center. https://www.usgs.gov/centers/ut-water/science/great-salt-lake-elevations?qt-science_center_objects=0#qt-science_center_objects. Accessed 22 Mar 2019

Gwynn JW (1980) Great Salt Lake: a scientific, historical, and economic overview. Print. Bulletin (Utah Geological and Mineral Survey) 116

Gwynn JW (1996) Commonly asked questions about Utah's Great Salt Lake & Lake Bonneville. Utah State Geological Survey, Salt Lake City, UT

Hasegawa PM, Bressan RA, Zhu JK, Bohnert HJ (2000) Plant cellular and molecular responses to high salinity. Annu Rev Plant Physiol Plant Mol Biol 51:463–499

Hingole SS, Pathak AP (2016) Saline soil microbiome: a rich source of halotolerant PGPR. JCSB 19:231–239

Hrusa GF, Wilken DH (2012) *Kochia americana*. In: Jepson Flora Project (eds) Jepson eFlora. http://ucjeps.berkeley.edu/eflora/eflora_display.php?tid=29939. Accessed 8 June 2019

Jennings DH (1968) Halophytes, succulence and sodium in plants - a unified theory. New Phytol 67:899–911. https://doi.org/10.1111/j.1469-8137.1968.tb06402.x. Accessed 21 Mar 2019
Ji H, Pardo JM, Batelli G, Van Oosten MJ, Bressan RA, Li X (2013) The salt overly sensitive (SOS) pathway: established and emerging roles. Mol Plant 6:275–286
Maathuis FJM, Ahmad I, Patishtan J (2014) Regulation of Na+ fluxes in plants. Front Plant Sci 5:467. https://doi.org/10.3389/fpls.2014.00467. Accessed 21 Mar 2019
Madison RJ (1970) Effects of a causeway on the chemistry of the brine in Great Salt Lake Utah. Water-Resources Bulletin 14. https://pubs.er.usgs.gov/publication/70179728
McNulty IB (1985) Rapid osmotic adjustment by a succulent halophyte to saline shock. Plant Physiol 78(1):100–103. https://doi.org/10.1104/pp.78.1.100
Minnesota Wildflowers (2019) *Salicornia rubra* (red saltwort). https://www.minnesotawildflowers.info/flower/red-saltwort. Accessed 21 Mar 2019
Munns R, Tester M (2008) Mechanisms of salinity tolerance. Annu Rev Plant Biol 59:651–681
Niu X, Bressan RA, Hasegawa PM, Pardo JM (1995) Ion homeostasis in NaCl stress environments. Plant Physiol 109:735–742
Oviatt CG, Thompson RS, Kaufman DS, Bright J, Forester RM (1999) Reinterpretation of the Burmester Core, Bonneville Basin, Utah. Quat Res 52:180–184
Parker WR (2019) University of Austin Plant Database. https://biodiversity.utexas.edu/collections. Accessed 4 Apr 2019
PLANTS Database (2019) https://plants.usda.gov/java/factSheet. Accessed 2 Apr 2019
Range Plants of Utah Database (2019) https://extension.usu.edu/rangeplants/. Accessed 13 Mar 2019
Rawley EV (1980) Plant life of the Great Salt Lake study area. In: Gwynn J (ed) Great Salt Lake: a scientific, historical and economic overview. USGS, Salt Lake City, UT
Rhodes D, Hanson AD (1993) Quaternary ammonium and tertiary sulfonium compounds in higher plants. Annu Rev Plant Physiol Plant Mol Biol 44:357–384
Sanders D (2000) Plant biology: the salty tale of Arabidopsis. Curr Biol 10:486–488. Available via SEINet Portal Network. http://swbiodiversity.org/seinet/index.php. Accessed 8 May 2019
Semenova GA, Fomina IR, Biel KY (2010) Structural features of the salt glands of the leaf of *Distichlis spicata* 'Yensen 4a' (Poaceae). Protoplasma 240:75–82
Shroder JF, Cornwell K, Oviatt CG, Lowndes TC (2016) Landslides, alluvial fans, and dam failure at Red Rock Pass: the outlet of Lake Bonneville. In: Oviatt CG, Shroder JF (eds) Lake Bonneville: a scientific update. Elsevier, Dordrecht
Stephens DW, Gardener J (2007) Water-resources investigations report, Great Salt Lake, Utah. USGS publication 1999-4189. https://pubs.usgs.gov/wri/wri994189/PDF/WRI99-4189.pdf. Accessed 2 Apr 2019
Tester M, Davenport R (2003) Na+ tolerance and Na+ transport in higher plants. Ann Bot 91:503–527
Thomson WW (1975) The structure and function of salt glands. In: Poljakoff-Mayber A, Gale J (eds) Plants in saline environments. Ecological studies (analysis and synthesis), vol 15. Springer, Berlin
Vest E (1962) Biotic communities in the Great Salt Lake Desert. Ecology and epizoology series, no. 73. Institute of Environmental Biological Research, Division of Biological Sciences, University of Utah, Salt Lake City
Welsh SL, Atwood ND, Goodrich S, Higgins LC (1987) A Utah flora. Brigham Young University, Provo, UT
Woodbury AM (ed) (1956) Ecological check lists: the Great Salt Lake Desert series. University of Utah, Salt Lake City
Wurtsbaugh WA, Miller C, Null SE, DeRose RJ, Wilcock P, Hahnenberger M, Howe F, Moore J (2017) Decline of the world's saline lakes. Nat Geosci 10:816–821
Wyckoff JW (1973) The effects of soil texture on species diversity in an arid grassland of the eastern Great Basin. Great Basin Nat 33(3):4

Yokoi S, Bressan RA, Hasegawa PM (2002) Salt stress tolerance of plants. JIRCAS Working Report, pp 25–33

Yuan Z, Druzhinina IS, Labbé J, Redman R, Qin Y, Rodriguez R, Zhang C, Tuskan GA, Lin F (2016) Specialized microbiome of a halophyte and its role in helping non-host plants to withstand salinity. Sci Rep 6:32467. https://doi.org/10.1038/srep32467. Accessed 4 Apr 2019

Photograph Credits

Gratwicke B. *Salicornia rubra*. https://calscape.org/photos/3312

Hagwood S. *Distichlis stricta/spacata.* https://plants.usda.gov/java/largeImage?imageID=disp_007_avp.jpg

Hagwood S. *Phlox longifolia.* https://plants.usda.gov/java/largeImage?imageID=phlo2_009_avp.jpg

Lavin M. *Atriplex confertifolia.* http://temperate.theferns.info/image/Atriplex+confertifolia

Lee G. *Suaeda erecta.* https://www.saskwildflower.ca/nat_Suaeda-calceoliformis.html

Licher M. *Aristida longiseta.* http://swbiodiversity.org/seinet/imagelib/imgdetails.php?imgid=219382

Maylett C. *Sarcobatus vermiculatus.* https://en.wikipedia.org/wiki/Sarcobatus#/media/File:Greasewood_01.jpg

Parrott D. *Kochia americana.* Personal photograph

Shebs S. *Allenrolfea occidentalis.* https://commons.wikimedia.org/wiki/User:Stan_Shebs/Gallery/Plants/A#/media/File:Allenrolfea_occidentalis_4.jpg

Sivinski R. *Salicornia utahensis.* https://calphotos.berkeley.edu/cgi/img_query?enlarge=0000+0000+0909+0417

Chapter 13
Invasive Plants of Great Salt Lake Wetlands: What, Where, When, How, and Why?

Karin M. Kettenring, Chad R. Cranney, Rebekah Downard, Keith R. Hambrecht, Emily E. Tarsa, Diane R. Menuz, and Christine B. Rohal

Abstract Great Salt Lake (GSL) and its wetlands are recognized around the world for the valuable habitat they provide for millions of migratory birds. GSL wetlands are threatened by a number of invasive plants, the most problematic of which is non-native phragmites (*Phragmites australis*) although there are a number of other species that are concerning and also a target of management. In this chapter, we describe the major invasive plants of and their distributions across GSL wetlands, detail the mechanisms driving these plant invasions and their historical context, discuss why different invasive species present unwanted impacts, and synthesize best practices for invasive plant control for these species in GSL wetlands. Managers of GSL wetlands face a daunting task to control these plants, particularly in the case of phragmites, where hundreds of hectares of infestations must be treated and retreated annually. Eradication of phragmites will not be possible given its intense propagule pressure and dense seed banks, thus strategic and prioritized management

K. M. Kettenring (✉) · E. E. Tarsa
Ecology Center and Department of Watershed Sciences, Utah State University, Logan, UT, USA
e-mail: karin.kettenring@usu.edu

C. R. Cranney
Utah Division of Wildlife Resources, Utah Department of Natural Resources, Salt Lake City, UT, USA

R. Downard
Utah Division of Water Quality, Utah Department of Environmental Quality, Salt Lake City, UT, USA

K. R. Hambrecht
Utah Division of Forestry, Fire & State Lands, Utah Department of Natural Resources, Salt Lake City, UT, USA

D. R. Menuz
Utah Geological Survey, Utah Department of Natural Resources, Salt Lake City, UT, USA

C. B. Rohal
Department of Environmental Horticulture, University of Florida, Gainesville, FL, USA

B. K. Baxter, J. K. Butler (eds.), *Great Salt Lake Biology*,
https://doi.org/10.1007/978-3-030-40352-2_13

approaches are critical. Future success for all invaders will be contingent upon continued cooperation between scientists and managers to develop robust treatment techniques and between managers to coordinate their management to reduce invader cover and impacts. Furthermore, future research and management priorities should include (1) limiting invader propagule pressure and seed bank densities, (2) optimizing native plant revegetation following invader removal, (3) early detection and control of new invaders that are likely to increase with climate change, (4) more refined hydrologic management to promote invader control, and (5) quantitative documentation of avian impacts from invaders, especially given the continental importance of this habitat to migratory birds. Despite the threats GSL and its wetlands face with anthropogenic development, water diversions, and climate change, we are optimistic that at least in the case of invasive species, collaborative and science-backed management can continue to be effective given current partnerships and practices.

Keywords Great Salt Lake · *Cardaria draba* · Alien species · Non-indigenous species · Disturbance · *Frankenia pulverulenta* · Invasion mechanisms · Invasive plant management · Invasive species · *Lepidium latifolium* · *Lythrum salicaria* · Non-native species · Nutrients · *Phragmites australis* · *Potamogeton crispus* · Revegetation · Seed sowing · Seed-based restoration · Seeding density · *Typha domingensis* · *Typha latifolia* · Wetland restoration

13.1 Introduction

Wetland plant communities are the food web foundation of Great Salt Lake (GSL) wetlands (Downard et al. 2017). However, native-dominated plant communities in these wetlands—like in many wetlands around the world—are being heavily impacted by invasive plants (Downard et al. 2017; Kettenring et al. 2012; Zedler and Kercher 2004). Invasive plants are usually non-native species (some notable native but undesirable species exist), likely introduced due to human activity, that have substantial negative ecological or economic impacts or cause harm to human health (Executive Presidential Order 1999). Invasive plants are considered particularly problematic in GSL wetlands because they reduce habitat quality for many wildlife species, including continentally important migratory birds, by converting native plant assemblages to monotypic plant stands often with little habitat value (Downard et al. 2017; Intermountain West Joint Venture 2013; Kettenring et al. 2012; Rohal et al. 2018; SWCA 2012). Understanding the mechanisms driving these plant invasions is critical for improving management and recovering the critical habitat of GSL wetlands. In this chapter, we describe the major invasive plants of and their distributions across GSL wetlands, detail the mechanisms driving these plant invasions and their historical context, discuss why different invasive species are undesirable in terms of their impacts, and synthesize best practices for control of these species in GSL wetlands. We conclude with our assessment of next steps for further enhancing our understanding of the mechanisms, impacts, and management of common plant invaders.

13.2 Overview

13.2.1 Significance of Great Salt Lake and Its Wetlands

Understanding the significance of GSL and its wetlands is important to grasp what is currently present and can be lost with widespread plant invasions (Fig. 13.1). The GSL ecosystem is a desert oasis for millions of birds that breed in and migrate through the Great Basin region (Weller 1999; Oring et al. 2000; Olson et al. 2004; Intermountain West Joint Venture 2013). GSL is recognized regionally, nationally, and hemispherically for its importance to more than 130 waterbird species that rely on the lake and its wetlands during certain stages of their life history (Neill and Sorensen, unpublished list 2020). For some of these bird species, a major proportion of their population can be found on GSL during certain times of the year. For example, the largest staging concentration of Wilson's phalarope (*Phalaropus tricolor*), a small wading bird, is found on GSL, numbering over 500,000 (Fig. 13.2) (Paul and Manning 2002). Some of the highest counts within the Pacific Flyway of American avocets (*Recurvirostra americana*) and black-necked stilts (*Himantopus mexicanus*), two iconic shorebirds of GSL wetlands, have been recorded on GSL—250,000 and 65,000, respectively (Fig. 13.2) (Barber and Cavitt 2012; Shuford et al. 1995). GSL wetlands are the only interior staging area in western North America for thousands of marbled godwits (*Limosa fedoa*), another shorebird, with peak counts of 43,000 birds (Fig. 13.2) (Paul and Manning 2002; Shuford et al. 1995). Over 60,000 tundra swans (75% of its western population; *Cygnus columbianus*) rely on GSL wetlands for staging and refueling during their fall migration (Fig. 13.2) (Aldrich and Paul 2002). In addition, an extensive 5-year survey estimated 86,752,258 mean bird use days (1 bird spending 24 h in the study area) for GSL (Paul and Manning 2002). The diversity and vast numbers of aquatic birds that use GSL and its associated wetlands are due to the diversity of habitats and abundance of resources within these habitats that ultimately fuel these birds through critical stages within their life cycle (Downard et al. 2017; Intermountain West Joint Venture 2013; Paul and Manning 2002).

The modern economic and cultural significance of these avian populations is also worth noting. Non-hunting resource users, namely birdwatchers and others who visit the lake and its wetlands for aesthetic, spiritual, and intellectual inspiration, likely contribute more than $50 million in direct spending to the Salt Lake City area economy (Bioeconomics, Inc. 2012). Cultural events that celebrate these bird populations including the annual GSL Bird Festival bring together hundreds of birdwatchers, hunters, and interested citizens for workshops, field trips, and family activities (Burr and Scott 2004; Great Salt Lake Bird Festival 2019). Also, a substantial proportion of these migratory bird populations include waterfowl species

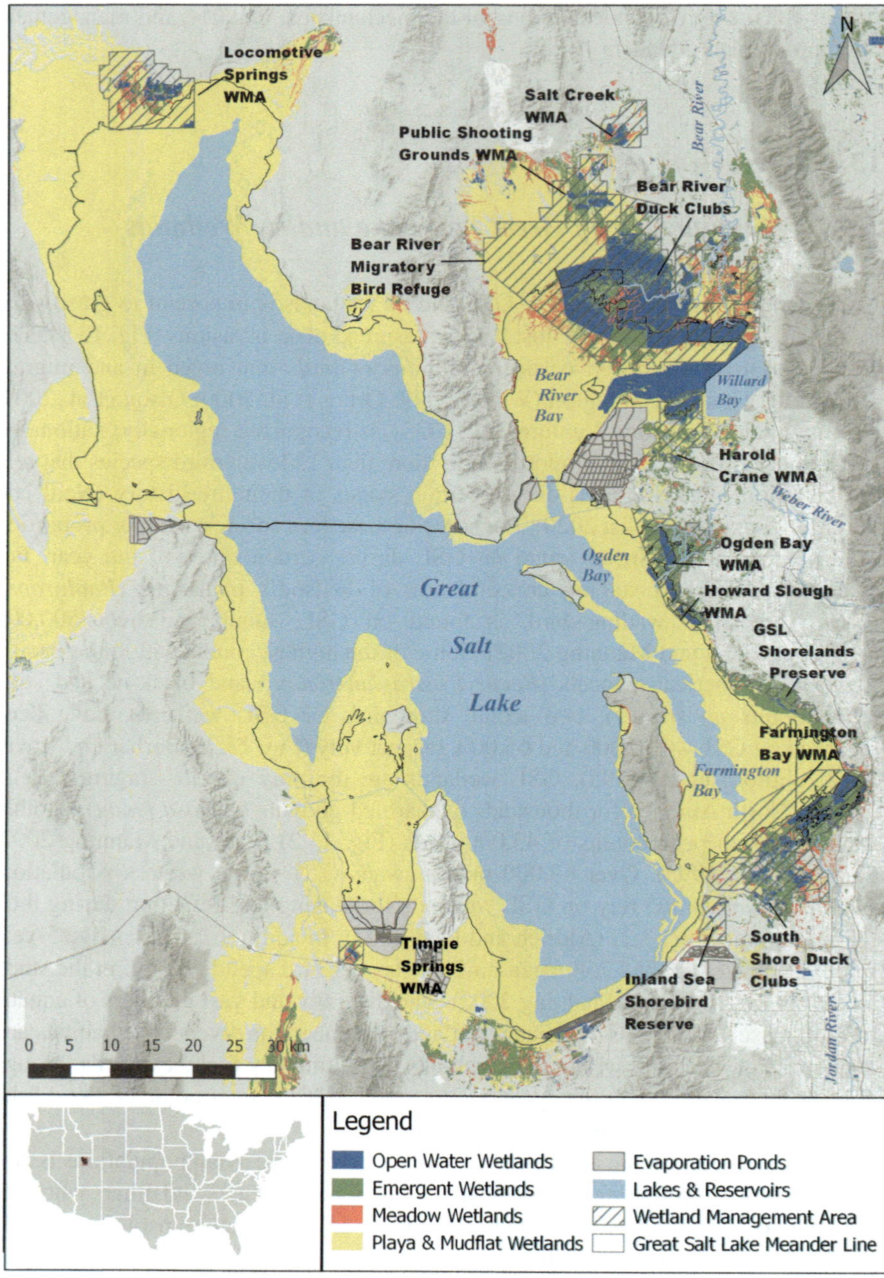

Fig. 13.1 The major federal, state, private, and nonprofit wetland complexes on Great Salt Lake. WMA = Waterfowl Management Area. GSL = Great Salt Lake. Wetland layer: US Fish & Wildlife Service 2017. National Wetlands Inventory website. US Department of the Interior, Fish & Wildlife Service, Washington, DC. http://www.fws.gov/wetlands/. Map tiles by Stamen Design, under CC BY 3.0. Data by OpenStreetMap, under ODbL

Fig. 13.2 Examples of bird species for whom Great Salt Lake and its wetlands provide significant habitat: (**a**) American avocet (*Recurvirostra americana*), (**b**) black-necked stilt (*Himantopus mexicanus*), (**c**) Wilson's phalarope (*Phalaropus tricolor*), (**d**) marbled godwit (*Limosa fedoa*), (**e**) tundra swans (*Cygnus columbianus*), (**f**) snowy plover (*Charadrius nivosus*), (**g**) northern shoveler (*Anas clypeata*), (**h**) cinnamon teal (*Anas cyanoptera*), (**i**) green-winged teal (*Anas carolinensis*), and (**j**) redhead (*Aythya americana*). All images courtesy of Mia McPherson, On the Wing Photography, with permission

that are popular for hunting. Waterfowl hunting has been a part of the Utah economy for generations (Thursby 2004) and contributes a hundred million dollars annually to the Salt Lake City area economy (Duffield et al. 2011; Bioeconomics, Inc. 2012). Interestingly, a survey of more than 550 GSL hunters in 2011 found that they considered invasive plants the top threat to their hunting hobbies and livelihoods (Duffield et al. 2011). Because of their economic and cultural significance, GSL wetlands are owned and managed for the public by state and federal agencies—US Fish & Wildlife Service, Utah's Division of Wildlife Resources (DWR) and Division of Forestry, Fire & State Lands (DFFSL)—nonprofit conservation groups like The Nature Conservancy and the Audubon Society for habitat conservation, privately managed mitigation wetlands like Inland Sea Shorebird Reserve, and private duck clubs that provide waterfowl hunting privileges to their members (Fig. 13.1).

13.2.2 Great Salt Lake Wetland Habitats Impacted by Invasives

The millions of birds that visit GSL and associated wetlands each year are dependent upon these unique habitats for food, nesting, shelter, loafing, and brood rearing (Aldrich and Paul 2002; Cox and Kadlec 1995; Roberts 2013; Vest and Conover 2011). Habitat types around GSL range from the hypersaline open lake, brackish emergent wetlands (with species like *Bolboschoenus maritimus*, alkali bulrush), mudflat, and playas (with species like *Salicornia rubra*, pickleweed), meadows (with species like *Eleocharis palustris*, common spikerush; *Distichlis spicata*, saltgrass; and *Juncus arcticus* ssp. *littoralis*, mountain rush) to fresh open water impoundments with productive submerged aquatic communities of native pondweeds (*Stuckenia* and *Potamogeton* species) surrounded by emergent wetlands (with species like *Schoenoplectus acutus* and *S. americanus,* hardstem and threesquare bulrush) (Figs. 13.1 and 13.3; Downard et al. 2017). The distribution and extent of each of these habitats are dependent on GSL elevations and the flux between saltwater and freshwater (Downard et al. 2017; SWCA 2013).

Over 70 non-native plant species have been documented in and near GSL wetlands and almost 40% of plant species listed in a recent flora for the region are non-native (Downard et al. 2017). There are numerous reasons why there are so many non-native plants inhabiting GSL wetlands, including high rates of spread from drainages; movement by livestock, birds, and humans; intentional planting by managers and landowners; and a high level of disturbance—such as nutrient enrichment and sedimentation—that lead to opportunities for invasive plants to establish (Long et al. 2017a; Zedler and Kercher 2004). In this chapter, we focus on the most prevalent invasive species in these wetland systems, the most common and impactful of which is *Phragmites australis* (Cav.) Trin. ex Steud. (common reed or phragmites) (Downard et al. 2017; Duffield et al. 2011; Kettenring et al. 2012) (Fig. 13.4). We bring particular attention to phragmites in this chapter because of its dominance in GSL wetlands and its high importance to managers. In addition, we focus on a

Fig. 13.3 Common native wetland plants that are important for wildlife habitat but can be replaced by non-native phragmites: (**a**) hardstem bulrush (*Schoenoplectus acutus* (Muhl. ex Bigelow) Á. Löve & D. Löve; image by Rachel Hager, with permission), (**b**) threesquare bulrush (*Schoenoplectus americanus* (Pers.) Volkart ex Schinz & R. Keller), (**c**) alkali bulrush (*Bolboschoenus maritimus* (L.) Palla), (**d**) saltgrass (*Distichlis spicata* (L.) Greene), (**e**) pickleweed (*Salicornia rubra* A. Nelson), (**f**) sago pondweed (*Stuckenia pectinata* (L.) Börner), (**g**) mountain rush (*Juncus arcticus* Willd. spp. *littoralis* (Engelm.) Hultén), (**h**) common spikerush (*Eleocharis palustris* (L.) Roem. & Schult.), (**i**) nodding beggartick (*Bidens cernua* L.), (**j**) rayless alkali aster (*Symphyotrichum ciliatum* (Ledeb.) G.L. Nesom), and (**k**) fringed willowherb (*Epilobium ciliatum* Raf.)

handful of other problematic species to management—*Lythrum salicaria* (purple loosestrife), *Frankenia pulverulenta* (European seaheath), *Cardaria draba* (whitetop), *Conium maculatum* (poison hemlock), *Lepidium latifolium* (perennial pepperweed), and *Typha latifolia* and *Typha domingensis* (cattails) (Downard et al. 2017). We also mention a number of plants that are non-native and may have a noxious weed classification, but foster differing levels of concern among managers (Table 13.1; Fig. 13.5).

Fig. 13.4 Phragmites phenology, reproduction, and growth: (**a**) phragmites towering above alkali bulrush, (**b**) immature phragmites inflorescences, (**c**) mature phragmites inflorescences, (**d**) phragmites inflorescences on a senesced stand (winter), (**e**) live phragmites towering over wetland researcher and manager, Chad Cranney, and (**f**) drought-stressed phragmites

13.3 Phragmites Distribution, Historical Context, Mechanisms of Invasion, and Impacts

13.3.1 Distribution and Historical Context of Invasion

Phragmites australis subsp. *australis* is a non-native plant (composed of multiple haplotypes; hereafter non-native or invasive phragmites) from Eurasia that is now widespread in coastal and inland wetlands and moist, disturbed habitats across North America (Chambers et al. 1999; Kettenring et al. 2012; Meyerson and Cronin 2013; Saltonstall 2002). Studying and managing the invasion of non-native phragmites is

Table 13.1 Non-native and invasive plant species of Great Salt Lake wetlands. The list includes native species considered to be aggressive (invasive) and non-native species that are the most common, known to be problematic (e.g., noxious weeds), or with high potential to invade wetlands (due to anaerobic tolerance). Species in bold are frequently the focus of management control efforts

Scientific name with author citation Common name	Growth duration and habit[a]	Wetland indicator status[a]	Description
Agrostis stolonifera L. Creeping bentgrass	Perennial grass	FACW	Introduced as a pasture grass, spreading into wetlands. High tolerance to salinity and anaerobic soils (Downard et al. 2017; USDA NRCS 2019)
Atriplex spp. Saltbush	Annual forb	N/A or FACW	*A. micrantha* Ledeb. (twoscale saltbush; non-native), *A. prostrata* Bouchér ex DC. (triangle orache; nativity uncertain), and *A. patula* L. (spear saltbush; non-native) are all found in GSL wetlands and are difficult to distinguish (Downard et al. 2017); the latter two species are FACW (USDA NRCS 2019)
Bassia hyssopifolia (Pall.) Kuntz Fivehorn bassia	Annual forb	FAC	Common along road shoulders, planted to prevent erosion (Downard et al. 2017). Has a high salinity tolerance and medium tolerance of anaerobic soils. Important habitat for wildlife (C. Cranney, *pers. obs.*)
***Cardaria draba* (L.) Desv.** **Whitetop**	**Perennial forb**	**N/A**	**Found in dry to moist, alkaline soil near agriculture and listed as a noxious weed in Utah** (Downard et al. 2017; Lowry et al. 2017)
***Carduus nutans* L.** **Musk thistle**	**Perennial forb**	**FACU**	**Biennial or winter annual that can grow 2 m tall; found in roadsides, pastures, and wetland and stream edges (Lowry et al.** 2017)
Cichorium intybus L. Chicory	Perennial forb	FACU	Considered noxious weed in Colorado (USDA NRCS 2019). No salinity tolerance and low anaerobic tolerance; often occurs on dry fringes (Downard et al. 2017; USDA NRCS 2019)
Cirsium spp. **Thistles**	**Perennial forb**	**FACU**	***C. vulgare* (Savi) Ten. (bull thistle) and *C. arvense* (L.) Scop. (Canada thistle) are nuisance species found in dry margins of GSL wetlands. *C. arvense* is listed as a Class 3 noxious weeds in Utah (Utah Department of Agriculture and Food** 2019)
***Conium maculatum* L.** **Poison hemlock**	**Biennial forb**	**FACW**	**Poisonous to people and livestock, a Class 3 noxious weed in Utah (Utah Department of Agriculture and Food** 2019)

(continued)

Table 13.1 (continued)

Scientific name with author citation Common name	Growth duration and habit[a]	Wetland indicator status[a]	Description
Dipsacus fullonum L. Fuller's teasel	Biennial forb	FAC	A noxious species in Colorado, capable of outcompeting and shading out desirable native species (Downard et al. 2017; USDA NRCS 2019)
Echinochloa crus-galli (L.) P. Beauv. Barnyardgrass	Annual grass	FACW	Native to Asia, this grass is planted as a food source for waterfowl. Able to accumulate a high concentration of nitrogen and phosphorus (Cao et al. 2020)
Elaeagnus angustifolia L. Russian olive	Perennial tree	FAC	A Class 4 noxious weed in Utah with high salinity tolerance and low anaerobic tolerance (Utah Department of Agriculture and Food 2019)
Frankenia pulverulenta L. European seaheath	Annual forb	N/A	Native to Europe, thrives in playa habitat, and very recently documented as widespread (D. Menuz, *pers. obs.*). Impacts unknown
Hordeum marinum Huds. Seaside barley	Annual grass	FAC	A grass that can cover significant wetland area, especially during dry conditions (Downard et al. 2017; USDA NRCS 2019)
Iris pseudacorus L. Pale yellow iris	Perennial forb	OBL	A noxious weed in 3 Western states, dispersed along canals via floating seeds (Downard et al. 2017; USDA NRCS 2019)
Lactuca serriola L. Prickly lettuce	Annual forb	FACU	Common weed in GSL wetlands, opportunistic late-season invader (Downard et al. 2017; USDA NRCS 2019). Important habitat for wildlife (C. Cranney, *pers. obs.*)
***Lepidium latifolium* L.** **Perennial pepperweed**	**Perennial forb**	**FAC**	**Roots are buoyant, can spread long distances by water, can remain dormant in the soil for years, and have been found more than 3 m deep in the soil profile (Fire Effects Information System** 2019)
Lepidium perfoliatum L. Clasping pepperweed	Annual forb	FACU	Found in a wide variety of ecosystems (open desert, alkaline flats, wetlands), though the area it covers is low (Downard et al. 2017)
***Lythrum salicaria* L.** **Purple loosestrife**	**Perennial forb**	**OBL**	**One individual plant can produce up to 2.7 million seeds, which are spread through waterways (DiTomaso et al.** 2013)
***Onopordum acanthium* L.** **Scotch thistle**	**Biennial forb**	**N/A**	**Can grow up to 12 ft tall; found along canals and streams in pastures and rangelands (Lowry et al.** 2017)

Phalaris arundinacea L. Reed canarygrass	Perennial grass	FACW	A cryptic species comprised of native and non-native, invasive genotypes likely introduced to the United States (Kettenring et al. 2019). It has medium salinity tolerance and high tolerance of anaerobic conditions (Downard et al. 2017; USDA NRCS 2019)
Polygonum persicaria L. Spotted ladysthumb	Annual forb	FACW	Moderately common in freshwater wetlands and looks a lot like native *P. lapathifolium* L. (Downard et al. 2017). Important habitat for wildlife (C. Cranney, *pers. obs.*)
Polypogon monspeliensis (L.) Desf. Annual rabbitsfoot grass	Annual grass	FACW	A common wetland-associated grass with low salinity tolerance but high anaerobic tolerance
Potamogeton crispus L. Curly pondweed	Perennial forb	OBL	Submerged aquatic species found in ponded brackish, alkaline, or eutrophic waters (Haynes and Hellquist 2000); not yet a nuisance in GSL wetlands, but listed as a noxious weed in at least six other states (USDA NRCS 2019)
Rumex stenophyllus Ledeb. Narrowleaf dock	Perennial forb	FACW	Very common wetland-associated species, similar to the also non-native *R. crispus* L. (FAC) (Downard et al. 2017). Important habitat for wildlife (C. Cranney, *pers. obs.*)
***Tamarix* L. spp.** **Tamarisk**	**Perennial tree**	**FAC**	**A Class 3 noxious weed in Utah. Deep taproots draw local water tables down while salty leaves make soils saltier (Utah Department of Agriculture and Food** 2019)
Thinopyrum intermedium (Host) Barkworth & D.R. Dewey Intermediate wheatgrass	Perennial grass	N/A	Common near wetlands, planted along road shoulders (Downard et al. 2017). Medium salinity tolerance and low anaerobic tolerance (USDA NRCS 2019). Important habitat for wildlife (C. Cranney, *pers. obs.*)
Trifolium fragiferum L. Strawberry clover	Perennial forb	FACU	High salinity and medium anaerobic tolerances. Likely spreads from pastures to wetlands (Downard et al. 2017; USDA NRCS 2019)
Typha L. spp. Cattail	Perennial forb	OBL	*T. domingensis* Pers. (southern cattail) and *T. latifolia* L. (broadleaf cattail) are native GSL wetland species but undesirable because they are considered poor waterfowl habitat and can push out more desirable wetland species (Downard et al. 2017; Ochterski 2003)

[a]Growth habit, duration, and wetland indicator status, which refers to the percentage of time the species is expected to be found in a wetland (OBL = obligate, >99%; FACW = facultative wetland, 67–99%; FAC = facultative, 34–66%; FACU = facultative upland, 1–33%; N/A = not rated) as determined by USDA NRCS (2019)

Fig. 13.5 Other common invasive and undesirable plant species in Great Salt Lake wetlands: (**a**) purple loosestrife (*Lythrum salicaria*), (**b**) European seaheath (*Frankenia pulverulenta*), (**c**) whitetop (*Cardaria draba*), (**d**) perennial pepperweed (*Lepidium latifolium*), (**e**) southern cattail (*Typha domingensis*), and (**f**) curly pondweed (*Potamogeton crispus*)

complicated by the fact that it is a cryptic invader, meaning that it is morphologically similar to a native subspecies and the two lineages cannot be easily distinguished without genetic analyses (Saltonstall 2002). Non-native phragmites co-occurs with a native subspecies (*Phragmites australis* subsp. *americanus* Saltonst., P.M. Peterson & Soreng; comprised of numerous haplotypes; hereafter native phragmites) in a number of places in North America including Utah (Kettenring and Mock 2012;

Kulmatiski et al. 2011; Lambert et al. 2016; Meadows and Saltonstall 2007; Meyerson et al. 2010a; Price et al. 2014; Saltonstall 2002, 2011, 2016). Native phragmites has been a part of the flora of North American inland and coastal wetlands for thousands of years (Goman and Wells 2000; Hansen 1978; Kiviat and Hamilton 2001; Niering et al. 1977; Orson 1999). In Utah and other parts of the American Southwest, native phragmites is broadly distributed but rarely abundant, reflecting the distribution of its habitats—isolated springs, riparian areas, and mesohaline and alkaline wetlands (Kettenring and Mock 2012; Kulmatiski et al. 2011; Meyerson et al. 2010a; Saltonstall et al. 2016). Native phragmites is mostly or entirely absent from the flora of GSL wetlands and the negative impacts associated with this species are due to the non-native lineage, the main focus of this chapter. Non-native phragmites has been in Utah for decades—it was first documented in herbarium records in 1993 (confirmed as non-native with genetic analyses) and flourished after the flooding of GSL in the 1980s (Kettenring et al. 2012; Kulmatiski et al. 2011; Smith and Kadlec 1983). In 2004, Kulmatiski et al. (2011) estimated that phragmites covered 86 km^2 of GSL wetlands (using methods based on visual inspection of 30-m resolution NAIP imagery and ground-truthing). Another census in 2011, using 1-m resolution imagery and supervised classification estimated that phragmites occupied 93 km^2 of GSL wetlands (Long et al. 2017a). A recent draft of updated National Wetlands Inventory (NWI) mapping for GSL, developed using visual interpretation of 2014 aerial imagery and ancillary data following NWI standards (Dahl et al. 2015), showed extensive spread of phragmites into previously unvegetated mudflats and 164 km^2 of wetland with $\geq$60% phragmites cover (US Fish & Wildlife Service 2019). Although the methods and exact boundaries of these surveys differ, they each indicate that phragmites is widespread and abundant in GSL wetlands, despite concerted efforts to limit its coverage. To encourage its further control, phragmites was recently placed on the state noxious weed list in Utah as a Class 3 containment species (Utah Department of Agriculture and Food 2019). In addition, the state spends hundreds of thousands of dollars each year on invasive phragmites control.

Replacement of native phragmites with non-native phragmites is a major concern across North America (Kettenring et al. 2012; Price et al. 2014; Taddeo and Blois 2012). However, in Utah it appears that many of the historical native phragmites populations still exist and there were likely few native phragmites populations along GSL due to potentially unsuitable (brackish) environmental conditions (Kettenring and Mock 2012; Kulmatiski et al. 2011). Hybridization between the lineages is also a conservation concern because of the increased competitiveness of native/non-native hybrids for other plants species and potential loss of the somewhat rare native phragmites (Galatowitsch et al. 1999; Meyerson et al. 2010b; Saltonstall 2011). Phragmites hybrids form under controlled pollination in experimental settings, and hybrids have been documented in a few places in North America including Las Vegas in the arid Southwest (Meyerson et al. 2010b, 2012; Paul et al. 2010; Saltonstall et al. 2014, 2016; Wu et al. 2015). However, in Utah and in GSL wetlands, phragmites hybrids have not been found (Kettenring and Mock 2012; Kulmatiski et al. 2011; Lambert et al. 2016).

13.3.2 Mechanisms of Invasion

Numerous factors contribute to the establishment and spread of phragmites. It was long believed that phragmites spreads mostly by asexual means—clonal expansion from established plants through stolons and rhizomes as well as dispersal (mostly by water) of stolon and rhizome fragments (Chambers et al. 1999; Keller 2000; Pellegrin and Hauber 1999; Saltonstall 2002). However, more recent research found that sexual reproduction is the predominant mechanism for phragmites dispersal and establishment (Albert et al. 2015; Belzile et al. 2010; Kettenring et al. 2011; Kettenring and Mock 2012; McCormick et al. 2010a, b). In Utah, comparisons between non-native and native phragmites showed that non-native phragmites relies much more on sexual reproduction and spread than does its native congener (Kettenring and Mock 2012). Seeds are important to the spread of non-native phragmites over moderate distances (up to approximately 100–500 m) but expansion of existing patches is mostly clonal with occasional seedling establishment (Fig. 13.4) (Kettenring et al. 2016; Kettenring and Mock 2012; McCormick et al. 2010a, b, 2016). Given that phragmites seeds require light for germination and seedlings are generally poor competitors, it is not surprising that phragmites thrives under disturbed conditions where vegetative cover is minimal and light and nutrients are abundant (Kettenring et al. 2015; Kettenring and Whigham 2018). Throughout its North American range, the occurrence of phragmites has correlated with increasing agricultural activities, suburban development, and highway networks; the presence of shoreline structures like riprap and docks; declines in water levels; and nutrient enrichment (Brisson et al. 2010; Chambers et al. 2008; Jodoin et al. 2008; King et al. 2007; Sciance et al. 2016; Tulbure and Johnston 2010). Along GSL, phragmites is more common at lower elevations with prolonged flooding and in areas closer to point sources of pollution and freshwater inflows likely to have elevated nutrients and moderate salinities (Long et al. 2017a). And it is likely that the massive disturbance of GSL wetlands in the 1980s—when hypersaline lake water completely inundated all wetlands—created the perfect high resource (exposed soil with high light) environment for phragmites seeds and seedlings to flourish (Kettenring et al. 2012, 2015).

13.3.3 Impacts on Native Plant Communities, Habitat Value, Wildlife, Human Use of Wetlands

The impact on avian habitat of non-native phragmites expansion into GSL wetlands is extremely concerning to wetland managers (Kettenring et al. 2012; Rohal et al. 2018). Native plants are not able to resist the clonal expansion of established phragmites stands. And, as described above, phragmites can initially get established via seeds in small (and large) disturbances due to sedimentation, herbivory, dike construction, and the like (Long et al. 2017a; Kettenring et al. 2015; Kettenring and Whigham 2018). Naturally unvegetated areas, such as mudflats and drawdown areas that are critical shorebird habitat, are also very susceptible to phragmites invasion. The fundamental alteration to wetland plant communities (composition and structure) and the macroinvertebrates they support have been observed with concern by GSL wetland stakeholders (37 individuals from 20 agencies and organizations) who in 2018 ranked phragmites as the #2 threat to these wetlands (Low and Downard 2018). The extent of the phragmites impact is so great that recent efforts to find "reference" (i.e., high quality) wetlands within the GSL wetland complex were unsuccessful and instead scientists had to seek out reference wetlands in the west desert of Utah (Utah Division of Water Quality 2015).

To understand the impact of phragmites on bird populations, we can look at research from other regions of North America where negative impacts of phragmites on avian habitats have been well-documented (Benoit and Askins 1999; Kessler et al. 2011; Kettenring et al. 2012; Meyerson et al. 2000; Robichaud and Rooney 2017; Whyte et al. 2015). There are robust data on bird usage of uninvaded, native plant-dominated wetlands (Fig. 13.1) including the types of habitats birds use, the season of use, and the significance of GSL wetlands to particular bird species (Table 13.2). From these bird-use data, we would expect substantial impacts of phragmites invasion on, for example, snowy plover (Fig. 13.2) summer breeding habitat when phragmites invades mudflat and playa areas replacing halophyte species such as pickleweed and saltgrass as well as open areas (Fig. 13.3). Or, in another example, as phragmites expands into deeper water habitats such as emergent wetlands dominated by bulrushes (hardstem, threesquare, and alkali) and submergent wetlands dominated by sago pondweed (Fig. 13.3), we would expect to see substantial declines in waterfowl species, such as swans, northern shovelers, redheads, green-winged teal, and cinnamon teal (Fig. 13.2), which use these habitats for some combination of breeding, staging, wintering, or migrating. Furthermore, GSL wetland managers who have observed their managed properties for decades, have noted highly productive native plant-dominated wetlands that supported abundant waterfowl and shorebirds become largely devoid of bird activity once phragmites dominates the area (Chad Cranney, Randy Berger, Rich Hansen, Jason Jones, *pers. comm.*). Given the critical importance of these GSL native plant-dominated habitats to North American populations of these birds, habitat loss to phragmites poses an enormous threat (Aldrich and Paul 2002; Intermountain West Joint Venture 2013; Paul and Manning 2002).

13.4 Phragmites Management in Great Salt Lake Wetlands

Widespread concern about phragmites impacts across many interest groups including hunters and hunting clubs, wetland managers, birdwatchers, the scientific community, and concerned citizens has led to significant and large-scale efforts to reduce the spread of phragmites and restore invaded areas to native plant-dominated wetlands. Throughout the relatively short period of the phragmites invasion in GSL wetlands, these interest groups have mobilized to recognize the scale of the problem

Table 13.2 Avian species that likely experience negative impacts associated with phragmites expansion as well as the significance of Great Salt Lake to the avian populations, preferred avian habitats, and season of avian use

Avian species common name Latin name	GSL significance to avian populations	Habitat type	Seasonal use
American avocet *Recurvirostra americana*	250,000 which is several times higher than any other Pacific Flyway wetland (Shuford et al. 1995)	Emergent Mudflat/playa	Spring: Migrating Summer: Breeding Fall: Migrating
Black-necked stilt *Himantopus mexicanus*	65,000 which is several times higher than any other Pacific Flyway wetland (Shuford et al. 1995)	Emergent Mudflat/playa	Spring: Migrating Summer: Breeding Fall: Migrating
Cinnamon teal *Anas cyanoptera*	Up to 60% of the US breeding population (Bellrose 1980)	Emergent Submergent Wet meadow	Spring: Migrating Summer: Breeding Fall: Migrating
Green-winged teal *Anas carolinensis*	600,000 migrating (Great Salt Lake Planning Team 2000)	Open lake Submergent Emergent Wet meadow Mudflat/playa	Fall: Migrating Winter: Wintering Spring: Migrating
Marbled godwit *Limosa fedoa*	30,000; only staging area in the interior of western North America (Shuford et al. 1995; Paul and Manning 2002)	Emergent Wet meadow	Fall: Staging
Northern shoveler *Anas clypeata*	>160,000 migrating and 10,000 breeding (Paul and Manning 2002)	Open lake Submergent Emergent Wet meadow	Spring: Migrating Summer: Breeding Fall: Staging Winter: Wintering
Redhead *Aythya americana*	20,000 breeding pairs and >150,000 migrating (Great Salt Lake Planning Team 2000)	Submergent Emergent	Spring: Migrating Summer: Breading Fall: Migrating
Snowy plover *Charadrius alexandrinus*	>5000 which is the world's largest assemblage representing 23% of breeding population (Thomas et al. 2012)	Mudflat/Playa	Summer: Breeding
White-faced ibis *Plegadis chihi*	>27,000 breeding adults that are 20% of western North American breeding population (Cavitt et al. 2014)	Emergent Wet meadow	Spring: Migrating Summer: Breeding Fall: Migrating

and coordinate across disciplines and management boundaries to address this challenge (Rohal et al. 2018). Managers and scientists have collaborated to evaluate the tools for phragmites control and containment and to continue to refine management practices to reduce the reproduction and spread of this species (Rohal et al. 2017, 2018, 2019a, b).

13.4.1 Great Salt Lake Phragmites Management History, Collective Problem Solving, and Science-Management Partnerships

The early establishment of phragmites throughout the GSL basin was met with mixed reactions among wetland managers and visitors. Before the invasion was officially documented in 1993 (or fully understood), some welcomed the new plant because of the greater structure it provided and the increase in cover for duck blinds. Others saw the new plant as an unwanted intruder, and they quickly acted to remove it. In many cases, early action was implemented by duck clubs with greater management resources and a long history of intensive wetland management (Rohal et al. 2018). Managers observed that methods previously used to encourage plant diversity and to create open wetland areas, such as water drawdowns and fire, were now encouraging the spread of phragmites (Rohal et al. 2018). Through trial and error, they found that maintaining deep water in some areas while drought stressing others could create conditions that could limit the spread of phragmites (Rohal et al. 2018). Nevertheless, despite early efforts to alter management, phragmites continued to spread. Managers realized that the problem was prominent across property lines, and collective action and coordination were necessary to fully address the issue.

Partnerships between agencies and property holders have developed and expanded as phragmites management efforts have increased to meet the scale of the problem. For example, some adjacent duck clubs have developed working groups to more effectively control phragmites that can easily spread seeds and rhizomes across property lines (e.g., the Southshore Wetlands & Wildlife Management, Inc. for the duck clubs on the south shore, near Farmington Bay, and partnerships among the Chesapeake, Bear River, and Ferry Duck Clubs north of GSL; Fig. 13.1). State agencies (DWR and DFFSL) now coordinate management plans at the interface of Waterfowl Management Areas (WMAs; DWR jurisdiction) and the GSL lakebed (i.e., Utah's Sovereign Lands that are under DFFSL jurisdiction). In some cases, agencies like the DWR and the US Fish & Wildlife Service pool resources to conduct aerial herbicide spraying.

The DWR has been one of the leaders in phragmites management efforts across the GSL watershed. Since 2006, the DWR has coordinated a long-term phragmites management plan that treats several thousand hectares in six state-owned WMAs covering $> 24{,}000$ wetland hectares over a month or more each year. To accomplish this management, they utilize the power of six full-time employees, 3–6 part-time technicians, and over 225 hours of volunteer time. Additionally, the DWR and DFFSL uses prviate contractors for aerial herbicide, ground herbicide, and mechanical removal treatments. One of the major goals of the DWR phragmites management plan is to disseminate phragmites education and management information to other agencies, county governments, private properties, and the public. They have facilitated proposal writing efforts for adjacent property holders and county governments to encourage regional phragmites management coordination and upstream phragmites control. The DFFSL has also worked to improve coordination among

GSL property holders, initiate "Phragmites Working Group" meetings, and work with other agencies to prioritize management in high need areas.

Many management agencies and property holders have also coordinated with researchers to facilitate phragmites management research (Cranney 2016; Rohal et al. 2017, 2018, 2019a, b; Rohal 2018; Duncan 2019). Managers have made properties accessible for research projects, provided feedback on relevant experimental questions, and assisted with research treatment implementation. Wildlife agencies and hunting organizations have also provided financial help to promote phragmites management research.

13.4.2 Methods for Phragmites Control

There are many methods that GSL managers use to control phragmites. The most commonly applied methods include herbicide, mowing, burning, and grazing (Figs. 13.6 and 13.7) (Rohal et al. 2018). Often these methods are used in combination with one another to achieve multiple goals (Figs. 13.6 and 13.7).

13.4.2.1 Herbicide

The primary tool for managing phragmites in GSL wetlands is herbicide. A 2011 survey of GSL managers showed 97% of managers used herbicide as their primary tool (Rohal et al. 2018). Glyphosate and imazapyr are the two herbicides approved for aquatic environments that are most frequently used to remove phragmites in North America (Hazelton et al. 2014). Each herbicide type has its own benefits and drawbacks. Glyphosate is a nonselective, broad-spectrum herbicide, which is absorbed into the plant through its leaves. Imazapyr is also a broad-spectrum herbicide, but it can be absorbed by both plant leaves and roots (Tu et al. 2001). Both herbicides have the potential for nontarget impacts, though imazapyr may be more damaging to nontarget plants due to its ability to impact plants through their roots. Imazapyr is used less frequently in GSL wetlands primarily due to its higher cost (Rohal et al. 2018). In GSL wetlands, glyphosate and imazapyr are equally effective at reducing phragmites cover and the resulting native plant recovery is similar (Cranney 2016; Rohal 2018). The timing of herbicide application can also influence phragmites removal success and native plant recovery. Herbicide applications in the fall (August–September) are more effective at reducing phragmites cover (Fig. 13.7; Cranney 2016; Rohal 2018, 2019a, b) because the herbicide is more effectively translocated to the roots and rhizomes where it has the greatest impact (Tu et al. 2001).

13.4.2.2 Mowing

Mowing phragmites without additional interventions is not effective at removing phragmites as it can actually stimulate its growth (Derr 2008). Nevertheless, in GSL,

Fig. 13.6 Phragmites management: (**a**) Loglogic Softrak, marsh-capable equipment driving through a vast expanse of phragmites, (**b**) Marsh Master amphibious equipment used for phragmites herbicide and mowing management, (**c**) Loglogic Softrak being used for herbicide application to phragmites, (**d**) Wilco amphibious equipment used by the US Fish & Wildlife Service for phragmites management at the Bear River Migratory Bird Refuge (see Fig. 13.4), (**e**) Loglogic Softrak dragging a roller/crusher to break down phragmites litter in winter, (**f**) phragmites litter from winter mowing with live phragmites emerging in spring, (**g**) phragmites litter waist to shoulder high on wetland researchers Brittany Duncan and Karin Kettenring, (**h**) phragmites mowed to provide cattle access for grazing research study by Brittany Duncan, (**i**) cattle grazing phragmites-invaded wetland, (**j**) loading hydroseeding tank with native seeds, a tackifier, and mulch prior to seeding by David England, Emily Tarsa, and Keith Hambrecht, (**k**) seeds in paint strainer mesh for cold stratification prior to hydroseeding, and (**l**) hydroseeding by David England and Chad Cranney at Farmington Bay WMA

mowing is frequently used as a tool in phragmites management programs to address the excess biomass that remains after herbicide applications (Figs. 13.4, 13.6, and 13.7). The large amount of dead phragmites biomass is a major impediment to native plant recovery because it shades the wetland surface. Many native wetland plants require high light levels to trigger germination (e.g., Marty and Kettenring 2017). Mowing and mulching this biomass accelerates its decomposition, opening up the light resources needed for native plant species to germinate. Mowing can be conducted in the summer as long as it does not impact bird nesting nor spread

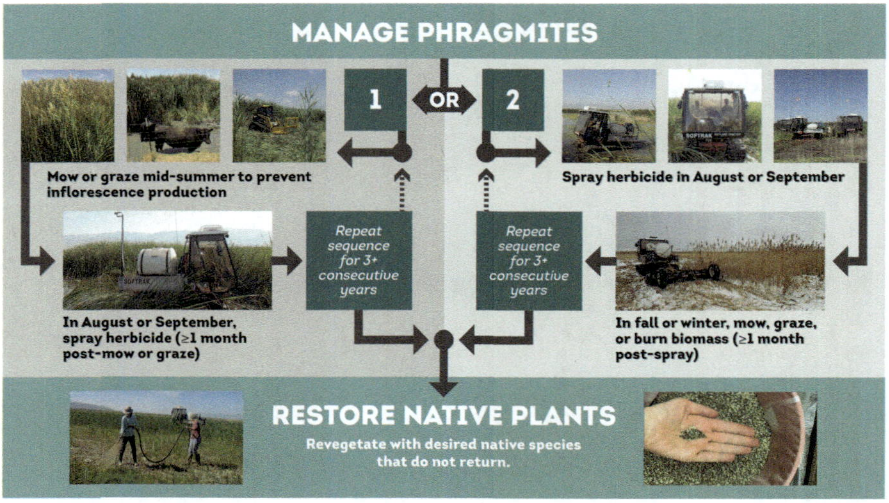

Fig. 13.7 Recommended treatment options for managing phragmites through mowing, grazing, burning, and herbicide application and reestablishing native plants through hydroseeding. Graphic design by Michael Wernert

plant fragments that may further its spread prior to a fall herbicide application. Alternatively, mowing can be done in the winter, following a fall herbicide application (Fig. 13.7). Summer mowing is not always feasible, however, because some mowing equipment can get stuck in the flooded and saturated soil conditions. Mowing in the winter is often more feasible due to the frozen ground conditions.

13.4.2.3 Burning

Burning, like mowing, is ineffective at controlling phragmites when used as a single management tool. However, it is commonly used to reduce phragmites biomass following herbicide applications (Hazelton et al. 2014). Burning is most frequently conducted in the spring in GSL wetlands (Rohal et al. 2018). Burning can be highly effective at biomass removal, and it does not require additional time for the biomass to degrade (unlike mowing), allowing native plant species to quickly germinate on the bare soil that remains. Unfortunately, burning is often infeasible in GSL wetlands due to air quality standards that make it difficult to obtain permission to burn (Rohal et al. 2018).

13.4.2.4 Grazing

Cattle grazing of phragmites is a management tool that is increasingly being used in GSL wetlands. While grazing is unlikely to kill phragmites over repeated grazing seasons, it has many other benefits. High intensity, short-term grazing of phragmites over 1–2 growing seasons can reduce phragmites living biomass, making phragmites-invaded wetlands more accessible, while not degrading water quality (Duncan 2019; Duncan et al. 2019). Over time, cattle also trample large amounts of standing dead phragmites, further opening up the wetland surface for native plant germination and accessibility (Duncan 2019; Duncan et al. 2019).

13.4.3 Managing Phragmites to Limit Seed Production

As mentioned earlier, the ability of phragmites to reproduce via seed is an important mechanism enabling its establishment and spread. Managing phragmites seed production is thus vital to both reduce the incidence of phragmites colonization into new areas and to reduce the chances for reinvasion following phragmites control. Phragmites reinvasion once control ceases is common, in part because the density of its seed in the soil of invaded patches can be very high (Cranney 2016; Rohal 2018). When soil seed densities are high, the propagule pressure of the invader increases the chances for establishment and increased competition with other species (Byun et al. 2015). Phragmites propagule pressure is particularly a concern in GSL wetlands, which have the highest recorded phragmites seed densities in North America (Rohal 2018). For example, a review of phragmites seed banks from other regions found the highest phragmites seed densities at ~700 seeds m^{-2} (Baldwin et al. 2010) whereas seed bank densities in GSL wetlands were as high as ~14,000 seeds m^{-2} (Rohal 2018).

There are a variety of management actions that can reduce the ability of phragmites to produce seed. Impacting the plant in the summer greatly reduces phragmites seed production during the reproductive season in fall (Cranney 2016; Duncan 2019; Rohal 2018; Rohal et al. 2019a, b; Duncan et al. 2019). Summer mowing and high-intensity summer livestock grazing both reduce phragmites seed production (Fig. 13.7; Duncan 2019; Rohal 2018). Summer herbicide applications can also greatly reduce phragmites seed production, but summer herbicide does not reduce phragmites cover as effectively as fall applications, so it is not recommended (Cranney 2016; Rohal 2018; Rohal et al. 2019a, b). In addition, drought-stressed phragmites rarely produces high densities of inflorescences (C. Rohal, *pers. obs.*). Managers with water control can intentionally drought stress phragmites patches that are not the target of other management actions to reduce the production of seeds that can spread into surrounding areas.

13.4.4 Managing Wetlands to Reduce Phragmites Spread

Managers have a number of methods available to reduce the spread of phragmites. One strategy is to minimize the conditions that promote phragmites seed germination. As discussed previously, physical disturbance of existing vegetation can create the high light conditions that are favorable for phragmites germination (Kettenring et al. 2015; Kettenring and Whigham 2018). Thus, minimizing disturbance to existing native vegetation (such as burning "decadent" vegetation) is an important tool for limiting phragmites seed germination and establishment. In addition, phragmites seeds need specific hydrologic conditions to germinate (moist to shallowly flooded mudflat conditions). In areas with water control, manipulating water depth to deeper flooding (>3.5 cm) or drought-stressing impounded areas can greatly minimize phragmites germination.

The outward spread of phragmites patches via rhizomes is another important mechanism for its expansion (Kettenring et al. 2016). Rhizome spread can be reduced by hydrologic manipulations. Flooding impoundments as deep as possible (>0.5 m) can restrict phragmites growth (Hudon et al. 2005), and this flooding is often used in GSL wetlands to prevent phragmites spread and to encourage open water and submerged aquatic vegetation habitat favorable to many waterfowl species. In contrast, shallow flooding with a sheet flow of water through the growing season (e.g., to mimic the historical hydrology of river deltas) creates conditions optimal for phragmites seed germination and seedling establishment, and therefore it should be avoided in areas where phragmites seeds are present. Finally, grazing is a low-cost tool for reducing phragmites biomass and stressing phragmites throughout the growing season, which can thus reduce its potential for outward spread (Duncan 2019).

13.4.5 Environmental Context of Management Areas Influences Management Success

The management of phragmites is not equally effective in all locations (Cranney 2016; Duncan 2019; Rohal 2018; Rohal et al. 2017, 2019b). There are a variety of factors, often outside of management control, that can influence management success, including abiotic conditions like site hydrology and nutrient conditions, biotic conditions like the condition of the native seed bank and site disturbance history, as well as the size of the phragmites patch (Brudvig et al. 2017; Zimmerman et al. 2018; Quirion et al. 2018; Rohal et al. 2019a, b). Hydrology is the predominant factor that can influence both phragmites control and native plant recovery (Rohal et al. 2019a, b). Herbicide is typically ineffective when sprayed on phragmites that has been drought stressed (Rohal 2018; Rohal et al. 2019a, b) because the herbicide is not effectively translocated to the rhizomes where it is needed to permanently kill the plant (Tu et al. 2001). Thus, counterintuitively, phragmites control with herbicide is most successful on the healthiest, greenest stands (Rohal 2018; Rohal et al. 2017, 2019a, b). These are typically in areas that have consistent moisture throughout the summer growing season. These conditions also favor more robust native plant recovery following control. More shallow flooding (<10 cm) allows for greater native plant germination, while deeper flooding (>10 cm) reduces germination opportunities for many desirable native plant species, and tends to favor the recruitment of cattail.

Another factor that can influence management success is the scale of the treated patch. In GSL wetlands, phragmites is present in both large, multi-hectare monocultures, which are often isolated from native plants, and small patches (<0.40 ha) that are still surrounded by a matrix of native plant species. Phragmites is more effectively controlled in small patches and native plant species return at a higher cover in these areas (Rohal et al. 2019b). Small patches tend to have more successful outcomes likely because the matrix of native plant species can provide higher densities of propagules to recolonize (Matthews et al. 2017; Rohal et al. 2019b), while large patches typically lack this source of native propagules. In addition, large patches are often in areas with a history of hydrologic manipulation and frequently have deeper flooding throughout the growing season. These conditions can prevent native plant germination and can favor the expansion of extant phragmites patches via rhizomes (Rohal et al. 2019a, b).

13.4.6 Revegetation Following Phragmites Control

Revegetation is an important tool following phragmites control, particularly in areas where native plant recovery is limited (Rohal et al. 2017, 2019a, b). Active revegetation (e.g., seeding, planting plugs) is often necessary as desirable native species rarely recruit at high densities following phragmites control (Cranney 2016; Rohal 2018). In GSL wetlands, revegetation is essential for restoring native species that provide high-quality food and habitat for waterfowl and shorebirds and deliver valuable ecosystem functions and services that are characteristic of wetland ecosystems (e.g., flood control, carbon sequestration). Furthermore, revegetating areas where phragmites has been treated can be an effective way to prevent phragmites reinvasion because it encourages the quick establishment of native plants, which limits the high light, high nutrient, and bare soil conditions that favor phragmites germination and growth (Byun et al. 2013, 2015; Kettenring et al. 2015; Kettenring and Whigham 2018; Peter and Burdick 2010).

For small sites or projects that have ample budgets, revegetation outcomes can be improved by planting native plugs or installing sod mats as these methods bypass the vulnerable seedling stage (Grubb 1977). When a site for revegetation is large, as are many restorations in GSL wetlands, it is logistically and financially more feasible to sow native seeds as compared to plugs or other forms of active revegetation (Hurd and Shaw 1992; Palmerlee and Young 2010). However, the seedling stage is the most limiting stage of a plant's life cycle and represents a bottleneck in recruitment (Barrett-Lennard et al. 2016; James et al. 2011). As such, several actions should be taken prior to seeding that increase the chance of native plant survival and improve restoration outcomes in GSL wetlands.

13.4.6.1 Preparing the Site for Revegetation

Preparation of the site prior to seeding is essential to create and maintain ideal conditions for native seedling recruitment. Phragmites litter left on the site should be removed so sown seeds have sufficient seed–soil contact and adequate light necessary to trigger germination (Cranney 2016; Lishawa et al. 2015; Rohal 2018). Additionally, maintaining ideal hydrological conditions is critical as many wetland seedlings are unable to survive in water depths greater than 0 cm above the soil surface (Fraser and Karnezis 2005). Therefore, the hydrology should be maintained as follows: (1) during seeding, water levels should be drawn down to the soil surface so that soil is exposed and saturated, thus preventing buoyant wetland seeds from floating away and encouraging seed–soil contact. In areas with unpredictable hydrologic regimes (i.e., flooding), a tackifier can be used while seeding to keep seeds in place through germination (Tilley and John 2013; England 2019); (2) the soil should remain waterlogged through germination and establishment for most species, although there are notable exceptions like saltgrass that performs well against phragmites with lower soil moisture levels (Webb et al. 2012; E.E. Tarsa, *pers. obs.*); and (3) flooding events that are deep or long in duration should be avoided in the first growing season, giving native plants enough time to develop adaptive structures (e.g., aerenchyma) necessary to withstand high water conditions (Cronk and Fennessy 2001). Unfortunately, these conditions are also ideal for phragmites germination and seedling establishment, which underscores the importance of depleting phragmites from the seed bank and removing nearby phragmites propagule sources prior to beginning revegetation while also being vigilant about spraying new phragmites as the native plant community becomes established. Native seedling recruitment can also be improved by creating "safe sites"—or small areas around a seed that have ideal environmental conditions for germination and establishment (Peach and Zedler 2006; Urbanska 1997). This variation in microtopography can be created by hand (e.g., using a shovel to rut the soil) or using large machinery (e.g., tractor rutting) (Moser et al. 2007).

13.4.6.2 Choosing Native Species for Revegetation

Which native species to sow in GSL wetlands is an important consideration that is based on the environmental conditions at a site, target wildlife habitat, and the ability for native species to resist phragmites reinvasion. In GSL wetlands, hardstem, threesquare, and alkali bulrush (Fig. 13.3) provide important habitat for waterfowl in this region and are often included in revegetation seeding mixes. However, sowing these species alone may not be an effective revegetation tool as they are slow-growing perennials and likely do not keep pace with the early emergence of phragmites from the seed bank (Gioria and Pyšek 2017; Downard et al. 2017). Therefore, adding annual (or otherwise fast growing or broadly environmentally tolerant) species to the seed mix that germinate quickly and preempt resources may be particularly effective at resisting phragmites reinvasion (Byun et al. 2013). Such species in GSL wetlands include nodding beggartick (*Bidens cernua*), rayless alkali aster (*Symphyotrichum ciliatum*), and fringed willowherb (Fig. 13.3) as well as (not pictured) Nuttall's alkaligrass (*Puccinellia nuttalliana* (Schult.) Hitchc.), golden dock (*Rumex maritimus* L.), and curlytop knotweed (*Polygonum lapathifolium* L.). Species with varying growth forms, such as mat-forming species (e.g., common spikerush), can also be particularly effective at preventing phragmites reinvasion in the long-term due to their ability to limit light (Fig. 13.3). Ensuring that the native seed mix sown at a restoration site can handle a diversity of environmental conditions (e.g., seeding saltgrass for drier site conditions with hardstem bulrush for wet conditions; Fig. 13.3) is a form of bet-hedging that increases the likelihood that native species will establish given the natural fluctuations and sometimes unpredictable conditions in GSL wetlands (Evans and Dennehy 2005).

13.4.6.3 When to Revegetate Great Salt Lake Wetlands?

Revegetation should occur after phragmites stands have been treated with herbicide and mowed for at least 3 years (Cranney 2016; Rohal 2018; Rohal et al. 2019a, b) (Fig. 13.7). In GSL wetlands, revegetation often occurs in the spring (May–June) when there is adequate moisture and the temperature is within optimal germination requirements of most GSL native species (approximately 28–35 °C) (Downard et al. 2017; Kettenring 2016; Marty and Kettenring 2017). Dormancy, or the ecological adaptation that prevents seeds from germinating during conditions that are sub-optimal for seedling survival (Willis et al. 2014), is present in many wetland species and must be broken prior to seeding in the spring. Breaking dormancy in only half of the seed lot and seeding the remaining seeds dormant can help build the native seed bank, thus ensuring native species are present on the site in future years and varying environmental conditions (Evans and Dennehy 2005). Alternatively, fall seeding of dormant seeds (i.e., a "dormant seeding") can serve to break dormancy naturally, but increases the likelihood of seed predation and germination of nondormant seeds during harsh winter conditions (Galatowitsch and van der Valk 1994; Kettenring and Galatowitsch 2011).

13.4.6.4 Seeding Density

Given the high phragmites propagule pressure in GSL wetlands (Rohal 2018), native seed mixes should be sown at high enough densities to competitively exclude phragmites reinvasion from the seed bank. The current recommended seeding rates (~1900 seeds m^{-2}), often based on adult plant distributions, are likely not high enough to prevent phragmites return from the seed bank (Tarsa and Kettenring, unpubl. data). Preliminary results suggest that seeding between 5800 and 9700 seeds m^{-2} can significantly reduce phragmites biomass at a site (Tarsa and Kettenring, unpubl. data). However, these results are contingent on phragmites seed density in the seed bank—thus it is important to prioritize restoration sites that have low phragmites seed densities in the seed bank or, as mentioned earlier, focus on depleting phragmites seeds through repeated years of mowing and herbicide (Tarsa and Kettenring, unpubl. data; Rohal 2018). Furthermore, competitive dynamics change across environmental conditions (e.g., water and nutrient availability; Tilman 1994; Wilson and Keddy 1986). For instance, saltgrass can outcompete phragmites under low soil moisture conditions even when saltgrass is sown at a low seeding density (Tarsa and Kettenring unpubl. data).

13.4.6.5 Monitoring and Maintenance

GSL wetlands, as with many wetlands, are highly invasion prone due to high invader propagule pressure and the lake's low position in the landscape that facilitates disturbance and nutrient enrichment (Rohal 2018; Zedler and Kercher 2004). Furthermore, restoration activities themselves create disturbances that result in high light and nutrient conditions, making sites highly prone to secondary invasion (Davis et al. 2000). As such, monitoring and maintaining sites in the years following revegetation is necessary to encourage native species establishment and survival (Rieger et al. 2014). This maintenance is especially critical in the native seedling stage (<1 year of growth) as mature stands have not yet formed to limit light availability for germinating phragmites seeds (Adams and Galatowitsch 2006; Kettenring et al. 2015). Phragmites that is returning from the seed bank or encroaching from nearby sites should be removed as quickly as possible by hand or spot sprayed with glyphosate to prevent expansion into the revegetated wetland (Adams and Galatowitsch 2006; Rohal et al. 2017, 2019a, b). It should be expected that a significant amount of time will be spent to control reinvading phragmites in the revegetation site for at least the first 2 years, with a large reduction in time spent doing these activities over time (Bohnen and Galatowitsch 2005). Revegetation sites should also be coarsely assessed for native species survival rates, which will inform whether seeds should be sown or plugs should be planted the following year.

13.5 Other Common Invasive and Undesirable Species: What Is Known and On-going Management

Little regionally specific research has been conducted on the majority of other invasive species in GSL wetlands, though wetland condition surveys have estimated disturbance and non-native species distribution and abundance around the lake. Data compiled from GSL wetland surveys show 79 non-native or undesirable species, though few of these species are both widely distributed and abundant in wetlands (Downard et al. 2017; Menuz et al. 2014, 2016; Menuz and McCoy-Sulentic 2019; Menuz and Sempler 2018; Utah Division of Water Quality 2016). A species' status as native or non-native is not necessarily a useful indicator of how problematic it is in managed wetlands. In wetlands that are managed for waterfowl habitat, the ability of a species to help meet management goals by providing food and cover is more important to whether it is considered an undesirable weed than where the species originated. Because of this, native species like cattail that grow dense and have little nutritional value are often considered less desirable, while Eurasian species like barnyard grass and sorghum (*Sorghum bicolor* (L.) Moench) are deliberately planted to provide food and shelter for waterfowl and forage kochia (*Bassia* All. spp.) is left untreated (Table 13.1).

A variety of factors determine whether non-native plants become established in GSL wetlands. Species with high anaerobic soil tolerance (e.g., species listed as facultative wetland or obligate species in Table 13.1) are more likely to be able to establish in a wetland than species that cannot grow in waterlogged conditions. Undesirable species like native cattails thrive in artificially stabilized hydroperiods that result from deliberate management or incidental impoundment by roads. On the other hand, drought and altered hydrology can open up habitat that was formerly wetland to a wide variety of upland invaders while annual species such as prickly lettuce (Table 13.1) can move into wetlands that dry out during the late summer (Downard et al. 2017; Zedler and Kercher 2004).

An invader's salinity tolerance will also determine its success (Cronk and Fennessy 2001). Many GSL wetland invaders are primarily found along canals and other sources of freshwater, especially purple loosestrife (Fig. 13.5) and annual rabbitsfoot grass (Table 13.1). Species from the Eurasian steppe, like tamarisk or kochia (forage kochia and related plants) often have high salinity tolerance (Table 13.1). Both those species were purposely introduced to the United States for windbreaks and to prevent erosion, respectively, but their tolerance of harsh environments has facilitated their expansion into wetlands (Downard et al. 2017). Lastly, nutrients also play a role in allowing some species to be successful, particularly species that can take advantage of nutrients to grow quickly, similar to phragmites (Cronk and Fennessy 2001). Barnyard grass is capable of concentrating high nitrogen and phosphorus in its tissues, and fast-growing, floating species like duckweed (*Lemna* L. spp. that are actually native) are indicators of eutrophic conditions (Table 13.1; Esser 1994; Penning et al. 2008).

Non-native species may displace native plant species and in some cases disrupt food webs dependent on those species (Cronk and Fennessy 2001). At least eight species listed as noxious weeds have been documented in the region; many of these are detrimental to grazing, such as Canada thistle, and some are poisonous to livestock and people, like poison hemlock (Table 13.1; Utah Department of Agriculture and Food 2019). Many non-native plant species have the potential to impact hydrology, such as the submerged aquatic plant curly pondweed, which can form dense mats that clog waterways (DiTomaso et al. 2013). Species can also alter biogeochemistry of soils by depleting or enhancing nitrogen in the soil or altering the salinity (e.g., Russian olive or tamarisk; Table 13.1).

While the most intensive invasive species management around GSL is focused on phragmites, some attention is being paid to the impacts and management of other undesirable species. Most control efforts are focused on species listed as noxious weeds in Utah (Table 13.1), including dyer's woad (*Isatis tinctoria* L.), Dalmatian toadflax (*Linaria dalmatica* (L.) Mill.), spotted knapweed (*Centaurea stoebe* L.), Russian knapweed (*Acroptilon repens* (L.) DC.), and yellow star-thistle (*Centaurea solstitialis* L.). Native cattail has also been the focus of management on some of the state-owned WMAs; see below for additional information on the treatment of this species. Most control efforts use chemical application in the spring, except for dyer's woad, which is hand pulled. Focal areas for control efforts in publicly managed areas include property adjacent to private land, highly visible areas, and high bird use areas. Details of some of the species of highest concern within wetlands are described below.

13.5.1 Purple Loosestrife (Lythrum salicaria)

Purple loosestrife is a noxious weed in 33 states, including Utah (USDA NRCS 2019), that was likely first introduced from Europe in shipping ballast, but also spread when it was purposely planted in gardens (Fig. 13.5; Munger 2002). This species grows as a 1.8–3.6 m tall bush with a deep root system and beautiful purple flowers. An individual purple loosestrife plant can produce millions of seeds, which are spread through waterways. Because purple loosestrife is so prolific and puts down such deep roots, controlling new, small patches is the most effective means for managing an invasion (DiTomaso et al. 2013). Young plants can be pulled out by hand or treated with an herbicide that is approved for use in aquatic environments. Two species of beetle native to Europe have also been used to manage large infestations of purple loosestrife (DiTomaso et al. 2013). Once established, purple loosestrife pushes out native plants, crowds out open water refuges, and clogs irrigation systems (Munger 2002).

13.5.2 European Seaheath (Frankenia pulverulenta)

European seaheath is a European plant species that has been introduced to Utah, a few states on the east and west coasts of the United States, and also South America, Australia, and elsewhere (Fig. 13.5; Whalen 2015). The species was first recorded in Utah in 1972 at a privately owned duck club near Salt Lake City International Airport, with two other collections nearby in the late 1970s and early 1980s (Intermountain Region Herbarium Network 2019). The species was then not observed for many years and was even speculated to be eradicated by floods or heavy equipment by the 2000s (Holmgren 2005). However, by 2018 it had been documented across the entire eastern side of the lake and found in 17% and 40% of playa sites in the Bear River and Farmington Bays, respectively, in recent surveys around GSL though always with low ($\leq$2%) cover (Menuz and Sempler 2018; Menuz and McCoy-Sulentic 2019). The species is associated with intermittently flooded and frequently sparsely vegetated areas with high salinity, including playas, mudflats, and greasewood stands. This species is of interest because it is clearly adapted to the harsh conditions of playas around the lake and has become very widespread relatively quickly, though its potential for negative impacts beyond replacing native species is unknown. A 2009 study in Pakistan proposed European seaheath as a newly emerging species of concern in saline areas (Waheed et al. 2009), but a brief literature search failed to find any documented impacts of the species in other areas where it has invaded.

13.5.3 Whitetop (*Cardaria draba*) and Perennial Pepperweed (*Lepidium latifolium*)

Whitetop and perennial pepperweed are both members of the Brassicaceae family and Class 3 noxious weeds in Utah associated with moist agricultural sites and other disturbed areas (Fig. 13.5; Downard et al. 2017; Utah Department of Agriculture and Food 2019). Both species are widespread along the eastern shore of GSL, though neither is particularly abundant within wetlands. Of the two species, perennial pepperweed is more strongly associated with wetlands and riparian areas, though some case studies have suggested the species is intolerant of prolonged inundation and may grow poorly under saturated conditions (Blank et al. 2002), which may explain why it has not become more of a nuisance in GSL wetlands. Asexual reproduction is important to both species, with spread via roots, buds, and rhizomes common, though sexual reproduction is common as well (Fire Effects Information System 2019). The roots of perennial pepperweed are buoyant, can spread long distances by water, can remain dormant in the soil for years, and have been found more than 3 m deep in the soil profile (Fire Effects Information System 2019). Both species have been documented to reduce crop or hay yields, displace native plant species, and reduce wildlife habitat (Fire Effects Information System 2019). More importantly for GSL's ecosystem, perennial pepperweed may negatively affect nesting habitat for waterfowl and other wildlife and displace important food grasses for waterfowl based on observations made in California, though this phenomenon has never been studied in Utah (Fire Effects Information System 2019).

13.5.4 Cattails (*Typha domingensis* and *T. latifolia*)

Cattails are a native GSL wetland species, but undesirable because they make for poor waterfowl habitat and can push out more desirable wetland species (Fig. 13.5; Downard et al. 2017; Ochterski 2003). Cattails are a common problem in wetlands managed for migratory bird habitat where wetland hydroperiods have been lengthened through diversions and dikes. Cattails have large underground rhizomes that transfer oxygen and nutrients between daughter ramets of the same plant (Cronk and Fennessy 2001). They also grow quickly, making them well adapted to consistently deep flooding (Cronk and Fennessy 2001). To control the expansion of cattails, managers must disrupt its robust root system, which requires more than one method of treatment. Small patches of cattail can be hand pulled. Once cattail becomes tall and dense, a combination of growing season mowing, burning, and herbicide use may help control cattail by stressing the plant when carbohydrate reserves in rhizomes are at their lowest (Gleason et al. 2012). Any method of cattail control should be followed by deep flooding ($\geq$0.3 m) to ensure the roots do not survive (DiTomaso et al. 2013). Prior to the expansion of phragmites, cattails were a primary concern of waterfowl managers around GSL and are re-emerging as weed control targets as phragmites cover decreases (C. Cranney, *pers. obs.*).

13.5.5 Curly Pondweed (Potamogeton crispus)

Curly pondweed is an invasive submerged aquatic species native to Eurasia that is found in ponded brackish, alkaline, or eutrophic waters in northern Utah, but has not yet become a major nuisance or focus of management (Fig. 13.5; Haynes and Hellquist 2000; Intermountain Region Herbarium Network 2019). The plant is unusual in that it produces fruit in late spring or early summer and then decays, leaving behind special leaf buds called turions that germinate in late summer or fall; the resulting plants, only a few centimeters tall, overwinter under ice and then resume growth in the spring (Haynes and Hellquist 2000). The species spreads vegetatively via turions along canals and potentially attached to boats, boots, or other equipment. Curly pondweed can deplete nutrients during periods of rapid growth (Brusati and DiTomaso 2005) and cause phosphorus to spike and dissolved oxygen to rapidly decline when it decomposes mid-summer (Haynes and Hellquist 2000; Thayer et al. 2019). Large infestations of curly pondweed can impede water flow and disrupt recreation (Brusati and DiTomaso 2005; Thayer et al. 2019). Despite the negative impacts, the species can provide food and cover for birds, fishes, and macroinvertebrates and may be an important food source in waters too turbid to support other submergent species (Brusati and DiTomaso 2005; Thayer et al. 2019).

13.6 Summary and Concluding Remarks

GSL and its wetlands are recognized around the world for the valuable habitat they provide for millions of migratory birds (Aldrich and Paul 2002; Evans and Martinson 2008; Paul and Manning 2002). The largest threat to these wetlands in terms of invasive plants is phragmites, although there are a number of species that are concerning and a target of management (Table 13.1). Managers of GSL wetlands face a daunting task to control these plants, particularly in the case of phragmites, where hundreds of hectares of infestations must be treated and retreated annually. Eradication will not be possible given the intense propagule pressure and dense seed banks (Rohal 2018; Rohal et al. 2019b), thus strategic and prioritized management approaches are critical (Long et al. 2017b). In addition, there have been exciting advancements in terms of cooperation between scientists and managers, in developing robust treatment techniques, and cooperation between managers, to coordinate their efforts to reduce phragmites cover and impacts (Rohal et al. 2017, 2018). These partnerships are the foundation for any future management programs should current or future invaders prove as formidable as phragmites. Given the threats GSL and its wetlands face with anthropogenic development, water diversions, and climate change (Downard et al. 2014; Downard and Endter-Wada 2013; Li et al. 2019; Wurtsbaugh et al. 2017), we are optimistic that at least in the case of invasive species, collaborative and science-backed management can continue to yield successes.

There are a number of research and management priorities that must be addressed in the near future to foster further invasive management success. First, revegetation following invasive species control, particularly after phragmites removal, is still in the initial stages of development. There are many opportunities for refining techniques to maximize native plant establishment and survival. Second, because phragmites propagule pressure is extraordinarily high in GSL wetlands (Rohal 2018), management efforts need to address this propagule pressure and focus on greatly reducing phragmites seed bank densities through multiple years of summer management prior to seed maturation (Fig. 13.7; Rohal et al. 2017). Third, looking into the future, what will be the next big invader? It is critical to recognize these new invaders—that are likely already problematic in other regions of North America—that may emerge, particularly as environmental conditions shift with climate change. Early detection, rapid response efforts will be essential because once an invader is well established and widespread, the cost of management increases substantially and the likelihood of management success declines markedly. Fourth, hydrologic management can be used to the advantage of managers (and many managers do so effectively already) but there are opportunities to further refine techniques to best prevent invasions and further facilitate successful management (e.g., Alminagorta et al. 2016). Finally, impacts of these invaders and especially phragmites have been documented qualitatively but there are few quantitative data on impacts to avian species from these invasions. These impacts should be a research priority considering the continental importance of this habitat to migratory birds.

Acknowledgments We thank the many agencies and organizations whose funding supported the primary research synthesized in this book chapter including the Community Foundation of Utah; Delta Waterfowl; Ducks Unlimited Canada; FRIENDS of Great Salt Lake; Intermountain West Joint Venture; Kennecott Utah Copper Charitable Foundation; the Salt Institute; Society of Wetland Scientists; Southshore Wetlands and Wildlife Management, Inc.; US Fish & Wildlife Service, Utah Department of Agriculture and Food; Utah Division of Forestry, Fire & State Lands; Utah Division of Water Quality; Utah Division of Wildlife Resources; Utah State University Ecology Center; Utah State University Research and Graduate Studies; U.S. Environmental Protection Agency; Utah Public Lands Initiative; Utah Waterfowl Association; Utah Wetlands Foundation; Wasatch Front Water Quality Council; Western Integrated Pest Management Center; and The Wetland Foundation. We greatly appreciate the many landowners along GSL who have been our partners in wetland research and management. M. Dunphy and N. Berg provided helpful feedback on an earlier version of this book chapter. This research was supported by the Utah Agricultural Experiment Station, Utah State University, and approved as journal paper number 9234.

References

Adams CR, Galatowitsch SM (2006) Increasing the effectiveness of reed canary grass (*Phalaris arundinacea* L.) control in wet meadow restorations. Restor Ecol 14:441–451

Albert A, Brisson J, Belzile F, Turgeon J, Lavoie C (2015) Strategies for a successful plant invasion: the reproduction of *Phragmites australis* in north-eastern North America. J Ecol 103:1529–1537

Aldrich TW, Paul DS (2002) Avian ecology of Great Salt Lake. In: Gwynn JW (ed) Great Salt Lake: an overview of change. Utah Department of Natural Resources and Utah Geological Survey Special Publication, Salt Lake City, pp 343–374

Alminagorta O, Rosenberg DE, Kettenring KM (2016) Systems modeling to improve the hydro-ecological performance of diked wetlands. Water Resour Res 52:7070–7085

Baldwin AH, Kettenring KM, Whigham DF (2010) Seed banks of *Phragmites australis*-dominated brackish wetlands: relationships to seed viability, inundation, and land cover. Aquat Bot 93:163–169

Barber B, Cavitt J (2012) Dietary review for aquatic birds utilizing Willard Spur, Great Salt Lake. Final Report, Utah Division of Water Quality, Salt Lake City, UT

Barrett-Lennard EG, Norman HC, Dixon K (2016) Improving saltland revegetation through understanding the "recruitment niche": potential lessons for ecological restoration in extreme environments. Restor Ecol 24:S91–S97

Bellrose FC (1980) Ducks, geese and swans of North America. Stackpole Books, Harrisburg, PA

Belzile F, Labbé J, LeBlanc MC, Lavoie C (2010) Seeds contribute strongly to the spread of the invasive genotype of the common reed (*Phragmites australis*). Biol Invasions 12:2243–2250

Benoit LK, Askins RA (1999) Impact of the spread of *Phragmites* on the distribution of birds in Connecticut tidal marshes. Wetlands 19:194–208

Bioeconomics, Inc (2012) Economic significance of the Great Salt Lake to the State of Utah. Missoula, MT. https://documents.deq.utah.gov/water-quality/standards-technical-services/great-salt-lake-advisory-council/Activities/DWQ-2012-006864.pdf. Accessed 8 Apr 2019

Blank RR, Qualls RG, Young JA (2002) *Lepidium latifolium*: plant nutrient competition-soil interactions. Biol Fertil Soils 35:458–464

Bohnen JL, Galatowitsch SM (2005) Spring peeper meadow: revegetation practices in a seasonal wetland restoration in Minnesota. Ecol Restor 23:172–181

Brisson J, de Blois S, Lavoie C (2010) Roadside as invasion pathway for common reed (*Phragmites australis*). Invasive Plant Sci Manag 3:506–514

Brudvig LA, Barak RS, Bauer JT, Caughlin TT, Laughlin DC, Larios L, Matthews JW, Stuble KL, Turley NE, Zirbel CR (2017) Interpreting variation to advance predictive restoration science. J Appl Ecol 54:1018–1027

Brusati E, DiTomaso J (2005) *Potamogeton crispus* plant assessment form *Cal-IPC Inventory*. https://www.cal-ipc.org/plants/paf/potamogeton-crispus-plant-assessment-form/

Burr SW, Scott D (2004) Application of the recreational specialization framework to understanding visitors to the Great Salt Lake Bird Festival. Event Manag 9:27–37

Byun C, de Blois S, Brisson J (2013) Plant functional group identity and diversity determine biotic resistance to invasion by an exotic grass. J Ecol 101:128–139

Byun C, de Blois S, Brisson J (2015) Interactions between abiotic constraint, propagule pressure, and biotic resistance regulate plant invasion. Oecologia 178:285–296

Cao L, Berent L, Fusaro A (2020) *Echinochloa crus-galli* (L.) P. Beauv. In: U.S. Geological Survey Nonindigenous Aquatic Species Database, Gainesville, FL, and NOAA Great Lakes Aquatic Nonindigenous Species Information System, Ann Arbor, MI. https://nas.er.usgs.gov/queries/GreatLakes/FactSheet.aspx?SpeciesID=2664

Cavitt JF, Jones SL, Wilson NM, Dieni JS, Zimmerman TS, Doster RH, Howe WH (2014) Atlas of breeding colonial waterbirds in the interior western United States. Final Report, US Department of the Interior, Fish & Wildlife Service, Denver, CO

Chambers RM, Meyerson LA, Saltonstall K (1999) Expansion of *Phragmites australis* into tidal wetlands of North America. Aquat Bot 64:261–273
Chambers RM, Havens KJ, Killeen S, Berman M (2008) Common reed *Phragmites australis* occurrence and adjacent land use along estuarine shoreline in Chesapeake Bay. Wetlands 28:1097
Cox RR, Kadlec JA (1995) Dynamics of potential waterfowl foods in Great Salt Lake marshes during summer. Wetlands 15:1–8
Cranney CR (2016) Control of large stands of *Phragmites australis* in Great Salt Lake, Utah wetlands. Graduate Theses and Dissertations. https://digitalcommons.usu.edu/etd/4988
Cronk JK, Fennessy MS (2001) Wetland plants: biology and ecology. CRC Press, Boca Raton, FL
Dahl TE, Dick J, Swords J, Wilen BO (2015) Data collection requirements and procedures for mapping wetland, deepwater and related habitats of the United States. USFWS Division of Habitat and Resource Conservation, National Standards and Support Team, Madison, WI, 92p
Davis MA, Grime JP, Thompson K (2000) Fluctuating resources in plant communities: a general theory of invasibility. J Ecol 88:528–534
Derr JF (2008) Common reed (*Phragmites australis*) response to mowing and herbicide application. Invasive Plant Sci Manag 1:12–16
DiTomaso JM, Kyser GB, Oneto SR, Wilson RG, Orloff SB, Anderson LW, Wright SD, Roncoroni JA, Miller TL, Prather TS, Ransom C, Beck KG, Duncan C, Wilson KA, Mann JJ (2013) Weed control in natural areas in the western United States. Weed Research and Information Center, University of California
Downard R, Endter-Wada J (2013) Keeping wetlands wet in the western United States: adaptations to drought in agriculture-dominated human-natural systems. J Environ Manag 131:394–406
Downard R, Endter-Wada J, Kettenring KM (2014) Adaptive wetland management in an uncertain and changing arid environment. Ecol Soc 19:23
Downard R, Frank M, Perkins J, Kettenring K, Larese-Casanova M (2017) Wetland plants of Great Salt Lake, a guide to identification, communities, and bird habitat. Current Publications. https://digitalcommons.usu.edu/extension_curall/1761
Duffield J, Neher C, Patterson D (2011) Utah waterfowl hunting: 2011 hunter survey: hunter attitudes and economic benefits. Bioeconomics, Missoula, MT
Duncan BL (2019) Impacts of cattle grazing as a tool to control *Phragmites australis* in wetlands on nitrogen, phosphorus, and carbon. Graduate Theses and Dissertations. https://digitalcommons.usu.edu/etd/7420
Duncan BL, Hansen R, Cranney C, Shah JJF, Veblen KE, Kettenring KM (2019) Cattle grazing for invasive *Phragmites australis* (common reed) management in Northern Utah wetlands. All Current Publications. Paper 2030. https://digitalcommons.usu.edu/extension_curall/2030
England DM (2019) Seeding treatments to enhance seedling performance of the bulrushes *Bolboschoenus maritimus*, *Schoenoplectus acutus* and *S. americanus* in wetland restorations. All Graduate Theses and Dissertations. 7659. https://digitalcommons.usu.edu/etd/7659/
Esser LL (1994) *Echinochloa crus-galli.* In: Fire Effects Information System. https://www.fs.fed.us/database/feis/plants/graminoid/echcru/all.html
Evans MEK, Dennehy JJ (2005) Germ banking: bet-hedging and variable release from egg and seed dormancy. Q R Biol 80:431–451
Evans KE, Martinson W (2008) Utah's featured birds and viewing sites: a conservation platform for IBAs and BHCAs. Sun Litho, Salt Lake City, UT
Executive Presidential Order (1999) Executive Order 13112 of February 3, 1999: invasive species. Fed Regist 64:6183–6186
Fire Effects Information System (2019). https://www.fs.fed.us/database/feis/plants/forb/lytsal/all.html
Fraser LH, Karnezis JP (2005) A comparative assessment of seedling survival and biomass accumulation for fourteen wetland plant species grown under minor water-depth differences. Wetlands 25:520–530

Galatowitsch SM, van der Valk AG (1994) Restoring prairie wetlands: an ecological approach. Iowa State University Press, Ames, IA

Galatowitsch SM, Anderson NO, Ascher PD (1999) Invasiveness in wetland plants in temperate North America. Wetlands 19:733–755

Gioria M, Pyšek P (2017) Early bird catches the worm: germination as a critical step in plant invasion. Biol Invasions 19:1055–1080

Gleason RA, Tangen BA, Laubhan MK, Lor S (2012) A multi-refuge study to evaluate the effectiveness of growing-season and dormant-season burns to control cattail. US Geological Survey, Reston, VA. https://pubs.usgs.gov/sir/2012/5143/sir2012-5143.pdf

Goman M, Wells L (2000) Trends in river flow affecting the northeastern reach of the San Francisco Bay Estuary over the past 7000 years. Quat Res 54:206–217

Great Salt Lake Bird Festival (2019) http://www.daviscountyutah.gov/greatsaltlakebirdfest. Accessed 19 May 2019

Great Salt Lake Planning Team (2000) Great Salt Lake comprehensive management plan and descision document. Utah Department of Natural Resources, Salt Lake City, UT

Grubb PJ (1977) The maintenance of species-richness in plant communities: the importance of the regeneration niche. Biol Rev 52:107–145

Hansen RM (1978) Shasta ground sloth food habits, Rampart Cave, Arizona. Paleobiology 4:302–319

Haynes RR, Hellquist CB (2000) *Potamogeton crispus* in Flora of North America @ eflorasorg. Flora N Am Editor Comm Eds Flora N Am North Mex 20 Vols N Y Oxf Vol 22. http://www.efloras.org/florataxonaspx?flora_id=1&taxon_id=200024690. Accessed 6 Apr 2019

Hazelton ELG, Mozdzer TJ, Burdick DM, Kettenring KM, Whigham DF (2014) *Phragmites australis* management in the United States: 40 years of methods and outcomes. AoB Plants 6. https://doi.org/10.1093/aobpla/plu001

Holmgren NH (2005) In: Holmgren NH, Holmgren PK, Cronquist A (eds) Frankeniaceae, volume two, Part B: Subclass Dilleniidae. Botanical Garden, New York

Hudon C, Gagnon P, Jean M (2005) Hydrological factors controlling the spread of common reed (*Phragmites australis*) in the St Lawrence River (Québec, Canada). Écoscience 12:347–357

Hurd EG, Shaw NL (1992) Seed technology for *Carex* and *Juncus* species of the Intermountain Region. In: Intermountain Nurseryman Association annual meeting. USDA, Forest Service, Rocky Mountain Forest and Range Experiment Station, Park City, UT, pp 74–83

Intermountain Region Herbarium Network (2019) http://intermountainbiota.org/portal/index.php. Accessed 3 Mar 2019

Intermountain West Joint Venture (2013) Implementation plan: strengthening science and partnerships. Missoula, MT

James JJ, Svejcar TJ, Rinella MJ (2011) Demographic processes limiting seedling recruitment in arid grassland restoration. J Appl Ecol 48:961–969

Jodoin Y, Lavoie C, Villeneuve P, Theriault M, Beaulieu J, Belzile F (2008) Highways as corridors and habitats for the invasive common reed *Phragmites australis* in Quebec, Canada. J Appl Ecol 45:459–466

Keller BEM (2000) Genetic variation among and within populations of *Phragmites australis* in the Charles River watershed. Aquat Bot 66:195–208

Kessler AC, Merchant JW, Allen CR, Shultz SD (2011) Impacts of invasive plants on sandhill crane (*Grus canadensis*) roosting habitat. Invasive Plant Sci Manag 4:369–377

Kettenring KM (2016) Viability, dormancy, germination, and intraspecific variation of *Bolboschoenus maritimus* (alkali bulrush) seeds. Aquat Bot 134:26–30

Kettenring KM, Galatowitsch SM (2011) *Carex* seedling emergence in restored and natural prairie wetlands. Wetlands 31:273–281

Kettenring KM, Mock KE (2012) Genetic diversity, reproductive mode, and dispersal differ between the cryptic invader, *Phragmites australis*, and its native conspecific. Biol Invasions 14:2489–2504

Kettenring KM, Whigham DF (2018) The role of propagule type, resource availability, and seed source in *Phragmites* invasion in Chesapeake Bay wetlands. Wetlands 38:1259–1268

Kettenring KM, McCormick MK, Baron HM, Whigham DF (2011) Mechanisms of *Phragmites australis* invasion: feedbacks among genetic diversity, nutrients, and sexual reproduction. J Appl Ecol 48:1305–1313

Kettenring KM, de Blois S, Hauber DP (2012) Moving from a regional to a continental perspective of *Phragmites australis* invasion in North America. AoB Plants 2012. https://doi.org/10.1093/aobpla/pls040

Kettenring KM, Whigham DF, Hazelton ELG, Gallagher SK, Weiner HM (2015) Biotic resistance, disturbance, and mode of colonization impact the invasion of a widespread, introduced wetland grass. Ecol Appl 25:466–480

Kettenring KM, Mock KE, Zaman B, McKee M (2016) Life on the edge: reproductive mode and rate of invasive *Phragmites australis* patch expansion. Biol Invasions 18:2475–2495

Kettenring KM, Menuz DR, Mock KE (2019) The nativity and distribution of the cryptic invader *Phalaris arundinacea* (reed canarygrass) in riparian areas of the Columbia and Missouri River Basins. Wetlands 39:55–66

King RS, Deluca WV, Whigham DF, Marra PP (2007) Threshold effects of coastal urbanization on *Phragmites australis* (common reed) abundance and foliar nitrogen in Chesapeake Bay. Estuar Coasts 30:469–481

Kiviat E, Hamilton E (2001) *Phragmites* use by native North Americans. Aquat Bot 69:341–357

Kulmatiski A, Beard KH, Meyerson LA, Gibson JR, Mock KE (2011) Nonnative *Phragmites australis* invasion into Utah wetlands. West North Am Nat 70:541–552

Lambert AM, Saltonstall K, Long R, Dudley TL (2016) Biogeography of *Phragmites australis* lineages in the southwestern United States. Biol Invasions 18:2597–2617

Li E, Endter-Wada J, Li S (2019) Dynamics of Utah's agricultural landscapes in response to urbanization: a comparison between irrigated and non-irrigated agricultural lands. Appl Geogr 105:58–72

Lishawa SC, Lawrence BA, Albert DA, Tuchman NC (2015) Biomass harvest of invasive *Typha* promotes plant diversity in a Great Lakes coastal wetland. Restor Ecol 23:228–237

Long AL, Kettenring KM, Hawkins CP, Neale CMU (2017a) Distribution and drivers of a widespread, invasive wetland grass, *Phragmites australis*, in wetlands of the Great Salt Lake, Utah, USA. Wetlands 37:45–57

Long AL, Kettenring KM, Toth R (2017b) Prioritizing management of the invasive grass common reed (*Phragmites australis*) in Great Salt Lake wetlands. Invasions Plant Sci Manag 10:155–165

Low G, Downard R (2018) Great Salt Lake Wetlands Water Quality Standards Workshops, 25

Lowry BJ, Ransom C, Whitesides R, Olsen H (2017) Noxious weed field guide for Utah. https://digitalcommons.usu.edu/cgi/viewcontent.cgi?article=2351&context=extension_curall

Marty JE, Kettenring KM (2017) Seed dormancy break and germination for restoration of three globally important wetland bulrushes. Ecol Restor 35:138–147

Matthews JW, Molano-Flores B, Ellis J, Marcum PB, Handel W, Zylka J, Phillippe LR (2017) Impacts of management and antecedent site condition on restoration outcomes in a sand prairie. Restor Ecol 25:972–981

McCormick MK, Kettenring KM, Baron HM, Whigham DF (2010a) Extent and reproductive mechanisms of *Phragmites australis* spread in brackish wetlands in Chesapeake Bay, Maryland (USA). Wetlands 30:67–74

McCormick MK, Kettenring KM, Baron HM, Whigham DF (2010b) Spread of invasive *Phragmites australis* in estuaries with differing degrees of development: genetic patterns, Allee effects and interpretation. J Ecol 98:1369–1378

McCormick MK, Brooks HEA, Whigham DF (2016) Microsatellite analysis to estimate realized dispersal distance in *Phragmites australis*. Biol Invasions 18:2497–2504

Meadows RE, Saltonstall K (2007) Distribution of native and introduced *Phragmites australis* in freshwater and oligohaline tidal marshes of the Delmarva peninsula and southern New Jersey. J Torrey Bot Soc 134:99–107

Menuz D, McCoy-Sulentic M (2019) Bear River watershed wetland assessment and landscape analysis: Salt Lake City, Utah Geological Survey contract deliverable for the US Environmental Protection Agency Wetland Program Development Grant #CD96851201

Menuz D, Sempler R (2018) Jordan River watershed wetland assessment and landscape analysis: Salt Lake City, Utah Geological Survey contract deliverable for the US Environmental Protection Agency Wetland Program Development Grant #CD96804801

Menuz D, Sempler R, Jones J (2014) Great Salt Lake wetland condition assessment: Salt Lake City, Utah Geological Survey contract deliverable for the US Environmental Protection Agency Wetland Program Development Grant #CD96811101

Menuz D, Sempler R, Jones J (2016) Weber River watershed wetland condition assessment: Salt Lake City, Utah Geological Survey contract deliverable for the US Environmental Protection Agency Wetland Program Development Grant #CD96812601-0

Meyerson LA, Cronin JT (2013) Evidence for multiple introductions of *Phragmites australis* to North America: detection of a new non-native haplotype. Biol Invasions 15:2605–2608

Meyerson LA, Saltonstall K, Windham L, Kiviat E, Findlay S (2000) A comparison of *Phragmites australis* in freshwater and brackish marsh environments in North America. Wetl Ecol Manag 8:89–103

Meyerson LA, Lambert AM, Saltonstall K (2010a) A tale of three lineages: expansion of common reed (*Phragmites australis*) in the US Southwest and Gulf Coast. Invasive Plant Sci Manag 3:515–520

Meyerson LA, Viola DV, Brown RN (2010b) Hybridization of invasive *Phragmites australis* with a native subspecies in North America. Biol Invasions 12:103–111

Meyerson LA, Lambertini C, McCormick MK, Whigham DF (2012) Hybridization of common reed in North America? The answer is blowing in the wind. AoB Plants. https://doi.org/10.1093/aobpla/pls022

Moser K, Ahn C, Noe G (2007) Characterization of microtopography and its influence on vegetation patterns in created wetlands. Wetlands 27:1081–1097

Munger GT (2002) *Lythrum salicaria*. In: Fire Effects Information System. https://www.fs.fed.us/database/feis/plants/forb/lytsal/all.html

Niering WA, Warren RS, Weymouth GC (1977) Our dynamic tidal marshes, vegetation changes as revealed by peat analysis [Connecticut]. Bulletin-Connecticut Arboretum (USA)

Ochterski J (2003) Controlling cattails. Cornell Coop Ext. http://albany.cce.cornell.edu/environment/ponds/controlling-cattails

Olson BE, Lindsey K, Hirschboeck V (2004) Habitat management plan: Bear River Migratory Bird Refuge. U.S. Fish & Wildlife Service, Brigham City, UT

Oring LW, Neel L, Oring KE (2000) Intermountain west regional shorebird plan. US Shorebird Conservation Plan

Orson RA (1999) A paleoecological assessment of *Phragmites australis* in New England tidal marshes: changes in plant community structure during the last few millennia. Biol Invasions 1:149–158

Palmerlee AP, Young TP (2010) Direct seeding is more cost effective than container stock across ten woody species in California. Nativ Plants J 11:89–102

Paul DS, Manning AE (2002) Great Salt Lake waterbird survey five-year report (1997–2001). Great Salt Lake Ecosystem Project, Utah Division of Wildlife Resource, Salt Lake City, UT

Paul J, Vachon N, Garroway CJ, Freeland JR (2010) Molecular data provide strong evidence of natural hybridization between native and introduced lineages of *Phragmites australis* in North America. Biol Invasions 12:2967–2973

Peach M, Zedler JB (2006) How tussocks structure sedge meadow vegetation. Wetlands 26:322–335

Pellegrin D, Hauber DP (1999) Isozyme variation among populations of the clonal species, *Phragmites australis* (Cav) Trin ex Steudel. Aquat Bot 63:241–259

Penning WE, Mjelde M, Dudley B, Hellsten S, Hanganu J, Kolada A, van den Berg M, Poikane S, Phillips G, Willby N, Ecke F (2008) Classifying aquatic macrophytes as indicators of eutrophication in European lakes. Aquat Ecol 42:237–251

Peter CR, Burdick DM (2010) Can plant competition and diversity reduce the growth and survival of exotic *Phragmites australis* invading a tidal marsh? Estuar Coasts 33:1225–1236

Price AL, Fant JB, Larkin DJ (2014) Ecology of native vs introduced *Phragmites australis* (common reed) in Chicago-area wetlands. Wetlands 34:369–377

Quirion B, Simek Z, Dávalos A, Blossey B (2018) Management of invasive *Phragmites australis* in the Adirondacks: a cautionary tale about prospects of eradication. Biol Invasions 20:59–73

Rieger JP, Stanley J, Traynor R (2014) Project planning and management for ecological restoration. Springer, New York

Roberts AJ (2013) Avian diets in a saline ecosystem: Great Salt Lake, Utah, USA. Hum Wildl Interact 7:158–168

Robichaud CD, Rooney RC (2017) Long-term effects of a *Phragmites australis* invasion on birds in a Lake Erie coastal marsh. J Gt Lakes Res 43:141–149

Rohal CB (2018) Invasive *Phragmites australis* management in Great Salt Lake wetlands: context dependency and scale effects on vegetation and seed banks. Graduate Theses and Dissertations. https://digitalcommons.usu.edu/etd/7228

Rohal CB, Hambrecht KR, Cranney C, Kettenring KM (2017) How to restore *Phragmites*-invaded wetlands. Utah Agric Exp Stn Res Rep 224:2

Rohal CB, Kettenring KM, Sims K, Hazelton ELG, Ma Z (2018) Surveying managers to inform a regionally relevant invasive *Phragmites australis* control research program. J Environ Manag 206:807–816

Rohal CB, Cranney C, Hazelton ELG, Kettenring KM (2019a) Invasive *Phragmites australis* management outcomes and native plant recovery are context dependent. Ecol Evol 9:13835–13849

Rohal CB, Cranney C, Kettenring KM (2019b) Abiotic and landscape factors constrain restoration outcomes across spatial scales of a widespread invasive plant. Front Plant Sci 10:481

Saltonstall K (2002) Cryptic invasion by a non-native genotype of the common reed, *Phragmites australis*, into North America. Proc Natl Acad Sci 99:2445–2449

Saltonstall K (2011) Remnant native *Phragmites australis* maintains genetic diversity despite multiple threats. Conserv Genet 12:1027–1033

Saltonstall K, Castillo HE, Blossey B (2014) Confirmed field hybridization of native and introduced *Phragmites australis* (Poaceae) in North America. Am J Bot 101:211–215

Saltonstall K, Lambert AM, Rice N (2016) What happens in Vegas, better stay in Vegas: *Phragmites australis* hybrids in the Las Vegas Wash. Biol Invasions 18:2463–2474

Sciance MB, Patrick CJ, Weller DE, Williams MN, McCormick MK, Hazelton ELG (2016) Local and regional disturbances associated with the invasion of Chesapeake Bay marshes by the common reed *Phragmites australis*. Biol Invasions 18:2661–2677

Shuford WD, Roy VL, Page GW, Paul DS (1995) Shorebird surveys in wetlands at Great Salt Lake, Utah. Point Reyes Bird Observatory Report no. 708

Smith LM, Kadlec JA (1983) Seed banks and their role during drawdown of a North American marsh. J Appl Ecol 20:673–684. https://doi.org/10.2307/2403534

SWCA (2012) Definition and assessment of Great Salt Lake health. SWCA Environmental Consultants, Salt Lake City, UT. https://documents.deq.utah.gov/water-quality/standards-technical-services/great-salt-lake-advisory-council/activities/DWQ-2012-006862.pdf. Accessed 8 Apr 2019

SWCA (2013) Final Great Salt Lake comprehensive management plan and record of decision. https://www.scribd.com/document/294220307/Final-Great-Salt-Lake-Comprehensive-Management-Plan-and-Record-of-Decision. Accessed 19 Mar 2019

Taddeo S, de Blois S (2012) Coexistence of introduced and native common reed (*Phragmites australis*) in freshwater wetlands. Écoscience 19:99–105

Thayer DD, Pfingsten IA, Cao L, Berent L (2019) *Potamogeton crispus* L: US Geological Survey, Nonindigenous Aquatic Species Database, Gainesville, FL. https://nas.er.usgs.gov/queries/factsheet.aspx?SpeciesID=1134. Accessed 6 Apr 2019

Thomas SM, Lyons JE, Andres BA, T-Smith EE, Palacios E, Cavitt JF, Royle JA, Fellows SD, Maty K, Howe WH, Mellink E, Melvin S, Zimmerman T (2012) Population size of snowy plovers breeding in North America. Waterbirds 35:1–14

Thursby J (2004) Ritual identity: hunting and feasting in Utah and southeastern Idaho. West Folk 63:101–121

Tilley DJ, St John L (2013) Hydroseeding improves field establishment of Nebraska sedge regardless of seed treatment. Native Plants J 14:89–94

Tilman D (1994) Competition and biodiversity in spatially structured habitats. Ecology 75:2–16

Tu M, Hurd C, Randall JM (2001) Weed control methods handbook: tools & techniques for use in natural areas. The Nature Conservancy

Tulbure MG, Johnston CA (2010) Environmental conditions promoting non-native *Phragmites australis* expansion in Great Lakes coastal wetlands. Wetlands 30:577–587

Urbanska KM (1997) Safe sites—interface of plant population ecology and restoration ecology. Restor Ecol Sustain Dev 81–110

US Fish & Wildlife Service (2019) National Wetlands Inventory website. U.S. Department of the Interior, Fish & Wildlife Service, Washington, DC. http://www.fws.gov/wetlands/

USDA NRCS (2019) PLANTS database. https://plants.sc.egov.usda.gov/java/

Utah Department of Agriculture and Food (2019) State of Utah noxious weed list. https://ag.utah.gov/farmers/plants-industry/noxious-weed-control-resources/state-of-utah-noxious-weed-list/

Utah Division of Water Quality (2015) Ecological characteristics of potential reference standard sites for Great Salt Lake impounded wetlands: 2014 & 2015. Survey: http://DWQ-2015-017187.pdf

Utah Division of Water Quality (2016) Ecological characteristics of Great Salt Lake fringe wetlands. Utah Division of Water Quality. https://documents.deq.utah.gov/water-quality/standards-technical-services/gsl-website-docs/wetlands-program/wetland-monitoring-assessment/DWQ-2015-017187.pdf

Vest JL, Conover MR (2011) Food habits of wintering waterfowl on the Great Salt Lake, Utah. Waterbirds 34:40–50

Waheed A, Qureshi R, Jakhar GS, Tareen H (2009) Weed community dynamics in wheat crop of district Rahim Yar Khan, Pakistan. Pak J Bot 41:247–254

Webb JA, Wallis EM, Stewardson MJ (2012) A systematic review of published evidence linking wetland plants to water regime components. Aquat Bot 103:1–14

Weller MW (1999) Wetland birds: habitat resources and conservation implications. Cambridge University Press, Cambridge

Whalen MA (2015) *Frankenia pulverulenta*. In: Flora of North America Editorial Committee, eds. 1993+. Flora of North America North of Mexico. 20+ vols. New York and Oxford. Vol 6, p 411. http://beta.floranorthamerica.org/Frankenia_pulverulenta

Whyte RS, Bocetti CI, Klarer DM (2015) Bird assemblages in *Phragmites* dominated and non-*Phragmites* habitats in two Lake Erie coastal marshes. Nat Areas J 35:235–245

Willis CG, Baskin CC, Baskin JM, Auld JR, Venable DL, Cavender-Bares J, Donohue K, de Casas RR, The NESCent Germination Working Group (2014) The evolution of seed dormancy: environmental cues, evolutionary hubs, and diversification of the seed plants. New Phytol 203:300–309

Wilson SD, Keddy PA (1986) Species competitive ability and position along a natural stress/disturbance gradient. Ecology 67:1236–1242

Wu CA, Murray LA, Heffernan KE (2015) Evidence for natural hybridization between native and introduced lineages of *Phragmites australis* in the Chesapeake Bay watershed. Am J Bot 102:805–812

Wurtsbaugh WA, Miller C, Null SE, DeRose RJ, Wilcock P, Hahnenberger M, Howe F, Moore J (2017) Decline of the world's saline lakes. Nat Geosci 10:816–821

Zedler JB, Kercher S (2004) Causes and consequences of invasive plants in wetlands: opportunities, opportunists, and outcomes. Crit Rev Plant Sci 23:431–452

Zimmerman CL, Shirer RR, Corbin JD (2018) Native plant recovery following three years of common reed (*Phragmites australis*) control. Invasive Plant Sci Manag 11:175–180

Chapter 14
Mercury Bioaccumulation and Biomagnification in Great Salt Lake Ecosystems

Abigail F. Scott and Frank J. Black

Abstract Historic and ongoing mercury (Hg) contamination of Great Salt Lake (GSL) has resulted in concern of Hg toxicity for both humans and wildlife. Although Hg levels in GSL surface waters are below Environmental Protection Agency (EPA)'s aquatic life standard, anoxic waters of the deep brine layer at the bottom of the stratified south arm of GSL contain some of the highest concentrations of methylmercury ever measured for a natural water body. High Hg concentrations in muscle from three species of ducks at GSL exceeded EPA screening levels and resulted in the world's first waterfowl consumption advisory due to Hg. Because the greatest concerns of Hg toxicity at GSL are its potential negative impacts to birds and hunters who target waterfowl, Hg research in this ecosystem has focused on birds and their common prey (brine shrimp and brine flies). Hg concentrations in brine shrimp and brine fly larvae pose a low to moderate toxicity risk for birds, but GSL has been identified as an avian Hg hotspot with Hg blood levels commonly exceeding toxicity benchmarks. Hg in waterfowl increases as they feed higher on the food chain.

Keywords Great Salt Lake · Methylmercury · Stratified · Methylation · Birds · Waterfowl · Brine flies · Brine shrimp · Bioaccumulation · Biomagnification · Bioadvection · Toxic

14.1 The Environmental Mercury Problem at Great Salt Lake

Great Salt Lake (GSL) is unique in a number of ways, as described elsewhere in this volume, yet one more includes the nature of the environmental mercury (Hg) problem here. While most human exposure to Hg occurs via the consumption of fish and seafood (Sunderland 2007), and thus most concerns of environmental Hg

A. F. Scott · F. J. Black (✉)
Department of Chemistry, Westminster College, Salt Lake City, UT, USA
e-mail: fblack@westminstercollege.edu

B. K. Baxter, J. K. Butler (eds.), *Great Salt Lake Biology*,
https://doi.org/10.1007/978-3-030-40352-2_14

problems revolve around the bioaccumulation of high levels of Hg by predatory fish, the high salinity of most regions of GSL precludes fish from living in the majority of the lake. Instead, concerns of Hg toxicity at GSL are primarily due to high Hg levels in waterfowl and other birds. GSL is designated as a site of hemispheric importance by the Western Hemisphere Shorebird Reserve Network, and the Hg accumulated here could have far-ranging effects on the 176 species of birds that reside at GSL and the millions of migratory birds that stop to feed at GSL during their annual migration (Aldrich and Paul 2002; Neill et al. 2016).

Recognition of the environmental mercury problem at GSL began with the measurement of high concentrations of Hg in three duck species (Cinnamon Teal (*Anas cyanoptera*), Northern Shoveler (*Anas clypeata*), and Common Goldeneye (*Bucephala clangula*)) at GSL, which resulted in the world's first waterfowl consumption advisories due to their muscle Hg concentrations exceeding the EPA screening level of 0.3 mg/kg wet weight (Utah DoH 2005). Subsequent studies found that the GSL waters contained some of the highest concentrations of methylmercury (MeHg) ever reported for a natural water body (Naftz et al. 2008; Johnson et al. 2015). These findings prompted additional studies on the cycling and bioaccumulation of Hg at GSL, which have since reported elevated concentrations of Hg in a range of other biota at GSL, including brine flies, brine shrimp, spiders, other waterfowl, and even some songbirds (Conover and Vest 2009; Vest et al. 2009; Wurtsbaugh et al. 2011; Saxton et al. 2013; Valdes et al. 2017). While it is clear that GSL is an important resource for both human and bird populations, both are threatened by elevated levels of Hg in this environment.

In addition to its role in the breeding, staging, and migration of millions of birds, GSL also supports industrial and recreational interests, some of which could be affected by the elevated levels of Hg in GSL. Estimates suggest that the total annual economic output and labor income from the GSL ecosystem is nearly $1.7 billion (Utah DWQ 2012). This includes revenue from the extraction and production of minerals including halite, potassium salts, magnesium, chlorine, titanium, potash, and nutritional supplements, recreation including boating, swimming, and birdwatching, waterfowl hunting, grazing leases, and oil and gas drilling (Utah DWQ 2012). The harvest of brine shrimp eggs accounts for over $80 million in total economic effect annually (Utah DWQ 2012). The brine shrimp eggs, or cysts, are collected for the aquaculture industry where they can be hatched on-demand to provide a highly nutritious food source for larval fish and other crustaceans in commercial aquaculture operations (Belovsky et al. 2011; Utah DNR 2011). Brine shrimp cysts collected from GSL account for 35–45% of the world's supply (Utah DNR 2011). During the last decade, 9–16 metric tons of brine shrimp cysts have been harvested annually (GSLEP 2019). One positive research finding has been that female brine shrimp minimize transfer of Hg to their eggs, resulting in much lower Hg concentrations in brine shrimp eggs and cysts compared with adults (Saxton et al. 2013), which is good news given that brine shrimp cysts from GSL are harvested and used extensively in aquaculture globally, including in the growth of fish for human consumption.

14.2 Mercury Toxicity

Mercury is a toxic heavy metal and environmental pollutant (Clarkson and Magos 2006; Mergler et al. 2007). Hg exists in nature as elemental Hg(0), inorganic divalent Hg(II), and organomercury compounds including methylmercury (MeHg). Although all forms of Hg are toxic, MeHg is the chemical species of most environmental concern because (1) ingested MeHg is more efficiently absorbed than other forms of Hg, (2) the rate of MeHg depuration is much slower than its rate of accumulation, and (3) MeHg is the only form of Hg that is consistently biomagnified up food webs, reaching toxic concentrations in apex predators (Wang and Wong 2003; Watras et al. 1998; Chen et al. 2008). Most human exposure to Hg occurs via the consumption of fish and seafood (Sunderland 2007), with large, long-lived piscivorous fish and other apex predators in aquatic ecosystems having the highest Hg concentrations (Bargagli et al. 1998; Francesconi and Lenanton 1992; Baeyens et al. 2003).

As a neurotoxin, the harmful effects of Hg to both humans and wildlife following consumption of contaminated prey are often greatest in young, developing organisms. The greatest concern for human health related to environmental Hg pollution is the consumption of fish by the most susceptible populations, specifically pregnant women and children (Clarkson and Magos 2006; Mergler et al. 2007). Hg exposure in pregnant women has been linked to neurological and developmental deficiencies in the developing fetus, while in adults it also results in increased cardiovascular disease (Mergler et al. 2007). As a result, elevated Hg concentrations in fish have resulted in consumption advisories being issued for multiple species of fish in areas of Europe, Canada, and all 50 states in the United States (e.g., US EPA 2007).

Ecosystems can be threatened by elevated environmental Hg levels, with higher trophic-level organisms considered most at risk. These include piscivorous birds and mammals, as well as predatory fish that often have high body burdens of Hg due to the biomagnification of MeHg up aquatic food chains (Brookens et al. 2008; Scheuhammer et al. 2007; Sonne et al. 2007). The deleterious effects of MeHg poisoning on wildlife include decreased growth, delayed development, decreased survival rate, and abnormal development of the central nervous system, gonads, and other organs (Scheuhammer et al. 2007). In birds, in particular, sublethal levels of MeHg can affect behavior, coordination, reproduction, and immune response (Scheuhammer et al. 2007).

14.3 Global Mercury Cycle

Mercury is a global pollutant of major concern due to (1) its interhemispheric atmospheric transport, allowing it to be emitted to the atmosphere and later deposited elsewhere globally far from its source (Fitzgerald et al. 1998), (2) its methylation to methylmercury (MeHg) by sulfate-reducing bacteria and other microbes in anoxic aquatic sediment and waters (Benoit et al. 2003), and (3) the ensuing bioaccumulation and biomagnification of that MeHg by organisms in aquatic food chains (Watras et al. 1998).

Roughly 90% of the Hg found in the atmosphere exists as volatile, elemental Hg (0), with the remaining 10% existing as a combination of gaseous Hg(II) (often called reactive gaseous mercury) and particulate bound Hg (Selin et al. 2008). Hg can be emitted to the atmosphere from both the ocean and the land via natural processes, such as volcanic eruptions, off-gassing of the crust, and evasion from the ocean. However, current Hg emissions to the atmosphere are dominated by anthropogenic sources, with estimates suggesting that roughly one-third of atmospheric emissions are natural and two-thirds anthropogenic (Selin et al. 2008). Anthropogenic sources of Hg include coal combustion, gold production, non-iron metal smelting, cement production, caustic soda manufacturing, waste incineration, and biomass burning (Pacyna et al. 2006; Selin et al. 2008).

Hg has an atmospheric residence time of roughly one year (Mason and Sheu 2002; Selin et al. 2008), which is sufficiently long to allow for the long-range transport of Hg from point sources to remote, pristine areas globally before being deposited. As a result, much of the atmospherically deposited Hg at most locations originates from global rather than local sources. For example, Selin et al. (2008) estimated that of the anthropogenic Hg that is deposited from the atmosphere to the United States each year, 50% more originates from global sources outside the United States than what originates from the United States itself. This highlights the truly global nature of the environmental Hg problem.

Atmospheric Hg(0) can be deposited directly to land or the oceans via dry deposition, but, the atmospheric deposition of Hg commonly first involves the oxidation of Hg(0) to Hg(II), which is very particle reactive and commonly becomes associated with aerosols or other particles, which then undergo wet or dry deposition. The reduction of Hg(II) and the reoxidation of Hg(0) in surface waters and the atmospheric boundary layer is exceedingly rapid, and Hg deposited to the oceans and other bodies of water can be quickly reemitted and recycled (Mason and Sheu 2002; Strode et al. 2007).

The amount of Hg stored in the earth's soils and surficial sediments is many orders of magnitude greater than that found in the ocean and atmosphere, although the residence time of Hg on land is also much longer. The predominant form of Hg found in these rocks, aquatic sediment, and soils is Hg(II) (Selin et al. 2008). Hg inputs to soils and land occur via atmospheric deposition and sediment burial. Hg deposited in terrestrial systems can be preferentially revolatilized to the atmosphere, while the Hg retained by vegetation and soils binds strongly with reduced sulfur groups, and remains in the soil for centuries to millennia (Skyllberg et al. 2003; Selin et al. 2008). Hg concentrations in uncontaminated soils are typically in the range 20–70 ng/g dry weight (dw), and vary with factors controlling Hg adsorption and deposition, including the type of soil and concentrations of certain species, including sulfide, chloride, and dissolved organic carbon. Hg is lost from land and soils via mineral weathering, runoff, and evasion to the atmosphere.

Hg(II) is the primary form of Hg found in the waters of aquatic ecosystems, both freshwaters and the ocean. This Hg(II) can exist either bound to particles, or dissolved in solution. The inorganic Hg(II) that exists in the dissolved phase is generally found in either inorganic complexes (bound to chloride, hydroxide, or sulfide) or organic complexes (bound to thiols and dissolved organic matter). The form of Hg and what it is bound to has a profound influence on this bioavailability and cycling in aquatic ecosystems (Hsu-Kim et al. 2013). A small portion of the Hg in natural waters exists as Hg(0) and organic forms of Hg, in particular, methylated species.

MeHg is formed in aquatic ecosystems by the methylation of inorganic mercury, a process carried out by sulfate and iron-reducing microbes that thrive in anoxic conditions (Benoit et al. 2003). Hg methylation most frequently occurs in aquatic sediment, but can also take place in anoxic regions of the water column (Eckley and Hintelmann 2006; Lamborg et al. 2008). Factors that control rates of MeHg production include sediment composition and structure, redox conditions, sulfide and sulfate concentrations, organic carbon content and composition, temperature, microbial community composition, rates of microbial respiration, as well as Hg speciation (the forms of Hg present) (Hsu-Kim et al. 2013). Once produced, MeHg can be demethylated via both biotic and abiotic mechanisms, with microbial demethylation being most important in sediment and deep waters, and photo-demethylation being more important in surface waters with abundant solar irradiance (Marvin-DiPasquale et al. 2000; Black et al. 2012).

Phytoplankton and algae, at the base of aquatic food webs, bioaccumulate MeHg to concentrations ~10,000 times greater than the natural waters in which they live (Pickhardt and Fisher 2007; Watras et al. 1998). This transfer from natural waters to phytoplankton represents the single greatest bioconcentration of MeHg that occurs at any trophic level in aquatic food chains, with smaller enrichment factors (~2–5×) characterizing trophic transfers higher in the food chain. The result of this process is that most of the Hg in fish at higher trophic levels exists as MeHg, which comes primarily from dietary sources (Pickhardt et al. 2006; Wang and Wong 2003). Concentrations of MeHg in apex, predatory fish in aquatic ecosystems can exceed 1 part per million (Bargagli et al. 1998; Francesconi and Lenanton 1992; Baeyens et al. 2003), concentrations high enough to constitute a health hazard to humans or wildlife that consume them (Clarkson and Magos 2006; Mergler et al. 2007; Scheuhammer et al. 2007).

14.4 Mercury Cycling in Great Salt Lake

GSL is the largest hypersaline lake in the western hemisphere, and one of the largest terminal lakes in the world. Railroad and other causeways constructed between 1959 and 1969 separated and compartmentalized GSL into four major embayments (Fig. 14.1): Bear River Bay, Farmington Bay, Gunnison Bay (the north arm), and Gilbert Bay (the south arm). Differences in freshwater inputs have resulted in significant salinity differences between regions of the lake (Fig. 14.2).

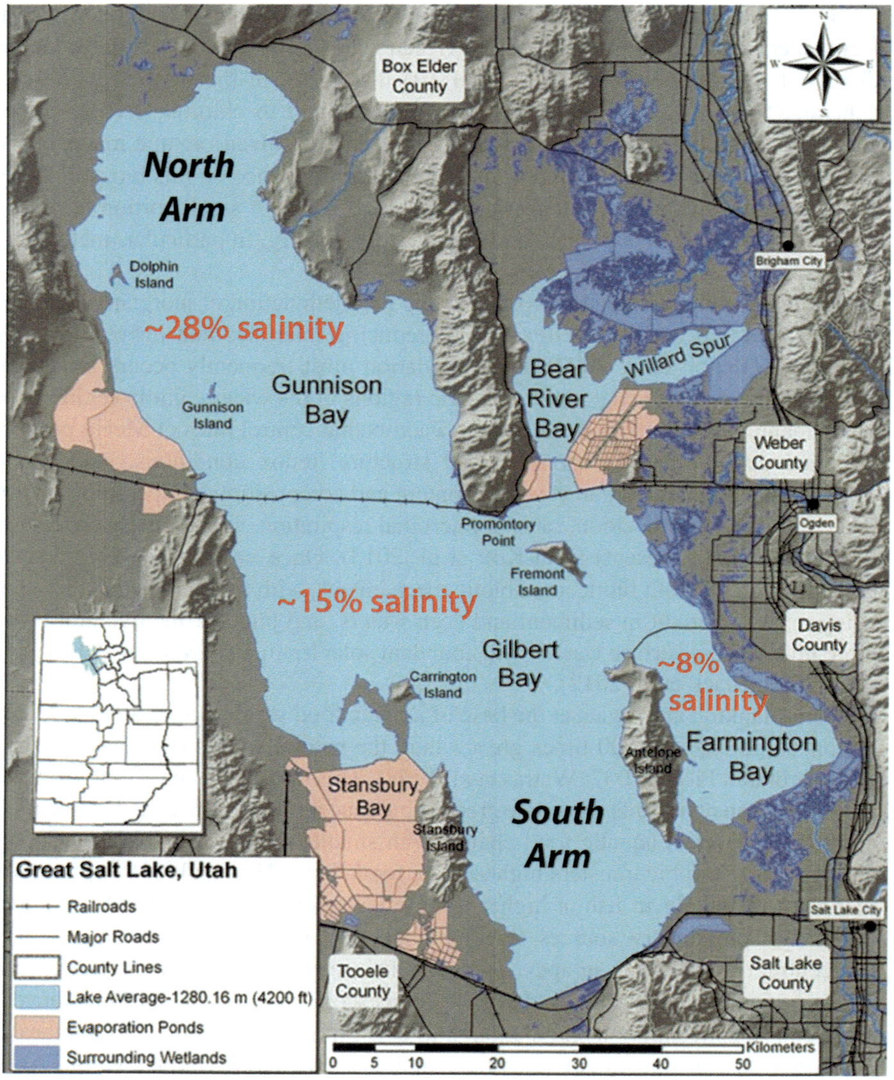

Fig. 14.1 Map of Great Salt Lake, Utah, showing its various embayments

Limited volumes of the more saline, denser water from the north arm (~28% salinity) is able to flow into the south arm, primarily via culverts in the railroad causeway, where it sinks and produces a distinct layer of water at the bottom of the south arm. This process has resulted in the south arm becoming perennially stratified, a condition known as meromixis. The south arm consists of an anoxic, deep brine layer (DBL) rich in dissolved organic carbon (~25% salinity, [DOC] > 70 mg/L) and an oxic upper brine layer (~15% salinity) (Loving et al. 2000). The DBL occupies the deepest portions of GSL, at depths greater than approximately 6–7 m

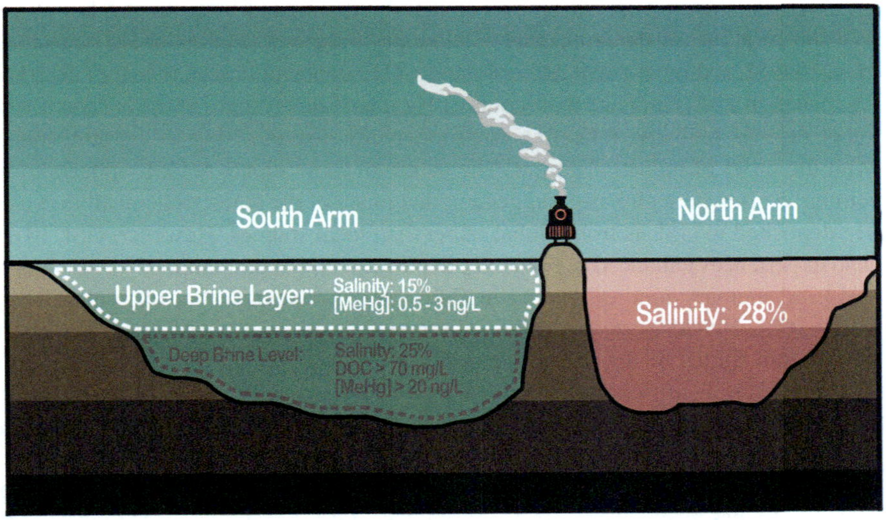

Fig. 14.2 Stylized cross section of Great Salt Lake, showing the perennial stratification of the south arm (Gilbert Bay) (Image credit: Johanna Bossart)

below the surface (Diaz et al. 2009; Valdes et al. 2017). The DBL is home to some of the highest MeHg concentrations ever reported in natural waters, and routinely exceed 20 ng/L (Naftz et al. 2008; Johnson et al. 2015). The DBL can persist for many annual cycles because it does not experience annual turnover, and microbial respiration of organic matter in the DBL has resulted in it becoming anoxic. There is at least some vertical mixing between the DBL and upper brine layer (UBL) due to internal waves created by sustained wind events, and roughly 50% of the DBL is estimated to mix up into the UBL each year (Jones and Wurtsbaugh 2014; Valdes et al. 2017).

Concentrations of total Hg in surface waters of GSL are typically 2–15 ng/L (Naftz et al. 2008; Valdes et al. 2017), and are thus usually below the US EPA's aquatic life standard for Hg of 12 ng/L. The highest concentrations of total Hg (HgT) in the GSL waters are found in the anoxic DBL, where they are typically 30–70 ng/L, can reach up to 100 ng/L, and consistently exceed the EPA's threshold for aquatic life (Naftz et al. 2008; Valdes et al. 2017). A substantial fraction of this Hg in the DBL (30–80%) exists as MeHg (Naftz et al. 2008; Johnson et al. 2015), which is much higher than the % MeHg typically found in surface waters of either freshwater or marine ecosystems.

Estimates of contemporary inputs of Hg to GSL are dominated by atmospheric deposition, which is true of most lakes globally not exposed to local point sources of Hg pollution (Fitzgerald et al. 1998). The most important current sources of Hg to GSL are estimated to be wet atmospheric deposition (47%), dry atmospheric deposition (37%), and riverine inputs (16%), with roughly half of the riverine inputs of Hg coming from outflows of Farmington Bay (Peterson and Gustin 2008; Naftz et al. 2009). Sinks or loss mechanisms for total Hg from GSL have been far less studied than Hg sources, but would be expected to be dominated by sediment burial and evasion of Hg(0) to the atmosphere following reduction. Studies of Hg concentrations in GSL sediment cores and estimates of Hg deposition rates (Naftz et al. 2008; Wurtsbaugh 2012) indicate that inputs of Hg increased around 1850, consistent with other regions globally, reflecting anthropogenic releases of Hg to the atmosphere with the onset of industrialization. The nature of these inputs at GSL appears to be similar to elsewhere globally, suggesting that much of the anthropogenically derived Hg deposited to GSL is from global rather than local sources. However, GSL likely also experiences enhanced regional Hg deposition due to contemporary and historic gold and mercury mining upwind in California and Nevada. For example, during 2001–2002, Nevada was one of the largest sources of atmospheric Hg emissions in the western United States, with the majority of those Hg emissions coming from gold ore processing facilities (Jones and Miller 2005).

The elevated Hg concentrations currently found in GSL, if assumed to be at near steady state, are the combined result of Hg sources and sinks, and thus may reflect either higher rates of Hg inputs to GSL than aquatic ecosystems elsewhere, or lower rates of Hg loss mechanisms. As a terminal lake, GSL has no loss of Hg via riverine outflows, the lack of which could allow for the accumulation of Hg over time. However, in many other large lakes for which mass balances of Hg have been created, sediment burial and volatilization to the atmosphere, not riverine outflows, were the dominant loss mechanism, and waters in those other lakes still have much lower Hg concentrations than GSL (Gao et al. 2006; Qureshi et al. 2009), suggesting this alone is not to blame. While little is known about rates of Hg loss from GSL, Hg concentrations in the GSL sediments, and estimated rates of Hg accumulation in sediments suggest that sediment burial of Hg is not dramatically lower than lakes elsewhere. Rates of Hg(0) volatilization to the atmosphere from GSL have never been directly measured, so it is unknown if the high salinity of GSL or other factors results in lower rates of evasive loss of Hg compared to marine waters or lakes elsewhere. Riverine inputs of Hg to GSL have been estimated (Naftz et al. 2008), and are not substantially greater than lakes of similar size elsewhere.

One possibility that could account for the high concentrations of total Hg in the GSL waters, apart from greater recycling of past inputs, include greater rates of atmospheric deposition than for lakes elsewhere. The shallow bathometry of GSL results in it having a high surface area-to-volume ratio, which increases with decreasing lake volume (Baskin 2005), which would allow for greater atmospheric deposition of Hg per unit volume compared to many other lakes of its size. One additional contributing factor could be that due to a combination of the high concentrations of chloride and bromine in the GSL waters, as well as the 900–3500 metric tons of chlorine gas emitted each year to the atmosphere along the western margin GSL by US Magnesium (Utah DEQ 2017), the atmosphere above GSL may be more effective at oxidizing Hg(0) to Hg(II) (reactive gaseous Hg), which would result in enhanced atmospheric deposition of Hg to GSL. However, the one study to date designed to test this hypothesis did not find evidence to support it, nor evidence of elevated rates of Hg atmospheric deposition in general at GSL (Peterson and Gustin 2008).

14.4.1 Reasons for the High Levels of Methylmercury in Great Salt Lake

While previous studies have documented the elevated concentrations of total Hg and MeHg in water, sediment, and biota at GSL (Naftz et al. 2008; Johnson et al. 2015; Vest et al. 2009; Saxton et al. 2013), the reasons for these elevated MeHg concentrations are as yet unclear. Because MeHg production is mediated primarily by sulfate and iron-reducing bacteria in anoxic environments (Benoit et al. 2003; Kerin et al. 2006), it has been hypothesized that the anoxic DBL is responsible. This could be the case if Hg methylation occurred not only in surficial sediments, as is true in most aquatic systems, but also extensively in the DOM (dissolved organic matter) rich water column of the anoxic DBL of GSL. However, rates of Hg methylation measured in DBL water and underlying sediment in incubation experiments suggest that most methylation likely occurs in the sediment, and that rates of Hg methylation in GSL are similar to in other aquatic ecosystems (Johnson et al. 2015; Boyd et al. 2017). Hg methylation rates in GSL have also been reported to be inversely related to salinity, with the highest rates of microbial MeHg production in sediments with low pore-water salinity, although the highest rates of microbial degradation of MeHg were also found in these same areas (Boyd et al. 2017). The abundance of 16S rRNA gene transcripts from the sulfate-reducing bacteria *Desulfobacterium* sp. was positively correlated with MeHg methylation rates and MeHg concentrations in GSL sediment, suggesting a possible role for this genus of bacteria in Hg methylation at GSL.

The DBL may be uniquely efficient at Hg methylation due to its anoxic conditions, high concentrations of sulfate and organic material, and high activity of sulfate-reducing bacteria, which are all factors associated with high rates of microbial Hg methylation (Benoit et al. 2003). The importance of Hg methylation in the water column of both anoxic and oxic saline marine waters and freshwaters (Monperrus et al. 2007; Cossa et al. 2009; Sunderland et al. 2009) has been highlighted elsewhere in recent years, and may contribute to Hg methylation at GSL. Alternatively, the presence of the anoxic DBL may enhance MeHg production in, or export flux from, the underlying sediments, as has been reported for some other systems (Covelli et al. 2008). These ideas have proven difficult to test at GSL (Johnson et al. 2015), but this has not prevented proposals being put forth at the state level to try and decrease Hg methylation at GSL by mechanically aerating the DBL, despite a lack of understanding of where Hg methylation primarily occurs in this ecosystem nor where most of the MeHg in the GSL biota is produced. Other potentially important sources of MeHg to GSL could include the fringing wetlands, some of which are hotspots of MeHg production and have high concentrations of MeHg (Johnson et al. 2015).

14.5 The Great Salt Lake Biota and Mercury Bioaccumulation

Due to the high salinity of the lake, no fish live in the saline bays of GSL, and birds are the primary predators of invertebrates in the lake. GSL sits at the intersection of the Pacific and Central flyways, and represents one of the most important locations in North America for migratory birds (Gill 1995). GSL's eastern and northern shores include 400,000 acres of freshwater and saltwater wetlands (US EPA 2010), which along with GSL are utilized by millions of shorebirds as breeding grounds, and millions more waterfowl stop at GSL to feed on brine shrimp and brine flies during their biannual migration (Aldrich and Paul 2002). Coupled with the elevated concentrations of Hg in brine shrimp and brine flies here, GSL may serve as a major source of not only food, but also Hg poisoning to millions of resident and migratory birds each year.

14.5.1 Mercury Bioaccumulation at the Base of the Great Salt Lake Food Web

Birds are often the organisms of concern for Hg toxicity at GSL, so research on the bioaccumulation of Hg at GSL has largely focused on birds and their food sources in this ecosystem. Brine flies (*Ephydra* spp.) and brine shrimp (*Artemia franciscana*) represent keystone species in the short food chains at GSL (Fig. 14.3), and the abundant biomass of these two invertebrates comprise a significant fraction of the diet of many of the 338 species of resident and migratory birds at GSL (Sorensen et al. 2020). As such, the mechanisms by which Hg is transferred from water and algae at the base of the GSL food chain to brine shrimp and brine flies play an important role in controlling Hg exposure to birds and other organisms at higher trophic levels (Wurtsbaugh et al. 2011; Johnson et al. 2015; Valdes et al. 2017).

GSL has two, loosely related food webs (Belovsky et al. 2011). The first is a phytoplankton-based web, involving phytoplankton and the brine shrimp that feed on them. The second is a benthic algae, microbialite, and particulate organic matter (detritus) based food web that includes microbialites and their associated periphyton, as well as the brine flies that feed on these. Bird species feeding on either brine flies or brine shrimp at GSL have both been shown to have elevated Hg concentrations (UDH 2005; Vest et al. 2009; Valdes et al. 2017), highlighting the key roles that both *Artemia* and *Ephydra* play in the transfer of Hg to higher-order biota.

While the south arm is home to over 70 species of phytoplankton that are the center of the plankton food web of GSL, only a few species of phytoplankton can tolerate the higher salinity (28%) of the north arm (Larson and Belovsky 2013). Phytoplankton communities in the south arm are dominated by Chlorophytes, Bacillariophytes, and Cyanophytes (Belovsky et al. 2011). Phytoplankton favored by brine shrimp include the green algae *Dunaliella* spp., which are a better food

Fig. 14.3 Stylized Great Salt Lake food web, with pathways of mercury bioaccumulation, trophic transfer, and exposure (Image credit: Johanna Bossart)

source for brine shrimp than Cyanophytes or Bacillariophytes. The abundance of phytoplankton in the south arm is frequently controlled by top-down grazing pressure by *Artemia*. Adult brine shrimp are present during the warmer months from roughly March to November. Predation by brine shrimp during this time of year results in phytoplankton abundance and chlorophyll A being highest during winter months (Belovsky et al. 2011). However, phytoplankton growth in the south arm is also influenced by bottom-up nutrient availability, which varies seasonally and interannually, including from the expansion and contraction of the DBL (Belovsky et al. 2011). Mixing of the underlying DBL up into the UBL where the phytoplankton live increase the availability of nutrients, but also facilitates the transfer of MeHg from the DBL to the UBL, where it can then be assimilated and enter the GSL aquatic food web. Hg bioaccumulation by phytoplankton has been little studied in GSL, so typical concentrations of HgT or MeHg in different species of phytoplankton in this ecosystem are not known, nor are how they vary spatially or temporally.

Brine shrimp (*Artemia franciscana*) distribution in GSL is largely controlled by salinity, while their abundance is controlled by seasonal changes in temperature and availability of their phytoplankton prey (Wurtsbaugh and Gliwicz 2001). With an optimal salinity range of perhaps 10–17%, brine shrimp thrive in the south arm, but are absent from the higher salinity north arm, as well as in the fresher waters of the Bear River Bay and wetlands (Wurtsbaugh and Gliwicz 2001). Farmington Bay, with its lower salinity and higher nutrients than the south arm, is able to support smaller, and more variable populations of brine shrimp. Brine shrimp originate as cysts or nauplii (first larval stage) through two modes of reproduction, but the cysts are able to overwinter while the nauplii cannot. Brine shrimp densities in the south arm of GSL can reach almost six shrimp/liter (Belovsky et al. 2011), and *Artemia* biomass in GSL can exceed 90,000 metric tons.

Concentrations of Hg in brine shrimp are generally 0.08–0.9 mg/kg dw (~0.02–0.12 mg/kg ww), with 80–100% of this existing as MeHg (Utah DWQ 2012, 2014). These data suggest that brine shrimp pose a low (0–0.05 mg/kg ww) to moderate (0.05–0.15 mg/kg ww) Hg toxicity risk to avian wildlife that consume them (Evers et al. 2004). Despite this, potentially toxic levels of Hg have been found in the tissue of birds that consume brine shrimp, as detailed below. Hg and MeHg concentrations in brine shrimp eggs and cysts are 90% lower than in the adult females that produce them (Saxton et al. 2013). A study designed to evaluate if the Hg found in brine shrimp at GSL is accumulated directly from the DBL found that, in laboratory experiments using microcosms to simulate the GSL water column, brine shrimp feed near the chemocline and interface between the DBL and the UBL. However, under such conditions, the brine shrimp accumulated less Hg than in a water column with no DBL present, which the authors attributed to the low Hg:POC (particulate organic carbon) ratio in the DOM rich DBL and a detrital dilution effect (Jones and Wurtsbaugh 2014).

The benthic algae- and detritus-based food web of GSL includes microbialites, organosedimentary structures formed by the microbially mediated precipitation of carbonates (Dupraz and Visscher 2005). Microbialites cover roughly 24% of the north and south arms of GSL, although only those in the south arm are active, and are limited to depths of roughly 0–4 m below the water surface where there is sufficient light for photosynthesis (Wurtsbaugh et al. 2011; Lindsay et al. 2017). Depending on the salinity and other factors, the microbialites are composed primarily of

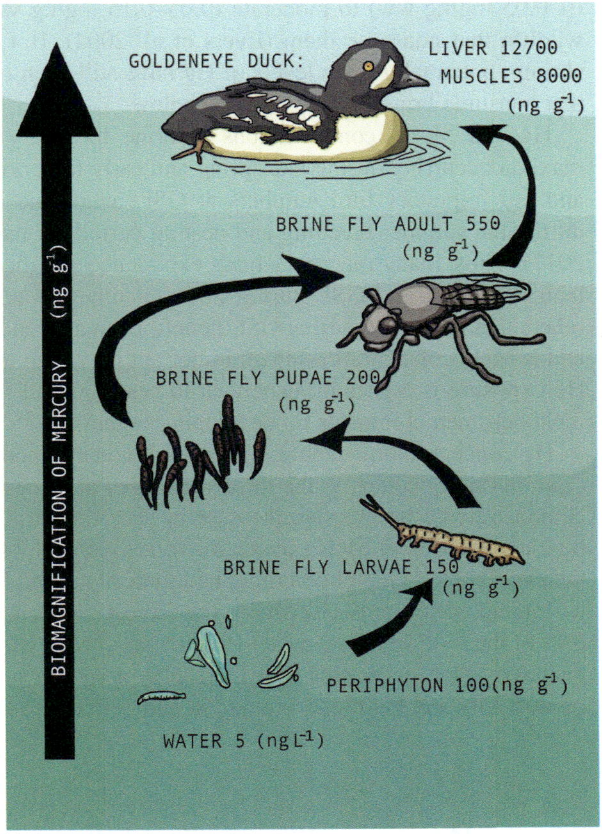

Fig. 14.4 Bioaccumulation of mercury in one of the Great Salt Lake food chains. Total mercury concentrations in biota are reported on a dry weight basis (Image credit: Johanna Bossart)

cyanobacteria of the genera *Aphanothece* and *Euhalothece*, or the bacilliarophyte *Navicula salinicola* (Carozzi 1962; Halley 1976; Lindsay et al. 2017). The bioaccumulation factor for Hg between water and periphyton in the microbialites of the south arm is roughly 20,000 (Wurtsbaugh et al. 2011), with additional biomagnification occurring up the GSL food chain (Fig. 14.4).

Brine flies (*Ephedra* spp.) are a keystone species at GSL, and the estimated 90,000 metric tons of brine flies that emerge from GSL each year (Collins 1980) are a primary food source, along with brine shrimp, for the millions of migratory birds that visit GSL each year. Brine fly larvae at GSL primarily feed on periphyton growing on hard substrates in shallow areas of the lake, which are dominated by the microbialites abundant near the shore of GSL (Collins 1980). Brine fly larval densities on the microbialites are highest in October and lowest in July, while pupal densities are highest in July and lowest in December (Wurtsbaugh et al. 2011). Hg concentrations in brine flies increase as they progress through their larval, pupal, and adult life stages, and are typically 0.1–0.2 mg/kg dw in brine fly larvae compared to 0.25–0.80 mg/kg dw (~0.03–0.10 mg/kg ww) in adults (Wurtsbaugh et al. 2011; Valdes et al. 2017). These data suggest that brine flies pose a low

(0–0.05 mg/kg ww) to moderate (0.05–0.15 mg/kg ww) Hg toxicity risk to avian wildlife that consume them (Evers et al. 2004). But similar to the case of brine shrimp, potentially toxic levels of Hg have still been reported in the tissue of birds that consume brine flies, as detailed below.

Hg and MeHg concentrations in brine flies vary seasonally, with an annual maxima occurring during late spring and early fall correspond to the peak in spring and fall migratory bird numbers at GSL, while the seasonal minimum occurred during the summer breeding and nesting period for many local birds (Valdes et al. 2017). Thus, many migratory birds experience significantly greater exposure to Hg than if they visited GSL during different times of year, while resident birds and others that breed and nest at GSL benefit from a serendipitous seasonal low in HgT and MMHg concentrations in an important prey item during this period. This lower Hg exposure is especially beneficial to nestlings and fledglings because Hg has its most detrimental impacts on developing organisms.

Hg concentrations moving up the GSL aquatic food chain from water to periphyton on microbialites, to the three life stages of brine flies, up to ducks (Fig. 14.4) demonstrated a progressive increase in Hg concentrations, which is typical of the biomagnification of MeHg in aquatic food chains globally. A similar trend exists in the parallel food chain involving planktonic algae and brine shrimp of GSL. Similar to aquatic ecosystems elsewhere, the largest bioaccumulation factor occurs at the base of the food web from water to either periphyton associated with microbialites or plankton algae, indicating that these play a significant role in both the biological productivity and Hg bioaccumulation and delivery to higher trophic levels at GSL.

14.5.2 Mercury in Waterfowl and Wading Birds at Great Salt Lake

Hg concentrations in GSL waterfowl are lowest in herbivores and detritivores, such as mallard, gadwall, northern pintail, Canada goose, and green winged teal, which generally have Hg concentrations in muscle tissue of less than 0.1 μg/g ww (Utah DoH 2005; Ackerman et al. 2015; Valdes et al. 2017). Waterfowl feeding higher in the food web have higher Hg concentrations, with cinnamon teal, northern shoveler, and common goldeneyes having some of the higher Hg muscle concentrations (Utah DoH 2005; Valdes et al. 2017), while Caspian and Foster's terns, snowy plovers, black-necked stilts, pied-billed grebes, and ring-billed gulls have some of the highest Hg levels in eggs (Ackerman et al. 2015). Total Hg concentrations measured in GSL waterfowl eggs and muscle tissue, the tissues for which by far the most Hg measurements in GSL birds have been reported, are summarized in Table 14.1.

While various studies have reported spatial, seasonal, and interannual variability in HgT concentrations in waterfowl at GSL (Conover and Vest 2009; Waddell et al. 2009; Vest et al. 2009; Valdes et al. 2017), differences in the avian species targeted,

Table 14.1 Concentrations of total mercury, HgT (mean ± SD; μg/g wet weight) in the Great Salt Lake waterfowl eggs and breast muscle tissue from 25 avian species

Bird common name	Scientific name	Eggs HgT mean ± SD (μg/g ww)	Eggs Sample year	Eggs Sample site	Muscle HgT mean ± SD (μg/g ww)	Muscle Sample year	Muscle Sample site	References
Caspian tern	*Hydroprogne caspia*	0.87 ± 0.53	2010–2012	BRMBR, FBWMA, AI				Ackerman et al. (2015)
Forster's tern	*Sterna forsteri*	0.64 ± 0.70	2010–2012	BRMBR, FBWMA, AI				Ackerman et al. (2015)
Common golden eye	*Bucephala clangula*				2.01 ± 1.56	2004	BRMBR, OBWMA, FBWMA	Utah DoH (2005)
Northern shoveler	*Anas clypeata*				1.99 ± 2.80	2004–2005	BRMBR, OBWMA, FBWMA	Utah DoH (2005)
Northern shoveler	*Anas clypeata*				0.78 ± 0.48	2014–2015	FBWMA, OBWMA, HSWMA, HSWMA, HCWMA	Valdes et al. (2017)
Snowy plovers	*Charadrius nivosus*	0.44 ± 0.25	2010–2012	BRMBR, FBWMA, AI				Ackerman et al. (2015)
Black-necked stilt	*Himantopus mexicanus*	0.30 ± 0.29	2010–2012	BRMBR, FBWMA, AI				Ackerman et al. (2015)
Black necked stilt	*Himantopus mexicanus*	0.14 ± 0.07	2012	OBWMA, AI				Cavitt et al. (2012)

(continued)

Table 14.1 (continued)

Bird common name	Scientific name	Eggs			Muscle			References
		HgT mean ± SD (μg/g ww)	Sample year	Sample site	HgT mean ± SD (μg/g ww)	Sample year	Sample site	
Black necked stilt	*Himantopus mexicanus*	0.32 ± 0.12	2017	AI				Black, unpublished data
Pied-billed grebe	*Podilymbus podiceps*	0.22 ± 0.13	2010–2012	BRMBR, FBWMA, AI				Ackerman et al. (2015)
Ring-billed gull	*Larus delawarensis*	0.22 ± 0.15	2010–2012	BRMBR, FBWMA, AI				Ackerman et al. (2015)
Double crested cormorant	*Phalacrocorax auritus*	0.18 ± 0.16	2010–2012	BRMBR, FBWMA, AI				Ackerman et al. (2015)
Northern pintail	*Anas acuta*	0.18 ± 0.10	2010–2012	BRMBR, FBWMA, AI	0.046 ± 0.038	2005–2006	BRMBR, OBWMA, FBWMA	Utah DoH (2006), Ackerman et al. (2015)
Black-crowned night heron	*Nycticorax nycticorax*	0.17 ± 0.14	2010–2012	BRMBR, FBWMA, AI				Ackerman et al. (2015)
Gadwall	*Anas strepera*	0.16 ± 0.09	2010–2012	BRMBR, FBWMA, AI	0.057 ± 0.055	2005	BRMBR, OBWMA, FBWMA	Utah DoH (2005), Ackerman et al. (2015)
Gadwall	*Anas strepera*				0.11 ± 0.17	2014–2015	FBWMA, OBWMA, HSWMA, HSWMA, HCWMA	Valdes et al. (2017)

Killdeer	*Charadrius vociferus*	0.15 ± 0.11	2010–2012	BRMBR, FBWMA, AI				Ackerman et al. (2015)
Cinnamon teal	*Anas cyanoptera*	0.15 ± 0.13	2010–2012	BRMBR, FBWMA, AI	0.60 ± 0.41	2014–2015	FBWMA, OBWMA, HSWMA, HSWMA, HCWMA	Valdes et al. (2017), Ackerman et al. (2015)
American avocet	*Recurvirostra americana*	0.12 ± 0.18	2010–2012	BRMBR, FBWMA, AI				Ackerman et al. (2015)
American avocet	*Recurvirostra americana*	0.08 ± 0.03	2012	OBWMA, AI				Cavitt et al. (2012)
American avocet	*Recurvirostra americana*	0.16 ± 0.04	2016–2017	AI				Black, unpublished data
Great blue heron	*Ardea herodias*	0.14 ± 0.08	2010–2012	BRMBR, FBWMA, AI				Ackerman et al. (2015)
Mallard	*Anas platyrhynchos*	0.13 ± 0.06	2010–2012	BRMBR, FBWMA, AI	0.28 ± 0.20	2005	BRMBR, OBWMA, FBWMA	Utah DoH (2005), Ackerman et al. (2015)
Mallard	*Anas platyrhynchos*				0.19 ± 0.23	2014–2015	FBWMA, OBWMA, HSWMA, HSWMA, HCWMA	Valdes et al. (2017)
Green winged teal	*Anas carolinensis*				0.21 ± 0.09	2004–2005	BRMBR, OBWMA, FBWMA	Utah DoH (2005)
Franklin's gull	*Leucophaeus pipixcan*	0.11 ± 0.11	2010–2012	BRMBR, FBWMA, AI				Ackerman et al. (2015)
American coot	*Fulica americana*	0.10 ± 0.12	2010–2012	BRMBR, FBWMA, AI				Ackerman et al. (2015)

(continued)

Table 14.1 (continued)

Bird common name	Scientific name	Eggs			Muscle			References
		HgT mean ± SD (μg/g ww)	Sample year	Sample site	HgT mean ± SD (μg/g ww)	Sample year	Sample site	
Eared grebes	*Podiceps nigricollis*	0.10 ± 0.07	2010–2012	BRMBR, FBWMA, AI				Ackerman et al. (2015)
Redhead	*Aythya americana*	0.10 ± 0.05	2010–2012	BRMBR, FBWMA, AI	0.017 ± 0.018	2005–2006	BRMBR, OBWMA, FBWMA	Utah DoH (2006), Ackerman et al. (2015)
White faced ibis	*Plegadis chihi*	0.08 ± 0.18	2010–2012	BRMBR, FBWMA, AI				Ackerman et al. (2015)
California gull	*Larus californicus*	0.08 ± 0.10	2010–2012	BRMBR, FBWMA, AI				Ackerman et al. (2015)
Canada geese	*Branta canadensis*	0.01 ± 0.04	2010–2012	BRMBR, FBWMA, AI				Ackerman et al. (2015)

Where necessary, egg total mercury concentrations have been converted from dry weight to wet weight assuming an egg moisture content of 75%. Site abbreviations: Bear River Migratory Bird Refuge (BRMBR), Ogden Bay Waterfowl Management Area (OBWMA), Howard Slough Wildlife Management Area (HSWMA), Harold Crane Waterfowl Management Area (HCWMA), Antelope Island and causeway (AI) tissues sampled, locations visited, and seasons or years sampled have prevented such differences or spatial and temporal trends from being elucidated or confirmed. That said, Antelope Island appears to be a local hotspot for mercury in birds, with HgT in eggs here generally being higher than those from the same bird species collected elsewhere around GSL (Waddell et al. 2009; Ackerman et al. 2015).

Abandoned and unsuccessful eggs in some GSL waterfowl had elevated Hg concentrations (Ackerman et al. 2015), suggesting Hg is negatively impacting the reproductive success of some waterfowl at GSL, while a synthesis of Hg exposure and toxicology to birds in the western United States identified GSL as a hotspot for elevated Hg in birds, with Hg blood equivalent concentrations commonly exceeding toxicity benchmarks, and birds at GSL frequently exhibited blood Hg concentrations above 3.0 μg/g ww (Ackerman et al. 2016). Previous research has reported that ibis, egrets, and herons are among the most sensitive bird species to Hg toxicity (Heinz et al. 2009) and that their median lethal concentration (LC_{50}) was less than 0.25 μg/g ww in eggs (Heinz et al. 2009). Ackerman et al. (2015) reported that 50% of great egrets, 40% of snowy egrets, 12% of black-crowned night herons, 8% of great blue herons, and 2% of white-faced ibis eggs at GSL exceeded that 0.25 μg/g ww threshold. On the other hand, ducks are believed to be less sensitive to Hg exposure, and no duck eggs at GSL exceeded a toxicity threshold of 0.5 μg/g ww suggested for HgT in duck eggs (Ackerman et al. 2015).

Studies have suggested that the intracellular binding of mercury by selenium can reduce the toxicity of Hg, a process most effective at a Se:Hg molar ratio greater than 1 (Scheuhammer et al. 2007). Over 99% of GSL waterfowl eggs analyzed for both Se and Hg have had a Se:Hg molar ratio greater than 1 (Ackerman et al. 2015), which may help alleviate Hg toxicity in birds at GSL, despite the high Hg concentrations found here, although this may contribute to Se toxicity in these birds.

14.5.3 Mercury Bioadvection from Great Salt Lake to Surrounding Terrestrial Ecosystems

Research on the bioaccumulation of Hg by organisms in GSL for many years focused on brine shrimp, brine flies, waterfowl, and wading birds; i.e., organisms that live or feed directly in GSL. This focus was based on the assumption that because MeHg is produced in the lake, only organisms that live or feed directly in the Lake will be subject to accumulating high levels of MeHg. However, brine flies are able to accumulate Hg from waters of GSL during their larval stage in the lake when they feed on periphyton, algae, and cyanobacteria, then as adults they can fly to surrounding terrestrial ecosystems, thus transferring their Hg load with them (Fig. 14.3). We estimate that the roughly 90,000 metric tons of brine flies that emerge from GSL each year (Collins 1980) would carry with them roughly 10 kg of HgT and 7 kg of MeHg out of GSL each year given typical HgT and MeHg concentrations in brine flies (Valdes et al. 2017). This annual flux would be equal to roughly 10% of the HgT in GSL, and 24% of the MeHg. Many of these brine flies are likely decomposed in the waters of GSL, so much of this would be recycled back into the water column. However, even if only 10% of these brine flies either died on land or were consumed by terrestrial predators, then brine flies would still be responsible for the bioadvection of 1 kg of HgT and 0.7 kg of MeHg annually from GSL to surrounding terrestrial ecosystems, which is substantial.

Brine flies are consumed by a number of terrestrial organisms along the shores of GSL, including lizards and numerous species of spiders, which in turn are prey for a number of birds (Fig. 14.3). Portions of the GSL shoreline with high densities of brine flies experience an explosion in the number of orb-weaving spiders (*Neoscona* spp.) every summer, with their prey composed primarily of brine flies (Black et al., unpublished data). Typical HgT levels in spiders along the shores of Gilbert Bay range from roughly 1.2 ppm dw in orb-weaving spiders (*Neoscona* spp.) up to almost 3.5 ppm dw in black widows (*Latrodectus hesperus*) with wolf spiders of the genera *Schizocosa* and *Alopecosa* having Hg levels intermediate of these (Black et al., unpublished data). Spatial and temporal variability in the concentrations of HgT in spiders at GSL reflect variations in the HgT levels in their brine fly prey.

The elevated Hg levels in the spiders at GSL pose a potential risk to the spiders themselves, as well as to organisms that consume them. Organisms that prey on spiders include reptiles, small mammals, and songbirds. Spiders represent an ideal food source for nestlings and fledgling birds, even for species of birds for which spiders do not typically represent an important part of their diet. A number of species of songbirds have been observed preying on spiders and feeding spiders to their fledglings around GSL, including loggerhead shrikes, a predatory songbird (Fig. 14.3) (Stracey et al. unpublished data). High levels of Hg can have negative effects on the behavior and reproductive success of songbirds resulting in significant decreases in reproductive output (Jackson et al. 2011). Given that aridland bird populations are experiencing widespread declines across the United States (NABCI 2009), this Hg contamination could represent another significant stressor on these populations.

Total blood Hg concentrations in shrikes at GSL range from 160 ppb to 4000 ppb ww, and average 1100 $\pm$ 1200 ppb ww. Blood HgT levels in 13% of the shrikes sampled exceeded 2000 ppb (Stracey et al. unpublished data), a threshold above which sublethal effects of Hg toxicity have been detected in other songbirds. Thus, some loggerhead shrikes at GSL may suffer from negative impacts of Hg toxicity. Birds closest to the lake had the highest blood HgT concentrations, supporting the hypothesis that much of the Hg in these birds originates from the lake and is transferred to the surrounding terrestrial ecosystem via bioadvection by brine flies, which are consumed by spiders, which are then consumed by the shrikes.

14.6 Mercury Cycling and Bioaccumulation in a Changing Great Salt Lake

The GSL terminal basin ecosystem is not static, and future changes in various coupled physical, chemical, and biological factors will undoubtedly influence the biogeochemical cycling and accumulation of Hg at GSL, although the nature, magnitude, and timing of these are uncertain. In particular, the GSL water budget and lake level are subject to climate variability and climate change, as well as changes in human diversions of freshwater inputs upstream of GSL (Wurtsbaugh et al. 2017).

Climate models for Utah and the broader Mountain West and southwestern United States, while not certain, predict that climate change in Utah will result in a longer growing season, an increase in evapotranspiration, a decrease in snowpack, earlier snowpack melting each year, a potential increase in precipitation during the fall and winter, but a decrease in summer precipitation, as well as an increase in the frequency and severity of droughts (Utah DWR 2007; Bardsley et al. 2013; Garfin et al. 2013; Scalzitti et al. 2016). Collectively these changes are expected to result in a decrease in freshwater runoff and inputs to GSL, as well as changes in the timing and magnitude of runoff events.

Stress on water resources and decreased freshwater inputs to GSL are expected to be further exacerbated by increased human water consumption. It has been estimated that the lake elevation is currently 11 ft lower, and the lake volume 48% smaller, than it would be without historic human water consumption since the arrivals of pioneers in Utah in the mid-1800s, with most of these diversions being for agriculture (Wurtsbaugh et al. 2017). A rapidly growing population in northern Utah and the Wasatch Front are expected to result in even greater water diversions in the future. For example, the Bear River Development Project that has been proposed to provide future water resources for a growing northern Utah population would dramatically decrease freshwater inputs to GSL, which would have effects similar to a permanent drought (Wurtsbaugh et al. 2017).

Decreased freshwater inputs and a shrinking GSL could affect the biogeochemical cycling of Hg in the lake in a number of ways. The shallow bathometry of GSL results in a substantial decrease in lake surface area for even a modest decrease in lake elevation (Baskin 2005). A smaller surface area would result in less atmospheric deposition of Hg, which is currently the largest source of Hg inputs to GSL (Peterson and Gustin 2008; Naftz et al. 2009). Such a decrease in lake elevation would also result in the exposure of a large surface area of the lake bed, and newly exposed lake sediments are highly prone to dust generation. For example, Owens Lake in California dried out completely in the 1920s due to water diversions. The dry Owens Lake bed, with a surface area of less than 280 km^2, became the largest single source of dust in the United States, accounting for 5% of all dust emissions in the country, and is home to the highest atmospheric particulate and PM 10 concentrations ever measured in the United States that are 25 times higher than the federal limit (Gill and Gillette 1991; Cahill et al. 1996). The GSL sediments are elevated in Hg as well as other trace elements, so there are concerns that decreasing lake levels could result in GSL becoming a source of metal-laden dust to the adjacent Wasatch Front.

Conversely, decreased freshwater runoff into GSL could also result in decreased inputs of Hg to GSL via riverine discharge if Hg concentrations in those waters remained constant. However, predictions from climate models suggest that while water runoff to GSL will decrease, it is likely to occur earlier in the spring and involve larger runoff events. An increase in the importance of high discharge events would likely increase riverine inputs of Hg to GSL given that Hg is very particle reactive and high discharge events transport much greater sediment and Hg loads, resulting in flood events being disproportionally important in Hg transport in riverine systems (David et al. 2009).

Yet another way in which decreasing lake level could influence Hg cycling and bioaccumulation in GSL is via its influence on the DBL at the bottom of the south arm. The waters of the DBL contain the highest concentrations of HgT and MeHg at GSL, and the role of the DBL in Hg methylation and as the source of MeHg found in the GSL biota is a topic of ongoing research (Johnson et al. 2015; Valdes et al. 2017). The DBL persists 6–7 m below the surface of the lake, with this depth controlled by top down wind-driven mixing (Naftz et al. 2008; Johnson et al. 2015; Valdes et al. 2017). A decrease in lake surface elevation would result in the top-down erosion of the DBL, and the elimination of the DBL in any areas of the south arm where the total water column depth drops below 6–7 m. Thus, a decrease in lake volume will result in a decrease in the extent of the DBL, or even its complete elimination if lake levels decrease sufficiently. Such a disappearance of the DBL is exactly what occurred in 2014–2015.

The DBL is created and maintained by denser, higher salinity water from Gunnison Bay (the north arm) flowing into Gilbert Bay (the south arm) via culverts in the railway causeway that separates the north and south arms (Fig. 14.2). In late 2013, the causeway culverts were closed. This closure coincided with a multiyear drought that led to a substantial decrease in the lake elevation, resulting in top-down erosion of the upper surface of the DBL. Together these events resulted in the disappearance of the DBL, the destratification of the south arm water column, and the oxygenation of the deep waters of the south arm. In the deep waters of the south arm, HgT concentration decreased by 81%, MeHg concentrations in deep waters decreased by 86%, and MeHg in underlying sediments decreased by 77% in response to the disappearance of the DBL (Valdes et al. 2017). A new bridge opened on the railroad causeway in late 2016 reestablished water flow from the north to south arm, leading to the reformation of the DBL. However, future decreases in freshwater inputs and lake elevation could result in the near permeant disappearance of the DBL, and with it the extremely high concentrations of HgT and MeHg in deep waters and sediments of the south arm. However, it remains unclear if this would lead to a decrease in the bioaccumulation of MeHg by GSL biota, but research on this topic suggests that it will not result in decreased Hg concentrations in at least some biota, including brine flies or some ducks (Valdes et al. 2017).

The anticipated increase in the population of the Wasatch Front in the coming years is likely to be accompanied by increases in anthropogenic pollution from growing urban and industrial sources of Hg, which could increase Hg loading to GSL. Some of this would occur via wastewater inputs to GSL given that the majority of wastewater effluent in northern Utah is discharged directly or indirectly into GSL, including via Farmington Bay, the Jordan River, and the Bear River. And because GSL is a terminal lake, these Hg inputs would not be lost via water outflows.

Changes in the water chemistry of GSL and other physical and biological factors could alter Hg methylation by sulfate-reducing bacteria or cause shifts in the microbial community. Alternatively, such changes could result in changes in phytoplankton assemblies, the diet of various organisms, food web structure, or other factors that influence the bioaccumulation and biomagnification of Hg in GSL ecosystems. Ongoing research aims to better understand the effects of such future changes at GSL.

References

Ackerman JT, Herzog MP, Hartman CA, Isanhart JP, Hering G, Vaughn S, Cavitt JF, Eagles-Smith CA, Browers H, Cline C, Vest J (2015) Mercury and selenium contamination in waterbird eggs and risk to avian reproduction at Great Salt Lake, Utah (No. 2015-1020). US Geological Survey

Ackerman JT, Eagles-Smith CA, Herzog MP, Hartman CA, Peterson SH, Evers DC, Jackson AK, Elliott JE, Vander Pol SS, Bryan CE (2016) Avian mercury exposure and toxicological risk across western North America: a synthesis. Sci Total Environ 568:749–769

Aldrich TW, Paul DS (2002) Avian ecology of Great Salt Lake. In: Gwynn JW (ed) Great Salt Lake: an overview of change. Utah Department of Natural Resources Special Publication, Salt Lake City, UT

Baeyens W, Leermakers M, Papina T, Saprykin A, Brion N, Noyen J, De Gieter M, Elskens M, Goeyens L (2003) Bioconcentration and biomagnification of mercury and methylmercury in North Sea and Scheldt Estuary fish. Arch Environ Contam Toxicol 45(4):498–508

Bardsley T, Wood A, Hobbins M, Kirkham T, Briefer L, Niermeyer J, Burian S (2013) Planning for an uncertain future: climate change sensitivity assessment toward adaptation planning for public water supply. Earth Interact 17(23):1–26

Bargagli R, Monaci F, Sanchez-Hernandez JC, Cateni D (1998) Biomagnification of mercury in an Antarctic marine coastal food web. Mar Ecol Prog Ser 169:65–76

Baskin RL (2005) Calculation of area and volume for the south part of Great Salt Lake, Utah. USGS Open-File Report 2005 1327. US Geological Survey, Salt Lake City, UT

Belovsky GE, Stephens D, Perschon C, Birdsey P, Paul D, Naftz D, Baskin R, Larson C, Mellison C, Luft J, Mosley R (2011) The Great Salt Lake ecosystem (Utah, USA): long term data and a structural equation approach. Ecosphere 2(3):1–40

Benoit JM, Gilmour CC, Heyes A, Mason RP, Miller C (2003) Geochemical and biological controls over methylmercury production and degradation in aquatic ecosystems. In: Biogeochemistry of environmentally important trace elements. American Chemical Society, Washington, DC

Black FJ, Poulin BA, Flegal AR (2012) Factors controlling the abiotic photo-degradation of monomethylmercury in surface waters. Geochim Cosmochim Acta 84:492–507

Boyd ES, Yu RQ, Barkay T, Hamilton TL, Baxter BK, Naftz DL, Marvin-DiPasquale M (2017) Effect of salinity on mercury methylating benthic microbes and their activities in Great Salt Lake, Utah. Sci Total Environ 581:495–506

Brookens TJ, O'Hara TM, Taylor RJ, Bratton GR, Harvey JT (2008) Total mercury body burden in Pacific harbor seal, *Phoca vitulina richardii*, pups from Central California. Mar Pollut Bull 56 (1):27–41

Cahill TA, Fill TE, Reid JS, Gearhart EA, Gillette DA (1996) Saltasting particles, playa crusts and dust aerosols at Owens (dry) Lake, California. Earth Surf Process Landf 21(7):621–639

Carozzi AV (1962) Observations on algal biostromes in the Great Salt Lake, Utah. J Geol 70:246–252

Cavitt JF, Hall J, Swenson L, Wilson N (2012) Mercury and selenium contamination in waterbird eggs and risk to avian reproduction at Great Salt Lake, Utah. Report to Utah Department of Environmental Quality, Division of Water Quality, Salt Lake City, UT

Chen CY, Pickhardt PC, Xu MQ, Folt CL (2008) Mercury and arsenic bioaccumulation and eutrophication in Baiyangdian Lake, China. Water Air Soil Pollut 190:115–127

Clarkson TW, Magos L (2006) The toxicology of mercury and its chemical compounds. Crit Rev Toxicol 36(8):609–662

Collins N (1980) Population ecology of *Ephydra cinerea* Jones (Diptera: Ephydridae), the only benthic metazoan of the Great Salt Lake, U.S.A. Hydrobiologia 68(2):99–112

Conover MR, Vest JL (2009) Selenium and mercury concentrations in California gulls breeding on the Great Salt Lake, Utah, USA. Environ Toxicol Chem 28(2):324–329

Cossa D, Averty B, Pirrone N (2009) The origin of methylmercury in open Mediterranean waters. Limnol Oceanogr 54(3):837–844

Covelli S, Faganeli J, De Vittor C, Predonzani S, Acquavita A, Horvat M (2008) Benthic fluxes of mercury species in a lagoon environment (Grado Lagoon, Northern Adriatic Sea, Italy). Appl Geochem 23(3):529–546

David N, McKee LJ, Black FJ, Flegal AR, Conaway CH, Schoellhammer DH, Ganju NK (2009) Mercury concentrations and loads in a large river system tributary to San Francisco Bay, California, USA. Environ Toxicol Chem 28:2091–2100

Diaz X, Johnson WP, Fernandez D, Naftz DL (2009) Size and elemental distributions of nano-to micro-particulates in the geochemically-stratified Great Salt Lake. Appl Geochem 24 (9):1653–1665

Dupraz C, Visscher PT (2005) Microbial lithification in marine stromatolites and hypersaline mats. Trends Microbiol 13:429–438

Eckley CS, Hintelmann H (2006) Determination of mercury methylation potentials in the water column of lakes across Canada. Sci Total Environ 368(1):111–125

Evers DC, Lane OP, Savoy L, Goodale W (2004) Assessing the impacts of methylmercury on piscivorous wildlife using a wildlife criterion value based on the Common Loon, 1998–2003. Report BRI, 5

Fitzgerald WF, Engstrom DR, Mason R, Nater EA (1998) The case for atmospheric mercury contamination in remote areas. Environ Sci Technol 32(1):1–7

Francesconi KA, Lenanton RC (1992) Mercury contamination in a semi-enclosed marine embayment: organic and inorganic mercury content of biota, and factors influencing mercury levels in fish. Mar Environ Res 33(3):189–212

Gao Y, Yan CH, Yu XD, Wu SH (2006) Effects of perinatal exposure to methylmercury on the structure of hippocampus and cerebellum in young rats. J Hyg Res 35(4):402–405

Garfin G, Jardine A, Merideth R, Black M, LeRoy S (2013) Assessment of climate change in the southwest United States: a report prepared for the national climate assessment. Southwest climate alliance. Island Press, Washington, DC

Gill FB (1995) Migration. In: Ornithology, 2nd edn. W.H. Freeman, New York, pp 287–309

Gill TE, Gillette DA (1991) Owens Lake: a natural laboratory for aridification, playa desiccation and desert dust. Geol Soc Am 23(5):462

Great Salt Lake Ecosystem Program: Historic Harvest Data (2019). https://wildlife.utah.gov/gsl/harvest/historic_harvest_data.php. Accessed 9 May 2019

Halley RB (1976) Textural variation within Great Salt Lake algal mounds. In: Walter MR (ed) Developments in sedimentology. Elsevier, Amsterdam, pp 435–446

Heinz GH, Hoffman DJ, Klimstra JD, Stebbins KR, Kondgrad SL, Erwin CA (2009) Species differences in the sensitivity of avian embryos to methylmercury. Arch Environ Contam Toxicol 56(1):129–138

Hsu-Kim H, Kucharzyk KH, Zhang T, Deshusses MA (2013) Mechanisms regulating mercury bioavailability for methylating microorganisms in the aquatic environment: a critical review. Environ Sci Technol 47(6):2441–2456. https://doi.org/10.1021/es304370g

Jackson AK, Evers DC, Folsom SB, Condon AM, Diener J, Goodrick LF, McGann AJ, Schmerfeld J, Cristol DA (2011) Mercury exposure in terrestrial birds far downstream of an historical point source. Environ Pollut 159(12):3302–3308

Johnson WP, Swanson N, Black B, Rudd A, Carling G, Fernandez DP, Luft J, Van Leeuwen J, Marvin-DiPasquale M (2015) Total- and methyl-mercury concentrations and methylation rates across the freshwater to hypersaline continuum of the Great Salt Lake, Utah, USA. Sci Total Environ 511:489–500

Jones G, Miller G (2005) Mercury and modern gold mining in Nevada: final report to US Environmental Protection Agency Region IX. University of Nevada, Reno, NV

Jones EF, Wurtsbaugh WA (2014) The great salt Lake's monimolimnion and its importance for mercury bioaccumulation in brine shrimp (*Artemia franciscana*). Limnol Oceanogr 59 (1):141–155

Kerin EJ, Gilmour CC, Roden E, Suzuki MT, Coates JD, Mason RP (2006) Mercury methylation by dissimilatory iron-reducing bacteria. Appl Environ Microbiol 72(12):7919–7921

Lamborg CH, Yiğiterhan O, Fitzgerald WF, Balcom PH, Hammerschmidt CH, Murray J (2008) Vertical distribution of mercury species at two sites in the Western Black Sea. Mar Chem 111:77–89

Larson CA, Belovsky GE (2013) Salinity and nutrients influence species richness and evenness of phytoplankton communities in microcosm experiments from Great Salt Lake, Utah, USA. J Plankton Res 35(5):1154–1166

Lindsay MR, Anderson C, Fox N, Scofield G, Allen J, Anderson E, Bueter L, Poudel S, Sutherland K, Munson-McGee JH, Van Nostrand JD (2017) Microbialite response to an anthropogenic salinity gradient in Great Salt Lake, Utah. Geobiology 15(1):131–145

Loving BL, Miller CW, Waddell KM (2000) Water and salt balance of Great Salt Lake, Utah, and simulation of water and salt movement through the causeway, 1987-98 (No. 2000-4221). US Geological Survey

Marvin-DiPasquale M, Agee J, McGowan C, Oremland RS, Thomas M, Krabbenhoft D, Gilmour CC (2000) Methyl-mercury degradation pathways: a comparison among three mercury-impacted ecosystems. Environ Sci Technol 34(23):4908–4916

Mason RP, Sheu GR (2002) Role of the ocean in the global mercury cycle. Glob Biogeochem Cycles 16(4):40–41

Mergler D, Anderson HA, Chan LHM, Mahaffey KR, Murray M, Sakamoto M, Stern AH (2007) Methylmercury exposure and health effects in humans: a worldwide concern. AMBIO J Hum Environ 36(1):3–12

Monperrus M, Tessier E, Amouroux D, Leynaert A, Huonnic P, Donard OFX (2007) Mercury methylation, demethylation and reduction rates in coastal and marine surface waters of the Mediterranean Sea. Mar Chem 107(1):49–63

Naftz D, Angeroth C, Kenney T, Waddell B, Darnall N, Silva S, Perschon C, Whitehead J (2008) Anthropogenic influences on the input and biogeochemical cycling of nutrients and mercury in Great Salt Lake, Utah, USA. Appl Geochem 23(6):1731–1744

Naftz D, Fuller C, Cederberg J, Krabbenhoft D, Whitehead J, Garberg J, Beisner K (2009) Mercury inputs to Great Salt Lake, Utah: reconnaissance-phase results. Nat Resour Environ Issues 15 (1):5

Neill J, Leite B, Gonzales J, Sanchez K, Luft J (2016) 2015 Great Salt Lake Eared Grebe aerial photo survey. Annual Report. Great Salt Lake Ecosystem Program, Utah Division of Wildlife Resources

North American Bird Conservation Initiative (NABCI) (2009) NABCI US Committee. The state of the birds, United States of America

Pacyna EG, Pacyna JM, Steenhuisen F, Wilson S (2006) Global anthropogenic mercury emission inventory for 2000. Atmos Environ 40(22):4048–4063

Peterson C, Gustin M (2008) Mercury in the air, water and biota at the Great Salt Lake (Utah, USA). Sci Total Environ 405(1-3):255–268

Pickhardt PC, Fisher NS (2007) Accumulation of inorganic and methylmercury by freshwater phytoplankton in two contrasting water bodies. Environ Sci Technol 41(1):125–131

Pickhardt PC, Stepanova M, Fisher NS (2006) Contrasting uptake routes and tissue distributions of inorganic and methylmercury in mosquitofish (*Gambusia affinis*) and redear sunfish (*Lepomis microlophus*). Environ Toxicol Chem 25(8):2132–2142

Qureshi A, MacLeod M, Scheringer M, Hungerbühler K (2009) Mercury cycling and species mass balances in four North American lakes. Environ Pollut 157(2):452–462

Saxton HJ, Goodman JR, Collins JN, Black FJ (2013) Maternal transfer of inorganic mercury and methylmercury in aquatic and terrestrial arthropods. Environ Toxicol Chem 32(11):2630–2636

Scalzitti J, Strong C, Kochanski A (2016) Climate change impact on the roles of temperature and precipitation in western U.S. snowpack variability. Geophys Res Lett 43:5361–5369

Scheuhammer AM, Meyer MW, Sandheinrich MB, Murray MW (2007) Effects of environmental methylmercury on the health of wild birds, mammals, and fish. AMBIO J Hum Environ 36 (1):12–20

Selin NE, Jacob DJ, Yantosca RM, Strode S, Jaeglé L, Sunderland EM (2008) Global 3-D land-ocean-atmosphere model for mercury: present-day versus preindustrial cycles and anthropogenic enrichment factors for deposition. Global Biogeochem Cycles 22(2):1

Skyllberg U, Qian J, Frech W, Xia K, Bleam WF (2003) Distribution of mercury, methyl mercury and organic sulphur species in soil, soil solution and stream of a boreal forest catchment. Biogeochemistry 64:53–76

Sonne C, Dietz R, Leifsson PS, Asmund G, Born EW, Kirkegaard M (2007) Are liver and renal lesions in East Greenland polar bears (*Ursus maritimus*) associated with high mercury levels? Environ Health 6:11

Sorensen ED, Hoven HM, Neill J (2020) Great Salt Lake shorebirds, their habitats and food base. In: Baxter BK, Butler JK (eds) Great Salt Lake biology: a terminal lake in a time of change. Springer, Cham

Strode SA, Jaeglé L, Selin NE, Jacob DJ, Park RJ, Yantosca RM, Mason RP, Slemr F (2007) Air-sea exchange in the global mercury cycle. Glob Biogeochem Cycles 21(1). https://doi.org/10.1029/2006GB002766

Sunderland EM (2007) Mercury exposure from domestic and imported estuarine and marine fish in the U.S. seafood market. Environ Health Perspect 115:235–242. https://doi.org/10.1289/ehp.9377

Sunderland EM, Krabbenhoft DP, Moreau JW, Strode SA, Landing WM (2009) Mercury sources, distribution, and bioavailability in the North Pacific Ocean: insights from data and models. Glob Biogeochem Cycles 23(2). https://doi.org/10.1029/2008GB003425

US Environmental Protection Agency (2007) National listing of fish advisories technical fact sheet, EPA-823-F-07-003. U.S. Environmental Protection Agency, Washington, DC

US Environmental Protection Agency (2010) Alternative futures analysis of Farmington Bay wetlands in the Great Salt Lake ecosystem, EPA-600-R-10-032. US Environmental Protection Agency, Cincinnati, OH

Utah Department of Environmental Quality (2017) Utah toxic release inventory, reporting year 2017 Data Summary Report, December 2018. https://deq.utah.gov/environmental-response-and-remediation/cercla-comprehensive-environmental-response-compensation-and-liability-act/toxic-release-inventory-tri-program

Utah Department of Environmental Quality, Water Quality (2014) 2012-2014 Final Integrated Report: Chapter 7: Great Salt Lake

Utah Department of Health (2005) An evaluation of mercury concentrations in ducks from the Great Salt Lake Utah for 2004 and 2005. Environmental Epidemiology Program, Utah Department of Health, Salt Lake City, UT

Utah Department of Health (2006) An evaluation of mercury concentrations in ducks from the Great Salt Lake Utah for 2005 and 2006. Environmental Epidemiology Program, Utah Department of Health, Salt Lake City, UT

Utah Department of Natural Resources (2011) Great salt Lake comprehensive management plan revision. May 2011

Utah Division of Water Quality (2012). https://documents.deq.utah.gov/water-quality/standards-technical-services/great-salt-lake-advisory-council/Activities/DWQ-2012-006864.pdf

Utah Divisions of Water Resources (2007) Drought in Utah: learning from the past - preparing for the future. Utah Division of Water Resources, Salt Lake City, UT

Valdes C, Black FJ, Stringham B, Collins JN, Goodman JR, Saxton HJ, Mansfield CR, Schmidt JN, Yang S, Johnson WP (2017) Total mercury and methylmercury response in water, sediment, and biota to destratification of the Great Salt Lake, Utah, United States. Environ Sci Technol 51 (9):4887–4896

Vest JL, Conover MR, Perschon C, Luft J, Hall JO (2009) Trace element concentrations in wintering waterfowl from the Great Salt Lake, Utah. Arch Environ Contam Toxicol 56 (2):302–316

Waddell B, Cline C, Darnall N, Boeke E, Sohn R (2009) Assessment of contaminants in the wetlands and open waters of the Great Salt Lake, Utah. US Fish and Wildlife Service Ecological Services, Salt Lake City, UT

Wang WX, Wong RS (2003) Bioaccumulation kinetics and exposure pathways of inorganic mercury and methylmercury in a marine fish, the sweetlips *Plectorhinchus gibbosus*. Mar Ecol Prog Ser 261:257–268

Watras CJ, Back RC, Halvorsen S, Hudson RJ, Morrison KA, Wente SP (1998) Bioaccumulation of mercury in pelagic freshwater food webs. Sci Total Environ 219(2-3):183–208

Wurtsbaugh WA (2012) Paleolimnological analysis of the history of metals contamination in the Great Salt Lake, Utah. Utah State University Watershed Sciences Faculty Publications. Paper 556. http://digitalcommons.usu.edu/wats_facpub/556

Wurtsbaugh WA, Gliwicz ZM (2001) Limnological control of brine shrimp population dynamics and cyst production in the Great Salt Lake, Utah. In: Saline lakes. Springer, Dordrecht, pp 119–132

Wurtsbaugh WA, Gardberg J, Izdepski C (2011) Biostrome communities and mercury and selenium bioaccumulation in the Great Salt Lake (Utah, USA). Sci Total Environ 409 (20):4425–4434

Wurtsbaugh WA, Miller C, Null SE, DeRose RJ, Wilcock P, Hahnenberger M, Howe F, Moore J (2017) Decline of the world's saline lakes. Nat Geosci 10(11):816

Chapter 15
The Rozel Point Tar Seeps and Their Impact on the Local Biology at Great Salt Lake, Utah

Kara L. Kornhauser, H. Gregory McDonald, Rebecca S. Dennis, and Jaimi K. Butler

Abstract Great Salt Lake (GSL), Utah, is the largest lake in the Great Basin and one of the primary migratory stops for many species of birds in North America. Located at Rozel Point, on the north arm of the lake, are natural tar seeps that have formed on the former lake bed resulting from the migration of oil to the surface along fault lines. Once the petroleum reaches the surface of the ground, usually at low pressure, it then spreads out from the seep. The resulting tar seeps are numerous and vary in size. During warm weather, the surface of the Rozel Point tar is sufficiently sticky, and it can lead to entrapment of animals.

There are many factors that may influence the entrapment, including lake levels, the type of animal that becomes stuck and the presence of potential prey, the appearance of the tar seeps as they reflect light, the temperature at which the tar seeps become tacky, which is dependent on the season and time of day. Using direct observations, motion-activated cameras, temperature monitoring devices, and paleontological methods for identification, we have located and documented animal species present in the Rozel seeps. Our data suggest that the most common species stuck is the American White Pelican, but other birds, insects, and other animals may also be trapped. We have collected photographic evidence of scavenging species, such as coyotes and ravens, which scavenge on animals entrapped in the tar. In this chapter, we present a review of our data regarding animal entrapment at Rozel Point. Using this site at the GSL as a modern analogue, we compare our analysis to other significant petroleum-enriched sites.

K. L. Kornhauser · J. K. Butler (✉)
Great Salt Lake Institute, Westminster College, Salt Lake City, UT, USA
e-mail: jbutler@westminstercollege.edu

H. G. McDonald
Bureau of Land Management, Colorado State Office, Lakewood, CO, USA

R. S. Dennis
Department of Biology, Westminster College, Salt Lake City, UT, USA

B. K. Baxter, J. K. Butler (eds.), *Great Salt Lake Biology*,
https://doi.org/10.1007/978-3-030-40352-2_15

Keywords Tar · Oil · Petroleum · Lake levels · Pelicans · Rozel Point · North arm · Seep · Great Salt Lake

15.1 Introduction

The Rozel Tar Seeps are located on the north arm of Great Salt Lake (GSL) in Box Elder County, Utah, near artist, Robert Smithson's iconic land art installation, *Spiral Jetty* (Fig. 15.1). These tar seeps are mostly naturally occurring (Bortz 2002; Damsté et al. 1987, 1989), although there are a few that are the by-product of oil exploration drilling that took place from the 1960s to 1970s (Patton and Lent 1980). Our observations indicate that these tar seeps have entrapped many types of birds including American White Pelicans, owls and other raptors, and various species of small birds and insects. Some of these entrapped animals are subsequently scavenged by coyotes and ravens and eventually decay on the surface of the seep. Surviving parts of the bird's skeleton may eventually become covered with tar and potentially may be preserved as fossils. A study of the processes of entrapment, scavenging, and preservation at these GSL tar seeps can provide a modern analogue of tar deposits with fossils around the world, such as the La Brea Tar Pits in California, but also document the contribution of the seeps to the mortality of the local fauna.

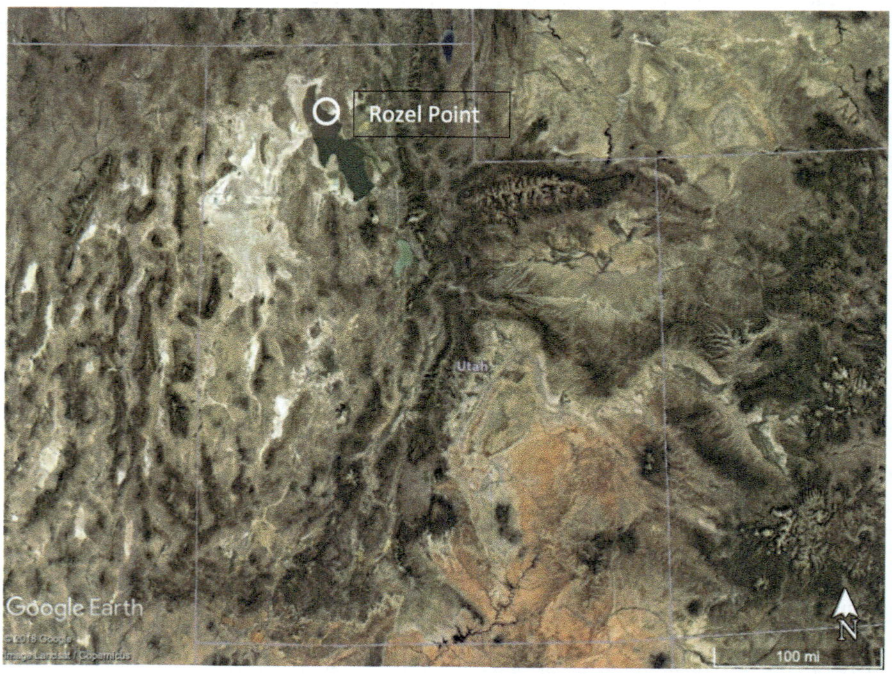

Fig. 15.1 Map showing the location of the Rozel Point Tar Seeps at Great Salt Lake (Google Earth 2019)

15.2 History of the Rozel Tar Seeps

15.2.1 Origin and Characteristics

The Rozel Point Tar Seeps are both naturally occurring and human-created areas where viscous oil oozes from an underground oil reservoir. The oil at Rozel Point formed during the Miocene to Pliocene (24–1.8 million years old) in organic-rich lake sediments deposited in a predecessor to GSL (Oviatt 2015). A volcanic tuff near the bottom of Amoco well No. 6 was determined to be 29.9 ± 1.3 m.y. based on a zircon fission-track date (Bortz 1987, 2002). At Rozel Point, this oil reservoir is in a 2- to 3-foot-thick porous basalt located 80 feet below the present lake bed. In addition to basalt flows, there are associated freshwater limestone beds, both of which may be tilted as a result of faults that have formed multiple grabens typical of the Basin and Range Geological Province. This includes the northwestward-trending Rozel graben, which is about 20 miles long and up to 5 miles wide, located on the northeastern margin of GSL (Cook et al. 1966). The faults extend through the oil reservoir, and the shattering of the rocks along the faults provides the pathway for the upward migration of the oil (Eardley 1963). Aerial photographs of the area show that many of the tar seeps form linear patterns reflecting the faults (Fig. 15.2).

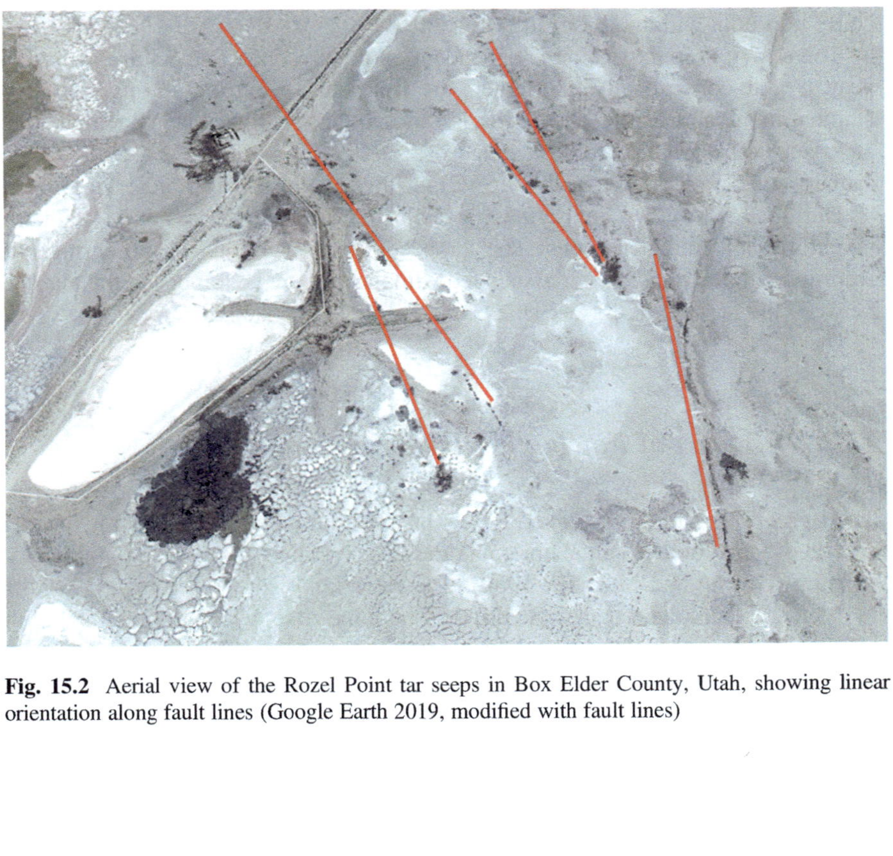

Fig. 15.2 Aerial view of the Rozel Point tar seeps in Box Elder County, Utah, showing linear orientation along fault lines (Google Earth 2019, modified with fault lines)

Once the petroleum reaches the surface, it flows out onto the former lake bed and creates a sticky layer of tar. The spreading oil forms relatively thin layers, often less than 10 cm thick. In some cases, there is sufficient pressure from methane gas to create a raised volcano-like mound that is built up of tar, commonly with a pit of liquid tar at the center. Often bubbles of methane erupt through this tar. The earliest mention of the tar seeps near the Rozel Hills was by Slentz and Eardley (1956) who described their occurrence as:

> The oil issues from craterlets whose cones are about 12 to 18 inches high and 20 to 40 feet in diameter. . .Each mound is formed of petroleum saturated oolitic sand and a smooth and sticky blue-gray or olive-green clay through which myriads of veinlets of petroleum ramify.

Significantly, none of the seeps form thick deposits that would result in the formation of tar “pits” such as those at Rancho La Brea in California. As such, they are closer to the description of the “fly paper” model for entrapment proposed by Akersten et al. (1983) rather than the “quicksand” description of tar pits at Rancho La Brea.

The continuous flow of oil onto the flat surface of the former lake bottom facilitates the expansion of the tar seeps. Usually, the younger oil will flow under the older oxidized tar, resulting in a “reversed stratigraphy.” Expansion of the seeps is by subsequent accretion along the seep margins once the newer oil is no longer under the older layer so instead of increasing in vertical thickness, the area of the seep increases horizontally since there are no barriers to restrict the flow of the tar or cause it to pool within a basin-like structure.

The rate of seep expansion is dependent on the seasonal temperature, with most growth and consequently entrapment of the fauna occurring during the warm summer months when the tar is least viscous and thus sticky. Seasonal temperatures at Rozel Point range from an average of −2.3 °C (range 2.4 °C to −7.1 °C) in January to 25.5 °C (range 33.4–17.6 °C) in July (National Weather Service 2019).

During the winter months, it is possible to walk on the tar, thus facilitating inventories of animal remains. The tar becomes less viscous in the spring with warming temperatures. Water associated with the tar may be trapped at the interface between the tar and the former lake beds.

15.2.2 Impact of Fluctuating Lake Elevation

The tar seeps at Rozel Point are on former lake bottom sediments consisting of clays, silts and sand, oolites, algal microbialites, fecal pellets, and minor amounts of gypsum (Gwynn and Murphey 1980). The elevation of the tar seeps at 1279.6 m (4198 feet) places them within the historic shoreline zone, when the shoreline of the GSL ranged from a historic high of 1284 m (4212 feet) in 1873 to a historic low of 1277 m (4191 feet) in 1963 (Curry 1980).

Before GSL existed, the basin was occupied by other lakes at various water levels (Atwood et al. 2016). Because of fluctuations in lake levels, the subaerial exposure

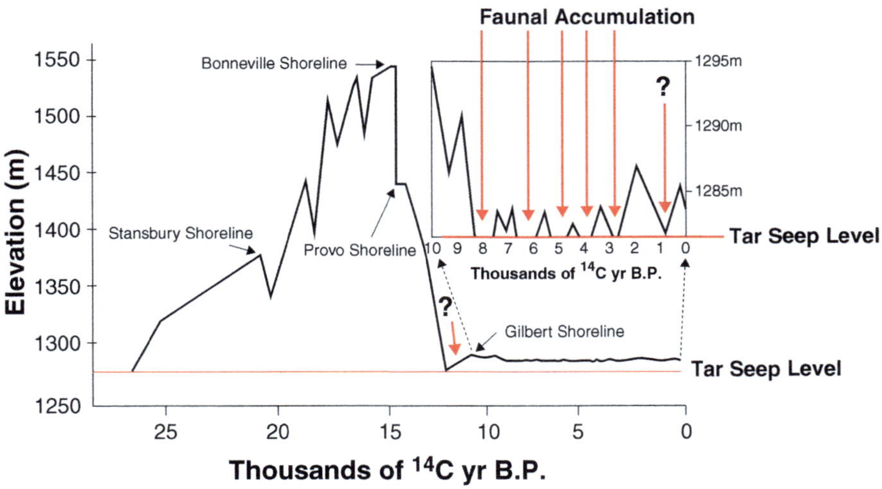

Fig. 15.3 History of lake level elevation changes during the transition from Pleistocene Lake Bonneville to modern Great Salt Lake, showing when lower lake levels in the Holocene would have permitted the formation of tar seeps at Rozel Point. Modified from Oviatt (2015)

has varied through time and allows the formation of tar seeps with the potential to entrap birds and other fauna. Changes in lake level result from long-term fluctuations in response to climate change and general warming from the Pleistocene to Holocene transition starting with the decline of Lake Bonneville and the drop in the lake level from the Bonneville to Provo Shorelines (Oviatt et al. 1999; Oviatt 2015; Atwood et al. 2016). During the Holocene at least six intervals are recorded when the lake level dropped below 1279.6 m (4198 feet) in elevation resulting in the subaerial exposure of the former lake bottom that permitted the creation of the seeps (Fig. 15.3).

Given the proximity of the seeps to the lake margin and the small difference in the surface elevation of the lake needed to cover them, the current surface seeps are also influenced by seasonal short-term fluctuations in lake level. Typically, there is a higher lake level in summer due to runoff from the nearby mountains and watershed. This rise, may cover the seeps for short intervals then re-expose them in the winter. In recent years, this fluctuation still occurs, yet there has been a consistent decrease in lake levels reflecting the impact of human activities such as diversion of water sources such as the Bear River for irrigation (Gwynn and Murphey 1980; Null and Wurtsbaugh 2020).

During periods when the seeps are submerged, tar seeping into the water forms tubular masses and threads that break into tar balls upon reaching the surface of the water and may be washed ashore. When the lake level is down and the seeps are exposed subaerially, the oil reaching the surface is able to spread out on the surface and form the animal entrapping seeps (Fig. 15.4).

Stansbury (1855), while camping in the vicinity of Rozel Point during his geologic survey of GSL (1849–1851) reported the presence of bitumen:

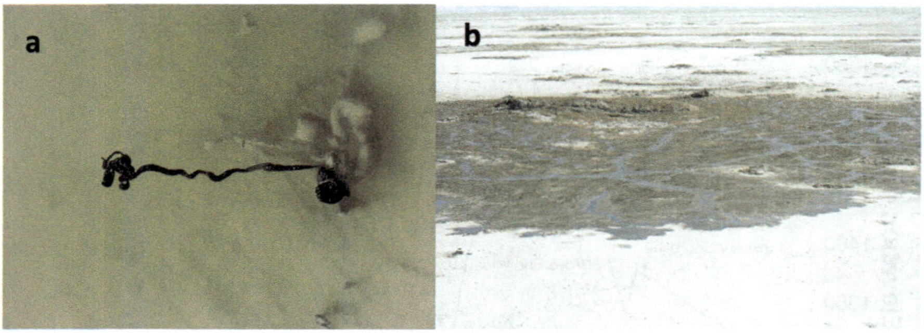

Fig. 15.4 (**a**) Petroleum seeping into standing water at the GSL forming a long thread, which will eventually break up to form tar balls. Photo courtesy of Steve Mulqueen, by permission. (**b**) Tar seeps that form subaerially allowing the oil to spread out on the lake bottom sediments

> Small quantities of bitumen were found on the shore, in masses of a substance which lined the beach to a depth of six inches, and resembled in appearance the brown dried seaweed of the ocean.

And:

> A piece of bitumen was found buried in the sand, which had adhered to it when softened by the sun, and completely frosted it over, so that it very much resembled one of the small chocolate lozenges of the shops, covered with miniature sugarplums.

He was not however able to identify the source of the tar balls he observed because at the time of his survey, the elevation of the lake was at about 1280 m (4201 feet) so the seeps at Rozel Point were submerged.

15.2.3 Oil Exploration

Despite the early report of bitumen in the vicinity of Rozel Point by Stansbury, oil exploration in the area and interest in its extraction lagged for nearly 100 years. The earliest recorded evidence of oil exploration at Rozel Point is in 1904 making it one of the oldest (if not the oldest) fields to produce oil in Utah (Boutwell 1904; MaGuire 1904). Drilling continued intermittently up to 1996 (Bortz 2002; Patton and Lent 1980).

During the time of exploration of the oil at Rozel Point, 30–50 wells were drilled and eventually abandoned and capped. Amoco drilled 13 exploratory wells in GSL, from June 1978 to December 1980 (Gwynn 2006). Because of the high lake level at that time, a floating barge (Fig. 15.5) was used that was disassembled and trucked from one side of the lake to the other because there was no access over or through the Southern Pacific Railroad causeway (Bortz 1987). The Amoco operation was discontinued in December 1980 because of the high water cut in the produced oil, and the high cost of operating an "offshore" field (Gwynn 2006).

Fig. 15.5 Historical photos of oil drilling at Rozel Point. (**a**) Asphalt Plant-Leonora Mining & Milling Co., 1929. Utah Division of State History Photo No. 3954. View looking southwest. (**b**) Looking northwest of the causeway and drilling platform. Photo by Charles Kelly, date unknown. Current tar seeps are found between the two structures and to the south of the end of the causeway. Utah Division of State History photo No. 10669. (**c**) Barge transporting Amoco's Parker drilling rig being into place for oil drilling on GSL, June 1978. View looking northeast toward the Rozel Hills. The Rozel Point tar seeps would be in the upper right of the picture, just offshore from the white shoreline. Image credit: Utah Division of State History photo No. 18556, public domain

The cleanup and capping process at Rozel Point began in 1996 (Hunt and Chidsey 2002) and has continued with capping of another well in January 2019. Much of the cleanup of the area was done by a cooperative effort of the Division of Oil Gas and Mining of the Utah Geological Survey and the Environmental Protection Agency from 1996 to 2005 (Gwynn 2006), although in recent years it was found that at least one of the caps from this cleanup potentially leaked and needed to be replaced (authors' observation confirmed by personnel at the Utah Division of Oil, Gas and Mining). Many of the structures constructed by multiple oil companies over the years today are often only represented by pilings that are still visible in the area.

The impact of the oil exploration is also still evident (Fig. 15.5). There are also metal remains of tools and structures. While some are sitting on the current shoreline, others are embedded into the sediment and have become a part of the lake bed. Along with random metal scraps, there are also metal pipes that protrude out of the ground. These pipes are likely the remains of old oil well heads that were drilled into the bed of the lake when the lake levels were higher and the water covered this area.

15.2.4 Exposed and Active Tar Seeps

Based on archived images in Google Earth, there has been a visible drop in lake elevation since June of 2009. While the lake level fluctuates naturally, in recent years the lake level has consistently declined (USGS 2019). Recent studies have raised an alarm about the declining levels for GSL and other saline lakes in the world (Wurtsbaugh et al. 2017). In fact, except for a short-term rise in 2011 (Fig. 15.6), much of the lake bed at Rozel Point has been exposed in the last decade, allowing the formation of tar seeps in the absence of lake water. Because GSL is so shallow (its current average depth is 4.3 m (14 feet) with a maximum depth of 10 m (33 feet), slight decreases in lake levels can reduce the surface area of the lake greatly, especially at the shallower shorelines such as at Rozel Point. The recent trend of low lake level has resulted in a longer exposure of the tar seeps at Rozel Point than has occurred historically. The extended subaerial exposure has resulted in a greater

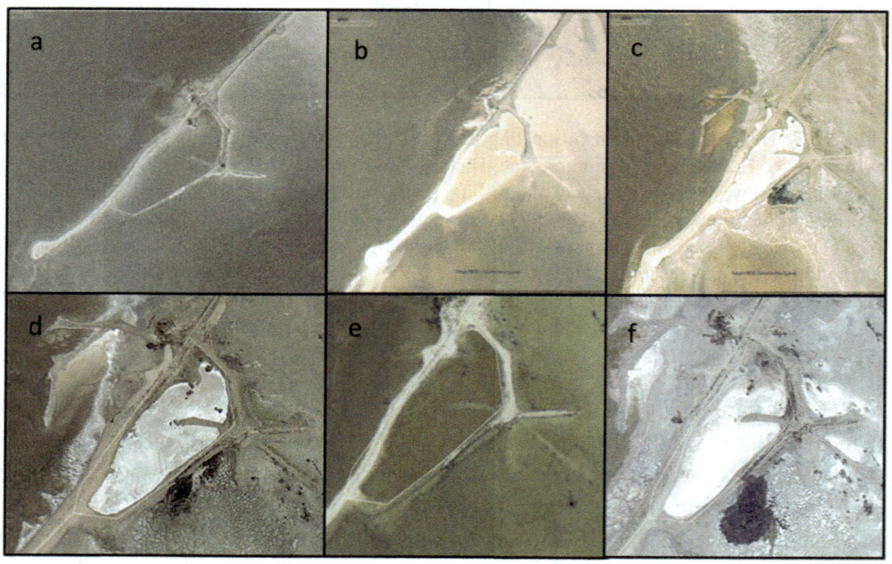

Fig. 15.6 Aerial views of the Rozel Point tar seeps showing changes in lake level over time and the growth of tar seeps following a decline in lake level. (**a**) June 1993, (**b**) July 1996, (**c**) June 2009, (**d**) May 2010, (**e**) Sept 2011, (**f**) August 2014 (Images from Google Earth Archives, public domain)

expansion of individual seeps so they cover a larger area and have allowed individual seeps to coalesce into larger seeps. Their large size increases their potential for entrapment of birds and other animals.

Despite ongoing interest in the petroleum resources at Rozel Point and observed tar seeps during lower lake levels (Slentz and Eardley 1956), there are no previous studies regarding the tar seeps trapping any of the local fauna. While the current iteration of the tar seeps has been exposed since 2014, there is little information on the entrapment of animals before 2017 when our investigation began. As seen in Fig. 15.2, there are a large number of seeps of varying size present at Rozel Point. While we have identified animal remains in a number of them, this is only a small percentage of the total number present and the majority of the seeps do not contain animal remains and have not acted as a trap. Given the large number of seeps, we have not been able to examine each one, but a general observation is that small seeps, those that lack active flow, have become oxidized so not become sufficiently soft even at warm temperatures, to function as an effective trap. For the purpose of this overview, we have focused on those seeps for which we have identified entrapped birds and other vertebrates.

At the beginning of the summer of 2018, there were many skeletons of entrapped birds from previous years. Although we do not know when these birds were entrapped, we visually monitored and photodocumented the decay of the previously entrapped birds to gain a better understanding of how birds decay on the tar seeps. Throughout the summer season of 2018 and 2019, when the seeps are most active there were many birds entrapped. While scavengers such as ravens were seen feeding on the carcasses in the motion-sensing cameras, none of them were entrapped. One coyote was entrapped while scavenging on a pelican carcass in 2019. Seepage of petroleum at Rozel Point is most active in the summer months, and trail camera images documenting this flow at the seeps show that the rate of seep expansion corresponds with increasing temperatures. Throughout the summer the seeps can be seen expanding, either from the edges when tar flows under the surface crust of the seep, or from the center when tar "volcanos" are formed. During this time, when it is warmer, the tar is less viscous and flows at a faster rate. In the colder winter months, the tar seeps are not very active; the tar does not expand as it does in the summer, and the seeps can seize up to a stiff surface on which one can walk.

15.3 Biology of the Rozel Point Tar Seeps

The avifauna found in the vicinity of GSL has 338 documented species including shorebirds, waterfowl, which along with raptors, are represented by both residents and migrants that inhabit the complex of wetlands and uplands associated with the lake (Sorensen et al. 2020). To date, only a small subset of this species diversity is represented by individuals entrapped in the tar seeps at Rozel Point; so the overview presented here should be considered preliminary, as further work should undoubtedly increase the overall diversity.

15.3.1 *American White Pelicans (Pelecanus erythrorhynchos)*

The American White Pelican (AWPE) is the most commonly trapped animal at the Rozel Tar Seeps during our study in 2018. Gunnison Island in GSL, 19 km from Rozel Point, is the location of one of the world's largest breeding populations of AWPE (Kijowski et al. 2020). These enormous birds, about a meter tall, spend their spring and summer, from March to September on the island nesting. While the location of the island in the remote north arm of GSL provides the solitude that the AWPE needs to raise their young, there is no food or freshwater near the island. The pelicans travel 50–60 km each way to the feeding areas that flank the eastern shore of the lake such as the Bear River Bird Refuge to get fish for their offspring on Gunnison Island.

Late in the summer, many of the young pelicans are ready to make the journey from Gunnison Island to the Bear River Bird Refuge on their own (Fig. 15.7). Rozel Point is located halfway between the island and the refuge, and is the first land area on which the young birds may encounter on their journey.

In 2018, we began documenting the pelicans observed entrapped in the seeps and recording those that have been marked by the Utah Division of Wildlife Resources

Fig. 15.7 Google Earth aerial view showing geographic relationship of Gunnison Island to Rozel Point and the Bear River Wildlife Refuge

Fig. 15.8 Aerial view taken by drone of largest seep showing the distribution of entrapped American White Pelicans. Other birds are also present but not visible at this resolution. This seep was found to be the result of an improperly capped well and was re-capped in 2019 by the State of Utah. Image credit: State of Utah, Department of Natural Resources, Utah Geological Survey

(UDWR) for research purposes (Kijowski et al. 2020). To date, we have documented 42 pelicans entrapped in the tar seeps, 31 of them in one seep (Fig. 15.8). Of these pelicans, 17 were marked by UDWR, three were tagged in 2015, four in 2016, one in 2017, seven in 2018, and two in 2019. Three individuals with tags were found outside the seeps. Since the Rozel Tar seeps became consistently exposed in 2014, with the most recent drop in the lake level, it appeared that entrapment began soon after their subaerial exposure.

The proximity of the tar seeps to the shoreline of GSL may also contribute to their "attractiveness" to birds and being mistaken for a water source. The flat surface upon which the oil flows is at a similar elevation as the lake level, and at times may be covered by water during high lake levels, or also have standing water after a rain or snow melt. Bernáth et al. (2001a) demonstrated that optical clues of the shiny surface of an open-air oil reservoir may be so strong that water-seeking birds may be visually compelled to remain in the area in spite of their other senses signaling that it is not water. Consequently, they may land directly on the surface of the tar or walk/wade onto it, which may result in their entrapment.

The experiment in Hungary by Bernáth et al. (2001a) documented that the reflectivity of the surface of the tar has a seasonal cycle, and in the summer, it is usually flat and shiny unless disturbed by rainfall or cool weather resulting in a lower viscosity. Beginning in the fall, the surface becomes dull, and with a decrease in temperature, the oil becomes more viscous. The winter surface looks matted and wrinkled and may have small pools of water. Exposed tar will oxidize and harden, which reduces its effectiveness at trapping animals. However, even oxidized tar, if it is smooth enough, is sufficiently reflective to produce a heat-induced shimmer commonly mistaken for water by birds (Kriska et al. 1998). For example, Brown Pelicans (*Pelecanus occidentalis*) have been reported crashing into asphalt roads in Arizona during the summer (Pitzl 2004). Flickinger (1981) reported on wildlife entrapment in artificial petroleum pits of styrene tar that like the seeps at Rozel Point were not very thick, 2.5–10.2 cm.

Heat-activated trail cameras have captured images of pelicans walking on the surface of the Rozel Tar seeps, which would indicate areas of the seeps that have an oxidized crust. The time of day and especially the temperature impact the entrapment of pelicans (Fig. 15.9). These trail cameras have also provided images of pelican entrapment in the tar seeps and subsequent scavenging of pelicans by coyotes (Fig. 15.10) and ravens.

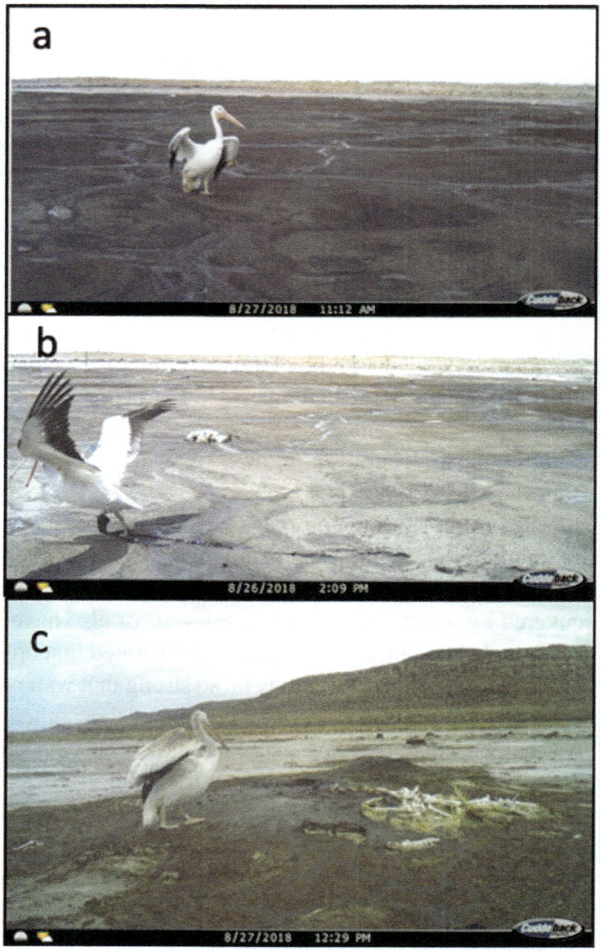

Fig. 15.9 Trail camera photos of pelicans on tar seeps. (**a**) American White Pelican walking on tar in the morning while it is still relatively firm and viscous. (**b**) American White Pelican walking on softer tar in the afternoon and leaving tracks on tar prior to entrapment, with a dead pelican in background. (**c**) American White Pelican on tar volcano with bones of dead pelicans

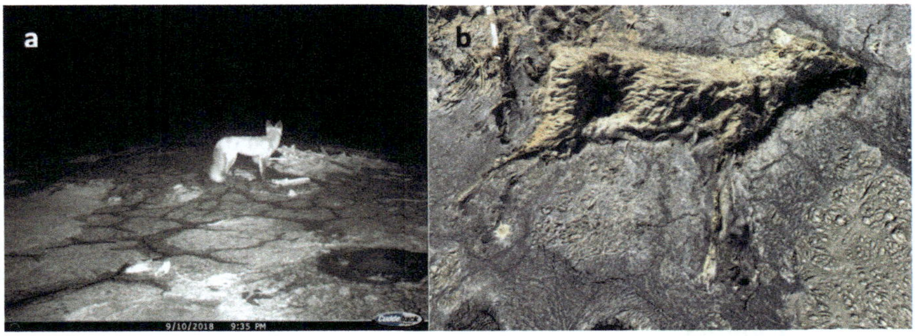

Fig. 15.10 (**a**) Night view from trail camera showing coyote scavenging pelican skeleton. (**b**) Coyote entrapped in the tar (Photo Credit: Mary Sanchez)

We observed a consistent pattern in the positioning of the deceased pelicans in the tar, which may reflect a common method in their attempts to escape. Initial entrapment is with the feet becoming stuck and immobile. In their attempts to free themselves, the pelican will put its wings down on the tar to push itself up. This results in the wings becoming stuck to the tar. Then the bird will use its beak to try and push itself up and away from the ground. The beak then becomes covered with tar and the bird may ingest tar during its efforts to extricate itself, entrapment contributes to the death of the animal along with heat exhaustion, dehydration, and starvation. In some cases, the beak becomes sufficiently coated with tar the bird cannot move its head. The result is a death position of the animal with the wings splayed, the head either outstretched or curved under their body and the legs and ventral part of the body firmly embedded in the tar (Fig. 15.8). This leaves only the dorsal part of the animal available for scavenging, so there is only a partial destruction of the skeleton as the majority of the body is immersed or covered with tar and not accessible for eating. This behavior was observed by KK and HGM when visiting the site. The entrapped pelican worked at attempting to free themselves as described above.

It is important to note that pelican mortality is not restricted to the tar seeps but also occurs in the lake shoreline area outside of the tar seeps. Scattered bones of pelicans can be found on the beach outside the petroleum area covered (USGS 2019; Kijowski et al. 2020), but the degree of disarticulation has made it difficult to determine the number of individuals, and none of these individuals represented by bones retained wing tags, with the exception of a few recently deceased individuals and decay had not yet started. In these soils, wing tags can be easily separated from the animal following the decomposition of soft tissues, unlike birds in the tar seeps where the adhesive nature of the tar often holds the tag in place next to the bird. Bones outside the tar seeps often are destroyed more quickly as well.

The sands in the GSL shoreline area are former lake bottom sediments and contain large amounts of halite and gypsum. Both are soluble and can be easily deposited in weathered bone during submersion. This combined with the intense solar ultraviolet radiation during the summer, which breaks down the bone collagen, facilitated the relatively rapid disintegration of bone. It is, therefore, possible that pelican mortality in the vicinity of Rozel Point is similar to or perhaps greater than that occurring due to tar entrapment and the tar seeps are simply providing a mechanism of better preservation. Bird bones on the shoreline of a saline-alkaline lake in Tanzania weathered quickly due to repeated episodic submersion and drying, coupled with mineral and salt uptake, both of which reduce the structural integrity of the bone and increased the rate of fragmentation (Prassack 2011). Additional research is needed on the overall mortality of pelicans both by tar entrapment and outside the seep area and the rate at which carcasses and bones disintegrate or are preserved.

15.3.2 Raptors

15.3.2.1 Hawks (*Buteo* sp.)

There are many species of hawks living in vicinity of GSL, including the Red-tailed Hawk (*Buteo jamaicensis*), Swainsons Hawk (*Buteo swainsoni*), and Sharp-shinned Hawk (*Accipiter striatus*). In the summer season of 2018, only one hawk was found to be entrapped, but the species is yet to be determined. The entrapped hawk was found with its wings spread out and its body and wings stuck face down in the tar. This hawk could have potentially spotted prey that had been entrapped in the seep or was crossing the tar. We have only detected one rodent crossing the tar and none have been found entrapped. There may, however, be limitations to rodent detection with motion-activated cameras due to their small size.

15.3.2.2 Owls

Two species of owl have been recovered from the Rozel tar seeps, the Barn Owl (*Tyto alba*) and Long-Eared Owl (*Asio otus*). The entrapment of 12 Barn Owls in one of the smaller seeps (Fig. 15.11) is puzzling as barn owls are not known to form "flocks." Since owls tend to be solitary hunters, it is not known why so many individuals should be concentrated in this one seep. At the time of the first observation, the stage of decay was very similar for all individuals suggesting entrapment occurred within a fairly narrow window of time.

The Long-Eared Owl (*Asio otus*) has been recovered from a different larger seep and is represented by two individuals, which were not found in close proximity to each other. Like the Barn Owls, they were preserved as essentially complete articulated skeletons and were well covered with tar.

Entrapped owls occur frequently in the asphalt deposits of Rancho La Brea (Campbell and Bochenski 2015) and are represented by nine species of which the Burrowing Owl is the most abundant, followed by the Barn Owl. The Long-Eared owl is also present in the fauna but is not as common. Burrowing Owls (*Athene cunicularia*) are seasonally present at Rozel Point and have been observed in the sagebrush habitat within close proximity to the tar seeps. Burrowing Owls form small and medium-sized colonies during the breeding season which lasts from early April through the first week of July in this part of the Great Basin. Fledged young

Fig. 15.11 Tar seep with 12 Barn Owls (*Tyto alba*), nine individuals are present in the picture

remain with the adults through July and part of August and they migrate out of the area before the end of September, with a few occasionally lingering in the areas until October (Smith and Murphey 1973). They are therefore active when the tar seeps are most effective in trapping but to date no individuals have been found entrapped.

Campbell and Bochenski (2015) assumed that entrapped owls were probably caught in the asphalt seeps while preying on entrapped animals struggling to free themselves. They noted that the number of owl specimens in the Rancho La Brea collections is larger than the number of true avian scavengers, suggesting that prey animals trapped, but still living, perhaps drew more attention from predators than dead animals did for scavengers.

In Utah Barn Owls, Smith et al. (1972) identified 21 different prey species in regurgitated pellets they examined. Mammalian remains comprised 90% of the total prey and 13 avian prey species were documented. For the mammals, *Microtis* spp. was the dominant food of barn owls throughout the year (81%). The frequency of avian prey, while comprising only 10% on an annual basis, doubled during autumn. While it might be surmised that the owls became entrapped while trying to catch mice running across the seeps, only one small rodent has been captured with the cameras. Given their small size they could be quickly covered by tar, when entrapped, and so are not readily visible. There is also likely limitations to the cameras detecting them. Adding to the challenge of determining the cause of entrapment of owls, it should be noted that the former lake bottom, where the seeps are located, is devoid of vegetation, reflecting the high amounts of salt in the sediments. When the amount of salt in the lake water exceeds 340 grams per liter of total dissolved solids halite is

Fig. 15.12 View looking east towards the Rozel Hills from the tar seeps showing the lack of vegetation on the former lake bottom in contrast to the sagebrush habitat outside the current lake margin

precipitated onto the lake bottom (Gwynn and Murphey 1980). The exposed lake bed thus lacks habitat and cover that might attract these prey species into the vicinity of the seeps, compared to the better cover nearby, on the sagebrush covered hills just outside the lake margin (Fig. 15.12).

The Rozel tar seeps not only caught numerous owls, but trapped them in large numbers. This is surprising given that they are primarily nocturnal hunters, a time when cooler temperatures reduce the effectiveness of the asphalt seeps as traps (Miller 1925). Owls are highly predaceous birds that feed almost exclusively on live-caught prey, and while some species will scavenge carrion, it is relatively rare. Campbell and Bochenski (2015) suggested that predaceous owls are more numerous in the Rancho La Brea avifauna than strict avian scavengers because of their hunting behavior, and this is evidence of the effectiveness of the asphalt seeps as baited traps. This observation appears to be valid at the Rozel Point tar seeps as well.

15.3.3 Non-raptoral Birds

We observed other avian specimens entrapped at Rozel Point. It is axiomatic, in fossil studies of tar pits, that the large percentage of raptors and scavengers in entrapped bird diversity reflects their attraction to live prey or carcasses stuck in the seep. Usually the presence of other taxa with different diets is attributed to accidental entrapment.

Two individuals of Gull (*Larus* sp.) were identified from the seeps. California Gull, Franklin's Gull, and Ring-Billed Gulls are common species at GSL (but there are also incidental occurrences of other species). It was not possible to determine the species trapped, but both individuals appear to be juveniles.

Three entrapped individuals of the Gray Flycatcher (*Empidonax wrightii*) were identified in our study of the Rozel tar seeps. The entrapment of such insectivorous birds may also be attributed to the presence of their prey, insects, which could be attracted to the reflective tar surface. This model parallels that of the more typical avian predators discussed above but with a different prey base. Oswald and Steadman (2015) reported two species of flycatcher, *Myiozetetes similis* and *Myiodynastes bairdii* as part of the Pleistocene avifauna from the Talara Tar Seeps in Peru and Steadman et al. (2015) reported six species of the Tyrannidae from the Pleistocene fauna recovered from the Mene de Inciarte tar pits in Venezuela.

15.3.4 Insects

The discussion of insectivorous birds leads to the question as to how insects are attracted to and entrapped in tar seeps. Many insects associated with water use polarotaxis, the plane of polarization of polarized light, to find aquatic habitat (Bernáth et al. 2001b). During warm weather, the shiny oil surface mimics such a water source, but during the winter, the surface is a matte texture and loses its polarization and hence attractiveness to insects. Some examples of polarotaxic insects include stoneflies, mayflies, caddish fly, and some scavenger beetles (Kriska et al. 1998). Bernáth et al. (2001b) observed dragonflies, mayflies, water bugs, water beetles, and butterflies were trapped en masse during their swarming and migration during the spring, summer, and autumn. Other insects such as dragonflies and beetles that do not swarm would be single entrapment events. Flickinger (1981) noted that trapped insects at a petroleum pit in Texas included grasshoppers, water scavenger beetles, predaceous diving beetles, dragonflies, butterflies, and moths that would have attracted insectivorous birds.

At Rozel Point, in the summer of 2019, both water boatmen (cf. *Sigara*) and brine flies (*Ephydra* spp.) were observed, associated with the tar seeps. Water boatmen (Corixidae) are predatory insects that are intolerant of salinities above 6% and thus are not found in large numbers in the open water of GSL but more often in brackish puddles where freshwater seeps dilute the salinity in the saline soils around the margins of the lake (Mellison 2000; Barrett and Belovsky 2020). *Ephydra* spp. spend their egg, larval, and pupal stages of their life cycle in the south arm of the lake, where the salinity is 12–15% (Collins 1980). Neither corixids nor *Ephydra* has been reported in the hypersaline (upwards of 30% salinity) north arm of GSL, but near the tar seeps are groundwater seeps, and brackish pools do form in this area near the lake's high-water line.

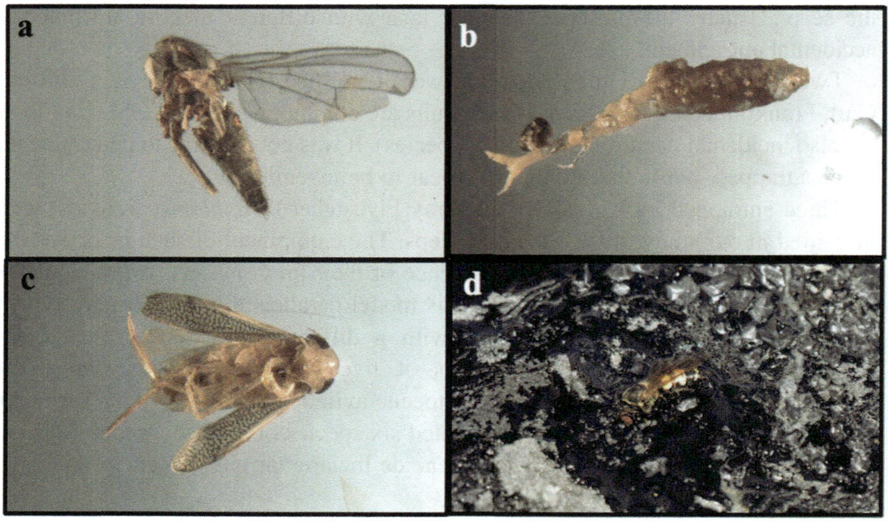

Fig. 15.13 Microscopy of insect samples taken from Rozel Point. (**a**) An *Ephydra* spp. fly taken from the surface of the tar. (**b**) An *Ephydra* spp. pupa taken from the freshwater seep. (**c**) A corixid taken from the surface of the tar. (**d**) cicada killer, cf. *Sphecius* (Image credit: Rebecca Dennis and Cayla Martin)

In early summer, brine flies (*Ephydra* sp.) were observed on the fresh tar, alive and not entrapped (Fig. 15.13). Live flies, dead flies, and pupae were also noted in a freshwater seep on the playa between two of the tar seeps (Fig. 15.13). On later trips to the site, we observed the brine flies, but in fewer numbers, and many that remained on the tar were dead. When the flies were sampled from the tar, the dead flies were not entrapped. The water boatmen (Corixidae), in contrast, were all dead and most of them were entrapped in the tar. An unidentified insect was observed hiding in a hole in the inner rim of a tar volcano. The hole it was going into was lined in fresh tar, but the insect flew away and was clearly not entrapped.

Only a single individual of a wasp, cf. *Sphecius*, or cicada killer has been observed trapped in the tar at the site. The specimen could not be collected and tar partially covered its markings limiting identification. However, as only one species is known from Utah, *S. grandis* and it has been documented in the vicinity of GSL (Coelho et al. 2011), we tentatively identified this specimen as that species. Wasps have been identified as prey for the Gray Flycatcher, which was entrapped in the seeps. Adults of the cicada killer emerge in summer, typically beginning around late June or early July and live for 60–75 days, usually until mid-September, so the timing of their maximum activity coincides with when the seeps are most effective as traps.

It appears it is possible for insects to be associated with the tar seeps and not get stuck, however, during periods of higher temperature the surface conditions of the tar could change quickly resulting in entrapment. We deduce, from this insect monitoring effort, that insects in the area of the Rozel Point tar seeps, may become trapped, and certainly may be an attractant to insectivorous birds, such as the gray flycatchers. Entrapment of the flycatcher may be accidental during foraging over the tar, or possibly direct, if the birds landed directly on the tar to catch an insect.

15.3.5 Scavenging Animals

The two scavengers documented by the trail cameras utilizing the carcasses, primarily pelicans, trapped in the tar are American Ravens (*Corvus corax*) and coyotes (*Canis latrans*) (Fig. 15.10). Interestingly only one coyote has been found entrapped. Meachen and Samuels (2012) noted that an extinct subspecies of coyote *Canis latrans orcutti* was one of the most common predators preserved in the Rancho La Brea, Maricopa and McKittrick tar pits. The ravens were documented during the day so they would have had a higher probability for entrapment given the warmer and less viscous state of the tar. Scavenging by the coyote was documented primarily at night, a time when the tar would have been cooler, firmer, and less sticky. The coyote that became entrapped was scavenging during the day when the tar was less viscous. Scavenging by coyotes accounts for the damage to the skeletons of the pelicans, particularly the vertebral column and synsacrum which are the parts of the body typically exposed above the tar and so is readily accessible. Smaller birds are more quickly covered by oil so are usually unavailable to scavengers and consequently are preserved as intact articulated skeletons.

While Turkey Vultures (*Cathartes aura*) have been observed in the vicinity of the tar seeps during the summer they have not been found entrapped. Neither have they been documented feeding on any carcasses in the tar by the trail cameras. Their absence was not expected given the predominance of vultures as fossils in tar pits like Rancho La Brea (Howard 1930). Their absence at the site is also puzzling since they are only seasonally present in the area during the summer when the seeps have the highest potential for entrapment. Given that the Turkey Vulture is one of the few birds with a sense of smell (Stager 1964), it is possible that the smell of the exposed oil masks the scent of the carcass. However, Smith and Paselk (1986) questioned the importance of smell in aiding Turkey Vultures in locating food sources.

15.3.6 Other Incidental Observations of Animals

A single gopher snake (*Pituophis catenifer*), was observed trapped in a seep at Rozel Point, caught along the margin near an active source of seepage from an abandoned drill hole. The animal was entirely covered within 2 months after it was originally spotted and was not scavenged.

Besides the coyote (*Canis latrans*) (Fig. 15.10) two other mammals were documented in the immediate vicinity of the tar seeps. A Black-tailed Jackrabbit

(*Lepus californicus*) was documented at night using the trail camera. No leporids, rabbits or hares, were found entrapped in the tar. A single unidentified rodent was documented crossing the tar at night and appears to be foraging on entrapped insects. It is likely that small rodents and other small mammals are so small that the heat-activated cameras are not triggered by them. Other methods of detection will need to be used in the future.

15.4 Concluding Remarks

The presence of tar seeps at Rozel Point is intimately tied to changes in the water level of GSL. Consequently, they only form when the lake level is below 1279.6 m (4198 feet), subaerially exposing the former lake bottom. When exposed, petroleum reaching the surface spreads out to form both thin layers of tar as well as tar volcanos. The current state of decline in lake level suggests that the exposed seeps will remain a prominent and stable feature of the modern GSL and will continue to have an impact on the wildlife, both vertebrate and invertebrate in the immediate vicinity.

In the summer and other intervals of higher temperatures, the tar forms an effective trap for animals that encounter them. To date, the primary vertebrate victims of entrapment are birds. Juvenile pelicans are commonly found in the tar seeps, but their remains are also found in the area around the seeps, reflecting the location of the tar seeps midway between the nesting area on Gunnison Island and the feeding area where the Bear River enters the GSL system.

During the Holocene, there were multiple times when the lake level has been low enough for tar seep formation and presumably they functioned as traps for birds and other wildlife in the past as well. This site is an excellent analogue for tar pit entrapment studies, providing at one site, key insight into both the modern process and a prehistoric view.

Acknowledgments Our investigations on the tar seep at Rozel Point were done under the Right of Entry No. 410-00654, issued by the Utah Division of Forestry, Fire and State Lands to Jaimi Butler of the Great Salt Lake Institute at Westminster College. We thank Jack M. Broughton of the Department of Anthropology, University of Utah for the use of the collections in the zooarchaeology laboratory to facilitate our identifications of bones from the birds. Arie Leeflang in Archaeology Records at the Utah Division of State History kindly provided historic photos of oil drilling operations at Rozel Point. John Neill of the Great Salt Lake Ecosystem Program, Utah Division of Wildlife Resources gave us access to information on the pelicans with wing tags. Christian Hardwick and Kellen Hatch of the Utah Geological Survey did the drone mapping of the tar seeps. Ryan Davis and Terry Griswold provided help on the wasp identification. Matt Coombs with the Utah Division of Forestry Fire and State Lands helped us with permits and site access for our fieldwork. We thank Mary Sanchez and Cayla Martin for their help monitoring the Rozel tar seeps. Funding was provided by the Stephen G. and Susan E. Denkers Family Foundation, W.M. Keck Foundation, and the Lawrence T. and Janet T. Family Foundation.

References

Akersten WA, Shaw CA, Jefferson GT (1983) Rancho La Brea: status and future. Paleobiology 9:211–217

Atwood G, Wambeam TJ, Anderson NJ (2016) The present as a key to the past: Paleoshoreline correlation insights from Great Salt Lake. In: Oviatt CG, Shroder JF (eds) Lake Bonneville a scientific update, 1st edn. Elsevier, Amsterdam, pp 1–25

Barrett KL, Belovsky GE (2020) Invertebrates and phytoplankton of Great Salt Lake: is salinity the driving factor? In: Baxter BK, Butler JK (eds) Great Salt Lake biology: a terminal lake in a time of change. Springer, Cham

Bernáth B, Szedenics G, Molnár G, Kriska G, Horváth G (2001a) Visual ecological impact of a peculiar waste oil lake on the avifauna: dual-choice field experiments with water-seeking birds using hug shiny black sheets. Archive für Natural Lands 40:1–28

Bernáth B, Szedenics G, Molnár G, Kriska G, Horváth G (2001b) Visual ecological impact of "shiny black anthropogenic products" on aquatic insects: oil reservoirs and plastic sheets as polarized traps for insects associated with water. Archive für Natural Lands 40:89–109

Bortz LC (1987) Heavy-oil deposit, Great Salt Lake, Utah. In: Meyer RF (ed) Exploration for heavy crude and natural bitumen, vol 25. American Association of Petroleum Geologists Studies in Geology, Tulsa, pp 555–563

Bortz LC (2002) Heavy oil deposits, Great Salt Lake, Utah. In: Gwynn W (ed) Great Salt Lake: an overview of change. Special Publication of Utah Department of Natural Resources, Salt Lake City, UT, pp 243–350

Boutwell JM (1904) Oil and asphalt prospects in Salt Lake Basin, Utah. In: Emmons SF, Hayes CW (eds) Contributions to economic geology: U.S. Geological Survey Bulletin, vol 260, pp 468–479

Campbell KE Jr, Bochenski ZM (2015) The owls (Aves:Strigiformes) of Rancho La Brea. In: Harris JM (ed) La Brea and beyond: the paleontology of asphalt-preserved biotas, vol 42. Natural History Museum of Los Angeles County Science Series Number, Los Angeles, CA, pp 5–21

Coelho JR, Holliday CW, Hastings JM (2011) The geographic distributions of Cicada Killers (*Sphecius*; Hymenoptera, Crabronidae) in the America. Open Entomol J 5:31–38

Collins N (1980) Population ecology of *Ephydra cinerea* Jones (Diptera: Ephydridae), the only benthic metazoan of the Great Salt Lake, U.S.A. Hydrobiologia 68:99–11

Cook KL, Berg Jr JW, Johnson WW, Novotny RT (1966) Some Cenozoic structural basins in the Great Salt Lake area, Utah, indicated by regional gravity surveys. In: Stokes, WL (ed) The Great Salt Lake: Utah Geological Society Guidebook to the Geology of Utah No 20, pp 57–76

Curry DR (1980) Coastal geomorphology of Great Salt Lake and vicinity. In: Gwynn JW (ed) Great Salt Lake: a scientific, historical and economic overview, vol 136. Utah Geological and Mineral Survey Bulletin, Salt Lake City, UT, pp 69–82

Damsté JSS, De Leeuw JW, Kock-van Dalen AC, De Zeeuw MA, De Lange F, Irene W, Rijpstra C, Schenck PA (1987) The occurrence and identification of series of organic sulphur compounds in oils and sediment extracts. I. A study of Rozel Point Oil (USA). Geochim Cosmochim Acta 51:2369–2391

Damsté JSS, Irene W, Rijpstra C, De Leeuw JW, Schenck PA (1989) The occurrence and identification of series of organic sulphur compounds in oils and sediment extracts: II. Their presence in samples from hypersaline and non-hypersaline palaeoenvironments and possible application as source, palaeoenvironmental and maturity indicators. Geochim Cosmochim Acta 53:1323–1341

Eardley AJ (1963) Oil seeps at Rozel Point. Utah Geological and Mineralogical Survey Special Studies No 5:5–26

Flickinger EL (1981) Wildlife mortality at petroleum pits in Texas. J Wildl Manag 45:560–564

Google Earth (2019) GSL map created by Google Earth. Accessed 25 Sept 2019

Gwynn JW (2006) Petroleum and gas occurrences and development history in Box Elder County, Utah. Geology of Northwest Utah Utah Geological Association Publication 34:1–13

Gwynn JW, Murphey PJ (1980) Recent sediments of the Great Salt Lake. In: Gwynn JW (ed) Great Salt Lake a scientific, historical and economic overview, vol 136. Utah Geological and Mineral Survey Bulletin, Salt Lake City, UT, pp 83–96

Howard H (1930) A census of the Pleistocene birds of Rancho la Brea from the collections of the Los Angeles Museum. Condor 32:81–88

Hunt GL, Chidsey TC Jr (2002) Rozel Point Oil Field, Box Elder County, Utah: geology, development history, and cleanup. In: Gwynn W (ed) Great Salt Lake: an overview of change. Special Publication of Utah Department of Natural Resources, Salt Lake City, UT, pp 251–258

Kijowski AK, Neill J, Wickline A, Swift J, Butler JK, Kimberly DA, VanLeeuwen J, Luft J, Stone K (2020) American White Pelicans of Gunnison Island. In: Baxter BK, Butler JK (eds) Great Salt Lake biology: a terminal lake in a time of change. Springer, Cham

Kriska G, Horváth G, Andrikovics S (1998) Why do mayflies lay their eggs en masse on dry asphalt roads? Water imitating polarized light reflected from asphalt attracts Ephemeroptera. J Exp Biol 201:2273–2286

MaGuire D (1904) Oil and asphaltum in the Great Salt Lake Basin. Lake Mining Rev 5:13–15

Meachen JA, Samuels JX (2012) Evolution in coyotes (Canis latrans) in response to the megafaunal extinctions. Proc Natl Acad Sci USA 9:4191–4196

Mellison SC (2000) Functional response of a water boatman (*Trichocorixa verticalis*) and environmental conditions that affect its distribution in the Great Salt Lake, Utah, USA. Thesis, Utah State University, Logan, UT

Miller LH (1925) The birds of Rancho La Brea, vol 349. Carnegie Institution of Washington, Washington, DC, pp 63–106

National Weather Service (2019) https://w2.weather.gov/climate/xmacis.php?wfo=slc. Accessed 20 Sep 2019

Null SE, Wurtsbaugh WA (2020) Water development, consumptive uses, and Great Salt Lake. In: Baxter BK, Butler JK (eds) Great Salt Lake biology: a terminal lake in a time of change. Springer, Cham

Oswald JA, Steadman DW (2015) The changing diversity and distribution of dry forest passerine birds in northwestern Peru since the last ice age. Auk 132:836–862

Oviatt CG (2015) Chronology of Lake Bonneville, 30,000 to 10,000 yr B.P. Quat Sci Rev 110:166–171

Oviatt CG, Thompson RS, Kaufman DS, Bright J, Forester RM (1999) Reinterpretation of the Burmester Core, Bonneville Basin, Utah. Quat Res 52:180–184

Patton TL and Lent RL (1980) Current hydrocarbon exploration activity in the Great Salt Lake. Great Salt Lake: A Scientific, Historical and Economic Overview: Utah Geological and Mineral Survey Bulletin 116, 115–124

Pitzl MJ (2004) Crash landing: pelicans mistake desert pavement for water. The Arizona Republic July 7, 2004

Prassack KA (2011) The effect of weathering on bird bone survivorship in modern and fossil saline-alkaline lake environments. Paleobiology 37:633–654

Slentz LW, Eardley AJ (1956) Geology of Rozel Hills. In: Eardley AJ, Hardy CT (eds) Geology of parts of northwestern Utah, vol 11. Utah Geological Society Guidebook to the Geology of Utah, Utah, pp 32–40

Smith DG, Murphey JR (1973) Late summer food habits of adult burrowing owls in central Utah. Raptor Res 7(3/4):112–115

Smith SA, Paselk RA (1986) Olfactory sensitivity of the Turkey Vulture (*Cathartes aura*) to three carrion-associated odorants. Auk 103:586–592

Smith DG, Wilson CR, Frost HH (1972) Seasonal food habits of Barn Owls in Utah. Great Basin Nat 32(4):229–234

Sorensen ED, Hoven HM, Neill J (2020) Great Salt Lake shorebirds, their habitats, and food base. In: Baxter BK, Butler JK (eds) Great Salt Lake biology: a terminal lake in a time of change. Springer, Cham

Stager KE (1964) The role of olfaction in food location by the Turkey Vulture (*Cathartes aura*). Los Angeles County Museum Contributions to Science No 81:1–63

Stansbury H (1855) An expedition to the valley of the Great Salt Lake of Utah. Lippincott, Grambo and Company, Philadelphia

Steadman DW, Oswald JA, Rincón AD (2015) The diversity and biogeography of late Pleistocene birds from the lowland Neotropics. Quat Res 83:555–564

United States Geologic Survey (2019) http://ut.water.usgs.gov/greatsaltlake/elevations. Accessed 5 Aug 2019

Wurtsbaugh WA, Miller C, Null SE, DeRose RJ, Wilcock P, Hahnenberger M, Howe F, Moore J (2017) Decline of the world's saline lakes. Nat Geosci 10:816. https://doi.org/10.1038/ngeo3052

Chapter 16
Great Salt Lake as an Astrobiology Analogue for Ancient Martian Hypersaline Aqueous Systems

Scott M. Perl and Bonnie K. Baxter

Abstract Great Salt Lake (GSL), Utah, is a thalassohaline terminal lake that currently occupies the Bonneville Basin, a depression in the larger Great Basin area of the western United States. Natural processes and climate conditions create a dynamic ecosystem with shifting salinity gradients and lake levels. The hypersaline north arm of GSL provides a model for exploring the limits of life on Earth and for potential life on other space bodies, especially the ancient closed-basin systems on Mars. The north arm water features hundreds of species of halophilic microorganisms with cellular strategies that allow them to live in hypersaline environments and high doses of ultraviolet light. These microbes also survive desiccation and can become entrapped in minerals as they are formed. The modern GSL evaporitic environment, generated by halite and gypsum precipitation events, illuminates the initial steps in preservation of biological material over geologic time. These minerals accumulate on the desiccated shores, in the sediment, and in the surrounding evaporite deposits and have been shown to have biopreservation abilities, protecting halophilic cells and their molecules inside brine fluid inclusions within the crystal structure. Entrapment allows in situ analyses of microbial diversity, which can be studied as a function of salt mineral assemblage. Globally across Mars these same types of evaporite precipitation events took place in closed-basin lake systems where surface waters have evaporated, leaving behind mineral vein structures composed of gypsum and other sulfate salts that have been modified or dissolved from later fluid shallow subsurface activity. We have chosen GSL as our analogue for Martian late Noachian/early Hesperian closed basin systems due to the overlapping evaporite mineralogy and fluid activity. Here we

S. M. Perl (✉)
NASA Jet Propulsion Laboratory, California Institute of Technology, Pasadena, CA, USA

Mineral Sciences, Los Angeles Natural History Museum, Los Angeles, CA, USA

NASA-JPL Origins and Habitability Laboratory, Pasadena, CA, USA
e-mail: scott.m.perl@jpl.nasa.gov

B. K. Baxter
Great Salt Lake Institute and Department of Biology, Westminster College, Salt Lake City, UT, USA

B. K. Baxter, J. K. Butler (eds.), *Great Salt Lake Biology*,
https://doi.org/10.1007/978-3-030-40352-2_16

explore the transference of biological material and organics from hypersaline GSL brine to the minerals as they form in the water. We draw parallels to the evaporites extensively mapped on Mars, which likely formed in a similar way. These observations and insights, taken together, suggest GSL is an appropriate analogue for the study of ancient salt lakes and evaporites discovered on Mars, and what is more, the halophilic archaea that live in Earth's salty lake may be good models for life elsewhere in our solar system.

Keywords Great Salt Lake · Hypersaline · Extremophile · Halophilic · Astrobiology · Biopreservation · Biosignature

16.1 Formation of Great Salt Lake Over Geologic Time

16.1.1 The Bonneville Basin

Great Salt Lake (GSL) fills one of the lowest depressions of the Great Basin in the western United States (Cohenour and Thompson 1966). This Bonneville Basin has hosted four deep lakes over the last 780,000 years including the Pleistocene Lake Bonneville, 30–13,000 years ago, covering greater than 32,000 square kilometers of western Utah, and including portions of eastern Nevada and southern Idaho (Oviatt et al. 1999; Shroder et al. 2016). GSL is exemplary of the shallow terminal lakes that primarily filled the Bonneville basin likely over the last several million years (Atwood et al. 2016). Indeed, evaporite shores and salt playa have dominated this region over that time despite the dramatic appearance of deep lakes.

The transition of Lake Bonneville to GSL occurred over just a few thousand years as water was lost in several phases (Atwood et al. 2016). As the last ice age thawed and the Earth warmed, an alluvial fan dam in southern Idaho burst, the water rushing out, leaving scars on the landscape that are still visible today (Shroder et al. 2016). The water evaporated and leaked out rapidly, dropping about 200 m in 2000 years, resulting in the current GSL margins by about 13,000 years ago (Atwood et al. 2016).

16.1.2 Modern Great Salt Lake

Though the precise margins of the current GSL vary with seasonal precipitation and drought cycles, it measures approximately 122 km in length and 50 km in width with an average depth of 4.3 m and a maximum depth of 9 m (Keck and Hassibe 1979; Stephens 1990). This lake experiences significant seasonal temperature variation from 0.5 °C in January to 26.7 °C in July (Crosman and Horel 2009) and up to 45 °C in the shallow margins (Post 1977) due to its elevation and desert setting.

This chapter describes the vibrant terminal lake that we see today (Fig. 16.1). GSL is the largest lake in the western United States, which supports a critical food

Fig. 16.1 Modern Great Salt Lake and its salinity gradients. (**a**) The hypersaline north arm with its pink water, colored by carotenoid-containing halophilic microorganisms, image credit: Great Salt Lake Institute. (**b**) The moderately saline south arm of the lake with the green and blue hues consistent with a diverse algae population, image credit: Great Salt Lake Institute. (**c**) The railroad causeway that separates Great Salt Lake is visible from space, producing segmented north and south arms with distinct coloration, image credit: NASA JPL, 2002. (**d**) The Union Pacific Railroad causeway that divides the lake, image credit: Great Salt Lake Institute

web, outlined in the preceding chapters (Baxter and Butler 2020). The simplicity of the food web of GSL (birds eat invertebrates) disguises the highly complex microbial consortia that inhabit this lake, particularly in the north arm which restricts that macro-level ecology (Almeida-Dalmet and Baxter 2020). Modern GSL is highly productive despite the reduced solubility of oxygen of hypersaline waters. Phototrophs power the system (Stephens 1974; Lindsay et al. 2017), anaerobic activities are prevalent (Boyd et al. 2017), and methanogenesis has been detected (Baxter et al. 2005). The metabolism of these microbial communities, living at salt saturation, is complex, but such reactions occur more slowly than at lower salinity levels (Ward and Brock 1978; Post and Stube 1988; Stube et al. 1976; Fendrich and Schink 1988).

With respect to astrobiology, this chapter is focused on the microbial communities of the lake (reviewed in Baxter and Zalar 2019) and the biosignatures that they may leave behind in the rock record. Diverse assemblages of halophilic (salt-thriving) microorganisms reside in even the most saline parts of GSL. We are particularly interested in the hypersaline region, the north arm of the lake (Fig. 16.1a), where the waters are saturated with sodium chloride. The less saline south arm may be of interest (Fig.16.1b) since it provides a future site for monitoring life in vacillating salinity, likely featuring changing communities as it becomes more saline over time. Prior molecular studies indicate that the microbial communities in the north arm of GSL are composed predominantly of halophilic archaea and to a lesser extent, bacteria (Baxter et al. 2005; Weimer et al. 2009; Parnell et al. 2011; Meuser et al. 2013; Tazi et al. 2014; Almeida-Dalmet et al. 2015; Boogaerts 2015; Perl 2019; Almeida-Dalmet and Baxter 2020). However, even in the hypersaline north arm, eukaryotic algae and fungi thrive (Baxter and Zalar 2019). Many of these microorganisms are pigmented; they have carotenoid compounds embedded in their cell membranes, which provide photoprotection for these cells (Jones and Baxter 2017). Owing its color to these tiny inhabitants, the north arm is colored pink, which can be seen from the shoreline to space (Fig. 16.1).

16.2 Salinity Gradients of Modern Great Salt Lake

GSL is a terminal lake, a closed system, subject to drought and flooding, dependent on temperature, groundwater inputs, evaporation, and precipitation cycles. The worst flooding of the modern lake on record occurred from 1983 to 1987 when the lake rose 4 m (Stephens 1990). Currently, we are in a long drought period (Wurtsbaugh et al. 2017). Such events cause the salinity to fluctuate as the salts are diluted or concentrated.

GSL is considered a thalassohaline lake, which is an inland saline body of water with ion proportions similar to the dissolved salts in seawater, indicating it formed from the evaporation of seawater in its history (Oren 1993). The modern evaporites and solid salts experience cycles of dissolution, bringing a continuous flow of minerals into the brine (Jones et al. 2009). GSL is rich in sodium chloride with an exceptionally high sulfate concentration, which distinguishes GSL from some salt lakes, such as the divalent-rich Dead Sea (Post 1977; Sturm 1980; Spencer et al. 1985; Domagalski et al. 1989; Baxter et al. 2005; Jones et al. 2009). The relative ratios of specific ions, for example, 1:1.7 Na^+ to Cl^-, remain somewhat consistent (Gwynn 1998; Rupke and McDonald 2012).

GSL boasts broad salinity ranges from 30 g/L of total dissolved salts to 340 g/L across the various regions of the lake and freshwater wetlands on the margins (e.g., USGS 2019; Naftz et al. 2011). Salinity gradients in the lake have been shown to markedly influence the structure and composition of the microbial communities in the water column and benthic regions (Boyd et al. 2014; Meuser et al. 2013). The most dramatic anthropomorphic impact is the railroad causeway that bisects GSL, constructed in the late 1950s (Fig. 16.1c, d) (Madison 1970; Cannon and Cannon 2002; Baxter et al. 2005). Soon after its construction, at least one study predicted the flow of lake water to the north, and the absence of freshwater input, would cause salt-saturation and formation of a permanent salt crust (Adams 1964). Within 7 years, distinct salinity differences were measured as the north arm was indeed approaching saturation, while the south arm, which received freshwater input from rivers flowing in, was less saline (Greer 1971). Currently, the open waters of south arm of GSL range from 110 to 150 g/L (USGS 2019) while the north arm is at saturation (280–340 g/L, dependent on the temperature) (e.g., Baxter et al. 2005; Almeida-Dalmet et al. 2015). Brine stratification also occurs, particularly near the causeway where the heavy hypersaline water seeps underneath, causing a halocline that does not readily mix (Naftz et al. 2008; Meuser et al. 2013). These hypersaline brine pockets provide niches for a variety of microbial communities.

16.3 Great Salt Lake as an Analogue for Studying Evaporites on Mars

16.3.1 Formation of Evaporites

Evaporite minerals, such as halite (NaCl) and gypsum ($CaSO_4 \cdot 2H_2O$) can precipitate on Earth and other planets following the evaporation of saline lakes. This mineralogy is a record of former fluvial histories and a diagnostic feature of the timing and chemistries of the ancient lake systems. Evaporites are key features in studying ecosystems, both modern and ancient (Sonnenfeld 1984). In addition, these minerals are capable of preserving signs of halophilic life (e.g., Lowenstein et al. 2011; Schopf et al. 2012).

16.3.2 Great Salt Lake Evaporites

The GSL north arm salinity is generally at saturation but the concentration varies as temperatures change and solubility is impacted, while other terrestrial closed basins might register higher concentrations (e.g., the Dead Sea) due to being warmer year-round (Baxter et al. 2005). At the north arm of GSL, both halite and gypsum have been observed (e.g., Eardley and Stringham 1952; Jones et al. 2009; Perl 2019) alongside other trace mineralogy. GSL is surrounded by the Bonneville Salt Flats, a halite deposit left from the evaporation of Lake Bonneville from the Pleistocene and other flooding/evaporation events (Turk 1970).

Our GSL north arm study site at Rozel Bay is a critical analogue as it provides a modern salt lake at the same location as an ancient evaporite site where life has thrived and adapted to changing hypersaline settings. As halite precipitates in the brine, we have observed terraces of halite layers on the lakeshores, desiccated in the summer season (Fig. 16.2a). The observable pink color of the halite is due to the enrichment of carotenoid-containing halophilic archaea and bacteria in these regions (Jones and Baxter 2017; Perl 2019). In the autumn, as the water cools, mineral precipitation is favored since the solubility is lowered. Evaporation of water at the

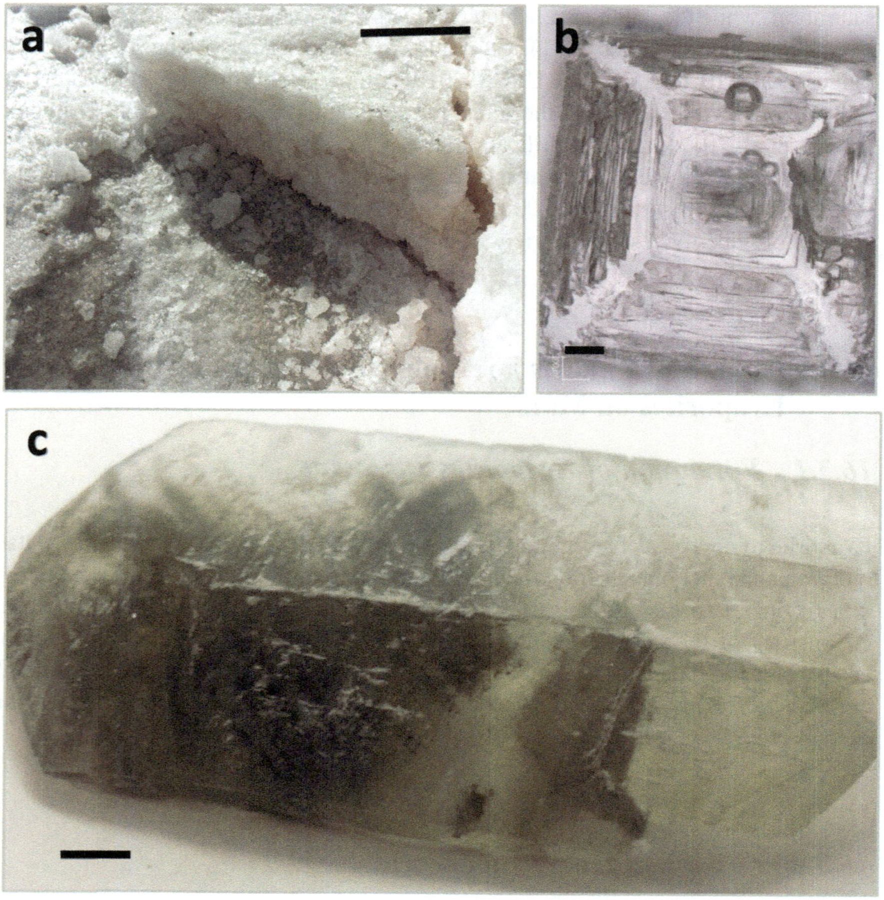

Fig. 16.2 Evaporite minerals at the north arm of Great Salt Lake. (**a**) Halite terrace forming as the north arm waters of Great Salt Lake recede in summer desiccating conditions. Evaporation may lead to entrapped pigmented halophilic microorganisms as the lakeshore is desiccated, which results in pink halite. Note the layered pigmented sections below non-pigmented material (Perl 2019). The scale bar represents ~12 cm. (**b**) A typical halite hopper crystal collected from the north arm of the lake. Fluid inclusions are evident in the crystalline structure. The scale bar represents 0.10 cm. (**c**) Gypsum ($CaSO_4 \cdot 2H_2O$) crystal retrieved from the sediment of evaporated brine from the north arm shore, showing entombed Fe-rich clays within the transparent mineral matrix. The scale bar represents ~1 cm. During precipitation, these minerals tend to align vertically and perpendicular to saturated sediments and fluids needed for crystal growth

surface of the hypersaline north arm of GSL pushes the molecules of NaCl closer together, forming small chevron crystals of halite floating on the surface of the water. The edges of these tiny crystals in the saturated brine, grow as salt molecules are layered on the nucleus. This salt crystal thus becomes heavier, and it sinks lower, creating a primary pyramid shaped, "hopper," crystal in which you can observe tiny fluid inclusions (Fig. 16.2b) (Gornitz and Schreiber 1981). Crystals eventually land on the lake floor, where they continue to precipitate halite and grow (Wardlaw and Schwerdtner 1966). Embedded in the sediment, both under the lake water and under the halite crust, are gypsum crystals (Fig. 16.2c) that can be retrieved by digging a few centimeters beneath the clays (Eardley and Stringham 1952).

Past work has demonstrated that halophilic microbial life from the GSL water column can become sequestered in halite and gypsum crystals, showing transfer of biology to the evaporitic mineralogy (Baxter et al. 2007; Kemp et al. 2018; Perl 2019). This preservation of cells occurs in these modern halite and gypsum evaporite samples at the lake–shoreline boundary and further inland within large-scale regolith megastructures, where gypsum has precipitated fairly close to the halite salt terraces (Perl 2019). Hopper halite crystals (Fig. 16.2b) were examined for the ability to preserve microorganisms. When surface sterilized, the hoppers reveled microorganisms stored inside the crystalline structure (Fig. 16.3a). Direct transmission electron microscopy of dissolved hopper crystals, that were surface sterilized, demonstrated the preservation of not only microorganisms, but also biological molecules, including cellulose and DNA (Fig. 16.3b). Inside hopper halite crystals are visible fluid inclusions (Fig. 16.3c) in which you can detect microorganisms with microscopy (Fig. 16.3d). Indeed, modern crystals can preserve the biology of the lake, and this also points to the fortitude of the microorganisms and the stability of the molecules stored in the salt-saturated fluid pockets (Pasteris et al. 2006).

Examination of the preserved microbial diversity in GSL minerals indicated that the community assemblages within the crystal structure are distinct dependent upon the in situ manner in which they are preserved (Perl 2019). The microbial community was assessed by amplification and DNA sequencing of the collection of 16S rRNA genes, which infers distinct genera and species. For example, communities that are preserved within pigmented halite (Fig. 16.2a) are different than communities that are preserved within nonpigmented crystals (Perl 2019). Moreover, microbial communities preserved in gypsum that contains Fe-bearing clays are distinct from halophilic microorganisms preserved in the non-clay (gypsum only) crystal matrix. When compared to regolith underneath the GSL salt terraces as an environmental control, these soil-only microorganisms are different than all of the aforementioned evaporite mineral matrices. This is consistent with carbonate minerals in GSL, which feature different microbial inhabitants than in the surrounding brine (Lindsay et al. 2017).

Our astrobiology studies of the modern GSL system (Baxter et al. 2007) aid in understanding ancient Mars lake systems and the evaporation of ancient aqueous settings that have left behind the same mineral assemblages (Perl 2019). Extreme life on our own planet in hypersaline settings gives us a wealth of knowledge for envisioning potentially extreme cellular life elsewhere in the solar system (Rothschild 1990).

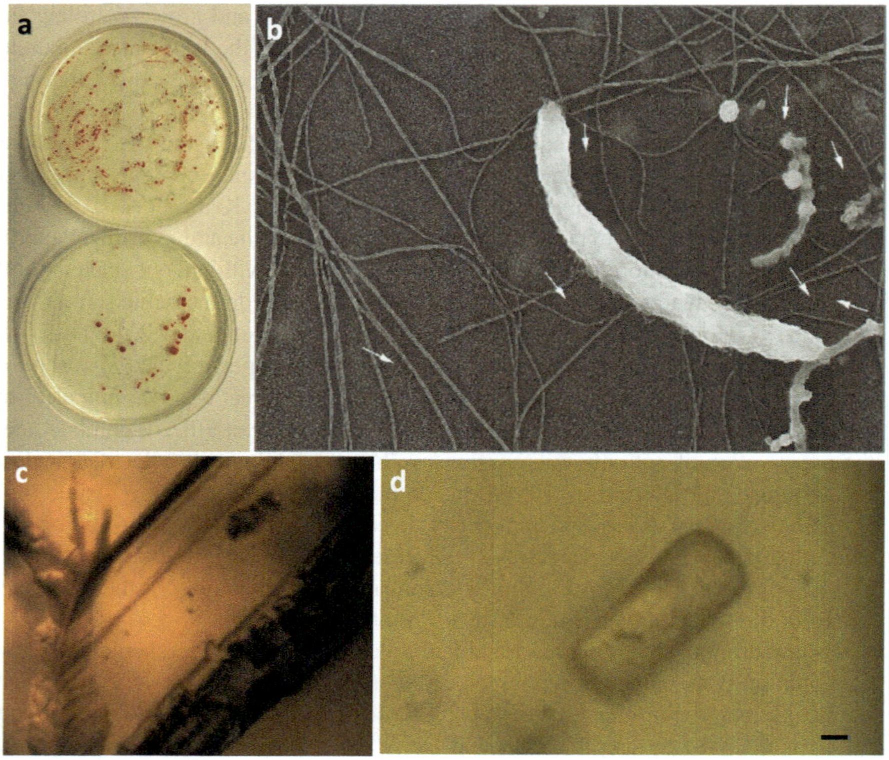

Fig. 16.3 Preservation of cells and molecules in modern Great Salt Lake halite. (**a**) Single halite hopper crystals (some treated with 1M HCl to sterilize the surface), as depicted in Fig. 16.2b, were each dissolved in 10 ml of sterile dionized water. 100 μl of the solution was plated on individual salt media agar plates, and the cells were allowed to grow for two weeks at 37 °C. The upper plate, from a hopper not surface sterilized, shows a higher colony count than the lower plate, representing a sterilized halite crystal. The microorganisms growing on the lower plate then are from within the crystal structure and not incidental to the hopper surface (Day et al. 2005). (**b**) Transmission electron microscopy of cells and molecules preserved in modern Great Salt Lake halite crystals. A curved microorganism [possibly *Salinibacter* sp. (Almeida-Dalmet and Baxter 2020)] , 3.5 μm long, fills the image on a background of cellular debris. Cellulose molecules are seen as thick ropes (Griffith et al. 2008), and arrows mark slender DNA strands, image credit: Jack D. Griffith. (**c**) Halite hopper crystal from Great Salt Lake at 10× resolution showing several rectangular fluid inclusions with zoomed-in area. (**d**) A single fluid inclusion, under 100× resolution, with microorganisms inside. Scale bar is 2 μm

16.3.3 Lessons from Ancient Evaporites on Earth

An example of an older terrestrial evaporite system is the Salado Formation in New Mexico, USA. This formation is a 600-m section of halite deposited ~253 Myr (Lowenstein 1988; Holt and Powers 1990; Powers et al. 2001; Beauheim and Roberts 2002; Satterfield et al. 2005). Due to the overburden rock (Lowenstein 1988), this halite deposit is protected from ionizing radiation and surface water contamination. This coupled with the fact that halite features low levels of naturally occurring radioactive isotopes means that the biological molecules in these halite fluid inclusions may be well-preserved. Indeed, our studies located preserved cellulose, a carbohydrate made by plants and algae (Griffith et al. 2008), and DNA (Griffith and Baxter unpublished) from the Permian salt deposit. If we compare modern GSL and the surrounding 13,000-year-old Bonneville Salt Flats to this ancient system, we can envision time points in the process of evaporite formation and the preservation of the biological molecules that are sheltered there.

16.3.4 Mars Evaporites

Globally across Mars, there have been orbital and in situ detections of evaporites (Murchie et al. 2009; Squyres et al. 2004; Ehlmann and Edwards 2014) showing a wide diversity of minerals precipitated primarily from water or secondary modifications by water (Ehlmann et al. 2011). The NASA Mars Exploration Rover, *Opportunity,* uncovered several outcrops in the Endurance Crater section of the Burns Formation where groundwater had upwelled and dissolved pore spaces (Fig. 16.4) of former Mg and Fe–sulfate minerals (McLennan et al. 2005; Perl et al. 2007). These groundwater recharge cycles observed in Eagle, Endurance, Erebus, and potentially Victoria craters showed how evaporite minerals can not only exist on the surface via groundwater breaching the top subsurface crustal layers (McLennan et al. 2005; Grotzinger et al. 2015; McLennan and Grotzinger 2008) but also be precipitated as buried evaporite mineral deposits. Hydrogeological modeling showed how widespread late Noachian/early Hesperian waters might have been in Meridiani Planum and the evaporation of these volumes of water would have yielded significant and buried mineralogy that were/are removed from the penetrating short-wave ultraviolet (UV) C light conditions on modern Mars (Andrews-Hanna and Lewis 2011).

More recent work by the Mars Science Laboratory (MSL) rover, *Curiosity,* delivered high-resolution photographs of detailed box-work structures of calcium sulfate veins in the "Garden City" outcrop within Gale Crater (Fig. 16.5a) (Milliken et al. 2010; Kronyak et al. 2019). Due to a lack of plate tectonics on Mars, mineral assemblages precipitated from late Noachian/early Hesperian surface and subsurface waters are still in the same relative place as they were during initial precipitation or modification by ancient Martian waters. Given the observation of salt mineral assemblages, similar to GSL, on Mars and their widespread diversity of chlorides, phyllosilicate clays, sulfate salts, and other hydrated minerals it is likely that many of the ancient lake systems had higher than seawater salt concentrations. On Earth, ocean salinity is ~34 g/L whereas in closed-basin lake systems salinities can range tenfold from this value (Hammer 1986). The widespread evaporite mineralogy and closed basin nature of many of the regions on Mars have been observed by digital elevation models from the High Resolution Imaging Science Experiment (HiRISE) and the Compact Reconnaissance Imaging Spectrometer for Mars (CRISM), which is onboard the Mars Reconnaissance Orbiter (MRO) (Viviano-Beck et al. 2014;

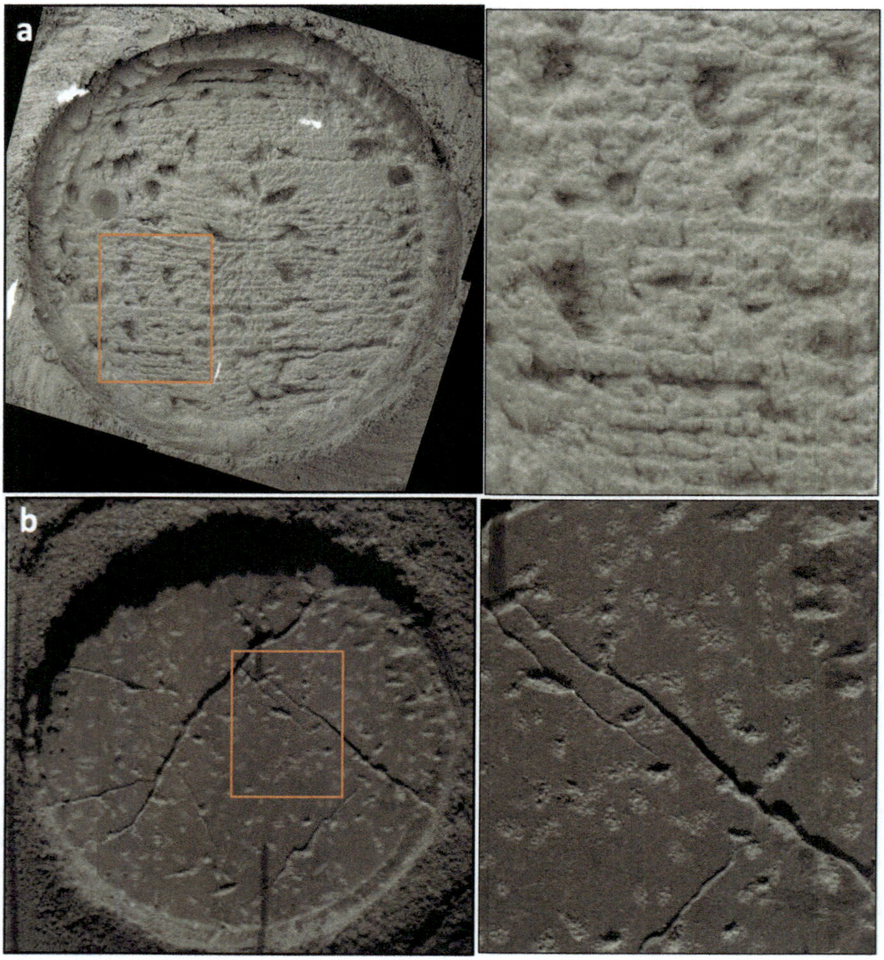

Fig. 16.4 Differences in secondary porosity from rocks within Endurance crater (**a**) and Erebus Crater (**b**). Rectangle on the left shows the enlarged region of each on the right. Scale bars are 3 mm each. Secondary pores in Endurance crater are sheet-like vugs that grade into channel pores whereas pores in Erebus are mostly cubic. The frequency of pores in Erebus shows more of a hopper-like crystal habit implying that the parent minerals may have been halite or other highly soluble mineral. Parent minerals in Endurance Crater were likely Mg/Fe sulfates (Perl et al. 2007)

Ehlmann and Edwards 2014; Weitz and Bishop 2019; Jolliff et al. 2019; Ye and Glotch 2018). A spectacular example of this is at the future landing site of the *Perseverance* rover to be deployed in the summer of 2020 to Mars, Jezero Crater (Fig. 16.5b), where smectite clay minerals and hydrated sulfate minerals (Goudge et al. 2015) show similar mineralogy to the north shores of GSL. These geochemical models and joint CRISM–HiRISE digital elevation models show how closed-basin hydrogeological settings store these hydrated minerals. This suggests the existence of potentially hundreds of ancient salt lake sites on the Martian surface.

Fig. 16.5 Evaporite minerals on Mars, image credit: NASA, JPL. (**a**) The box-work structure shows calcium sulfate veins (Kronyak et al. 2019) in the "Garden City" outcrop within Gale Crater, Mars (March 18, 2015). (**b**) Jezero Crater, Mars, the chosen landing site for the upcoming Mars rover mission, *Perseverance*, and where the first samples will be collected for Mars sample return. High Resolution Imaging Science Experiment (HiRISE) and the Compact Reconnaissance Imaging Spectrometer for Mars (CRISM) map of a delta within Jezero Crater shows smectite clay minerals (green) and hydrated sulfate minerals in (warmer colors). The mineral composition (Goudge et al. 2015) is analogous to that of Great Salt Lake evaporites

On Earth, studies on GSL have demonstrated transfer of biological material or cells from the lake to the evaporitic mineralogy (Fig. 16.3) (Day et al. 2005; Baxter et al. 2007; Kemp et al. 2018; Perl 2019). This process uses the hypersaline waters as the source for both the mineral precipitation and the transference mechanism. On Mars, these similar minerals from late-Noachian waters would have remained behind as the atmosphere dissipated over millions of years (Mancinelli et al. 2004). In addition, the halite and gypsum deposits on Mars are favorable to radiation protection that supports photosynthesis by allowing access to visible light but screening lower wavelengths (Cockell and Raven 2004).

Recent evidence of seasonally flowing brine fluids have been observed by CRISM in the form of the Recurring Slope Lineae (RSL) (Ojha et al. 2015). These dark slope streaks seem to become elongated over several Martian years on crater walls of steep angles of repose leading into the possibility of these streaks being sourced by a highly viscous brine that extends during the warmer Martian seasons and remains at its previous length during the colder months. The survivability and somewhat stable nature of potential surface brines bodes well for subsurface fluidic flow where more recent evaporite mineralogy may be precipitated on modern Mars. Significant comparisons between modern saline lake deposits on Earth with those of the late Noachian/ early Hesperian lakes on Mars have been published (Benison 2006).

16.4 Great Salt Lake Halophilic Archaea as Potential Astrobiological Models

The halophilic microorganisms that thrive in the salt-saturated regions of GSL teach us about the limits of life on Earth, and these lessons maybe applied elsewhere in the Universe (Rothschild 1990; Litchfield 1998; DasSarma 2006; Baxter et al. 2007, 2013). GSL halophilic archaea can survive high doses of UV light, desiccation of their environment, and osmotic challenges. These polyextremophile microorganisms may be excellent life forms to study when considering a search for potential extant (existing) or extinct life in a Martian evaporite formation.

16.4.1 Overcoming the Challenge of High Salinity

If microbial life were ever present on Mars during a time when water was disappearing from the planet's surface, then that life would likely have been halophilic in these areas since these cells could tolerate the hypersaline waters resulting from that evaporation and mineral concentration. These microorganisms would have had the opportunity for evolution over the slow-changing geologic time to manage the osmotic stress as their once aqueous ecosystem dried up (Fig. 16.6). On Earth, we know that rapid micro-evolution, in certain genetic pathways, can give microorganisms the ability to survive harsh environmental conditions, including adaptations to high salinity, within a few generations (Chen et al. 2018).

In a saturated brine, water molecules that interact with ions are less available to support life, and some have theorized that life cannot tolerate the saturated acidic Martian brines (Tosca et al. 2008). However, there is ample evidence from several hypersaline environments where halophilic organisms do maintain homeostasis and thrive despite the lower water activity (a_{w}) and acidic saline lake environs (Grant 2004; Fendrihan et al. 2012; Benison et al. 2008; Stevenson et al. 2015). Terrestrial salt lakes include more neutral or basic fluids (such as GSL) and those remain strong analogues to understanding how life can thrive in high salt settings with many of these overlapping features occurring in the Burns Formation fluids when surface waters were stable.

To live in salt-saturated brine, halophilic archaea must balance osmotically such that their cells do not shrivel up due to water loss. This is accomplished in part by the intracellular accumulation of osmotica, which balances against the salt on the outside of the cell membrane (Brown 1976). Halophilic archaea are shown to accumulate organic compatible solutes (e.g., ectoine or glycerol) in addition to ions (Larsen 1967; Galinski and Trüper 1982; Galinski 1993, 1995; Oren 1999), which explains their success in salty environments like GSL. These extremophiles also have modifications in their proteins that help them function at high salt (Litchfield 1998). Another strategy that has been characterized in halophilic archaea, Na^+ pumps, which will push toxic Na^+ ions out of the cell, while concentrating K^+ ions within the cell in order to balance osmotic pressure (Gilmour 1990; Litchfield 1998) (Fig. 16.6). As Martian lakes evaporated, the water would have become more and

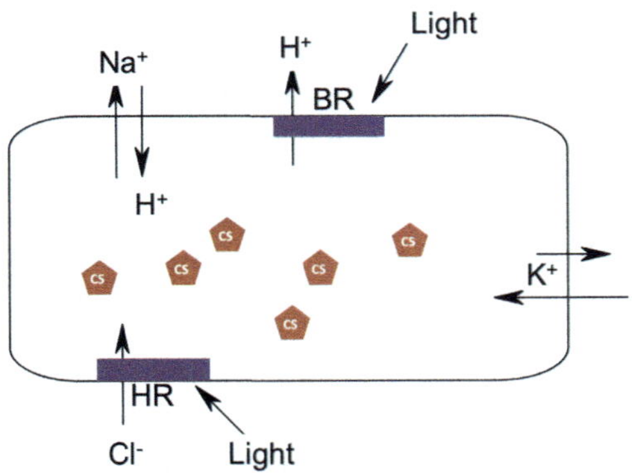

Fig. 16.6 The ability of halophilic archaea to survive high osmotic stress is related to their cellular biochemistry. Some species have a membrane protein with a retinal component, bacteriorhodopsin (BR). This light-mediated cellular pump creates a proton gradient and pumps Na+ out of the cell to avoid accumulating the ion, which is toxic at high levels. For each photon absorbed, BR pushes one proton across the membrane, and the proton gradient generates ATP, giving free energy to these pseudo-photosynthetic microorganisms. K^+ is brought into the cell to balance osmotic pressure created from the salt outside the cell membrane. Light-mediated Halorhodopsin (HR) pumps Cl^- into the cell, which balances the K^+ ions inside. In addition, intracellular accumulation of osmotica, or compatible solutes (CS), aid is balancing with the high salt solution outside the membrane (Adapted from Gilmour 1990; Litchfield 1998; Allred and Baxter 2019)

more saline. The (potential) last life flourishing in this brine would have been halophilic and may have had physiology related to modern GSL microorganisms.

16.4.2 Overcoming the Challenge of Desiccation

As GSL waters recede in drought cycles, precipitated minerals become a part of the matrix of the evaporite left behind (Fig. 16.2). Desiccation may continue for years at GSL (Wurtsbaugh et al. 2017; Null and Wurtsbaugh 2020; Baxter and Butler 2020). Can halophilic archaea survive this drying process? In fact, such biota on Earth is very resistant to drought: endolithic halophilic cyanobacteria can survive in and around halite crystals in the hot, dry climate of the Atacama Desert (Wierzchos et al. 2006). Recent studies have shown the same GSL halophilic archaea surviving in evaporite minerals (Baxter et al. 2007; Kemp et al. 2018; Perl 2019), and Fig. 16.3 demonstrates halophilic archaea seen directly from GSL halite.

As the halite crystals form, small pockets of brine are trapped within the salt structure, and this may be the secret to the survival of life, which has evolved to live in salt-saturated fluid. Studies of modern halophilic archaea that were desiccated and embedded in salt crystals indicate that the cells move toward the fluid inclusions and they can survive cold temperatures, but they suffer DNA damage (Fendrihan and Stan-Lotter 2004). The mechanism of dormancy is unknown but may involve low-grade metabolism and an effective DNA repair process (Johnson et al. 2007). As the rate of crystal growth increases, the quantity of fluid inclusions also increases, being greatest in the center of the crystal (Roedder 1984). Studies suggest that halophilic archaea become trapped within these fluid pockets of the halite crystals.

Beyond modern examples of halophilic archaea surviving in halite crystals, several studies point to the possibility of the survival of halophilic archaea in salt over geologic time (Norton and Grant 1988; Norton et al. 1993; Denner et al. 1994; Grant et al. 1998; Stan-Lotter et al. 1999; McGenity et al. 2000; Vreeland et al. 2000; Stan-Lotter et al. 2002a, b; Kminek et al. 2003; Mormile et al. 2003; Gruber et al. 2004; Park et al. 2009; Schubert et al. 2009, 2010; Lowenstein et al. 2011; Sankaranarayanan et al. 2011). However, the viability of the entombed cells is hypothesized to decrease over time as resources are depleted (Norton and Grant 1988). For example, crystals from a 253-million-year-old deposit found that two of 52 examined halite samples contained viable cells (Vreeland et al. 2000). Although the effect of dormancy is not understood, we do expect cells could experience some molecular stress from starvation. Rod-shaped halophilic archaea transitioned to a spherical morphology within a few weeks following crystal formation in the laboratory (Norton and Grant 1988) showing the pleomorphism typical of halophilic archaea in a starved state.

We do know in the GSL system, halophilic microorganisms can survive desiccation in their environment (Baxter et al. 2007; Kemp et al. 2018; Perl 2019) and may even color the halite left behind (Fig. 16.2). Since GSL halophilic archaea can survive desiccation inside minerals on Earth, there are two important lessons for Mars studies here: (1) Halophilic life are appropriate models for cells that survived evaporite formation on Mars, and (2) Evaporites are an excellent place to look for extant or extinct life on the red planet.

16.4.3 Overcoming the Challenge of High Ultraviolet Exposure

In addition to the high salt levels and desiccation, UV light exposure is intense in GSL. UV radiation penetrates more deeply in saline water (Huovinen et al. 2003). Wind activity and shallow waters increase UV penetration, even in areas of high dissolved organic carbon (Arts et al. 2000). GSL is in an elevated desert biome, and the high altitude (1280 m) results in increased UV exposure. Depending on the wavelength of UV light measured, the increase of UV exposure (300–370 nm) ranges between 9 and 24% per one thousand meters (Blumthaler et al. 1997). Also, salt in and around such lakes causes mobilization of atmospheric chlorine, which has depleted ozone concentrations, leading to more UV exposure (Stutz et al. 2002). And yet, GSL halophilic archaea thrive in this extreme environment to which they have adapted.

The damaging effects of UV light exposure result in helix-distorting damage to the DNA and also create oxidative damage to other cellular components (Friedberg 2003). All known life forms on Earth have DNA repair mechanisms, as the consequence of DNA lesions, for any organism is mutation or cell death. However, halophilic archaea are highly resistant to UV light (Dundas and Larsen 1963; Baxter et al. 2007; Jones and Baxter 2016; Jones and Baxter 2017). When embedded in salt, they are particularly resistant (Fendrihan et al. 2009). They do have efficient DNA repair (McCready and Marcello 2003; Kish and DiRuggiero 2012; Jones and Baxter 2017), but extreme halophilic microorganisms also typically contain carotenoid pigments in their membranes, which mitigate their photobiology by preventing DNA damage (White and Jahnke 2002; Baxter et al. 2007; Jones and Baxter 2017). In addition, halophilic archaea possess genome strategies to limit DNA damage such as bipyrimidine limitation (Jones and Baxter 2016) and polyploidy (Jones and Baxter 2017).

Mars features intense surface exposure of UV radiation, due to the planet lacking a significant ozone layer in combination with a lower total atmospheric pressure than Earth (reviewed in Cockell et al. 2000). The Martian UV radiation environment has high UVB and UVC fluxes, which are lower wavelengths than UVA and more damaging to life. In fact, the accumulated damage to potential cellular life on Mars could be amplified due to longer days. Dust on Mars may mitigate UV exposure on the surface (Hagen et al. 1970; Green et al. 1971; Horneck et al. 2001; Fendrihan et al. 2009), but evaporites below the surface may be more protective (Fendrihan et al. 2009). GSL halophilic microorganisms could likely survive these extreme UV conditions, especially if they were sheltered in the mineral deposits.

16.5 The Preservation of Biosignatures by Halite and Gypsum

Considering timescales of geological changes are magnitudes longer than the adaptation of halophilic and other extreme life, survivability of biological evidence in ever-changing hypersaline settings depends on both physical and chemical features, based on the carbon-based chemistry of life and preservation potential of molecules (Summons et al. 2008; Hays et al. 2017). Our knowledge of biological systems informs the search for metabolites or biochemicals in the environment that may indicate if life is present or once was. Physical *biosignatures* in this case (e.g., pigments, organic layering, and fossilization) may be seen with the naked eye or a visualization technique. Chemical *biomarkers* (e.g., amino acids, hopanes, fatty acids, and other long-chained biological macromolecules) cannot be detected visually but through chemical analyses, and their presence alone may not be diagnostic of life as some can be produced abiotically (e.g., amino acids).

Both biosignatures and biomarkers can be indicative of extinct or extant biology (Chan et al. 2019). These are useful tools with which to validate the former or current existence of the cell-based metabolisms of life, especially if analyzed in conjunction with knowledge of the ancient ecosystem. Terrestrial evaporite minerals halite and gypsum can preserve labile biological and cellular products over geologic time, creating a micron-scale ecosystem for life to continue post-preservation (e.g., Perl 2019). This post-preservation microcosm varies and can allow for modification of the evaporitic fluids and mineral substrates by the biological components. The authors demonstrated the variation of preserved microbial communities solely based on the evaporite mineral set from where they were entombed. Halite and gypsum salt-saturated fluid inclusions may offer solar protection from UV radiation and desiccation as well as shielding from potentially damaging enzymatic processes. While we have studied these fluid inclusions in our work at GSL, their importance is also noted in prior work by others (e.g., Jehlička et al. 2014; Roedder 1984; Van den Kerkhof and Hein 2001; Fendrihan et al. 2009; Winters 2013). Moreover, within these micron-scale fluid inclusion settings, motile halophilic GSL microorganisms have been observed over multiple hour-long timescales, suggesting non-Brownian motion. While the aforementioned macromolecules are well preserved, the flagellated microorganisms are also preserved (Perl 2019).

Isolated pockets of brine trapped in halite crystalline structures have been used to study the ancient chemical environments of the original fluid, in addition to the microorganisms and molecules from ancient waters (e.g., Griffith et al. 2008; Satterfield et al. 2005; Benison 2006; Lowenstein et al. 2011). Even if cells are no longer present, the chemistry of this ancient fluid, and the biological molecules stored inside, can provide clues about extinct life on early Earth or potentially on other space bodies. Consideration should be made for the stability of each type of molecule and the environmental radiation exposure over time (Kminek et al. 2003; Fendrihan et al. 2009). Which types of macromolecules can survive and for how long? This will inform work on Mars as we search for organics (Wiens et al. 2017).

Chemical biomarkers, such as nucleic acids (RNA and DNA) are obvious signatures for biology, with DNA being far more stable than RNA. Predictions of the rates of spontaneous DNA depurination, backbone cleavage, and other processes that degrade DNA suggest that large molecules should be degraded into small fragments within tens of thousands of years (Lindahl and Nyberg 1972; Lindahl 1993; Schroeder et al. 2006; Pääbo et al. 2004; Willerslev et al. 2004). In addition, these DNA molecules would lose some molecular integrity due to spontaneous chemical cross-linking (Hansen et al. 2006; Pääbo 1989). However, a number of studies have reported the presence of ancient DNA from minerals. For example, DNA molecules were isolated from a number of Permian halite samples (Radax et al. 2001), and even from salt that is 425 million years old (Fish et al. 2002). Ancient fluid inclusions in gypsum have also been found to harbor DNA (Panieri et al. 2010). Ancient DNA can likely remain in the environment over much longer geologic times if entrapped in minerals. It is important to note then that the measurements of DNA degradation over time do not address DNA in saturated salts, where these reactions would occur more slowly. Also, there is evidence of DNA repair within these ancient cells (Johnson et al. 2007). Considering DNA as an information storage molecule, we should keep in mind that molecular evolution of our own planet may not be the same for Mars or other space bodies. If life ever was present outside of Earth, it is likely that it had a separate evolutionary path than that of what we know of on our own planet (Woese 1987; Hug et al. 2016).

Carbohydrates can be preserved in halite. The polysaccharide, cellulose, is highly stable and relatively resistant to harsh conditions (Okuda et al. 1993; Kvien et al., 2005; Morán et al. 2008). Cellulose may be one of the most ancient biopolymers, and it is abundant within the biosphere of Earth (Cox et al. 2000). It is produced by cyanobacteria, perhaps one of the earliest known life forms, as well as other prokaryotic and eukaryotic species (Nobles et al. 2001). Cellulose can remain intact for a quarter of a billion years in halite fluid inclusions, as detected by electron microscopy (Griffith et al. 2008). Given that cellulose also has an infrared signature (Kondo and Sawatari 1996), it could potentially be detectable from orbit. This would take an extreme volume/quantity of such a sample and that does not take into account atmospheric opacity, dust coverage on the surface, among many other natural obstacles. Cellulose does not contain genetic information like nucleic acids, nor could a code be inferred as in the case of proteins. It would, however, provide a biosignature as it is not known to be produced abiotically.

Proteins may have high preservation potential and robustness over geologic time. For example, 68 Ma collagen fragments were isolated from dinosaur bones (Schweitzer et al. 2007). The building blocks of proteins, amino acids, may be key biomarkers, however, they can be produced by biotic or by abiotic processes (Bada 2001). Methods have been developed to detect the amino acid chirality (biotic processes produce only the L-form while abiotic reactions produce a 50:50 ratio of L and D forms), which can distinguish amino acids from biological sources versus abiotic ones (Hutt et al. 1999; Skelley and Mathies 2003).

Lipids may be fossilized in the rock record, especially in ancient marine and semiaquatic microbial mats (Pawlowska et al. 2013). Lipid fossils do not contain genetic information, but they are valuable biosignatures because specific types may be indicative of particular categories of microorganisms. The biopreservation of the chemical structure of lipids over geologic time has been demonstrated in halite fluid inclusions (Winters 2013) and in GSL sediment cores as well (Collister and Schamel 2002). Given that halophilic archaea contain carotenoid pigments in their membranes (Jones and Baxter 2017), it is relevant here that cyclic and branched hydrocarbons are stable over geologic time (Simoneit 2004). These compounds provide protection against high exposures to UV radiation (White and Jahnke 2002; Baxter et al. 2007; Jones and Baxter 2017) and at GSL, these pigments point to regions of halite that are enriched with microbial life (Perl 2019).

There has been much controversy over the study of ancient biomolecules in halite due to the high probability of contamination, for example the minerals may not provide a truly protected environment, allowing modern bacteria and archaea to contaminate the samples. This is a concern in the in the natural environment as well as in the laboratory where modern DNA or cells could be introduced (e.g., Pääbo et al. 2004; Hebsgaard et al. 2005). Of particular concern are analyses of environmental samples utilizing amplification methods such as the polymerase chain reaction (PCR) (e.g., Fish et al. 2002) or microbial cultivation techniques (e.g., Vreeland et al. 2000), which have the caveat of contamination possibilities. Other studies may avoid this with extensive surface sterilization (Sankaranarayanan et al. 2011) or by employing more direct methods like electron microscopy of the fluid from the inclusions, coupled with biochemical assays to identify the molecules (Griffith et al. 2008). Raman analyses of in situ minerals can provide a nondestructive methodology to demonstrate the presence of halophilic organisms inside fluid inclusions (Perl 2019) and monitor their biological processes over short-time periods. Moreover, Raman may be a valuable tool for future astrobiology and planetary landed campaigns to determine sites of interest, to examine the survivability of organics, and perhaps to find potential life with different chemistry.

16.6 Upcoming Planetary Rover and Recommendations for Returned Sample Analysis

The MSL Mars 2020 rover, *Perseverance*, will be the first part of a multistep process in returning Martian regolith and mineral samples to Earth for analyses and coring sites will in part be determined with Raman spectroscopic signature data (Wiens et al. 2017). The science objectives and the return sample science team will ensure that soil and mineral samples are collected and cached properly such that they remain representative of the Noachian sediments likely containing Fe/Mg smectites, Mg-carbonates, and other biosignature-relevant samples yet to be discovered. Explored sites in the Burns Formation set of abraded rocks revealed secondary pore spaces previously occupied by Fe/Mg sulfate salts. These minerals were later dissolved due to groundwater recharge events (McLennan et al. 2005; McLennan and Grotzinger 2008) leaving behind the aforementioned porous (and in the case of the Karatepe section, permeable) set of rocks in a ~7 m vertical section. Clark et al. (2005) note the fluid chemistry that needed to dissolve these original minerals and, due to the orientation of the secondary porosity (Perl et al. 2008, Fig. 16.4) generated post-fluid activity, these sections would be ideal for collection by *Perseverance* and a return sample analysis. The ancient fluid interactions within these rocks, as well as the specific mm-scale sections that have had specific ratios of water–rock interactions, should be taken into account when the next mission provides future Martian samples for analysis back on Earth. The Mars rover's coring ability will not be able to preserve microtextures but will provide the Noachian sediment that will be eventually returned to Earth in later sample return mission steps.

Mars mission strategies for the future return sample science campaign, following collection by *Perseverance,* will include analyses of Martian material that may have entrapped ancient material in its matrix. As Perl (2019) demonstrated, halite and gypsum mineralogy in the modern GSL environment at Rozel Bay have entombed carotenoid biomarkers within the NaCl mineral matrices and clays within the $CaSO_4{\cdot}2H_2O$ crystal shard structures. The evaporites of GSL show how modern biosignature preservation occurs from the lake water column and is transferred into precipitating minerals using the same fluid sources (e.g., the shallow lake waters) and annual evaporation–precipitation cycles. This process on Earth provides us with a perspective into how evaporite minerals on Mars may have entombed biosignatures or biomarkers, and this should be applied to the future sample analyses of Mars minerals, studied in situ by *Perseverance*, when they are returned to Earth for study in terrestrial laboratories.

Due to the lack of plate tectonics on Mars, mineral veins that were formed from late-Noachian/early-Hesperian surface and ground waters are still in roughly the same location as they were ~3.5 Gyr ago. The static nature of former aqueous features, mineral precipitation, and mineral transformation processes yields us the chance to potentially discover well-preserved organics and perhaps biology from an independent abiogenesis event if life began independently on the Red Planet. GSL provides an analogous system to apply to study Mars evaporites.

16.7 Life as We Do Not Know It

When comparing Earth life to potential life on Mars, an important caveat is that we are looking for life as defined on our planet. We have only a single sample of life arising in the Universe, as all life forms on Earth are related to the Last Universal Common Ancestor (LUCA) (Lazcano and Forterre 1999; Penny and Poole 1999; Koonin 2003; Weiss et al. 2016). Scientists have made considerable progress in pursing this concept and completing comparative genetic experiments with the goal of reconstruction of LUCA. We know that all examples of life on Earth use DNA as genetic material and ribosomes to produce proteins, therefore, we assume LUCA performed these functions in the same way with the same molecules. But what if there was a second origin of life event in our Solar System? Would it necessarily be based on the same molecular mechanisms or follow Earth's same evolutionary path? The original series Star Trek character, Spock, once described a (fictional) extraterrestrial silicon-based life form as "no life as we know it," meaning it was not carbon-based as we understand life to be. This quote and mindset have been adapted and expanded on by scientists in astrobiology to open our scientific thinking to life as we do not know it (e.g., Ward 2007). In our discussions herein, we have considered entrapment of life and/or life's molecules in the mineral record of Mars, but we have been limited in our scope, looking at the molecules of life only as we know them on Earth.

16.8 Insights

GSL is a hypersaline ecosystem, a model extreme environment, and an analogue for salty sites on other space bodies in the universe. The microbial life here may give us clues to potential Martian life from ancient salty seas. Clearly, halophilic archaea can handle the multiple extremes of osmotic challenges, desiccation, and the high solar radiation of their environment. All of these are potential challenges in space, and GSL microorganisms are critical to understanding life at its limits. Halophile communities that can survive in fluid inclusions over geologic time provide a model for life that could be remaining on Mars.

If aquatic life was on Mars, but it is now gone, will we find evidence of it in the rock record? Our work and that of others suggest that biopreservation in evaporite minerals can retain biosignatures over *hundreds of millions* of years. While this provides optimism, on Mars we assume the molecules or cells may have been entombed in the environment for *billions* of years due to the timeline of surface water loss on that planet. Mars and Earth formed around the same time, ~4.5 billion years ago (Bouvier and Wadhwa 2010). But Mars went through climatic changes, losing its atmosphere, surface water, and tectonic activity around an estimated 3.5 billion years ago (Carr and Wänke 1992; Jakosky and Phillips 2001; Maltagliati et al. 2011). During this time, prokaryotic life on Earth was flourishing, but more complicated eukaryotic cells would not evolve for another 1.5 billion years (Grosberg and Strathmann 2007). Since conditions on Earth that favored biological evolution toward complexity were altered on Mars, we could argue that any potential life on Mars would be microbial. The calculation of possibilities of life off our planet must also consider the proximity of Mars and Earth in the "architecture" of our Solar System, as these bodies could have exchanged life forms (Scharf and Cronin 2016). In this panspermia model, if we discover complex organics or cells on Mars that are related to Earth's biosignatures, we may not know which appeared first.

Direct environmental sampling on Earth and in situ analyses of minerals is critical to inform methodologies for missions on Mars. Work on biosignature preservation in gypsum and halite inform mission work on methodologies, but also on determining molecular targets for current studies as we prepare for sample return. GSL plays an important role as an analogue for ancient Martian salt lakes and evaporitic systems.

Acknowledgments Funding for some of the work reported here was from the NASA Utah Space Grant Consortium to B.K.B. (NNX15A124H, Sub-Award 10037896WEST), and the Caltech/JPL Presidents and Directors Research Fund award to S.M.P. We would like to thank Jaimi Butler for her assistance with sample collection efforts in 2014. These samples led to the original proof of concept for in situ modern biogenic preservation and its application to potential extant halophilic life. We are also indebted to Aaron Celestian for his analyses in mineral–microbe systems, mineral precipitation experiments, and his continued collaborations in our ongoing investigations. Finally, we would like to thank Frank Corsetti for investigation guidance and support since the beginning of this investigation.

References

Adams TC (1964) Salt migration to the northwest body of Great Salt Lake, Utah. Science 143 (3610):1027–1029

Allred A, Baxter BK (2019) Microbial life in hypersaline environments, Wood's Hole Microbial Life Series. http://serc.carleton.edu/microbelife/extreme/hypersaline/index.html Accessed 25 May 2019

Almeida-Dalmet S, Baxter BK (2020) Unexpected complexity at salinity saturation: microbial diversity of the north arm of Great Salt Lake, Utah. In: Baxter BK, Butler JK (eds) Great Salt Lake biology: a terminal lake in a time of change. Springer, Cham

Almeida-Dalmet S, Sikaroodi M, Gillevet PM, Litchfield CD, Baxter BK (2015) Temporal study of the microbial diversity of the north arm of Great Salt Lake, Utah, US. Microorganisms 3 (3):310–326

Andrews-Hanna JC, Lewis KW (2011) Early Mars hydrology: 2. Hydrologic evolution in the Noachian and Hesperian epochs. J Geophys Res 116:E02007. https://doi.org/10.1029/2010JE003709

Arts MT, Robarts RD, Kasai F, Waiser MJ, Tumber VP, Plante AJ, Rai H, De Lange HJ (2000) The attenuation of ultraviolet radiation in high dissolved organic carbon waters of wetlands and lakes on the northern Great Plains. Limnol Oceanogr 45:292–299

Atwood G, Wambeam TJ, Anderson NJ (2016) The present as a key to the past: paleoshoreline correlation insights from Great Salt Lake. In: Oviatt CG, Shroder JF (eds) Lake Bonneville a scientific update. Elsevier, Dordrecht

Bada JL (2001) State-of-the-art instruments for detecting extraterrestrial life. Proc Natl Acad Sci 98 (3):797–800

Baxter BK, Butler JK (eds) (2020) Great salt Lake biology: a terminal lake in a time of change. Springer, Cham

Baxter BK, Zalar P (2019) The extremophiles of Great Salt Lake: complex microbiology in a dynamic hypersaline ecosystem. In: Seckbach J, Rampelotto PH (eds) Model ecosystems in extreme environments. Elsevier, Dordrecht

Baxter BK, Litchfield CD, Sowers K, Griffith JD, Dassarma PA, Dassarma S (2005) Microbial diversity of Great Salt Lake. Microbial diversity of Great Salt Lake. In: Gunde-Cimerman N, Oren A, Plemenitaš A (eds) Adaptation to life at high salt concentrations in archaea, bacteria, and eukarya. Cellular origin, life in extreme habitats and astrobiology, vol 9. Springer, Dordrecht, Netherlands, pp 9–25

Baxter BK, Eddington B, Riddle MR, Webster TN, Avery BJ (2007) Great Salt Lake halophilic microorganisms as models for astrobiology: evidence for desiccation tolerance and ultraviolet radiation resistance. In: Hoover RB, Levin GV, Rozanov AY, Davies PCW (eds) Instruments, methods, and missions for astrobiology X, vol 6694. SPIE, Bellingham, WA, p 669415

Baxter BK, Butler JK, Kleba B (2013) Worth your salt: halophiles in education. In: Vreeland RH (ed) Advances in understanding the biology of halophilic microorganisms. Springer, Dordrecht, Netherlands

Beauheim RL, Roberts RM (2002) Hydrology and hydraulic properties of a bedded evaporite formation. J Hydrol 259:66–88

Benison KC (2006) A Martian analogue in Kansas: comparing Martian strata with Permian acid saline lake deposits. Geology 34:385–388

Benison KC, Jagniecki EA, Edwards TB, Mormile MR, Storrie-Lombardi MC (2008) "Hairy blobs:" microbial suspects preserved in modern and ancient extremely acid lake evaporites. Astrobiology 8(4):807–821

Blumthaler M, Ambach W, Ellinger R (1997) Increase in solar UV radiation with altitude. J Photochem Photobiol B Biol 39:130–134

Boogaerts GL (2015) Preliminary characterization of the microbial community in the Bonneville Salt Flats. Thesis, The University of Alabama at Birmingham

Bouvier A, Wadhwa M (2010) The age of the solar system redefined by the oldest Pb–Pb age of a meteoritic inclusion. Nat Geosci 3(9):637

Boyd ES, Hamilton TL, Swanson KD, Howells AE, Baxter BK, Meuser JE, Posewitz MC, Peters JW (2014) [FeFe]-hydrogenase abundance and diversity along a vertical redox gradient in Great Salt Lake, USA. Int J Mol Sci 15:21947–21966

Boyd ES, Yu R-Q, Barkay T, Hamilton TL, Baxter BK, Naftz DL, Marvin-DiPasquale M (2017) Effect of salinity on mercury methylating benthic microbes and their activities in Great Salt Lake, Utah. Sci Total Environ 581–582:495–506. https://doi.org/10.1016/j.scitotenv.2016.12.157

Brown AD (1976) Microbial water stress. Bacteriol Rev 40:803–846

Cannon JS, Cannon MA (2002) The southern Pacific railroad trestle - past and present. In: Gwynn JW (ed) Great Salt Lake: an overview of change. Utah Geological Survey, State of Utah Department of Natural Resources, Salt Lake City, UT, pp 283–294

Carr MH, Wänke H (1992) Earth and Mars: water inventories as clues to accretional histories. Icarus 98(1):61–71

Chan MA, Hinman NW, Potter-McIntyre SL, Schubert KE, Gillams RJ, Awramik SM, Boston PJ, Bower DM, Des Marais DJ, Farmer JD, Jia TZ, King PL, Hazen RM, Léveillé RJ, Papineau D, Rempfert KR, Sánchez-Román M, Spear JR, Southam G, Stern JC, Cleaves HJ (2019) Deciphering biosignatures in planetary contexts. Astrobiology 19(9):1075–1102

Chen Y, Shenkar N, Ni P, Lin Y, Li S, Zhan A (2018) Rapid microevolution during recent range expansion to harsh environments. BMC Evol Biol 18(1):187

Clark BC, Morris RV, McLennan SM, Gellert R, Jolliff B, Knoll AH, Squyres SW, Lowenstein TK, Ming DW, Tosca NJ, Yen A (2005) Chemistry and mineralogy of outcrops at Meridiani Planum. Earth Planet Sci Lett 240(1):73–94

Cockell CS, Raven JA (2004) Zones of photosynthetic potential on Mars and the early Earth. Icarus 169(2):300–310

Cockell CS, Catling DC, Davis WL, Snook K, Kepner RL, Lee P, McKay CP (2000) The ultraviolet environment of Mars: biological implications past, present, and future. Icarus 146(2):343–359

Cohenour RE, Thompson KC (1966) Geologic setting of Great Salt Lake. Utah Geological Society, Salt Lake City, UT

Collister JW, Schamel S (2002) Lipid composition of recent sediments from the Great Salt Lake. In: Gwynn JW (ed) Great Salt Lake: an overview of change. Special Publication of Utah Department of Natural Resources, Salt Lake City, UT, pp 127–142

Cox PM, Betts RA, Jones CD, Spall SA, Totterdell IJ (2000) Acceleration of global warming due to carbon-cycle feedbacks in a coupled climate model. Nature 408:184–187

Crosman ET, Horel JD (2009) Modis-derived surface temperature of the Great Salt Lake. Remote Sens Environ 113:73–81

DasSarma S (2006) Extreme halophiles are models for astrobiology. Microbe 1(3):121–123

Day J, Griffith JD, Baxter BK (2005) The revival of Halophilic Archaea from recently formed halite crystals obtained from Great Salt Lake, Utah. In: Proceedings of the National Conference on Undergraduate Research Lexington, Virginia, April 21–23

Denner EB, McGentity TJ, Busse HJ, Grant WD, Wanner G, Stan-Lotter H (1994) *Halococcus salifodinae* sp. nov., an archaeal isolate from an Austrian salt mine. Int J Syst Evol Microbiol 44 (4):774–780

Domagalski JL, Orem WH, Eugster HP (1989) Organic geochemistry and brine composition in Great Salt, mono, and Walker Lakes. Geochim Cosmochim Acta 53(11):2857–2872

Dundas ID, Larsen H (1963) A study on the killing by light of photosensitized cells of *Halobacterium salinarium*. Archices Microbiol 46:19–28

Eardley AJ, Stringham B (1952) Selenite crystals in the clays of Great Salt Lake. J Sediment Petrol 22(4):234–238

Ehlmann BL, Edwards CS (2014) Mineralogy of the Martian surface. Annu Rev Earth Planet Sci 42:291–315

Ehlmann BL, Mustard JF, Murchie SL, Bibring JP, Meunier A, Fraeman AA, Langevin Y (2011) Subsurface water and clay mineral formation during the early history of Mars. Nature 479 (7371):53–60

Fendrich C, Schink B (1988) Degradation of glucose, glycerol, and acetate by aerobic bacteria in surface water of Great Salt Lake, Utah, USA. Syst Appl Microbiol 11:94–96

Fendrihan S, Stan-Lotter H (2004) Survival of halobacteria in fluid inclusions as a model of possible biotic survival in Martian halite. In: Teodorescu HN, Griebel HS (eds) Mars and planetary science and technology. Performantica Press, Iasi, Romania, pp 9–18

Fendrihan S, Bérces A, Lammer H, Musso M, Rontó G, Polacsek TK, Holzinger A, Kolb C, Stan-Lotter H (2009) Investigating the effects of simulated Martian ultraviolet radiation on *Halococcus dombrowskii* and other extremely halophilic archaebacteria. Astrobiology 9 (1):104–112

Fendrihan S, Dornmayr-Pfaffenhuemer M, Gerbl FW, Holzinger A, Grösbacher M, Briza P, Erler A, Gruber C, Plätzer K, Stan-Lotter H (2012) Spherical particles of halophilic archaea correlate with exposure to low water activity–implications for microbial survival in fluid inclusions of ancient halite. Geobiology 10(5):424–433

Fish SA, Shepherd TJ, McGenity TJ, Grant WD (2002) Recovery of 16S ribosomal RNA gene fragments from ancient halite. Nature 420:432–436
Friedberg EC (2003) DNA damage and repair. Nature 421:436–440
Galinski EA (1993) Compatible solutes of halophilic eubacteria: molecular principles, water-solute interaction, stress protection. Experientia 49(6–7):487–496
Galinski EA (1995) Osmoadaptation in bacteria. Adv Microb Physiol 37:273–328
Galinski EA, Trüper HG (1982) Betaine, a compatible solute in the extremely halophilic phototrophic bacterium *Ectothiorhodospira halochloris*. FEMS Microbiol Lett 13(4):357–360
Gilmour D (1990) Halotolerant and halophilic microorganisms. In: Edwards C (ed) Microbiology of extreme environments. McGraw Hill, New York, pp 147–177
Gornitz VM, Schreiber BC (1981) Displacive halite hoppers from the Dead Sea; some implications for ancient evaporite deposits. J Sediment Res 51(3):787–794
Goudge TA, Mustard JF, Head JW, Fassett CI, Wiseman SM (2015) Assessing the mineralogy of the watershed and fan deposits of the Jezero crater paleolake system, Mars. J Geophys Res 120 (4):775–805
Grant WD (2004) Life at low water activity. Philos Trans R Soc Lond Ser B Biol Sci 359 (1448):1249–1267
Grant WD, Gemmell RT, McGenity TJ (1998) Halobacteria: the evidence for longevity. Extremophiles 2(3):279–287
Green RH, Taylor DM, Gustan EA, Fraser SJ, Olson RL (1971) Survival of microorganisms in a simulated Martian environment. Space Life Sci 3:12–24
Greer DC (1971) Annals map supplement fourteen: Great Salt Lake, Utah. Ann Assoc Am Geogr 61:214–215
Griffith JD, Willcox S, Powers DW, Nelson R, Baxter BK (2008) Discovery of abundant cellulose microfibers encased in 250 Ma Permian halite: a macromolecular target in the search for life on other planets. Astrobiology 8(2):215–228
Grosberg RK, Strathmann RR (2007) The evolution of multicellularity: a minor major transition? Annu Rev Ecol Evol Syst 38:621–654
Grotzinger JP, Gupta S, Malin MC, Rubin DM, Schieber J, Siebach K et al (2015) Deposition, exhumation, and paleoclimate of an ancient lake deposit, Gale crater, Mars. Science 350(6257): aac7575
Gruber C, Legat A, Pfaffenhuemer M, Radax C, Weidler G, Busse HJ, Stan-Lotter H (2004) *Halobacterium noricense* sp. nov., an archaeal isolate from a bore core of an alpine Permian salt deposit, classification of *Halobacterium* sp. NRC-1 as a strain of H. salinarum and emended description of H. salinarum. Extremophiles 8(6):431–439
Gwynn JW (1998) Great Salt Lake, Utah: chemical and physical variations of the brine and effects of the SPRR causeway, 1966-1996. In: Modern and ancient lake systems: new problems and perspectives. AAPG, Tulsa, OK, pp 71–90
Hagen CA, Hawrylewicz EJ, Anderson BT, Cephus ML (1970) Effect of ultraviolet on the survival of bacteria airborne in simulated Martian dust clouds. Life Sci Space Res 8:53–58
Hammer UT (1986) Saline lake ecosystems of the world, vol. 59. Springer, Dordrecht
Hansen AJ, Mitchell DL, Wiuf C, Paniker L, Brand TB, Binladen J, Gilichinsky DA, Rønn R, Willerslev E (2006) Crosslinks rather than strand breaks determine access to ancient DNA sequences from frozen sediments. Genetics 173(2):1175–1179
Hays LE, Graham HV, Des Marais DJ, Hausrath EM, Horgan B, McCollom TM, Parenteau N, Potter-McIntyre SL, Williams AJ, Lynch KL (2017) Biosignature preservation and detection in Mars analog environments. Astrobiology 17(4):363–400
Hebsgaard MB, Phillips M, Willerslev E (2005) Geologically ancient DNA: fact or artefact? Trends Microbiol 13:212–220
Holt RM, Powers DW (1990) Geological and hydrological studies of evaporites in the Northern Delaware Basin for the waste isolation pilot plant (WIPP), New Mexico: guidebook 14. In: Powers DW, Holt RM, Beauheim RL, Rempe N (eds) Geological Society of America Annual Meeting. Dallas Geological Society, Dallas, TX, pp 45–78

Horneck G, Rettberg P, Reitz G, Wehner J, Eschweiler U, Strauch K, Panitz C, Starke V, Baumstark-Khan C (2001) Protection of bacterial spores in space, a contribution to the discussion on Panspermia. Origins Life Evol Biosph 31:527–547

Hug LA et al (2016) A new view of the tree of life. Nat Microbiol 1:16048. https://doi.org/10.1038/nmicrobiol.2016.48

Huovinen PS, Penttilä H, Soimasuo MR (2003) Spectral attenuation of solar ultraviolet radiation in humic lakes in Central Finland. Chemosphere 51:205–214

Hutt LD, Glavin DP, Bada JL, Mathies RA (1999) Microfabricated capillary electrophoresis amino acid chirality analyzer for extraterrestrial exploration. Anal Chem 71(18):4000–4006

Jakosky BM, Phillips RJ (2001) Mars' volatile and climate history. Nature 412(6843):237

Jehlička J, Edwards HG, Oren A (2014) Raman spectroscopy of microbial pigments. Appl Environ Microbiol 80(11):3286–3295

Johnson SS, Hebsgaard MB, Christensen TR, Mastepanov M, Nielsen R, Munch K, Rønn R (2007) Ancient bacteria show evidence of DNA repair. Proc Natl Acad Sci 104(36):14401–14405

Jolliff BL, Mittlefehldt DW, Farrand WH, Knoll AH, McLennan SM, Gellert R (2019) Chapter 10 - Mars exploration rover opportunity: water and other volatiles on ancient Mars in volatiles in the Martian crust 2019, 285–328. https://doi.org/10.1016/B978-0-12-804191-8.00010-6

Jones DL, Baxter BK (2016) Bipyrimidine signatures as a photoprotective genome strategy in G+C-rich halophilic archaea. Life 6(3):37

Jones DL, Baxter BK (2017) DNA repair and photoprotection: mechanisms of overcoming environmental ultraviolet radiation exposure in halophilic archaea. Front Microbiol 8:1882

Jones BF, Naftz DL, Spencer RJ, Oviatt CG (2009) Geochemical evolution of Great Salt Lake, Utah, USA. Aquat Geochem 15(1–2):95–121

Keck W, Hassibe W (1979) The Great Salt Lake. U.S. Geological Survey, 25

Kemp B, Tabish E, Wolford A, Jones D, Butler J, Baxter B (2018) The biogeography of Great Salt Lake halophilic archaea: testing the hypothesis of avian mechanical carriers. Diversity 10:124

Kish A, DiRuggiero J (2012) DNA replication and repair in halophiles. In: Vreeland RH (ed) Advances in understanding the biology of halophilic microorganisms. Springer, Dordrecht, pp 163–198

Kminek G, Bada JL, Pogliano K, Ward JF (2003) Radiation-dependent limit for the viability of bacterial spores in halite fluid inclusions and on Mars. Radiat Res 159:722–729

Kondo T, Sawatari C (1996) A Fourier transform infra-red spectroscopic analysis of the character of hydrogen bonds in amorphous cellulose. Polymer 37(3):393–399

Koonin EV (2003) Comparative genomics, minimal gene-sets and the last universal common ancestor. Nat Rev Microbiol 1(2):127

Kronyak RE, Kah LC, Edgett KS, VanBommel SJ, Thompson LM, Wiens RC et al (2019) Mineral-filled fractures as indicators of multigenerational fluid flow in the Pahrump Hills member of the Murray formation, Gale crater, Mars. Earth Space Sci 6:238–265. https://doi.org/10.1029/2018EA000482

Kvien I, Tanem BS, Oksman K (2005) Characterization of cellulose whiskers and their nanocomposites by atomic force and electron microscopy. Biomacromolecules 6:3160–3165

Larsen H (1967) Biochemical aspects of extreme halophilism. Adv Microb Physiol 1:97–132

Lazcano A, Forterre P (1999) The molecular search for the last common ancestor. J Mol Evol 49:411–412

Lindahl T (1993) Instability and decay of the primary structure of DNA. Nature 362(6422):709–715

Lindahl T, Nyberg B (1972) Rate of depurination of native deoxyribonucleic acid. Biochemistry 11 (19):3610–3618

Lindsay MR, Anderson C, Fox N, Scofield G, Allen J, Anderson E, Bueter L, Poudel S, Sutherland K, Munson-McGee JH, Van Nostrand JD, Zhou J, Spear JR, Baxter BK, Lageson DR, Boyd ES (2017) Microbialite response to an anthropogenic salinity gradient in Great Salt Lake, Utah. Geobiology 15(1):131–145

Litchfield CD (1998) Survival strategies for microorganisms in hypersaline environments and their relevance to life on early Mars. Meteorit Planet Sci 33(4):813–819

Lowenstein TK (1988) Origin of depositional cycles in a Permian "saline giant": the Salado (McNutt zone) evaporites of New Mexico and Texas. Geol Soc Am Bull 100:592–608
Lowenstein TK, Schubert BA, Timofeeff MN (2011) Microbial communities in fluid inclusions and long-term survival in halite. Geol Soc Am Today 21(1):4–9
Madison RJ (1970) Effects of a causeway on the chemistry of the brine in Great Salt Lake Utah. Water Resour Bull 14
Maltagliati L, Montmessin F, Fedorova A, Korablev O, Forget F, Bertaux JL (2011) Evidence of water vapor in excess of saturation in the atmosphere of Mars. Science 333(6051):1868–1871
Mancinelli RL, Fahlen TF, Landheim R, Klovstad MR (2004) Brines and evaporites: analogues for Martian life. Adv Space Res 33(8):1244–1246
McCready S, Marcello L (2003) Repair of UV damage in Halobacterium salinarum. Biochem Soc Trans 31(3):694–698. https://doi.org/10.1042/bst0310694
McGenity TJ, Gemmell RT, Grant WD, Stan-Lotter H (2000) Origins of halophilic microorganisms in ancient salt deposits. Environ Microbiol 2(3):243–250
McLennan SM, Grotzinger JP (2008) The sedimentary rock cycle of Mars. In: The Martian surface: composition, mineralogy, and physical properties, vol 541. Cambridge University Press, Cambridge
McLennan SM, Bell Iii JF, Calvin WM, Christensen PR, Clark BD, De Souza PA, Farmer J, Farrand WH, Fike DA, Gellert R, Ghosh A (2005) Provenance and diagenesis of the evaporite-bearing burns formation, Meridiani Planum, Mars. Earth Planet Sci Lett 240(1):95–121
Meuser JE, Baxter BK, Spear JR, Peters JW, Posewitz MC, Boyd ES (2013) Contrasting patterns of community assembly in the stratified water column of Great Salt Lake, Utah. Microb Ecol 66 (2):268–280
Milliken RE, Grotzinger JP, Thomson BJ (2010) Paleoclimate of Mars as captured by the stratigraphic record in Gale Crater. Geophys Res Lett 37(4). https://doi.org/10.1029/2009GL041870
Morán JI, Alvarez VA, Cyras VP, Vázquez A (2008) Extraction of cellulose and preparation of nanocellulose from sisal fibers. Cellulose 15(1):149–159
Mormile MR, Biesen MA, Gutierrez MC, Ventosa A, Pavlovich JB, Onstott TC, Fredrickson JK (2003) Isolation of *Halobacterium salinarum* retrieved directly from halite brine inclusions. Environ Microbiol 5(11):1094–1102
Murchie SL, Mustard JF, Ehlmann BL, Milliken RE, Bishop JL, McKeown NK, Noe Dobrea EZ, Seelos FP, Buczkowski DL, Wiseman SM, Arvidson RE (2009) A synthesis of Martian aqueous mineralogy after 1 Mars year of observations from the Mars reconnaissance orbiter. J Geophys Res Planets 114(E2). https://doi.org/10.1029/2009JE003342
Naftz DL, Angeroth C, Kenney T, Waddell B, Darnall N, Silva S, Perschon C, Whitehead J (2008) Anthropogenic influences on the input and biogeochemical cycling of nutrients and mercury in Great Salt Lake, Utah, USA. Appl Geochem 23(6):1731–1744
Naftz DL, Millero FJ, Jones BF, Green WR (2011) An equation of state for hypersaline water in Great Salt Lake, Utah, USA. Aquat Geochem 17:809–820
Nobles DR, Romanovicz DK, Brown RM Jr (2001) Cellulose in cyanobacteria: origin of vascular plant cellulose synthase? Plant Physiol 127:529–542
Norton CF, Grant WD (1988) Survival of halobacteria within fluid inclusions in salt crystals. J Gen Microbiol 134:1365–1373
Norton CF, McGenity TJ, Grant WD (1993) Archaeal halophiles (halobacteria) from two British salt mines. Microbiology 139(5):1077–1081
Null SE, Wurtsbaugh WA (2020) Water development, consumptive uses, and Great Salt Lake. In: Baxter BK, Butler JK (eds) Great Salt Lake biology: a terminal lake in a time of change. Springer, Cham
Ojha L, Wilhelm MB, Murchie SL, McEwen AS, Wray JJ, Hanley J, Massé M, Chojnacki M (2015) Spectral evidence for hydrated salts in recurring slope lineae on Mars. Nat Geosci 8 (11):829–832

Okuda K, Kudlicka K, Kuga S, Brown RM Jr (1993) [beta]-Glucan synthesis in the cotton Fiber (I. identification of [beta]-1,4- and [beta]-1,3-glucans synthesized in vitro). Plant Physiol 101:1131–1142

Oren A (1993) Ecology of extremely halophilic microorganisms. In: Vreeland RH, Hochstein LI (eds) The biology of halophilic bacteria. CRC, Boca Raton, FL, pp 25–53

Oren A (1999) Bioenergetic aspects of halophilism. Microbiol Mol Biol Rev 63(2):334–348

Oviatt CG, Thompson RS, Kaufman DS, Bright J, Forester RM (1999) Reinterpretation of the Burmester Core, Bonneville Basin, Utah. Quat Res 52:180–184

Pääbo S (1989) Ancient DNA: extraction, characterization, molecular cloning, and enzymatic amplification. Proceedings of the National Academy of Sciences of the United States of America (1939-1943). National Academy of Sciences, Washington, DC

Pääbo S, Poinar H, Serre D, Jaenicke-Després V, Hebler J, Rohland N, Kuch M, Krause J, Vigilant L, Hofreiter M (2004) Genetic analyses from ancient DNA. Annu Rev Genet 38:645–679

Panieri G, Lugli S, Manzi V, Roveri M, Schreiber BC, Palinska KA (2010) Ribosomal RNA gene fragments from fossilized cyanobacteria identified in primary gypsum from the late Miocene, Italy. Geobiology 8(2):101–111

Park JS, Vreeland RH, Cho BC, Lowenstein TK, Timofeeff MN, Rosenzweig WD (2009) Haloarchaeal diversity in 23, 121 and 419 MYA salts. Geobiology 7(5):515–523

Parnell JJ, Rompato G, Crowl TA, Weimer BC, Pfrender ME (2011) Phylogenetic distance in Great Salt Lake microbial communities. Aquat Microb Ecol 64:267–273

Pasteris JD, Freeman JJ, Wopenka B, Qi K, Ma Q, Wooley KL (2006) With a grain of salt: what halite has to offer to discussions on the origin of life. Astrobiology 6(4):625–643

Pawlowska MM, Butterfield NJ, Brocks JJ (2013) Lipid taphonomy in the proterozoic and the effect of microbial mats on biomarker preservation. Geology 41(2):103–106

Penny D, Poole A (1999) The nature of the last universal common ancestor. Curr Opin Genet Dev 9 (6):672–677

Perl SM (2019) Quantifying the threshold of biogenic detection in evaporites: constraining potential martian biomarker preservation via terrestrial analogues. University of Southern California, Ph.D. dissertation, Department of Earth Sciences

Perl SM, McLennan SM, The Athena Science Team (2008) Comparison of secondary porosity and permeability from Eagle to Victoria craters, Meridiani Planum, Mars. Lunar and Planetary Science XXXIX, Abstract #2246, Lunar and Planetary Institute, Houston, TX

Perl SM, McLennan SM, Grotzinger JP, Herkenhoff KE, the Athena Science Team (2007) Sedimentological constraints on an infiltrating paleowater table in the Burns Formation, Meridiani Planum, Mars. In: Seventh International Conference on Mars, Abstract #3298, Lunar and Planetary Institute, Houston (CD-ROM). http://www.lpi.usra.edu/meetings/7thmars2007/pdf/3298.pdf

Post FJ (1977) The microbial ecology of the Great Salt Lake. Microb Ecol 3:143–165

Post FJ, Stube JC (1988) A microcosm study of nitrogen utilization in the Great Salt Lake, Utah. Hydrobiolgia 158:89–100

Powers DW, Vreeland RH, Rosenzweig WD (2001) How old are bacteria from the Permian age? Nature 411:155–156

Radax C, Gruber C, Stan-Lotter H (2001) Novel haloarchaeal 16S rRNA gene sequences from Alpine Permo-Triassic rock salt. Extremophiles 5:221–228

Roedder E (1984) The fluids in salt. Am Mineral 69:413–439

Rothschild LJ (1990) Earth analogues for Martian life. Icarus 88(1):246–260. https://doi.org/10.1016/0019-1035(90)90188-F

Rupke AL, McDonald A (2012) Great Salt Lake Brine Chemistry Database, 1966–2011. Utah Geological Survey, State of Utah Department of Natural Resources, Salt Lake City, UT

Sankaranarayanan K, Timofeeff MN, Spathis R, Lowenstein TK, Lum JK (2011) Ancient microbes from halite fluid inclusions: optimized surface sterilization and DNA extraction. PLOS One 6 (6):e20683. https://doi.org/10.1371/journal.pone.0020683

Satterfield CL, Lowenstein TK, Vreeland RH, Rosenzweig WD (2005) Paleobrine temperatures, chemistries, and paleoenvironments of Silurian Salina Formation F-1 salt, Michigan Basin, U.S. A., from petrography and fluid inclusions in halite. J Sediment Res 75:534–546
Scharf C, Cronin L (2016) Quantifying the origins of life on a planetary scale. Proc Natl Acad Sci 113(29):8127–8132
Schopf JW, Farmer JD, Foster IS, Kudryavtsev AB, Gallardo VA, Espinoza C (2012) Gypsum-permineralized microfossils and their relevance to the search for life on Mars. Astrobiology 12 (7):619–633
Schroeder GK, Lad C, Wyman P, Williams NH, Wolfenden R (2006) The time required for water attack at the phosphorus atom of simple phosphodiesters and of DNA. Proc Natl Acad Sci USA 103(11):4052–4055
Schubert BA, Lowenstein TK, Timofeeff MN (2009) Microscopic identification of prokaryotes in modern and ancient halite, Saline Valley and Death Valley, California. Astrobiology 9 (5):467–482
Schubert BA, Lowenstein TK, Timofeeff MN, Parker MA (2010) Halophilic archaea cultured from ancient halite, Death Valley, California. Environ Microbiol 12(2):440–454
Schweitzer MH, Suo Z, Avci R, Asara JM, Allen MA, Arce FT, Horner JR (2007) Analyses of soft tissue from Tyrannosaurus rex suggest the presence of protein. Science 316(5822):277–280
Shroder JF, Cornwell K, Oviatt CG, Lowndes TC (2016) Landslides, alluvial fans, and dam failure at Red Rock pass: the outlet of Lake Bonneville. In: Oviatt CG, Shroder JF (eds) Lake Bonneville: a scientific update. Elsevier, Dordrecht
Simoneit BR (2004) Biomarkers (molecular fossils) as geochemical indicators of life. Adv Space Res 33(8):1255–1261
Skelley AM, Mathies RA (2003) Chiral separation of fluorescamine-labeled amino acids using microfabricated capillary electrophoresis devices for extraterrestrial exploration. J Chromatogr A 1021(1):191–199
Sonnenfeld P (1984) Brines and evaporites. University of California, Academic Press, Oakland, CA
Spencer RJ, Eugster HP, Jones BF, Rettig SL (1985) Geochemistry of Great Salt Lake, Utah I: Hydrochemistry since 1850. Geochim Cosmochim Acta 49(3):727–737
Squyres SW, Grotzinger JP, Arvidson RE, Bell JF, Calvin W, Christensen PR, Clark BC, Crisp JA, Farrand WH, Herkenhoff KE, Johnson JR (2004) In situ evidence for an ancient aqueous environment at Meridiani Planum, Mars. Science 306(5702):709–1714
Stan-Lotter H, McGenity TJ, Legat A, Denner EB, Glaser K, Stetter KO, Wanner G (1999) Very similar strains of Halococcus salifodinae are found in geographically separated Permo-Triassic salt deposits. Microbiology *145*(12):3565–3574
Stan-Lotter H, McGenity TJ, Legat A, Denner EB, Glaser K, Stetter KO, Stutz J, Ackermann R, Fast JD, Barrie L (2002a) Atmospheric reactive chlorine and bromine at the Great Salt Lake, Utah. Geophys Res Lett 29(10). https://doi.org/10.1029/2002GL014812
Stan-Lotter H, Pfaffenhuemer M, Legat A, Busse H-J, Radax C, Gruber C (2002b) *Halococcus dombrowskii* sp. nov., an archaeal isolate from a Permo-Triassic alpine salt deposit. Int J Syst Bacteriol 52:1807–1814
Stephens DW (1974) A summary of biological investigations concerning the Great Salt Lake, Utah (1861–1973). Great Basin Nat 34(3):Article 7
Stephens DW (1990) Changes in lake levels, salinity and the biological community of Great Salt Lake (Utah, USA), 1847–1987. Dev Hydrobiol 59:139–146. https://doi.org/10.1007/BF0002694.
Stevenson A, Cray JA, Williams JP, Santos R, Sahay R, Neuenkirchen N, McClure CD, Grant IR, Houghton JD, Quinn JP, Timson DJ, Patil SV, Singhal RS, Antón J, Dijksterhuis J, Hocking AD, Lievens B, Rangel DEN, Voytek MA, Gunde-Cimerman N, Oren A, Timmis KN, McGenity TJ, Hallsworth JE (2015) Is there a common water-activity limit for the three domains of life? ISME J 9(6):1333–1351
Stube JC, Post FJ, Procella DB (1976) Nitrogen cycling in microcosms and application to the biology of the north arm of Great Salt Lake (Publication No. PRJSBA-016-1). Utah Water Research Laboratory, Utah State University, Logan, UT

Sturm PA (1980) The Great Salt Lake brine system. In: Gwynn JW (ed) Great Salt Lake: a scientific, historical and economic overview. Utah Geological Survey, State of Utah Department of Natural Resources, Salt Lake City, UT, pp 147–162

Stutz J, Ackermann R, Fast JD, Barrie L (2002) Atmospheric reactive chlorine and bromine at the Great Salt Lake, Utah. Geophys Res Lett *29*(10):18–11

Summons RE, Albrecht P, McDonald G, Moldowan JM (2008) Molecular biosignatures. In: Strategies of life detection. Springer, Dordrecht, pp 133–159

Tazi L, Breakwell DP, Harker AR, Crandall KA (2014) Life in extreme environments: microbial diversity in Great Salt Lake, Utah. Extremophiles 18:525–535

Tosca NJ, Knoll AH, McLennan SM (2008) Water activity and the challenge for life on early Mars. Science 320(5880):1204–1207

Turk LJ (1970) Evaporation of Brine: A Field Study on the Bonneville Salt Flats, Utah. Water Resour Res 6(4):1209–1215. https://doi.org/10.1029/WR006i004p01209

United States Geological Survey (2019). https://ut.water.usgs.gov/greatsaltlake/salinity/. Accessed 04 Mar 2019

Van den Kerkhof AM, Hein UF (2001) Fluid inclusion petrography. Lithos 55(1):27–47

Viviano-Beck CE, Seelos FP, Murchie SL, Kahn EG, Seelos KD, Taylor HW, Taylor K, Ehlmann BL, Wiseman SM, Mustard JF, Morgan MF (2014) Revised CRISM spectral parameters and summary products based on the currently detected mineral diversity on Mars. J Geophys Res Planets 119(6):1403–1431

Vreeland RH, Rosenzweig WD, Powers DW (2000) Isolation of a 250 million-year-old halotolerant bacterium from a primary salt crystal. Nature 407:897–900

Ward P (2007) Life as we do not know it: the NASA search for (and synthesis of) alien life. Penguin, New York

Ward DM, Brock TD (1978) Hydrocarbon biodegradation in hypersaline environments. Appl Environ Microbiol 35:353–359

Wardlaw NC, Schwerdtner WM (1966) Halite-anhydrite seasonal layers in the middle Devonian Prairie evaporite formation, Saskatchewan, Canada. Geol Soc Am Bull 77(4):331–342

Weimer BC, Rompato G, Parnell J, Gann R, Ganesan B, Navas C, Gonzalez M, Clavel M, Albee-Scott S (2009) Microbial biodiversity of Great Salt Lake, Utah. Nat Resour Environ Issues 15:15–22

Weiss MC, Sousa FL, Mrnjavac N, Neukirchen S, Roettger M, Nelson-Sathi S, Martin WF (2016) The physiology and habitat of the last universal common ancestor. Nat Microbiol *1*(9):16116

Weitz CM, Bishop JL (2019) Formation of clays, ferrihydrite, and possible salts in Hydrae Chasma, Mars. Icarus. https://doi.org/10.1016/j.icarus.2018.09.007

White AL, Jahnke LS (2002) Contrasting effects of UV-A and UV-B on photosynthesis and photoprotection of β-carotene in two *Dunaliella* spp. Plant Cell Physiol 43(8):877–884

Wiens RC, Newell R, Clegg SM, Sharma SK, Misra A, Bernardi P, Maurice S, McCabe K, Cais P (2017, March) The SuperCam Remote Raman Spectrometer for Mars 2020. In Proceedings of the 48th Lunar and Planetary Science Conference.

Wierzchos J, Ascaso C, McKay CP (2006) Endolithic cyanobacteria in Halite Rocks from the hyperarid core of the Atacama Desert. Astrobiology 6(3):415–422. https://doi.org/10.1089/ast.2006.6.415

Willerslev E, Hansen AJ, Rønn R, Brand TB, Barnes I, Wiuf C, Gilichinsky D, Mitchell D, Cooper A (2004) Long-term persistence of bacterial DNA. Curr Biol 14(1):R9–R10

Winters YD (2013) Haloarchaeal survival and preservation of biomaterials (carotenoids) in ancient halite (Doctoral Dissertation). State University of New York at Binghamton, Binghamton, NY

Woese CR (1987) Bacterial evolution. Microbiol Rev 51(2):221

Wurtsbaugh WA, Miller C, Null SE, DeRose RJ, Wilcock P, Hahnenberger M, Howe F, Moore J (2017) Decline of the world's saline lakes. Nat Geosci 10(11):816

Ye C, Glotch TD (2018) Spectral properties of chloride salt-bearing assemblages: implications for detection limits of minor phases in chloride-bearing deposits on Mars. JGR Planets 124:209–222. https://doi.org/10.1029/2018JE005859

Index

A

Acremonium, 128
Acroptilon repens, 423
Actitis macularius, 298
Aerobic, 98, 101, 103, 126, 137
Agriculture, 7, 13, 30, 31, 33, 35, 42, 44, 56, 68, 69, 72, 80, 82, 90, 255, 258, 405–407, 409, 423, 425, 455
Air quality, 13, 36, 42, 43, 67, 416
Albedo, 36, 37
Algae, 2, 11, 30, 31, 39, 56, 94, 98, 99, 105, 112, 121, 128, 147, 149, 150, 152, 154–156, 158, 159, 200, 203, 209, 277, 372, 439, 444, 446, 448, 453, 489, 490, 495
Alkali bulrush, *see Bolboschoenus maritimus*
Alkaline lakes, 240–242, 254
Alkalinity, 102–104
Alkali sacaton, *see Sporobolus airoides*
Allenrolfea occidentalis, 271, 273, 378, 380, 382, 383, 385, 389
Alopecosa, 454
Alternaria, 128
Alteromonadales, 98, 101
American avocet, *see Recurvirostra americana*
American bullfrog, *see Rana catesbeiana*
American golden plover, *see Pluvialis dominica*
American raven, *see Corvus corax*
American white pelican (AWPE), *see Pelecanus erythrorhynchos*
Ammonia, 194, 202–204, 214–216
Amniotes, 346
Amoco, 465, 468, 469
Amoeba, 128
Amphipoda, 274
Anaeromyxobacter, 135, 138
Anas acuta, 450
Anas clypeata, 206, 252, 401, 412, 436
Anas crecca, 206, 252
Anas platyrhynchos, 252
Annual rabbitsfoot grass, *see Polypogon monspeliensis*
Anoxia, 179, 185, 186
Antelope Island, 4, 27, 30, 55, 61, 62, 75, 78, 92, 93, 129, 251, 288, 313, 345–364, 452, 453
Antelope Island Causeway, 61, 62, 71, 76, 255, 286, 287, 289, 292–295, 299, 301, 302
Anthropogenic, 32, 34–42, 45, 89, 107–112, 148, 159, 166, 167, 184, 213, 219, 223, 251, 252, 257, 258, 427, 438, 442, 456
Anthropomorphic, 24, 28–33, 38, 42, 126, 490
Apanothece, 96, 101
Aragonite, 92, 93, 102–106, 112, 270, 277
Aral Sea, 13–15, 168, 204, 205, 219
Archaea, 2, 11, 56, 95, 96, 99, 121, 122, 124–127, 129–132, 136, 138, 209, 212, 503
Arenaria interpres, 291
Arid, 15, 24, 43, 44, 72, 82, 146, 167, 313, 346, 352, 361, 363, 409
Aristida longiseta, 382, 383, 386, 392
Artemia, *see Artemia franciscana*
Artemia franciscana, 9, 11, 32, 39, 147, 176, 222, 240, 315, 444, 446
Artemisia spinescens, 377, 385, 389
Artemisia tridentate, 385, 390
Arthropod, 176, 222
Ascomycota, 128
Ascpidocelis tigris tigris, 358–359, 364
Asexual, 410, 425
Asio otus, 476

B. K. Baxter, J. K. Butler (eds.), *Great Salt Lake Biology*,
https://doi.org/10.1007/978-3-030-40352-2

Aspergillus, 128
Asphalt, 469, 474, 476–478
Astrobiology, 487–506
Athalassohaline, 24
Athene cunicularia, 476
Atmosphere, 438, 442, 497, 506
Atriplex confertifolia, 377, 382, 385, 389
Atriplex prostrata, 405
Avian mechanical carriers, 124, 136

B
Bacillales, 101
Bacillariophytes, 147, 444, 445
Bacillus, 122, 138
Bacteria, 2, 11, 30, 31, 56, 95, 98, 99, 101–104, 121, 122, 124–127, 129–138, 158, 209, 212, 217, 381, 443, 457, 490, 491, 503
Bacteroides, 132, 133, 137
Bacteroidetes, 125
Baird's sandpiper, *see Calidris bairdii*
Baltic rush, *see Juncus arcticus*
Barn owl, *see Tyto alba*
Barnyardgrass, *see Echinochloa crus-galli*
Bar-tailed godwit, *see Limosa lapponica*
Bartramia longicauda, 289
Basidiomycota, 128
Bathometry, 442, 455
Bear River, 3, 6, 16, 17, 27, 30, 44, 55, 69, 72, 73, 79, 81, 240, 296, 316, 413, 424, 455, 457, 467, 472, 482
Bear River Bay, 4, 5, 9–11, 244, 439, 446
Bear River Migratory Bird Refuge, 31, 283, 291, 295, 297, 325, 415, 452
Bear River Project, 72
Benthic, 10, 39, 88, 89, 98, 101, 102, 104, 106, 107, 147–149, 152, 154, 155, 157–159, 164–166, 184, 209, 214, 267, 278, 315, 316, 444, 446, 490
Bicarbonate, 102, 124, 270
Bidens cernua, 403, 420
Big sagebrush, *see Artemisia tridentate*
Bioaccumulation, 339, 436–439, 441–448, 453–457
Bioadvection, 454
Biodiversity, 4, 5, 9, 12, 34, 36, 95, 375
Biofilm, 88, 128, 148, 149
Biogeography, 124, 126, 136–138, 347–352, 354
Bioherm, *see* Microbialite
Biomagnification, 33, 436–439, 441–444, 446–448, 453–457
Biomarkers, 106, 501–505
Biomass, 39, 56, 88, 92, 94, 95, 99–102, 104–106, 108, 110–112, 258, 315, 415–418, 421, 438, 444, 446
Biopreservation, 503, 506
Biosignatures, 112, 501–504, 506
Bitumen, 467, 468
Black-bellied plover, *see Pluvialis squatarola*
Black-necked grebe, *see Podiceps nigricollis*
Black-necked stilt, *see Himantopus mexicanus*
Black-tailed jackrabbit, *see Lepus californicus*
Black widow, *see Latrodectus hesperus*
Bolboschoenus maritimus, 269, 272, 402, 403
Bonneville Basin, 25, 26, 37, 42, 321, 369, 488
Bonneville Salt Flats, 491, 495
Bonneville shoreline, 375, 467
Box Elder County, 66, 69, 70, 79, 464, 465
Branta canadensis, 448, 452
Breeding, 40, 245, 248, 252, 255, 265, 266, 280–288, 290, 291, 293, 297–301, 313, 316, 318–321, 325–327, 330–333, 339, 340, 357, 358, 411, 412, 436, 444, 448, 472
Bridger Bay, 92, 93, 355
Brine fly, *see Ephydra* spp.
Brine shrimp fecal pellets, 92, 277
Brine shrimp, *see Artemia franciscana*
Brine stratification, 491
Bromus tectorum, 363, 386, 391
Brown pelican, *see Pelecanus occidentalis*
Bucephala clangula, 206, 252, 436
Budsage, *see Artemisia spinescens*
Buff-breasted sandpiper, *see Calidris subruficollis*
Burrowing owl, *see Athene cunicularia*
Business, *see* Industry
Buteo, 476

C
Calcium (Ca^{2+}), 24, 25, 92, 102, 103, 124, 125, 270, 275–277, 495, 497
Calidris acuminata, 292
Calidris alba, 293
Calidris alpina, 293
Calidris bairdii, 294
Calidris canutus, 292
Calidris ferruginea, 293
Calidris fuscicollis, 295
Calidris himantopus, 292
Calidris mauri, 296
Calidris minutilla, 294
Calidris pugnax, 292
Calidris pusilla, 295

Calidris subruficollis, 295
California gull, *see Larus californicus*
Canada goose, *see Branta canadensis*
Canis latrans, 253, 481
Canyon treefrog, *see Hyla arenicolor*
Capillary fringe, 275–277
Capnodiales, 128
Carbon dioxide, 96, 101, 102
Carbon flow, 209
Carbonate, 24, 25, 39, 92, 100, 103–106, 112, 125, 126, 128, 203, 247, 275–277, 316, 493
Cardaria draba, 403, 405, 408, 425
Carex nebrascensis, 272
Carex praegracilis, 272
Carotenoid, 94, 120, 490, 501, 503, 504
Carp, *see Cyprinus carpio*
Carrington, Albert, 317
Carrington Island, 149, 240, 244
Cathartes aura, 481
Cattails, *see Typha* spp.
Causeway, 4, 30–31, 59–62, 82, 107, 111, 112, 126, 127, 148–153, 184, 199, 212, 214, 219, 240, 301, 302, 313, 362, 452, 469, 491
Cedar City, 248
Centaurea solstitialis, 423
Centaurea stoebe, 423
Central Utah Project, 6
Chaetoceros sp., 199
Charadrius montanus, 287
Charadrius nivosus, 288, 401
Charadrius semipalmatus, 287
Charadrius vociferus, 287
Cheatgrass, *see Bromus tectorum*
Chemocline, 152, 203, 213–215, 217, 218, 446
Chewaucan River, 255
Chilophyra (*Prorodon*) utahensis, 128
Chironomids, 10, 151, 274, 301, 302
Chlorine, 64, 436, 442, 500
Chlorophyll-α, 155, 157, 164, 179, 180, 187, 191–195, 200–202, 216, 217
Chlorophyte, 128, 155, 198, 200, 210, 444
Chroococcales, 98, 211
Cicada killer, *see Sphecius speciosus*
Cinnamon teal, *see Spatula cyanoptera*
Cirsium arvense, 405
Cladocerans, 10, 207
Cladosporium, 128
Clay, 67, 92, 277, 303, 375–378, 466, 492, 493, 495–497, 504
Climate, 17, 24, 25, 28, 29, 34, 35, 38, 42, 68, 69, 79, 111, 146, 159, 160, 166, 168, 191, 313, 349, 371, 393, 455, 499
Climate change, 6, 16, 23–45, 78–80, 89, 107, 111–112, 148, 159–161, 163, 165–168, 258, 303, 325, 326, 363, 393, 427, 455, 467
Climate models, 34, 159, 455, 456
Climate projections, 25, 35–37, 160, 166
Clustered field sedge, *see Carex praegracilis*
Coccochloris elebans, 149, 150, 152
Coleoptera, 274, 284
Colorado Plateau, 287, 346, 350
Colorado River Compact, 17
Coluber constrictor mormon, 353, 359–360, 364
Columbia spotted frog, *see Rana luteiventris*
Common goldeneye, *see Bucephala clangula*
Common spikerush, *see Eleocharis palustris*
Community structure, 39, 125, 176, 202, 363
Coniochaeta, 128
Conium maculatum, 403, 405
Conservation, 14–17, 44, 69, 72, 81, 166–168, 181, 265, 276, 313, 354, 402, 409
Consumer, 39, 99–101, 106, 108, 110, 112, 154, 158, 166, 200
Consumption advisory, 436, 437
Contamination, 42, 339, 454, 494, 503
Copepods, 10, 110, 181, 207
Corixid, *see Trichocorixa verticalis*
Corixidae, 244, 274, 284, 285, 479, 480
Corvus corax, 481
Coyote, *see Canis latrans*
Cretaceous, 89, 348–350
Crystal, 76, 92, 93, 106, 122, 129, 138, 362, 380, 381, 492–494, 496, 499, 500, 504
Curlew sandpiper, *see Calidris ferruginea*
Curly pondweed, *see Potamogeton crispus*
Cyanobacteria, 11, 88, 91, 94, 99, 100, 102, 104, 112, 125, 147–149, 152, 154–156, 179, 198, 200–202, 207, 209, 210, 247, 446, 453, 499, 503
Cyanophytes, 444, 445
Cyclotella, 210
Cygnus columbianus, 399, 401
Cyprinus carpio, 253
Cysts, 32, 39, 40, 63, 92, 99, 107, 108, 110, 112, 113, 128, 148, 152, 153, 157, 161, 165–167, 176–178, 182, 183, 185–192, 194, 195, 198, 199, 204, 205, 207–209, 216–222, 252, 253, 256, 257, 315, 436, 446
Cytophaga-Flavobacterium Bacteroides, 137
Cytophagales, *see Cytophaga-Flavobacterium Bacteroides*

D

Dactylococcopsis, 210–212
Dalmatian toadflax, *see Linaria dalmatica*
Damming, *see* Dams
Dams, 16, 59, 69, 72, 258, 488
Davidiellaceae, 128
Dead Sea, 14, 15, 24, 138, 219, 490, 491
Decomposition, 5, 103, 415, 475
Deep brine layer (DBL), 152, 164, 201, 440, 441, 443, 446, 456
Denitrification, 103, 202, 203, 216, 217
Desert saltgrass, *see Distichlis* spp.
Dessication, 110, 111
Desulfobacterium sp., 443
Desulfohalobium utahense, 132
Desulfosalsimonas propionicica, 132
Detrital, 88, 92, 102, 103, 105, 158, 446
Development, 1–18, 31, 42, 44, 45, 58, 59, 66, 67, 70–73, 78–82, 98, 106, 181, 183, 186–189, 208, 256, 258, 275, 302, 303, 313, 324, 336, 340, 346, 410, 427, 437, 455
Devonian, 348
Diapause, 165, 177, 187–191
Diatoms, 39, 91, 98, 100, 101, 147–158, 166, 198, 199, 202, 210, 211
Diet, 148, 181, 199, 206, 207, 209–212, 247, 248, 251, 252, 274, 283, 285, 301, 317, 321–323, 444, 454, 457, 479
Dissolved inorganic nitrogen (DIN), 202, 215–217
Dissolved organic matter (DOM), 443, 446
Distichlis spicata, 269, 380–383, 402, 403
Distichlis spicata/stricta, 269, 377, 380–383, 386, 391, 402, 403
Distichlis stricta, 377, 382, 383, 386, 391
DNA, 99, 122–125, 129, 130, 209, 212, 348, 493–495, 499–503, 505
DNA sequencing, 124, 125, 134, 493
Dolomite, 93, 102–106, 112
Dormancy, 122, 185–191, 421, 499, 500
Dropseed, *see Sporobolus cryptandrus*
Drought, 5–7, 9–11, 34–36, 38, 41, 67, 68, 91, 159, 303, 319, 324, 338, 340, 363, 413, 417, 418, 423, 455, 488, 490, 499
Dunaliella salina, 128
Dunaliella viridis, 149, 150, 198, 200
Dunlin, *see Calidris alpina*
Dust, 10, 13–15, 17, 36, 37, 42–44, 58, 67, 76, 78, 80, 81, 220, 455, 456, 501, 503
Dyer's woad, *see Isatis tinctoria*

E

Eared grebe, *see Podiceps nigricollis*
East Lake Fault, 91
Echinochloa crus-galli, 406
Ecology, 34, 38, 88, 99, 159, 165, 167, 186, 285, 374–381, 427, 489
Economic, 2, 14, 17, 58, 59, 63–65, 81, 82, 89, 111–113, 166–168, 177–178, 220, 257, 258, 398, 399, 402, 436
Ectotherms, 346
Egg Island, 64, 318
Eleocharis palustris, 272, 402, 403
Elevation, 2–7, 9–17, 27–29, 31, 33–39, 41, 42, 54, 55, 58–60, 65–73, 77–80, 82, 107, 121, 127, 130, 178, 184, 197, 213, 255, 257, 258, 269, 271, 302, 315, 324, 325, 339, 355, 357, 402, 410, 455, 456, 466–468, 470, 473, 488, 495, 496
Emergent wetlands, 402, 411
Empidonax wrightii, 479–481
Endorheic saline lakes, 192
Environmental Protection Agency (EPA), 13, 14, 64, 159–161, 436, 437, 441, 444
Eocene, 89, 350
Ephemeroptera, 274
Ephydra spp., 10, 11, 39, 444, 479, 480
Ephydridae, 284, 285
Epigenetic responses, 189
Epilimnetic water, 201
Epilimnion, 202, 204, 215, 218, 219, 221
Euhalothece sp., 98
Eukaryota, 121
European seaheath, *see Frankenia pulverulenta*
European settlement, 63, 253, 324
Evaporation, 2, 4, 7, 24, 25, 28–32, 35, 37, 64, 65, 120, 146, 164, 168, 178, 179, 196, 240, 257, 258, 269, 275–278, 288, 313, 490–493, 495, 498, 504
Evaporite, 275, 276, 488, 490–502, 504–506
Extracellular Polymeric Substances (EPS), 103–105, 112, 125
Extremophiles, 122, 498

F

Farmington Bay, 4, 5, 30, 31, 62, 65, 129, 151, 207, 208, 244, 247, 248, 251, 313, 413, 415, 424, 439, 441, 446, 457
Faults, 89, 91, 465
Fecal pellets, 165, 277, 466
Federal Emergency Management Agency (FEMA), 67, 70

Fendler threeawn, *see Aristida longiseta*
Fermentation, 102, 104, 112
Ferrithrix, 135
Ferry Duck Club, 413
Fission-track date, 465
Flavobacteriales, 101
Flooding, 14, 15, 27, 28, 30, 34, 36, 55, 59, 61, 65–70, 78, 191, 269, 270, 278, 284, 285, 409, 410, 418–420, 425, 426, 490, 491
Flowers, Seville, 372, 393
Fluctuations/variations, 24, 25, 27–29, 34, 37, 38, 55, 67, 105, 146–148, 152, 153, 183, 194, 196, 200, 202, 216, 257, 269, 272, 276, 303, 363, 379, 393, 419–421, 454, 466, 467, 488, 502
Fluid inclusions, 129, 492–495, 499, 500, 502–505
Food chain, 33, 99, 147, 159, 166, 216, 339, 437, 439, 444, 447, 448
Food web, 2, 12, 30, 32, 96, 99, 100, 106, 107, 110–112, 147, 148, 151–154, 156, 158–160, 165, 166, 168, 178, 179, 191, 200, 213, 220, 222, 274, 398, 423, 437, 439, 444–446, 448, 457, 488, 489
Forage, 181, 206, 244–246, 250–253, 256, 266–268, 273, 280, 286–288, 290–300, 319–321, 323, 329, 330, 423
Fossil, 88, 312, 348, 464, 478, 481, 503
Frankenia pulverulenta, 403, 406, 408, 424
Franklin's gull, *see Leucophaeus pipixcan*
Frémont, J.C. (John Charles), 25, 27, 29, 55, 67, 74, 149, 316
Freshwater, 2, 4, 9, 26, 28–33, 39, 41, 43, 59, 76, 89, 91, 105, 107, 111, 112, 120, 147–149, 151, 153, 160, 167, 179, 191, 219, 240, 242–244, 251–253, 257, 258, 264, 271, 273, 278, 284, 293, 296, 298–300, 303, 313, 316, 319, 320, 324, 330, 340, 361, 370, 402, 407, 410, 423, 439, 441, 443, 444, 455, 456, 465, 479, 480, 490, 491
FRIENDS of Great Salt Lake, 45, 61, 113
Fungi, 56, 121, 128, 490

G

Gadwall, *see Mareca strepera*
Gallinago delicata, 298
Gastropoda, 274
Geochemical, 89, 120, 496
Geology, 25, 88, 91, 112, 275, 353, 374, 376
Gilbert Bay, 4, 9, 10, 149, 151–154, 176, 178–180, 183–185, 188, 190, 196–208, 212–217, 219, 248, 439, 441, 454
Gilbert, G.K. (Grove Karl), 25, 27
Global plate tectonics, 348
Glutamate, 127
Glycerol, 122, 127, 187, 379, 498
Glyphosate, 414, 422
Gopher snake, *see Pituophis catenifer*
Goshute, 29
Graben, 89, 465
Gray flycatcher, *see Empidonax wrightii*
Grazing, 29, 67, 148, 152, 156–158, 165, 179, 180, 194, 200, 204, 206, 211, 212, 216, 414–418, 423, 436, 445
Gray molly, *see Kochia americana*
Greasewood, *see Sarcobatus vermiculatus*
Great Basin, 25, 29, 35, 36, 38, 54, 89, 146, 167, 240, 252, 253, 350–352, 354, 358–363, 399, 488
Great Basin gopher snake, *see Pituophis catenifer deserticola*
Great Basin spadefoot, *see Spea intermontana*
Great Basin whiptail, *see Ascpidocelis tigris tigris*
Greater yellowlegs, *see Tringa melanoleuca*
Great Salt Lake Advisory Council, 44
Great Salt Lake Basin, 68, 71, 72, 79, 80
Great Salt Lake Ecosystem Program, 32, 44, 57, 153, 247, 257, 314, 482
Great Salt Lake Institute, 24–39, 41–44, 329, 332, 334, 482, 489
Great Salt Lake Water Quality Strategy, 220
Green frog, *see Rana clamitans*
Green molly, *see Kochia americana*
Green-winged teal, *see Anas crecca*
Groundwater, 3, 16, 28, 29, 31, 34, 35, 37, 39, 72, 271, 275–277, 479, 495, 504
Groundwater connection, 276
Guano extraction, 318
Gulf of California, 248, 288
Gunnison Island, 13, 40, 251, 253, 311–340, 472, 482
Gypsum, 275, 466, 475, 491–493, 497, 501–504, 506

H

Haliscomenobacter, 136
Halite, 24, 275, 436, 475, 477, 491–497, 499–504, 506
Haloanaerobium praevalens, 131
Haloarchaea, *see Halophilic archaea*
Haloarchaeon, *see Halophilic archaea*
Haloarcula, 130
Halobacteriaceae, 131
Halobacterium, 129, 130
Haloferacales, 96
Haloferax, 129

Halomonas, 130, 132, 133, 135, 137, 138
Halomonas utahensis, 132
Halonotius pteroides str., 129
Halophile, 98, 121, 158, 498, 501, 505
Halophilic archaea, 31, 60, 94, 122, 127, 129, 136–137, 490, 491, 498–501, 503, 505
Halophyte, 271, 372, 379, 380, 393, 411
Haloquadratum, 129, 130
Haloquadratum walsbyi, 96, 129
Halorubrum, 130
Halorubrum sp., 96
Halotolerant, 122, 146, 147, 153, 381
Halovibrio variabilis, 131
Hardstem bulrush, *see Schoenoplectus acutus*
Harmful algal bloom (HAB), 201
Harold Crane Waterfowl Management Area, 288, 293, 297, 452
Hat Island, 64, 251, 318–320
Health, 11, 13, 14, 33, 43, 45, 54, 57, 61, 63, 65, 80–83, 100, 112, 128, 186, 197, 220, 257, 321, 337, 398, 437, 439
Heat-shock protein (HSP), 186, 187
Heavy metals, 13, 33, 42, 168, 339, 437
Hemiptera, 274
Hemlock, *see Conium maculatum*
Hemolymphs, 185, 196
Herbicide, 413–418, 421, 424, 425
Herpetofauna, 348, 349, 352–363
Herpetology, 345
Heterotrophic, 98, 99, 101–103, 105, 158
Highly unsaturated fatty acids (HUFA), 181
Himantopus mexicanus, 206, 284, 399, 401, 412
Holocene, 25, 34, 467, 482
Holomixis, 216
Howard Slough Wildlife Management Area, 452
Hudsonian godwit, *see Limosa haemastica*
Hungary, 473
Hydroclimate, 37
Hydrogen sulfide, 5, 58, 102
Hydrology, 2–6, 24, 27, 30, 35, 158, 269, 275, 277, 278, 418, 420, 423
Hydrophilidae, 284, 285
Hyla arenicolor, 349, 350
Hypereutrophic, 146, 179, 180
Hyperosmotic solutions, 185
Hypersaline, 9, 11, 24, 39, 60, 96, 105, 106, 120–129, 131, 134, 137–139, 146, 166, 187, 194, 199, 211, 240–242, 245, 247, 254, 269, 270, 315–317, 335, 379, 439, 479, 487–506
Hypersaline lakes, 126, 137, 138, 146, 203, 240, 244, 245, 253–256, 410
Hypolimnia, 215

I

Idaho, 16, 26, 159, 253, 320, 321, 369, 488
Imazapyr, 414
Impounded wetlands, 7, 269, 278, 280, 324
Industry, 7, 14, 30–33, 45, 57, 63–65, 72, 80, 112, 113, 148, 150–153, 167, 168, 176, 178–183, 190, 192, 208, 209, 219, 222, 256, 315, 318, 436
Inputs, 24, 30–32, 37, 38, 41–45, 62, 89, 105, 107, 112, 120, 126, 202, 203, 216, 275, 303, 313, 316, 326, 438, 439, 441, 442, 455–457, 490, 491
International Space Station, 60, 120
Invasive plant management, 397–427
Invasive species, 398, 402, 422, 423, 427
Inversion, 43
Invertebrates, 2, 9, 30, 39–41, 56, 100, 138, 146–154, 156–159, 162, 164–168, 176, 185, 196, 206, 207, 222, 240, 244, 246, 250, 251, 315, 444, 482, 489
Iodine bush, *see Allenrolfea occidentalis*
Iron, 201, 202, 339
Isatis tinctoria, 423

J

Juncus arcticus, 272, 402, 403

K

Kesterson Reservoir, 257
Killdeer, *see Charadrius vociferus*
Knapweed, *see Centaurea stoebe*
Kochia americana, 377, 382, 383, 385, 389

L

La Brea Tar Pits, 464
Lactuca serriola, 406
Lacustrine, 88
Lake Abert, 168, 255
Lake Bonneville, 25–27, 29, 34, 89, 176, 312, 313, 369, 375, 467, 488, 491
Lake-effect, 37
Lake level, 2, 4–7, 9, 11–17, 25, 29, 31, 39, 42, 61, 66, 68, 73, 91, 107, 108, 149, 198, 255, 269, 303, 319, 325, 339, 355, 375,

393, 455, 456, 466–468, 470, 471, 473, 482
Lake Park Resort, 61
Lake Urmia, 37, 111, 137, 168, 219, 220
Lambourne, Alfred, 64, 318
Land bridge, 40, 41, 255, 319, 324, 326, 332, 334, 338, 339, 348, 362
Landforms, 269, 275
Lanius ludovicianus, 454
Larus californicus, 206, 245, 251, 252, 255, 257, 320, 326, 452, 479
Larus delawarensis, 251, 450
Late-embryogenesis abundant (LEA) Proteins, 186
Latrodectus hesperus, 454
Least sandpiper, *see Calidris minutilla*
Lepidium latifolium, 403, 406, 408, 425
Lepus californicus, 481, 482
Lesser yellowlegs, *see Tringa flavipes*
Leucophaeus pipixcan, 272
Limestone, 11, 465
Limnodromus griseus, 296
Limnodromus scolopaceus, 297
Limosa fedoa, 291, 399, 401, 412
Limosa haemastica, 290
Limosa lapponica, 290
Linaria dalmatica, 423
Lithified, 88, 112
Live feed, 179–182
Loggerhead shrike, *see Lanius ludovicianus*
Long-billed curlew, *see Numenius americanus*
Long-billed dowitcher, *see Limnodromus scolopaceus*
Long-eared owl, *see Asio otus*
Longleaf phlox, *see Phlox longifolia*
Loosestrife, *see Lythrum salicaria*
Lucin Cutoff, 59
Lythrum salicaria, 403, 406, 408, 424

M
Macroinvertebrates, 2, 10–12, 56, 264, 267, 268, 273–276, 278, 284, 285, 301–303, 315, 411, 426
Magnesium (Mg^{2+}), 102
Mallard, *see Anas platyrhynchos*
Management, 16, 17, 41, 43, 44, 57, 58, 80–82, 166–168, 176, 178, 181–185, 190, 199, 201, 204, 208, 209, 219, 220, 222, 223, 265, 278, 313, 315, 323, 324, 354, 398, 399, 403, 405, 412–418, 422–427
Marbled godwit, *see Limosa fedoa*
Mareca strepera, 252
Marinimicrobium haloxylanilyticum, 98, 101
Mark-resight, 327, 332
Mars, 491–498, 500–502, 504–506
Mars Exploration Rover (MER), 495
Masticophis taeniatus, 351, 360, 364
Mephitis mephitis, 246
Mercury (Hg), 221, 339, 435
Meromictic, 89, 124, 212, 215
Meromixis, 152, 212–219
Metabarcoding, 209
Metabolic, 123, 125, 127, 129, 137, 187–189, 222, 380
Metabolic processes, 88, 222
Metagenome, 123, 125
Meta-nauplii, 181
Methane, 96, 126, 466
Methanogens, 96, 127
Methanosarcinales, 96
Methylation, 439, 443, 456, 457
Methylmercury (MeHg), 436, 437, 439, 441, 443, 446, 448, 453, 454, 456
Mexican spadefoot, *see Spea multiplicata*
Microbial communities, 11, 33, 88, 94–96, 101, 106, 121–127, 210, 439, 457, 489–491, 493, 502
Microbial diversity, 30, 39, 121, 122, 126–136, 139, 493
Microbialite, 11, 39, 88–112, 128, 148–150, 152, 155, 157–159, 161, 164, 165, 184, 209–212, 247, 253, 315, 444, 446–448, 466
Microbial mat, 2, 11, 39, 89, 91, 94, 98–101, 103, 105, 108, 110, 125, 126, 128, 247, 503
Microbiology, 121
Microcosm, 108–110, 127, 128, 197, 201, 203, 222, 446, 502
Microorganisms, 30, 33, 39, 56, 99, 100, 102, 103, 105–107, 112, 120, 122–124, 126–130, 132, 138, 209, 316, 489, 490, 492–494, 498–503, 505
Microscopy, 92, 93, 96, 121, 123, 211, 480, 493, 494, 503, 504
Microtis spp., 477
Migrant, 264, 280–282, 284–286, 290–297, 299–302
Migration, 40, 41, 200, 248, 256, 265, 275, 278, 280, 281, 284–287, 289, 291–301, 325, 331, 332, 399, 436, 444, 465, 479
Migratory birds, 2, 56, 80, 99, 112, 113, 138, 167, 398, 399, 425–427, 436, 444, 447, 448

Mineral extraction, 2, 7, 14, 31–33, 63, 80, 159, 160, 167, 168
Miocene, 89, 349, 465
Mirabilite (sodium sulfate), 91
Mitochondria, 186
Molecular chaperones, 186
Molybdenum, 202
Monimolimnetic water, 201
Monimolimnion, 5, 202, 212–218, 220, 221
Mono Lake, 14–17, 37, 38, 43, 168, 191, 200, 203, 204, 206, 215, 216, 220, 241, 247, 253–256
Monomixis, 212–219
Morphometric data, 336–338
Mountain plover, *see Charadrius montanus*
Mountain rush, *see Juncus arcticus*
Mucronate, 357
Mudflat, 25, 39, 178, 264, 267, 268, 271, 272, 275–278, 280, 285–287, 289, 291–297, 302, 303, 329, 369, 402, 409–412, 418, 424
Muscidae, 284

N
National Atmospheric Deposition Program (NADP), 202
National Audubon Society, 41, 43, 72, 100
National Wetlands Inventory (NWI), 409
Native American, 56
Nature Conservancy, 72, 320, 402
Nauplii, 165, 176, 177, 181, 185, 191–193, 199, 211, 221, 446
Navicula, 100, 105, 108, 110, 111, 157, 211, 212
Navicula salinicola, 98, 447
Neanthes succinea, 256
Nebraska sedge, *see Carex nebrascensis*
Negit Island, 255
Neocamarosporium, 128
Neoscona spp., 454
Nesting, 41, 64, 67, 148, 166, 167, 243, 245, 246, 248, 249, 251, 257, 264, 267, 272, 273, 275, 276, 278–281, 285, 287, 288, 290, 291, 293, 296, 298, 300–302, 313, 316, 318–321, 326, 327, 332–334, 402, 425, 448, 472, 482
Neurotoxin, 437
Nevada, 15, 26, 35, 37, 241, 254, 299, 321, 322, 354, 369, 442, 488
New World, 249, 350
Nitzschia, 210
Nodding beggartick, *see Bidens cernua*
N-limitation, 200
Nodularia spumigena, 151
Non-native species, 398, 405, 422, 423
North America, 13, 41, 44, 57, 253–256, 258, 280, 282, 283, 289, 291–295, 298, 299, 301, 312, 313, 348–350, 352, 382, 404, 408, 409, 411, 414, 417, 427, 444
North Arm, 4, 11–13, 24, 28, 30, 31, 39, 55, 59, 60, 66, 79, 89, 91, 93–97, 120–123, 125–139, 151, 152, 212, 313, 316, 324, 325, 370, 393, 440, 446, 464, 472, 479, 489–492
Northern Basin, 346, 350
Northern leopard frog, *see Rana pipiens*
Northern pintail, *see Anas acuta*
Northern shoveler, *see Anas clypeata*
Numenius americanus, 289
Numenius phaeopus, 289
Nutrient cycling, 191, 200, 209, 217
Nutrient dynamics, 176

O
Oceanospirillales, 98, 101
Odiel marshes, 206
Odonata, 274, 284
Ogden, 31, 55, 59, 61, 63, 66, 176–221, 240, 244, 251, 252, 316
Ogden Bay Waterfowl Management Area, 452
Ogden, P.S. (Peter Skene), 316
Oligochaeta, 274
Omnivorous, 165, 274
Ooid(s), *see* Oolitic sand
Oolites, *see* Oolitic sand
Oolitic sand, 76, 92, 128, 355, 466
Operational Taxonomic Units (OTU), 96, 98, 123, 125
Orb-weaving spider, *see Neoscona* spp.
Organic carbon, 101, 103, 106, 438–440, 446, 500
Organosedimentary, 316, 446
Oscillatoria, 149
Osmophily, 122
Osmoregulatory, 146, 185
Osmoregulatory exertion, 196
Osmotica, 121, 122, 498, 499
Outreach, 328
Overharvesting, 219
Oviparity, 148, 187, 189, 195
Oviparous, 177, 187, 194, 216, 217
Ovoviviparity, 148, 194
Ovoviviparous, 177, 186, 187, 192, 193
Owens Lake, 13, 15, 44, 219, 220, 455

Oxidation, 104, 185, 438
Oxygen, 101, 110, 126, 130, 187, 192, 194, 195, 208, 214, 425, 426, 489

P

Pacific golden-plover, *see Pluvialis fulva*
Pacific treefrog, *see Pseudacris regilla*
Paleocene, 349
Parengydontium, 128
Parthenogenetic, 197
Pathology, 186
PCR, 95, 99, 125, 503
Pelecanus erythrorhynchos, 245, 312, 472–476
Pelecanus occidentalis, 312, 474
PELIcam, 332–334, 336, 340
Pelican Management Act, 320
PELI project, 332–338
Penicillium, 128
Perennial pepperweed, *see Lepidium latifolium*
Perexilibacter, 135
Periphyton, 94, 100, 106, 111, 209, 210, 274, 444, 447, 448, 453
Permian, 348, 495, 502
Pesticide, 318, 319, 339
Petroleum, 466, 468, 471, 474, 475, 479, 482
pH, 11, 96, 102–105, 112, 130, 133, 146, 187, 216, 240, 253, 379
Phalarope, 12, 41, 100, 167, 178, 206, 239–258, 264–268, 278, 280–283, 300–302, 315, 316, 399, 401
Phalaropus fulicarius, 280, 281, 302
Phalaropus lobatus, 41, 206, 248, 301, 316
Phalaropus tricolor, 41, 206, 248, 300, 315, 399, 401
Phlox longifolia, 382, 383, 386, 392
Photobiology, 501
Photoperiod, 189, 195
Photoprotection, 490
Photoreceptor, 357
Photosynthetic, 39, 88, 91, 92, 98–100, 102–106, 108, 112, 125, 126, 209, 212, 316, 380, 387
Phototrophic, 91, 94, 98–100, 102, 107, 110
Phragmites, 246, 269, 273, 303, 402–404, 408–423, 426, 427
Phragmites australis, 269, 273, 402, 404, 408
Phragmites spp., 246
Phylogeny, 124, 189
Physiology, 88, 123–125, 499
Phytoplankton, 2, 11, 12, 30, 39, 44, 89, 94, 106, 107, 128, 146–159, 161, 162, 164–168, 191, 192, 194–196, 198–200, 202, 204, 210–212, 216, 222, 315, 439, 444–446, 457
Pigmentation, 60, 120
Pileworm, *see Neanthes succinea*
Pituophis catenifer, 351, 355, 364, 481
Pituophis catenifer deserticola, 360–361
Planctomycetaceae, 135
Playa, 10, 13, 25, 37, 39, 42, 242, 271, 275–278, 284, 285, 288, 289, 302, 303, 314, 369, 372, 375, 402, 406, 411, 412, 424, 480, 488
Plegadis chihi, 272, 412, 452, 453
Pleistocene, 25, 26, 29, 34, 176, 271, 349, 352, 362, 467, 479, 488, 491
Pliocene, 349, 465
Pluvialis dominica, 286
Pluvialis fulva, 286
Pluvialis squatarola, 285
Podiceps nigricollis, 12, 148, 206, 315
Poison hemlock, *see Conium maculatum*
Polarotaxis, 479
Pollutant(s), 33, 42, 44, 220, 437
Polypogon monspeliensis, 407
Potamogeton crispus, 407, 408, 426
Potash, 436
Potassium, 32, 122, 168, 339, 380, 436
Precipitation, 2, 6, 24, 28, 29, 32, 34–37, 55, 79, 88, 102–106, 112, 120, 146, 148, 159, 160, 162, 164–166, 168, 196, 197, 248, 257, 269, 271, 275, 276, 315, 316, 324, 363, 375, 446, 455, 488, 490–492, 495, 497, 504–506
Preservation, 72, 80, 81, 88, 212, 222, 464, 476, 493–495, 501–504, 506
Prickly lettuce, *see Lactuca serriola*
Primary production, 88, 99–101, 108, 111, 155, 157–159, 164, 166, 176, 200, 201, 203, 204, 209, 216, 247, 257
Procyon lotor, 246
Productivity, 89, 95, 108, 110, 111, 146, 164, 167, 168, 179, 203, 217, 219, 254, 327, 381, 448
Promontory Mountains, 63
Promontory Point, 79, 92, 151
Proteins, 40, 122, 176, 178–181, 185–189, 222, 273, 498, 499, 503, 505
Proteobacteria, 125, 138
Protists, 128
Protodolomite, 102
Protozoa, 56, 128
Provo River Project, 6
Provo Shoreline, 467
Pseudacris regilla, 350

Pseudacris triseriata, 349, 350
Pseudomonas, 131, 132, 136–138
Pseudo-oligotrophic, 179
Psychroflexus, 136, 138
Public Trust Doctrine, 16, 43, 45
Purple loosestrife, *see Lythrum salicaria*
Pyramid Lake, 37, 219, 241, 321

Q
Quasi-decadal oscillation (QDO), 35
Quiescence, 188–191

R
Racoon, *see Procyon lotor*
Railroad, 59–61, 107, 112, 151, 439
Railroad causeway, 3, 4, 12, 30, 31, 55, 59, 60, 62, 89, 105, 111, 120, 151, 176, 184, 212, 219, 240, 255, 258, 313, 393, 440, 456, 468, 489, 490
Rana catesbeiana, 350
Rana clamitans, 350
Rana luteiventris, 350
Rana onca, 350
Rana pipiens, 350
Rayless alkali aster, *see Symphyotrichum ciliatum*
Recreation, 2, 10, 14, 30, 31, 45, 58, 61–64, 74–75, 80, 178, 340, 426, 436
Recurvirostra americana, 206, 284, 285, 315, 399, 401, 412
Red fox, *see Vulpes vulpes*
Red knot, *see Calidris canutus*
Red-necked phalarope, *see Phalaropus lobatus*
Red phalarope, *see Phalaropus fulicarius*
Red swampfire, *see Salicornia rubra*
Regional Screening Levels, 13
Regulation, 33, 40, 57, 58, 122, 130, 137, 182, 187
Relict leopard frog, *see Rana onca*
Revegetation, 419–422, 427
Rhizome, 384, 410, 413, 414, 418, 419, 425, 426
Rhizosphere, 381
Ricker curve, 183
Ring-billed gull, *see Larus delawarensis*
Riverine input, 27, 107, 203, 441, 442, 456
Rivers, 2–4, 6, 7, 10, 17, 25–28, 33, 44, 55, 59, 62, 66, 74, 75, 79, 80, 89, 107, 112, 120, 185, 208, 212, 255, 269, 271, 277, 284, 288, 312, 313, 318, 350, 370, 372, 418, 457, 491
RNA, 125, 502
Roseovarius, 136, 138
Rotifers, 110, 181, 207
Rozel point, 120, 129, 130, 463–482
Ruddy turnstone, *see Arenaria interpres*
Ruff, *see Calidris pugnax*
Ruppia cirrhosa, 246
Russian knapweed, *see Acroptilon repens*

S
Saccharospirillum salsuginis, 101
Sagebrush lizard, *see Sceloporus graciosus*
Sago pondweed, *see Stuckenia pectinate*
Salicola, 132
Salicornia rubra, 271, 377, 380, 383–385, 389, 402, 403
Salicornia utahensis, 377, 383–385, 389
Salinibacter, 121, 122, 130, 132–134, 137
Salinibacter ruber, 98, 133, 137
Salinity, 2, 4, 7–12, 14, 16, 17, 24, 28–31, 33, 37–41, 54–56, 59, 60, 62, 66, 67, 80, 89–91, 94–96, 98–101, 105–112, 119–139, 146–159, 161, 162, 164–168, 176, 178, 183–185, 189, 192, 195–200, 202, 206–208, 212, 214, 216, 218, 219, 222, 240, 242, 244, 253–258, 267, 269, 271, 274, 278, 301, 302, 315, 316, 325, 362, 370–373, 375, 377, 379–381, 384, 393, 405–407, 410, 423, 424, 436, 439, 440, 442–444, 446, 479, 489–491, 495, 498–499
Salinity gradient, 9, 30, 39, 112, 120, 126, 129, 132, 196, 201, 207, 212, 315, 489–491
Saltgrass, *see Distichlis* spp.
Salton Sea, 14, 15, 220, 241, 256, 288, 291, 325
San Francisco Bay, 189, 300
San Joaquin Valley, 257
Sanderling, *see Calidris alba*
Sanders Brine Shrimp Company, 152, 179
Sarcobatus vermiculatus, 377, 378, 383, 385, 386, 392
Sceloporus graciosus, 352, 353, 357, 363
Schizocosa, 454
Schlesneria, 135
Schoenoplectus acutus, 269, 402, 403
Schoenoplectus americanus, 403
Seasonal temperature, 214, 466, 488
Sea surface temperatures (SST), 35
Secondary production, 101–103, 148
Sedge, 272, 301
Sediment, 4, 13, 38, 88, 91, 94, 102, 104–106, 112, 126–128, 131, 138, 176, 200, 203,

217–220, 267, 269, 271, 277, 438, 439, 442, 443, 455, 456, 465, 466, 468, 470, 475, 477, 492, 493, 503, 504
Seepweed, *see Suaeda* spp.
Seismic, 91
Selenium, 33, 220, 248, 257, 453
Semiarid, 17, 167, 363
Semipalmated plover, *see Charadrius semipalmatus*
Semipalmated sandpiper, *see Calidris pusilla*
Sevier orogenic system, 89
Sewage, 30, 64, 315
Sexual reproduction, 410, 425
Shadscale, *see Atriplex confertifolia*
Sharp-tailed sandpiper, *see Calidris acuminata*
Shorebirds, 10, 41, 56, 67, 99, 100, 106, 167, 178, 206, 264–303, 315, 399, 402, 411, 419, 444
Shorelands Preserve, 272, 293
Short-billed dowitcher, *see Limnodromus griseus*
Shoshonean-Paiute, 29
Silica, 98, 154
Small subunit (SSU) rRNA genes, 94–98
Snowpack, 28, 34–36, 79, 160, 326, 363, 393, 455
Snowy plover, *see Charadrius nivosus*
Sodium chloride, 32, 91, 105, 120, 276, 277, 490
Solar evaporation ponds, 243, 257
Solitary sandpiper, *see Tringa solitaria*
Sorghum, *see Sorghum bicolor*
Sorghum bicolor, 423
South arm, 4, 5, 8, 11, 12, 28, 30–32, 39, 55, 59, 60, 89, 91, 93–97, 100, 102, 120, 126–129, 131, 147, 149, 151, 176, 212, 244, 313–315, 370, 440, 444, 446, 447, 479, 489–491
Southern basin, 214, 346
Spatial, 106, 112, 126–127, 129, 206, 269, 448, 453, 454
Spatula cyanoptera, 252
Spea intermontana, 349, 350, 362
Spea multiplicata, 350
Sphecius, 480
Sphecius speciosus, 480
Spiders, 436, 454
Spiral ditchgrass, *see Ruppia cirrhosa*
Sporobolus airoides, 377, 386, 392
Sporobolus cryptandrus, 377, 386, 392
Spotted knapweed, *see Centaurea stoebe*
Spotted sandpiper, *see Actitis macularius*
16S rRNA gene, 95, 96, 98, 104, 108, 109, 122, 123, 127, 129, 130, 132–135, 211, 381, 443
23S rRNA gene, 209
Stansbury, Howard (Captain), 25, 316, 353
Stansbury Island, 27, 108
Star-thistle, *see Centaurea solstitialis*
Stemphylium, 128
Stenohaline, 151
Stilt sandpiper, *see Calidris himantopus*
Stratification, 185, 191, 212–219, 415, 441
Streamflow, 2–4, 6, 7, 11, 14, 16, 17, 29, 35, 38, 42
Striped skunk, *see Mephitis mephitis*
Striped whipsnake, *see Masticophis taeniatus*
Stuckenia pectinata, 246, 403
Suaeda calceoliformis, 271
Suaeda erecta, 377, 378, 383, 385, 389
Suaeda salsa, 381
Submergent wetlands, 411
Sulfate (SO_4^{2}-), 32, 91, 112, 125, 126, 276, 439, 443, 490, 495
Sulfate (SO_4^{2}-) reducing bacteria (SRB), 30, 101–105, 437, 443, 457
Sulfide, 102, 104, 105, 438, 439
Sulfonolipids, 122
Surveys, 27, 35, 41, 57, 58, 63, 66, 68, 73, 75–77, 81, 126, 151, 153, 166, 191, 204, 213, 218, 247, 248, 256, 258, 264, 276–278, 283, 285, 291, 313, 318, 325, 327, 328, 332, 338, 340, 353, 355, 356, 372–374, 393, 399, 402, 409, 414, 422, 424, 467, 468, 490, 491
Symphyotrichum ciliatum, 403, 420
Syrphidae, 284

T

Tar seep, 463–482
Technical Advisory Group (TAG), 183, 329, 473, 475
Temperature, 2, 28, 30, 32, 34–38, 40, 42, 43, 96, 98, 106, 110, 120, 121, 127, 128, 130, 133, 146–148, 151, 154–162, 164–166, 185, 187, 189, 192–195, 197, 199, 201, 207, 208, 214–216, 246, 248, 326, 346, 363, 421, 439, 446, 466, 471, 473, 474, 478, 480, 482, 490, 491, 499
Temporal, 112, 126–127, 129–132, 187, 203, 204, 283, 453, 454
Terminal lake, 2, 3, 17, 24–29, 33, 37, 38, 42, 43, 54, 89, 111, 120, 138, 269, 369, 393, 457, 488, 490

Terrestrial ecosystem, 453, 454
Terrestrial garter snake, *see Thamnophis elegans*
Thalassiosira, 210
Thalassohaline, 24, 490
Thamnophis elegans, 352, 356
Thamnophis elegans vagrans, 361, 362, 364
Thistle, *see Cirsium arvense*
Threats, 12, 24, 41, 42, 168, 219–220, 253–258, 303, 318, 322, 324, 326, 402, 411, 426, 427
Threesquare bulrush, *see Schoenoplectus americanus*
Thrombolites, 247
Tipping point, 108, 111, 112
Titanium, 436
Tourism, 45, 60, 61, 64, 112, 113, 178, 318
Toxic, 5, 33, 168, 257, 379, 437, 446, 448, 498, 499
Toxicity, 33, 186, 220, 257, 436, 437, 444, 446, 448, 453, 454
Transcontinental Railroad, 59
Triangle orache, *see Atriplex prostrata*
Triassic, 348
Trichocorixa verticalis, 196, 206, 208
Tringa flavipes, 299
Tringa incana, 299
Tringa melanoleuca, 300
Tringa semipalmata, 299
Tringa solitaria, 299
Trophic, 38, 39, 89, 96, 100, 107, 108, 110–111, 178, 202, 216, 220, 222, 274, 439, 444, 448
Trophic cascade, 100, 199, 219
Trophic transfer, 220, 439, 445
Tundra swan, *see Cygnus columbianus*
Turkey vulture, *see Cathartes aura*
Typha domingensis, 403, 408, 425–426
Typha latifolia, 403, 407, 425–426
Typha spp., 246, 269, 273, 407
Tyto alba, 336, 476, 477

U

Ultraviolet (UV), 121, 185, 495, 498, 500–503
Union Pacific Railroad, 4, 255, 313, 489
Upland sandpiper, *see Bartramia longicauda*
Uronema, 128
Utah Artemia Association (UAA), 183
Utah Division of Water Resources, 7
Utah Division of Wildlife Resources (UDWR), 182, 183, 191, 192, 315, 316, 318, 320, 322, 323, 325–328, 330–332, 336, 338, 340
Utah Geological Survey, 90, 257, 374, 376, 469, 473, 482
Utah pickleweed, *see Salicornia utahensis*
Utah State Division of Forestry, Fire, and State Lands (FFSL), 10, 13, 14
Utah State Legislature, 44, 65, 81
Uta stansburiana elegans, 357–358

V

Vegetation, 13, 16, 25, 66, 246, 266–269, 271–273, 280, 284, 285, 287–289, 292, 295, 297–300, 302, 315, 316, 320, 346, 362, 372, 378, 417, 418, 438, 477, 478
Vertebrata, 346
Vibrio spp., 181
Viruses, 121, 138, 139
Volcanic tuff, 465
Vulgatibacter, 136
Vulpes vulpes, 246

W

Walker Lake, 14, 15, 17, 219
Wallemia, 128
Wandering gartersnake, *see Thamnophis elegans vagrans*
Wandering tattler, *see Tringa incana*
Wasatch Front, 2, 6, 10, 13, 15, 16, 36–37, 56, 160, 275, 313, 340, 427, 455, 456
Wasatch Range, 2
Wasatch-Uinta Mountains, 346
Water activity, 129, 139, 498
Water banking, 16
Water boatmen, *see* Corixidae
Water budget, 28, 455
Water column, 39, 91, 104, 108, 126, 128, 129, 189, 190, 194, 195, 198, 209–213, 216, 250, 277, 284, 439, 443, 446, 454, 456, 490, 493, 504
Water depth, 41, 246, 267–269, 274, 278, 302, 418, 420
Water diversions, 2, 5, 17, 41, 42, 45, 79, 107, 159, 160, 166–168, 240–258, 303, 324, 427, 455
Water residence time, 176
Water right(s), 16, 27, 43–45, 258
Water Strategy Advisory Team, 81
Waterfowl, 10, 56, 58, 74, 75, 178, 206, 264, 278, 302, 399, 402, 406, 407, 411,

418–420, 422, 423, 425–427, 436, 444, 448, 449, 453
Waterfowl management, 58, 272, 278, 290, 297, 400, 413
Watershed, 2, 3, 6–8, 16, 24–26, 29–32, 34, 36, 37, 41–43, 56, 146, 154, 159–161, 164, 165, 168, 201, 223, 257, 258, 269, 324, 413, 467
Weber River Project, 6
West Desert Pumping Project, 66
West Layton Marsh, 272, 273
West Nile Virus, 339
Western chorus frog, *see Pseudacris triseriata*
Western Hemisphere Shorebird Reserve, 41, 167, 178, 436
Western sandpiper, *see Calidris mauri*
Western side-blotched lizard, *see Uta stansburiana elegans*
Western yellow-bellied racer, *see Coluber constrictor mormon*
Wetland habitats, 2, 17, 178, 264, 300, 362
Wetlands, 2, 9–10, 14, 15, 29, 39–42, 55, 56, 58, 63, 70, 72, 79–81, 178, 242, 245, 246, 248, 252, 264, 266, 269–275, 278, 280, 281, 283–285, 288, 289, 291, 293, 295–298, 300, 301, 303, 313–315, 320, 323, 325, 330, 332, 340, 362, 398–427, 443, 444, 446, 490
Whimbrel, *see Numenius phaeopus*
White-faced ibis, *see Plegadis chihi*
White pelican, *see Pelecanus erythrorhynchos*
White Rock Bay, 355
White-rumped sandpiper, *see Calidris fuscicollis*
Whitetop, *see Cardaria draba*
Wildlife Management, 31, 72, 320, 413, 427
Willard Bay, 31
Willet, *see Tringa semipalmata*
Wilson's phalarope, *see Phalaropus tricolor*
Wilson's snipe, *see Gallinago delicata*
Winogradsky column(s), 128
Wyoming Plateau, 346

Y
Yellow star-thistle, *see Centaurea solstitialis*

Z
Zooniverse, 334, 340
Zooplankton, 191, 198, 200, 206–208, 212

9783030403515